Plant Secondary Metabolites

Occurrence, Structure and Role in the Human Diet

Edited by

Alan Crozier
Professor of Plant Biochemistry and Human Nutrition
Institute of Biomedical and Life Sciences
University of Glasgow, UK

Michael N. Clifford
Professor of Food Safety
Centre for Nutrition and Food Safety
School of Biomedical and Life Sciences
University of Surrey, UK

Hiroshi Ashihara
Professor of Plant Biochemistry
Department of Biology
Ochanomizu University, Tokyo, Japan

Blackwell
Publishing

© 2006 by Blackwell Publishing Ltd

Editorial Offices:
Blackwell Publishing Ltd, 9600 Garsington Road, Oxford OX4 2DQ, UK
   Tel: +44 (0)1865 776868
Blackwell Publishing Professional, 2121 State Avenue, Ames, Iowa 50014-8300, USA
   Tel:+1 515 292 0140
Blackwell Publishing Asia Pty Ltd, 550 Swanston Street, Carlton, Victoria 3053, Australia
   Tel: +61 (0)3 8359 1011

The right of the Author to be identified as the Author of this Work has been asserted in accordance with the Copyright, Designs and Patents Act 1988.

All rights reserved. No part of this publication may be reproduced, stored in a retrieval system, or transmitted, in any form or by any means, electronic, mechanical, photocopying, recording or otherwise, except as permitted by the UK Copyright, Designs and Patents Act 1988, without the prior permission of the publisher.

Designations used by companies to distinguish their products are often claimed as trademarks. All brand names and product names used in this book are trade names, service marks, trademarks or registered trademarks of their respective owners. The Publisher is not associated with any product or vendor mentioned in this book.

This publication is designed to provide accurate and authoritative information in regard to the subject matter covered. It is sold on the understanding that the Publisher is not engaged in rendering professional services. If professional advice or other expert assistance is required, the services of a competent professional should be sought.

First published 2006 by Blackwell Publishing Ltd

2   2007

ISBN: 978-1-4051-2509-3

Library of Congress Cataloging-in-Publication Data
Plant secondary metabolites: occurrence, structure and role in the human diet/edited by Alan Crozier, Michael N. Clifford, Hiroshi Ashihara.
      p.;cm.
   Includes bibliographical references and index.
   ISBN-13: 978-1-4051-2509-3 (hardback: alk.paper)
   ISBN-10: 1-4051-2509-8 (hardback: alk.paper)
   1. Plants–Metabolism. 2. Metabolism, Secondary. 3. Botanical chemistry.
   I. Crozier, Alan. II. Clifford, M. N. (Michael N.) III. Ashihara, Hiroshi.
   [DNLM: 1. Plants, Edible–metabolism. 2. Food Analysis–methods. 3. Heterocyclic Compounds–chemistry.
   4. Heterocyclic Compounds–metabolism. 5. Plants, Edible–chemistry. QK 887 P713 2006]

   QK881.P55 2006
   $572'.2$–dc22

2006004363

A catalogue record for this title is available from the British Library

Set in 10/12 pt Minion
by Newgen Imaging Systems (P) Ltd., Chennai, India
Printed and bound in Singapore
by COS Printers Pte Ltd

The publisher's policy is to use permanent paper from mills that operate a sustainable forestry policy, and which has been manufactured from pulp processed using acid-free and elementary chlorine-free practices. Furthermore, the publisher ensures that the text paper and cover board used have met acceptable environmental accreditation standards.

For further information on Blackwell Publishing, visit our website:
www.blackwellpublishing.com

Dedication

To Diego Hermoso Borges – a very special, brave boy

Contents

Contributors xi

1 Phenols, Polyphenols and Tannins: An Overview *Alan Crozier, Indu B. Jaganath and Michael N. Clifford* 1
   1.1 Introduction 1
   1.2 Classification of phenolic compounds 2
      1.2.1 Flavonoids 2
         1.2.1.1 Flavonols 4
         1.2.1.2 Flavones 4
         1.2.1.3 Flavan-3-ols 5
         1.2.1.4 Anthocyanidins 8
         1.2.1.5 Flavanones 8
         1.2.1.6 Isoflavones 9
      1.2.2 Non-flavonoids 11
         1.2.2.1 Phenolic acids 11
         1.2.2.2 Hydroxycinnamates 12
         1.2.2.3 Stilbenes 12
   1.3 Biosynthesis 14
      1.3.1 Phenolics and hydroxycinnamates 16
      1.3.2 Flavonoids and stilbenes 17
         1.3.2.1 The pathways to flavonoid formation 17
         1.3.2.2 Isoflavonoid biosynthesis 18
         1.3.2.3 Flavone biosynthesis 18
         1.3.2.4 Formation of intermediates in the biosynthesis of flavonols, flavan-3-ols, anthocyanins and proanthocyanidins 19
         1.3.2.5 Stilbene biosynthesis 19
   1.4 Genetic engineering of the flavonoid biosynthetic pathway 19
      1.4.1 Manipulating flavonoid biosynthesis 20
      1.4.2 Constraints in metabolic engineering 21
   1.5 Databases 21

		Acknowledgements	21
		References	22
2	Sulphur-Containing Compounds *Richard Mithen*		25
	2.1	Introduction	25
	2.2	The glucosinolates-myrosinase system	26
	2.3	Chemical diversity of glucosinolates in dietary crucifers	27
	2.4	Biosynthesis	29
	2.5	Genetic factors affecting glucosinolate content	31
	2.6	Environmental factors affecting glucosinolate content	31
	2.7	Myrosinases and glucosinolate hydrolysis	32
	2.8	Hydrolytic products	33
	2.9	Metabolism and detoxification of isothiocyanates	34
	2.10	The Alliin-alliinase system	34
	2.11	Biological activity of sulphur-containing compounds	37
	2.12	Anti-nutritional effects in livestock and humans	38
	2.13	Beneficial effects of sulphur-containing compounds in the human diet	38
		2.13.1 Epidemiological evidence	38
		2.13.2 Experimental studies and mechanisms of action	39
		2.13.2.1 Inhibition of Phase I CYP450	39
		2.13.2.2 Induction of Phase II enzymes	39
		2.13.2.3 Antiproliferative activity	40
		2.13.2.4 Anti-inflammatory activity	40
		2.13.2.5 Reduction in *Helicobacter pylori*	40
	References	41
3	Terpenes *Andrew J. Humphrey and Michael H. Beale*	47
	3.1	Introduction	47
	3.2	The biosynthesis of IPP and DMAPP	49
		3.2.1 The mevalonic acid pathway	49
		3.2.2 The 1-deoxyxylulose 5-phosphate (or methylerythritol 4-phosphate) pathway	52
		3.2.3 Interconversion of IPP and DMAPP	54
		3.2.4 Biosynthesis of IPP and DMAPP in green plants	55
	3.3	Enzymes of terpene biosynthesis	55
		3.3.1 Prenyltransferases	55
		3.3.2 Mechanism of chain elongation	56
		3.3.3 Terpene synthases (including cyclases)	58
	3.4	Isoprenoid biosynthesis in the plastids	59
		3.4.1 Biosynthesis of monoterpenes	59
		3.4.2 Biosynthesis of diterpenes	65
		3.4.3 Biosynthesis of carotenoids	74
	3.5	Isoprenoid biosynthesis in the cytosol	78
		3.5.1 Biosynthesis of sesquiterpenes	78
		3.5.2 Biosynthesis of triterpenes	85

		3.6	Terpenes in the environment and human health: future prospects	90
		References		94
4	Alkaloids *Katherine G. Zulak, David K. Liscombe, Hiroshi Ashihara and Peter J. Facchini*			102
	4.1	Introduction		102
	4.2	Benzylisoquinoline alkaloids		102
	4.3	Tropane alkaloids		107
	4.4	Nicotine		111
	4.5	Terpenoid indole alkaloids		113
	4.6	Purine alkaloids		118
	4.7	Pyrrolizidine alkaloids		122
	4.8	Other alkaloids		125
		4.8.1	Quinolizidine alkaloids	125
		4.8.2	Steroidal glycoalkaloids	127
		4.8.3	Coniine	129
		4.8.4	Betalains	130
	4.9	Metabolic engineering		130
	Acknowledgements			131
	References			131
5	Acetylenes and Psoralens *Lars P. Christensen and Kirsten Brandt*			137
	5.1	Introduction		137
	5.2	Acetylenes in common food plants		138
		5.2.1	Distribution and biosynthesis	138
		5.2.2	Bioactivity	147
			5.2.2.1 Antifungal activity	147
			5.2.2.2 Neurotoxicity	149
			5.2.2.3 Allergenicity	150
			5.2.2.4 Anti-inflammatory, anti-platelet-aggregatory and antibacterial effects	151
			5.2.2.5 Cytotoxicity	152
			5.2.2.6 Falcarinol and the health-promoting properties of carrots	153
	5.3	Psoralens in common food plants		155
		5.3.1	Distribution and biosynthesis	155
		5.3.2	Bioactivity	159
			5.3.2.1 Phototoxic effects	159
			5.3.2.2 Inhibition of human cytochrome P450	162
			5.3.2.3 Reproductive toxicity	162
			5.3.2.4 Antifungal and antibacterial effects	162
	5.4	Perspectives in relation to food safety		163
	References			164
6	Functions of the Human Intestinal Flora: The Use of Probiotics and Prebiotics *Kieran M. Tuohy and Glenn R. Gibson*			174

6.1	Introduction			174
6.2	Composition of the gut microflora			174
6.3	Successional development and the gut microflora in old age			177
6.4	Modulation of the gut microflora through dietary means			178
	6.4.1	Probiotics		179
		6.4.1.1	Probiotics in relief of lactose maldigestion	180
		6.4.1.2	Use of probiotics to combat diarrhoea	180
		6.4.1.3	Probiotics for the treatment of inflammatory bowel disease	182
		6.4.1.4	Impact of probiotics on colon cancer	183
		6.4.1.5	Impact of probiotics on allergic diseases	184
		6.4.1.6	Use of probiotics in other gut disorders	184
		6.4.1.7	Future probiotic studies	185
	6.4.2	Prebiotics		186
		6.4.2.1	Modulation of the gut microflora using prebiotics	186
		6.4.2.2	Health effects of prebiotics	189
	6.4.3	Synbiotics		192
6.5	In vitro and in vivo measurement of microbial activities			193
6.6	Molecular methodologies for assessing microflora changes			194
	6.6.1	Fluorescent in situ hybridization		195
	6.6.2	DNA microarrays – microbial diversity and gene expression studies		195
	6.6.3	Monitoring gene expression – subtractive hybridization and in situ PCR/FISH		196
	6.6.4	Proteomics		196
6.7	Assessing the impact of dietary modulation of the gut microflora – does it improve health, what are the likelihoods for success and what are the biomarkers of efficacy?			197
6.8	Justification for the use of probiotics and prebiotics to modulate the gut flora composition			198
References				199

7	Secondary Metabolites in Fruits, Vegetables, Beverages and Other Plant-Based Dietary Components *Alan Crozier, Takao Yokota, Indu B. Jaganath, Serena Marks, Michael Saltmarsh and Michael N. Clifford*		208
	7.1	Introduction	208
	7.2	Dietary phytochemicals	209
	7.3	Vegetables	211
		7.3.1 Root crops	212
		7.3.2 Onions and garlic	214
		7.3.3 Cabbage family and greens	217
		7.3.4 Legumes	219
		7.3.5 Lettuce	222

		7.3.6	Celery		223
		7.3.7	Asparagus		223
		7.3.8	Avocados		224
		7.3.9	Artichoke		224
		7.3.10	Tomato and related plants		225
			7.3.10.1	Tomatoes	225
			7.3.10.2	Peppers and aubergines	227
		7.3.11	Squashes		228
	7.4	Fruits			229
		7.4.1	Apples and pears		229
		7.4.2	Apricots, nectarines and peaches		231
		7.4.3	Cherries		231
		7.4.4	Plums		231
		7.4.5	Citrus fruits		232
		7.4.6	Pineapple		235
		7.4.7	Dates		235
		7.4.8	Mango		236
		7.4.9	Papaya		237
		7.4.10	Fig		238
		7.4.11	Olive		238
		7.4.12	Soft fruits		240
		7.4.13	Melons		245
		7.4.14	Grapes		245
		7.4.15	Rhubarb		248
		7.4.16	Kiwi fruit		249
		7.4.17	Bananas and plantains		250
		7.4.18	Pomegranate		251
	7.5	Herbs and spices			252
	7.6	Cereals			258
	7.7	Nuts			260
	7.8	Algae			262
	7.9	Beverages			263
		7.9.1	Tea		263
		7.9.2	Maté		271
		7.9.3	Coffee		273
		7.9.4	Cocoa		277
		7.9.5	Wines		278
		7.9.6	Beer		281
		7.9.7	Cider		285
		7.9.8	Scotch whisky		287
	7.10	Databases			288
	References				288

8 Absorption and Metabolism of Dietary Plant Secondary Metabolites
  Jennifer L. Donovan, Claudine Manach, Richard M. Faulks and Paul A. Kroon   303
  8.1   Introduction   303

8.2	Flavonoids			303
	8.2.1	Mechanisms regulating the bioavailability of flavonoids		304
		8.2.1.1	Absorption	304
		8.2.1.2	Intestinal efflux of absorbed flavonoids	308
		8.2.1.3	Metabolism	309
		8.2.1.4	Elimination	310
	8.2.2	Overview of mechanisms that regulate the bioavailability of flavonoids		311
	8.2.3	Flavonoid metabolites identified in vivo and their biological activities		311
		8.2.3.1	Approaches to the identification of flavonoid conjugates in plasma and urine	312
		8.2.3.2	Flavonoid conjugates identified in plasma and urine	315
	8.2.4	Pharmacokinetics of flavonoids in humans		317
8.3	Hydroxycinnamic acids			321
8.4	Gallic acid and ellagic acid			323
8.5	Dihydrochalcones			324
8.6	Betalains			324
8.7	Glucosinolates			325
	8.7.1	Hydrolysis of glucosinolates and product formation		327
	8.7.2	Analytical methods		329
	8.7.3	Absorption of isothiocyanates from the gastrointestinal tract		330
	8.7.4	Intestinal metabolism and efflux		330
	8.7.5	Distribution and elimination		331
8.8	Carotenoids			332
	8.8.1	Mechanisms regulating carotenoid absorption		334
	8.8.2	Effects of processing		335
	8.8.3	Measuring absorption		335
	8.8.4	Transport		337
	8.8.5	Tissue distribution		338
	8.8.6	Metabolism		339
	8.8.7	Toxicity		340
	8.8.8	Other metabolism		340
8.9	Conclusions			341
	References			341

Index   353

Contributors

Hiroshi Ashihara	Department of Biology, Ochanomizu University, Otsuka, Bunkyo-ku, Tokyo, 112-8610, Japan
Michael H. Beale	CPI Division, Rothamsted Research, West Common, Harpenden, Hertfordshire, AL5 2JQ, UK
Kirsten Brandt	School of Agriculture, Food and Rural Development, University of Newcastle upon Tyne, King George VI Building, Newcastle upon Tyne NE1 7RU, UK
Lars P. Christensen	Department of Food Science, Danish Institute of Agricultural Sciences, Research Centre Aarslev, Kirstinebjergvej 10, DK-5792 Aarslev, Denmark
Michael N. Clifford	Food Safety Research Group, Centre for Nutrition and Food Safety, School of Biomedical and Molecular Sciences, University of Surrey, Guildford, Surrey GU2 7XH, UK
Alan Crozier	Graham Kerr Building, Division of Biochemistry and Molecular Biology, Institute of Biomedical and Life Sciences, University of Glasgow, Glasgow G12 8QQ, UK
Jennifer L. Donovan	Laboratory of Drug Disposition and Pharmacogenetics, 173 Ashley Ave., Medical University of South Carolina, Charleston, SC 29425, USA
Peter J. Facchini	Department of Biological Sciences, University of Calgary, Calgary, Alberta, T2N 1N4, Canada
Richard M. Faulks	Nutrition Division Institute of Food Research, Colney Lane, Norwich NR4 7UA, UK
Glenn R. Gibson	Food Microbial Sciences Unit, School of Food Biosciences, The University of Reading, Whiteknights, PO Box 226, Reading, RG6 6AP, UK
Andrew J. Humphrey	CPI Division, Rothamsted Research, West Common, Harpenden, Hertfordshire AL5 2 JQ, UK
Indu B. Jaganath	Graham Kerr Building, Division of Biochemistry and Molecular Biology, Institute of Biomedical and Life Sciences, University of Glasgow, Glasgow G12 8QQ, UK

Paul A. Kroon	Nutrition Division, Institute of Food Research, Colney Lane, Norwich NR4 7UA, UK
David K. Liscombe	Department of Biological Sciences, University of Calgary, Calgary, Alberta T2N 1N4, Canada
Claudine Manach	Unite des Maladies Metaboliques et Micronitriments, INRA de Clermont-Ferand/Theix, 63122 St Genes-Champanelle, France
Serena C. Marks	Graham Kerr Building, Division of Biochemistry and Molecular Biology, Institute of Biomedical and Life Sciences, University of Glasgow, Glasgow G12 8QQ, UK
Richard Mithen	Nutrition Division, Institute of Food Research, Colney Lane, Norwich NR4 7UA, UK
Michael Saltmarsh	Inglehurst Foods, 53 Blackberry Lane, Four Marks, Alton, Hampshire GU35 5DF, UK
Kieran M. Tuohy	Food Microbial Sciences Unit, School of Food Biosciences, University of Reading, Whiteknights, PO Box 226, Reading, RG6 6AP, UK
Takao Yokota	Department of Biosciences, Teikyo University, Utsunomiya 320-85551, Japan
Katherine G. Zulak	Department of Biological Sciences, University of Calgary, Calgary, Alberta T2N 1N4, Canada

Chapter 1
Phenols, Polyphenols and Tannins: An Overview

Alan Crozier, Indu B. Jaganath and Michael N. Clifford

1.1 Introduction

Plants synthesize a vast range of organic compounds that are traditionally classified as primary and secondary metabolites although the precise boundaries between the two groups can in some instances be somewhat blurred. Primary metabolites are compounds that have essential roles associated with photosynthesis, respiration, and growth and development. These include phytosterols, acyl lipids, nucleotides, amino acids and organic acids. Other phytochemicals, many of which accumulate in surprisingly high concentrations in some species, are referred to as secondary metabolites. These are structurally diverse and many are distributed among a very limited number of species within the plant kingdom and so can be diagnostic in chemotaxonomic studies. Although ignored for long, their function in plants is now attracting attention as some appear to have a key role in protecting plants from herbivores and microbial infection, as attractants for pollinators and seed-dispersing animals, as allelopathic agents, UV protectants and signal molecules in the formation of nitrogen-fixing root nodules in legumes. Secondary metabolites are also of interest because of their use as dyes, fibres, glues, oils, waxes, flavouring agents, drugs and perfumes, and they are viewed as potential sources of new natural drugs, antibiotics, insecticides and herbicides (Croteau *et al.* 2000; Dewick 2002).

In recent years the role of some secondary metabolites as protective dietary constituents has become an increasingly important area of human nutrition research. Unlike the traditional vitamins they are not essential for short-term well-being, but there is increasing evidence that modest long-term intakes can have favourable impacts on the incidence of cancers and many chronic diseases, including cardiovascular disease and Type II diabetes, which are occurring in Western populations with increasing frequency.

Based on their biosynthetic origins, plant secondary metabolites can be divided into three major groups: (i) flavonoids and allied phenolic and polyphenolic compounds, (ii) terpenoids and (iii) nitrogen-containing alkaloids and sulphur-containing compounds. This chapter will provide a brief introduction to the first group, the flavonoids, and polyphenolic and related phenolic compounds, including tannins and derived polyphenols. Sulphur-containing compounds are covered in Chapter 2, terpenes in Chapter 3, alkaloids in Chapter 4 and acetylenes and psoralens in Chapter 5.

Table 1.1 Structural skeletons of phenolic and polyphenolic compounds (hydroxyl groups not shown)

Number of carbons	Skeleton	Classication	Example
7	$C_6–C_1$	Phenolic acids	Gallic acid
8	$C_6–C_2$	Acetophenones	Gallacetophenone
8	$C_6–C_2$	Phenylacetic acid	p-Hydroxyphenyl-acetic acid
9	$C_6–C_3$	Hydroxycinnamic acids	p-Coumaric acid
9	$C_6–C_3$	Coumarins	Esculetin
10	$C_6–C_4$	Naphthoquinones	Juglone
13	$C_6–C_1–C_6$	Xanthones	Mangiferin
14	$C_6–C_2–C_6$	Stilbenes	Resveratol
15	$C_6–C_3–C_6$	Flavonoids	Naringenin

1.2 Classification of phenolic compounds

Phenolics are characterized by having at least one aromatic ring with one or more hydroxyl groups attached. In excess of 8000 phenolic structures have been reported and they are widely dispersed throughout the plant kingdom (Strack 1997). Phenolics range from simple, low molecular-weight, single aromatic-ringed compounds to large and complex tannins and derived polyphenols. They can be classified based on the number and arrangement of their carbon atoms (Table 1.1) and are commonly found conjugated to sugars and organic acids. Phenolics can be classified into two groups: the flavonoids and the non-flavonoids.

1.2.1 Flavonoids

Flavonoids are polyphenolic compounds comprising fifteen carbons, with two aromatic rings connected by a three-carbon bridge (Figure 1.1). They are the most numerous of

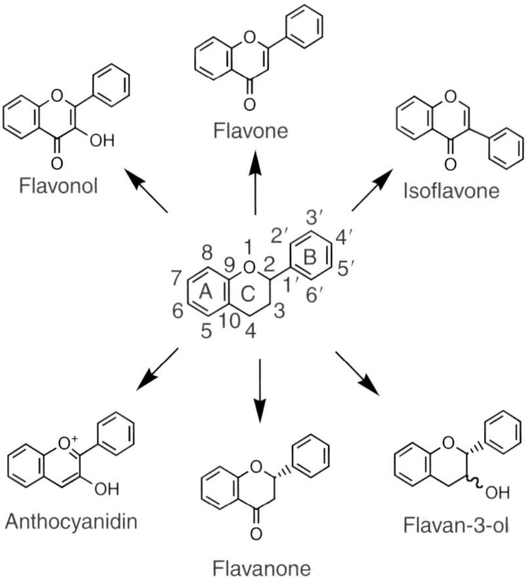

Figure 1.1 Generic structures of the major flavonoids.

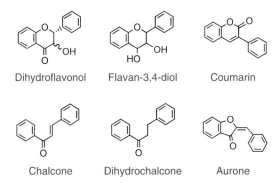

Figure 1.2 Structures of minor flavonoids.

the phenolics and are found throughout the plant kingdom (Harborne 1993). They are present in high concentrations in the epidermis of leaves and the skin of fruits and have important and varied roles as secondary metabolites. In plants, flavonoids are involved in such diverse processes as UV protection, pigmentation, stimulation of nitrogen-fixing nodules and disease resistance (Koes et al. 1994; Pierpoint 2000).

The main subclasses of flavonoids are the flavones, flavonols, flavan-3-ols, isoflavones, flavanones and anthocyanidins (Figure 1.1). Other flavonoid groups, which quantitatively are in comparison minor components of the diet, are dihydroflavonols, flavan-3,4-diols, coumarins, chalcones, dihydrochalcones and aurones (Figure 1.2). The basic flavonoid skeleton can have numerous substituents. Hydroxyl groups are usually present at the 4′, 5 and 7 positions. Sugars are very common with the majority of flavonoids existing

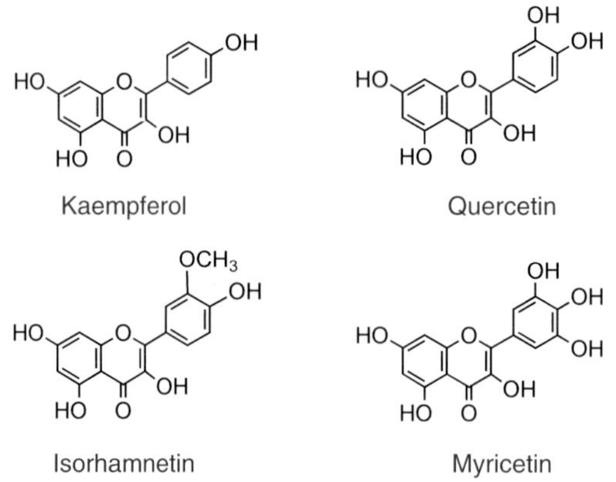

Figure 1.3 The flavonol aglycones kaempferol, quercetin, isorhamnetin and myricetin.

naturally as glycosides. Whereas both sugars and hydroxyl groups increase the water solubility of flavonoids, other substituents, such as methyl groups and isopentyl units, make flavonoids lipophilic.

1.2.1.1 Flavonols

Flavonols are arguably the most widespread of the flavonoids, being dispersed throughout the plant kingdom with the exception of fungi and algae. The distribution and structural variations of flavonols are extensive and have been well documented. Flavonols such as myricetin, quercetin, isorhamnetin and kaempferol (Figure 1.3) are most commonly found as *O*-glycosides. Conjugation occurs most frequently at the 3 position of the C-ring but substitutions can also occur at the 5, 7, 4', 3' and 5' positions of the carbon ring. Although the number of aglycones is limited there are numerous flavonol conjugates with more than 200 different sugar conjugates of kaempferol alone (Strack and Wray 1992). There is information on the levels of flavonols found in commonly consumed fruits, vegetables and beverages (Hertog *et al.* 1992, 1993). However, sizable differences are found in the amounts present in seemingly similar produce, possibly due to seasonal changes and varietal differences (Crozier *et al.* 1997). The effects of processing also have an impact but information on the subject is sparse.

1.2.1.2 Flavones

Flavones have a very close structural relationship to flavonols (Figure 1.1). Although flavones, such as luteolin and apigenin, have A- and C-ring substitutions, they lack oxygenation at C3 (Figure 1.4). A wide range of substitutions is also possible with flavones, including hydroxylation, methylation, *O*- and *C*-alkylation, and glycosylation. Most flavones occur as 7-*O*-glycosides. Unlike flavonols, flavones are not distributed widely with significant occurrences being reported in only celery, parsley and some herbs. In addition, polymethoxylated flavones, such as nobiletin and tangeretin, have been found in citrus species. Flavones in millet have been associated with goitre in west Africa (Gaitan *et al.* 1989).

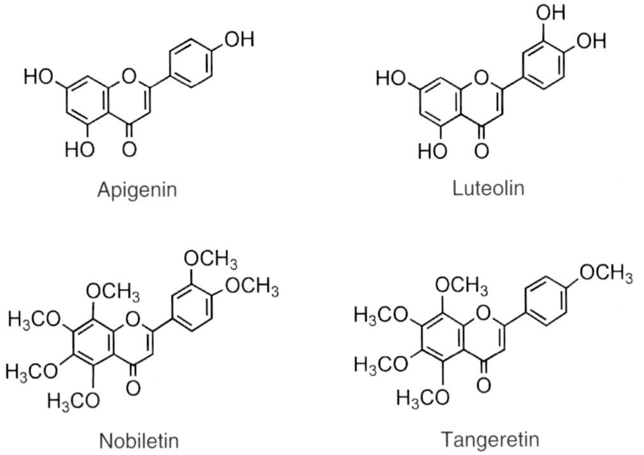

Figure 1.4 The flavones apigenin and luteolin, and the polymethoxylated flavones nobiletin and tangeretin.

1.2.1.3 Flavan-3-ols

Flavan-3-ols are the most complex subclass of flavonoids ranging from the simple monomers (+)-catechin and its isomer (−)-epicatechin, to the oligomeric and polymeric proanthocyanidins (Figure 1.5), which are also known as condensed tannins.

Unlike flavones, flavonols, isoflavones and anthocyanidins, which are planar molecules, flavan-3-ols, proanthocyanidins and flavanones have a saturated C3 element in the heterocyclic C-ring, and are thus non-planar. The two chiral centres at C2 and C3 of the flavan-3-ols produce four isomers for each level of B-ring hydroxylation, two of which, (+)-catechin and (−)-epicatechin, are widespread in nature whereas (−)-catechin and (+)-epicatechin are comparatively rare (Clifford 1986). The oligomeric and polymeric proanthocyanidins have an additional chiral centre at C4; the flavanones have only one chiral centre, C2. Pairs of enantiomers are not resolved on the commonly used reversed phase HPLC columns, and so are easily overlooked. Although difficult to visualize, these differences in chirality have a significant effect on the 3-D structure of the molecules as illustrated in Figure 1.6 for the (epi)gallocatechin gallates. Although this has little, if any, effect on their redox properties or ability to scavenge small unhindered radicals (Unno *et al.* 2000), it can be expected to have a more pronounced effect on their binding properties and hence any phenomenon to which the 'lock-and-key' concept is fundamental, for example, enzyme–substrate, enzyme–inhibitor or receptor–ligand interactions. It is interesting to note that humans fed (−)-epicatechin excrete some (+)-epicatechin indicating ring opening and racemization, possibly in the gastrointestinal tract (Yang *et al.* 2000). Transformation can also occur during food processing (Seto *et al.* 1997).

Type-B proanthocyanidins are formed from (+)-catechin and (−)-epicatechin with oxidative coupling occurring between the C-4 of the heterocycle and the C-6 or C-8 positions of the adjacent unit to create oligomers or polymers (Figure 1.5). Type A proanthocyanidins have an additional ether bond between C-2 and C-7. Proanthocyanidins can occur as polymers of up to 50 units. In addition to forming such large and

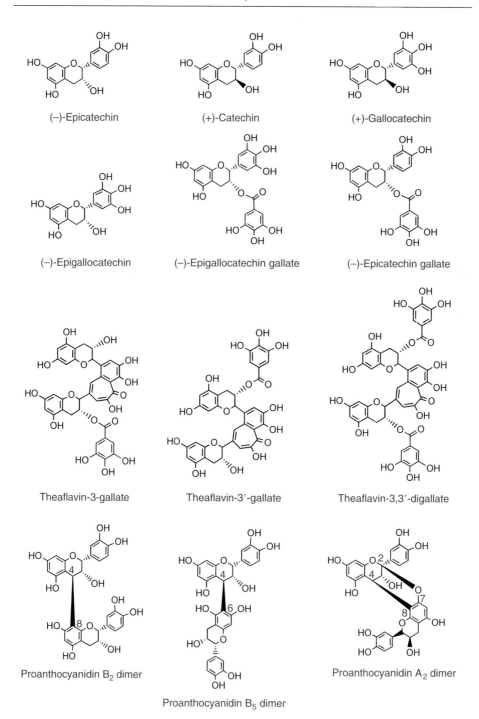

Figure 1.5 Flavan-3-ol structures.

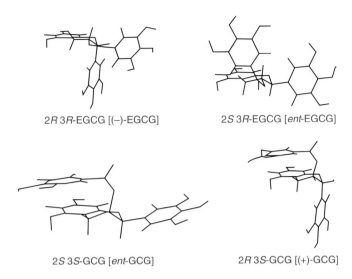

Figure 1.6 Computer-generated stereochemical projections for flavan-3-ol diastereoisomers. EGCG, epi-gallocatechin gallate; GCG, gallocatechin gallate. Three dimensional structures computed by Professor David Lewis, School of Biomedical and Molecular Sciences, University of Surrey, Guildford, Surrey, GU2 7XH, United Kingdom.

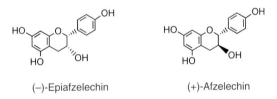

Figure 1.7 (−)-Epiafzelechin and (+)-afzelechin are less common flavan-3-ol monomers which form polymeric proanthocyanidins known as propelargonidins.

complex structures, flavan-3-ols are hydroxylated to form gallocatechins and also undergo esterification with gallic acid (Figure 1.5).

Proanthocyanidins that consist exclusively of (epi)catechin units are called procyanidins, and these are the most abundant type of proanthocyanidins in plants. The less common proanthocyanidins containing (epi)afzelechin (Figure 1.7) and (epi)gallocatechin (Figure 1.5) subunits are called propelargonidins and prodelphinidins, respectively (Balentine et al. 1997).

Red wines contain oligomeric procyanidins and prodelphinidins originating mainly from the seeds of black grapes (Auger et al. 2004) whereas dark chocolate is a rich source of procyanidins derived from the roasted seeds of cocoa (*Theobroma cacao*) (Gu et al. 2004). Green tea (*Camellia sinensis*) contains high levels of flavan-3-ols, principally (−)-epigallocatechin, (−)-epigallocatechin gallate and (−)-epicatechin gallate (Figure 1.5). The levels of catechins decline during fermentation of the tea leaves, and the main components in black tea are the high molecular-weight thearubigins and smaller quantities of theaflavins (Figure 1.5) (Del Rio et al. 2004). Although theaflavins

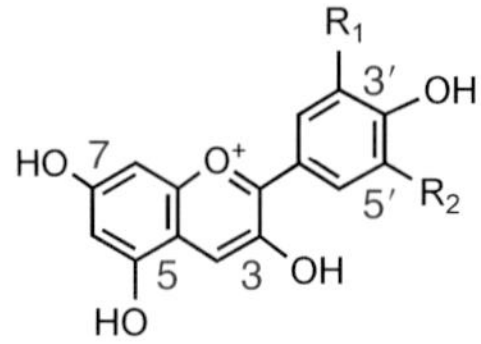

Anthocyanidin	R$_1$	R$_2$	MW
Pelargonidin	H	H	271
Cyanidin	OH	H	287
Delphinidin	OH	OH	303
Peonidin	OCH$_3$	H	301
Petunidin	OCH$_3$	OH	317
Malvidin	OCH$_3$	OCH$_3$	331

Figure 1.8 Structures of major anthocyanidins.

are derived from two flavan-3-ol monomer subunits, they are not strictly dimers although they are often referred to as such. Whereas thearubigins can reasonably be described as flavonoid-derived, their structures are largely unknown. These are often referred to as tannins although it is inappropriate since thearubigins will not convert hides to leather (see Section 1.2.2.1). Accordingly they are better referred to as 'derived polyphenols' until such time as their structures are elucidated and a more precise chemical name can be applied.

1.2.1.4 Anthocyanidins

Anthocyanidins, principally as their conjugated derivatives, anthocyanins, are widely dispersed throughout the plant kingdom, being particularly evident in fruit and flower tissue where they are responsible for red, blue and purple colours. In addition they are also found in leaves, stems, seeds and root tissue. They are involved in the protection of plants against excessive light by shading leaf mesophyll cells and also have an important role to play in attracting pollinating insects.

The most common anthocyanidins are pelargonidin, cyanidin, delphinidin, peonidin, petunidin and malvidin (Figure 1.8). In plant tissues these compounds are invariably found as sugar conjugates that are known as anthocyanins. The anthocyanins also form conjugates with hydroxycinnamates and organic acids such as malic and acetic acids. Although conjugation can take place on carbons 3, 5, 7, 3′ and 5′, it occurs most often on C3 (Figure 1.9). In certain products, such as matured red wines and ports, chemical and enzymic transformations occur and an increasing number of 'anthocyanin-derived polyphenols' that contribute to the total intake of dietary phenols are now known.

1.2.1.5 Flavanones

The flavanones are characterized by the absence of a $\Delta^{2,3}$ double bond and the presence of a chiral centre at C2 (Figure 1.1). In the majority of naturally occurring flavanones, the C-ring is attached to the B-ring at C2 in the α-configuration. The flavanone

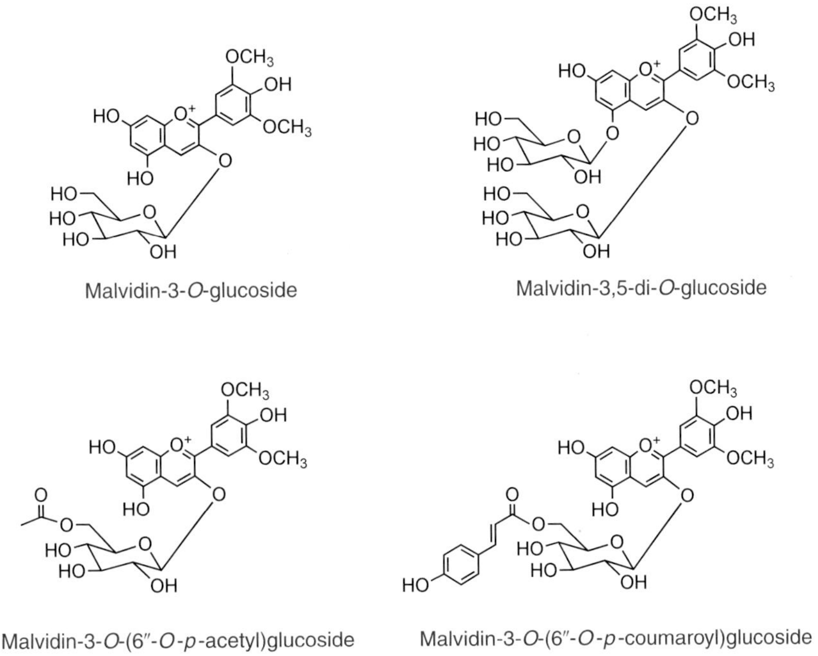

Figure 1.9 Anthocyanin structures: different types of malvidin-3-*O*-glucoside conjugates.

structure is highly reactive and have been reported to undergo hydroxylation, glycosylation and *O*-methylation reactions. Flavanones are dietary components that are present in especially high concentrations in citrus fruits. The most common flavanone glycoside is hesperetin-7-*O*-rutinoside (hesperidin) which is found in citrus peel. Flavanone rutinosides are tasteless. In contrast, flavanone neohesperidoside conjugates such as hesperetin-7-*O*-neohesperidoside (neohesperidin) from bitter orange (*Citrus aurantium*) and naringenin-7-*O*-neohesperidoside (naringin) (Figure 1.10) from grapefruit peel (*Citrus paradisi*) have an intensely bitter taste. The related neohesperidin dihydrochalcone is a sweetener permitted for use in non-alcoholic beers.

1.2.1.6 Isoflavones

Isoflavones are characterized by having the B-ring attached at C3 rather than the C2 position (Figure 1.1). They are found almost exclusively in leguminous plants with highest concentrations occurring in soyabean (*Glycine max*) (US Department of Agriculture, Agricultural Research Service, 2002). The isoflavones – genistein and daidzein – and the coumestan – coumestrol (Figure 1.11) – from lucerne and clovers (*Trifolium* spp) have sufficient oestrogenic activity to seriously affect the reproduction of grazing animals such as cows and sheep and are termed phyto-oestrogens. The structure of these isoflavonoids is such that they appear to mimic the steroidal hormone oestradiol (Figure 1.11) which blocks ovulation. The consumption of legume fodder by animals must therefore be restricted or low-isoflavonoid-producing varieties must be selected. This is clearly

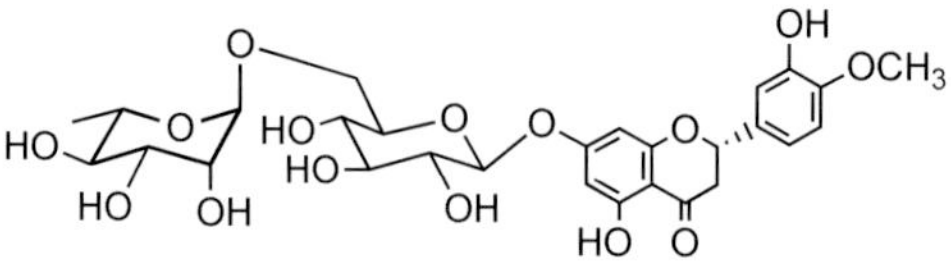

Hesperetin-7-*O*-rutinoside
(Hesperidin)

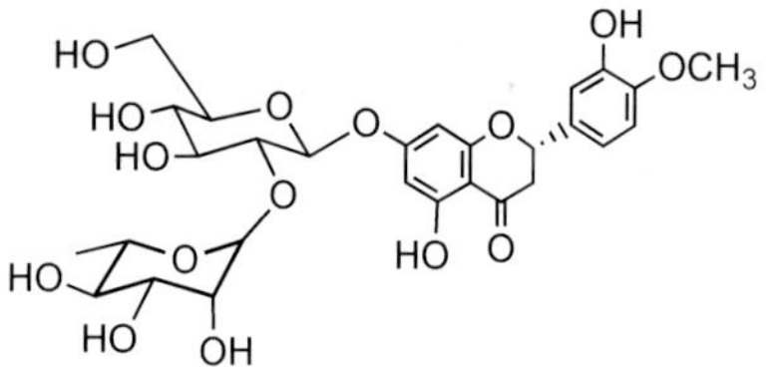

Hesperetin-7-*O*-neohesperidoside
(Neohesperidin)

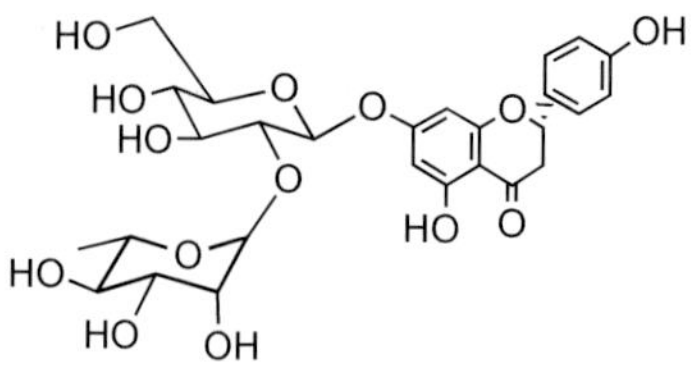

Naringenin-7-*O*-neohesperidoside
(Naringin)

Figure 1.10 Structures of the flavanones hesperidin, neohesperidin and naringin.

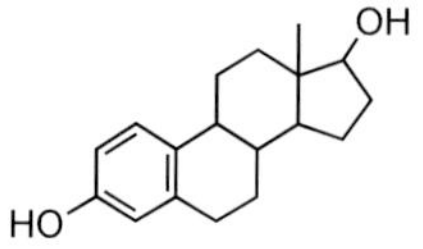

Oestradiol

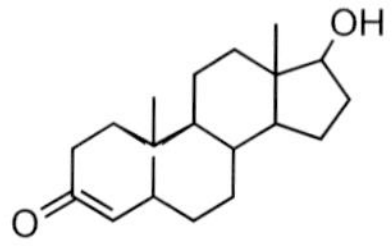

Testosterone

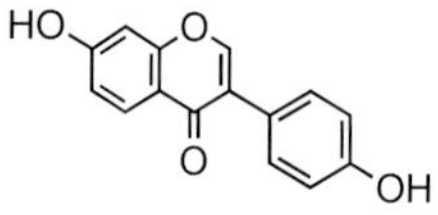

Daidzein

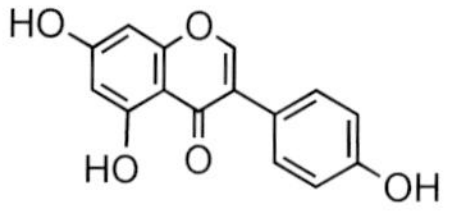

Genistein

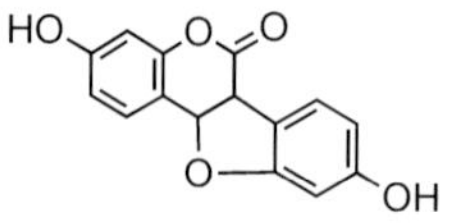

Coumestrol

Figure 1.11 Structures of the oestrogen oestradiol, the androgen testosterone and the isoflavonoids daidzein, genistein and coumestrol.

an area where it would be beneficial to produce genetically modified isoflavonoid-deficient legumes.

Dietary consumption of genistein and daidzein from soya products is thought to reduce the incidence of prostate and breast cancers in humans. However, the mechanisms involved are different. Growth of prostate cancer cells is induced by and dependent upon the androgen testosterone (Figure 1.11), the production of which is suppressed by oestradiol. When natural oestradiol is insufficient, the isoflavones can lower androgen levels and, as a consequence, inhibit tumour growth. Breast cancers are dependent upon a supply of oestrogens for growth especially during the early stages. Isoflavones compete with natural oestrogens, restricting their availability thereby suppressing the growth of cancerous cells.

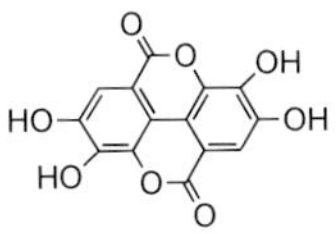

Ellagic acid 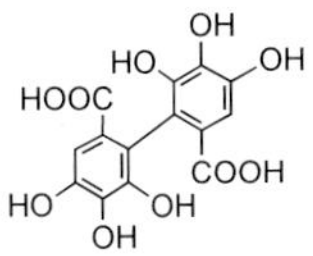   Hexahydroxydiphenic acid

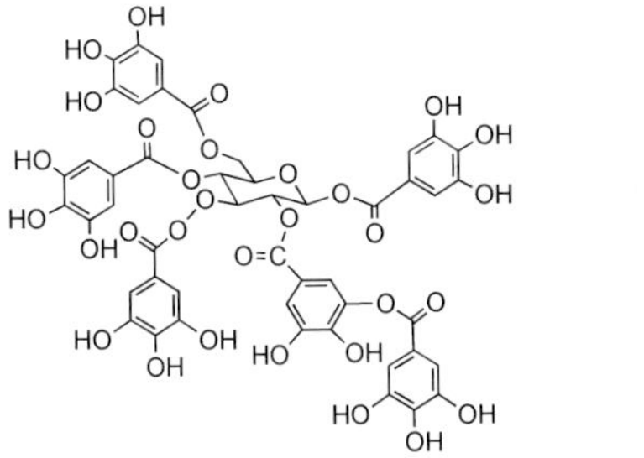

2-O-Digalloyl-tetra-O-galloylglucose
(Simple gallotannin)

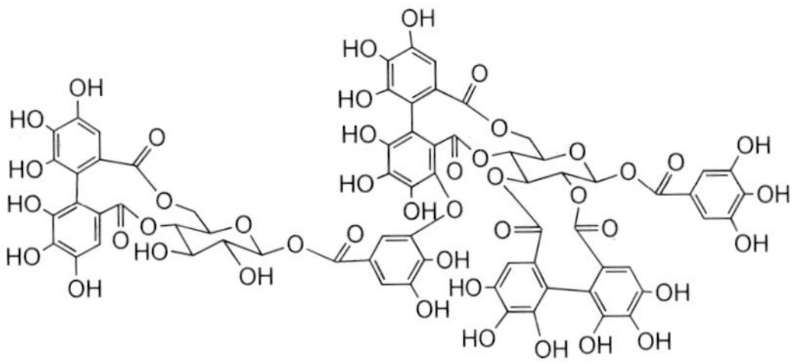

Sanguiin H-10
(Ellagitannin)

Figure 1.12 Structures of ellagic acid, hexahydroxydiphenic acid, 2-O–digalloyl-tetra-O-galloylglucose, a gallotannin and sanguiin H-10, a dimeric ellagitannin.

1.2.2 Non-flavonoids

The main non-flavonoids of dietary significance are the C_6–C_1 phenolic acids, most notably gallic acid, which is the precursor of hydrolysable tannins, the C_6–C_3 hydroxycinammates and their conjugated derivatives, and the polyphenolic C_6–C_2–C_6 stilbenes (Table 1.1).

1.2.2.1 Phenolic acids

Phenolic acids are also known as hydroxybenzoates, the principal component being gallic acid (Figure 1.12). The name derives from the French word *galle*, which means a swelling in the tissue of a plant after an attack by parasitic insects. The swelling is from a build up of carbohydrate and other nutrients that support the growth of the insect larvae. It has been reported that the phenolic composition of the gall consists of up to 70% gallic acid esters (Gross 1992). Gallic acid is the base unit of gallotannins whereas gallic acid and hexahydroxydiphenoyl moieties are both subunits of the ellagitannins (Figure 1.12). Gallotannins and ellagitannins are referred to as hydrolysable tannins, and, as their name suggests, they are readily broken down, releasing gallic acid and/or ellagic acid, by treatment with dilute acid whereas condensed tannins are not.

Condensed tannins and hydrolysable tannins are capable of binding to and precipitating the collagen proteins in animal hides. This changes the hide into leather making it resistant to putrefaction. Plant-derived tannins have, therefore, formed the basis of the tanning industry for many years. Tannins bind to salivary proteins, producing a taste which humans recognize as astringency. Mild astringency enhances the taste and texture of a number of foods and beverages, most notably tea and red wines. Clifford (1997) has reviewed the substances responsible for and the mechanisms of astringency.

Many tannins are extremely astringent and render plant tissues inedible. Mammals such as cattle, deer and apes characteristically avoid eating plants with high tannin contents. Many unripe fruits have a very high tannin content, which is typically concentrated in the outer cell layers. Tannin levels and/or the associated astringency decline as the fruits mature and the seeds ripen. This may have been an evolutionary benefit delaying the eating of the fruit until the seeds are capable of germinating.

It has been suggested that lack of tolerance to tannins may be one reason for the demise of the red squirrel. The grey squirrel is able to consume hazelnuts before they mature, and to survive on acorns. In contrast, the red squirrel has to wait until the hazelnuts are ripe before they become palatable, and it is much less able to survive on a diet of acorns which are the only thing left after the grey squirrels have eaten the immature hazelnuts (Haslam 1998).

Tannins can bind to dietary proteins in the gut and this process can have a negative impact on herbivore nutrition. The tannins can inactivate herbivore digestive enzymes directly and by creating aggregates of tannins and plant proteins that are difficult to digest. Herbivores that regularly feed on tannin-rich plant material appear to possess some interesting adaptations to remove tannins from their digestive systems. For instance, rodents and rabbits produce salivary proteins with a very high proline content (25–45%) that have a high affinity for tannins. Secretion of these proteins is induced by ingestion of food with a high tannin content and greatly diminishes the toxic effects of the tannins (Butler 1989).

1.2.2.2 Hydroxycinnamates

Cinnamic acid is a C_6–C_3 compound that is converted to range of hydroxycinnamates which, because they are products of the phenylpropanoid pathway, are referred to collectively as phenylpropanoids. The most common hydroxycinnamates are *p*-coumaric, caffeic and ferulic acids which often accumulate as their respective tartrate esters, coutaric, caftaric and fertaric acids (Figure 1.13). Quinic acid conjugates of caffeic acid, such as 3-, 4- and 5-*O*-caffeoylquinic acid, are common components of fruits and vegetables. 5-*O*-Caffeoylquinic acid is frequently referred to as chlorogenic acid, although strictly this term is better reserved for a whole group of related compounds. Chlorogenic acids form ~10% of leaves of green maté (*Ilex paraguariensis*) and green robusta coffee beans (processed seeds of *Coffea canephora*). Regular consumers of coffee may have a daily intake in excess of 1 g.

1.2.2.3 Stilbenes

Members of the stilbene family which have the C_6–C_2–C_6 structure (Table 1.1), like flavonoids, are polyphenolic compounds. Stilbenes are phytoalexins, compounds produced by plants in response to attack by fungal, bacterial and viral pathogens. Resveratrol is the most common stilbene. It occurs as both the *cis* and the *trans* isomers and is present in plant tissues primarily as *trans*-resveratrol-3-*O*-glucoside which is known as piceid and polydatin (Figure 1.14). A family of resveratrol polymers, viniferins, also exists.

The major dietary sources of stilbenes include grapes, wine, soya and peanut products. *Trans*-resveratrol and its glucoside are found in especially high amounts in the Itadori plant (*Polygonum cuspidatum*), which is also known as Japanese knotweed (Burns *et al.* 2002). It is an extremely noxious weed that has invaded many areas of Europe and North America.

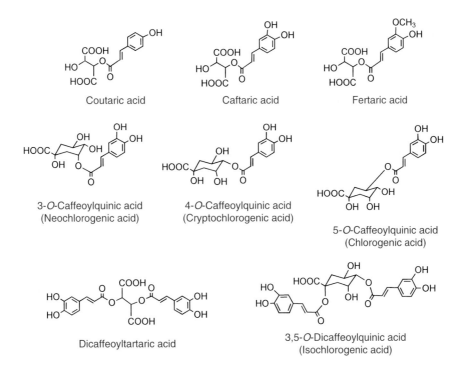

Figure 1.13 Structures of conjugated hydroxycinnamates.

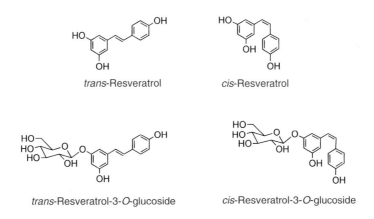

Figure 1.14 Structures of the stilbenes *trans*- and *cis*-resveratrol and their glucosides.

In its native Asia, the Itadori root is dried and infused to produce a tea. Itadori means 'well-being' in Japanese and Itadori tea has been used for centuries in Japan and China as a traditional remedy for many diseases including heart disease and stroke (Kimura *et al.* 1985). The active agent is believed to be *trans*-resveratrol and its glucoside which have also been proposed as contributors to the cardioprotective effects of red wine as it has been shown that *trans*-resveratrol can inhibit LDL oxidation, the initial stage of pathenogenesis of atherosclerosis (Soleas *et al.* 1997).

1.3 Biosynthesis

The biosynthesis of flavonoids, stilbenes, hydroxycinnamates and phenolic acids involves a complex network of routes based principally on the shikimate, phenylpropanoid and flavonoid pathways (Figures 1.15–1.17). An overview of these pathways will be discussed

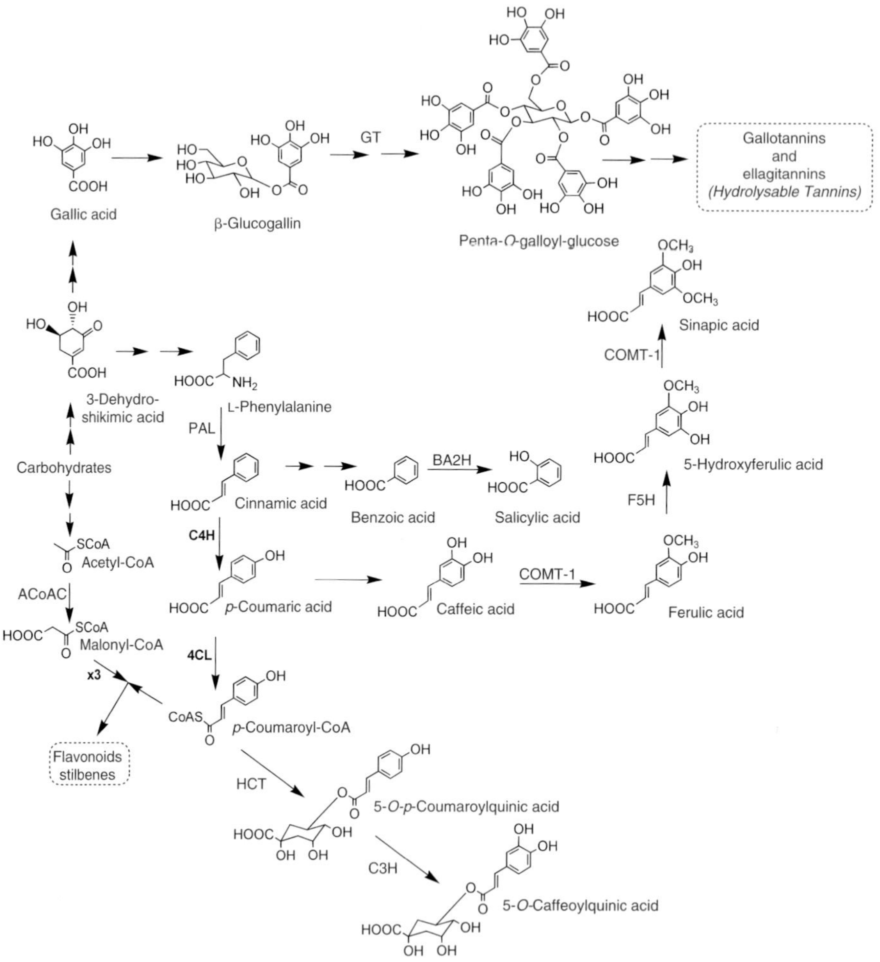

Figure 1.15 Schematic of the main pathways and key enzymes involved in the biosynthesis of hydrolysable tannins, salicylic acid, hydroxycinnamates and 5-caffeoylquinic acid. Enzyme abbreviations: PAL, phenylalanine ammonia-lyase; BA2H, benzoic acid 2-hydroxylase; C4H, cinnamate 4-hydroxylase; COMT-1, caffeic/5-hydroxyferulic acid O-methyltransferase; 4CL, p-coumarate:CoA ligase; F5H, ferulate 5-hydroxylase; GT, galloyltransferase; ACoAC, acetylCoA carboxylase.

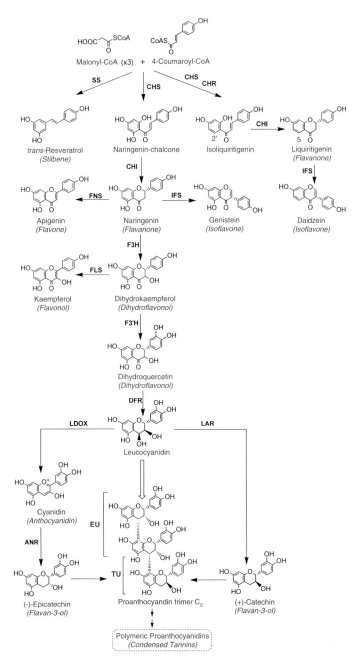

Figure 1.16 Schematic of the main pathways and enzymes involved in the production of stilbenes and flavonoids. Enzyme abbreviations: SS, stilbene synthase; CHS, chalcone synthase; CHR, chalcone reductase; CHI, chalcone isomerase; IFS, isoflavone synthase; FNS, flavone synthase; FLS, flavonol synthase; DFR, dihydroflavonol 4-reductase; ANS, anthocyanidin 4-reductase; F3H, flavanone 3-hydroxylase; F3′H, flavonol 3′-hydroxylase; LAR, leucocyanidin 4-reductase; LDOX, leucocyanidin deoxygenase; ANR, anthocyanidin reductase; EU, extension units; TU, terminal unit.

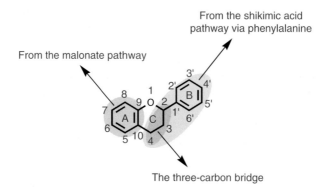

Figure 1.17 Biosynthetic origin of the flavonoid skeleton.

with particular emphasis on the production of secondary metabolites that are of dietary interest as they are significant components in commonly consumed fruits, vegetables and beverages. It should be pointed out that much of the recent information on these pathways, the enzymes involved and the encoding genes, has come from molecular biology-based studies that have utilized *Arabidopsis thaliana* as a test system. More comprehensive information on the network of pathways that are responsible for the synthesis of numerous secondary metabolites can be found in articles by Shimada *et al.* (2003), Tanner *et al.* (2003), Hoffmann *et al.* (2004), Dixon *et al.* (2005), Niemetz and Gross (2005) and Xie and Dixon (2005).

1.3.1 Phenolics and hydroxycinnamates

Gallic acid appears to be formed primarily via the shikimic acid pathway from 3-dehydroshikimic acid (Figure 1.15) although there are alternative routes from hydroxybenzoic acids. Enzyme studies with extracts from oak leaves have shown that gallic acid is converted to β-glucogallin which, in turn, is converted via a series of position-specific galloylation steps to penta-O-galloyl-glucose. Penta-O-galloyl-glucose is a pivotal intermediate that is further galloylated resulting in the synthesis of gallotannins and ellagitannins, the hydrolysable tannins (Niemetz and Gross 2005). Ellagitannins have an enormous structural variability forming dimeric and oligomeric derivatives (Figure 1.12). They also have a much more widespread distribution than gallotannins. The exact origin of ellagic acid, which is found in relatively low amounts in plant tissues, is unclear. Rather than being produced directly from gallic acid, it may be derived from ellagitannins, which upon hydrolysis liberate hexahydroxydiphenoyl residues as free hexahydroxydiphenic acid which undergoes spontaneous conversion to ellagic acid (Figure 1.12).

An alternate fate of the products of photosynthesis that are channeled through the shikimate pathway is for 3-dehydroshikimic acid to be directed to L-phenylalanine and so enter the phenylpropanoid pathway (Figure 1.15). Phenylalanine ammonia-lyase catalyses the first step in this pathway, the conversion of L-phenylalanine to cinnamic acid, which in a reaction catalysed by cinnamate 4-hydroxylase is converted to *p*-coumaric acid which in turn is metabolized to *p*-coumaroyl-CoA by *p*-coumarate:CoA ligase. Cinnamic acid is also metabolized to benzoic acid and salicylic acid although the flux through this latter step, which is catalysed by benzoic acid 2-hydroxylase, appears only to be significant in disease-resistant plants where infection induces the accumulation of salicylic acid. This acts as a trigger initiating events that restrict the spread of fungal, bacterial or viral pathogens by producing necrotic lesions around the initial point of infection (Crozier *et al.* 2000). *p*-Coumaric acid is also metabolized via a series of hydroxylation and methylation reactions leading to caffeic, ferulic, 5-hydroxyferulic and sinapic acids. Sinapic and ferulic acids are precursors of lignins. It was originally thought that caffeic acid was the immediate precursor of 5-*O*-caffeoylquinic acid, a common component of fruits and vegetables. However, recent molecular biology studies indicate that the main route to 5-*O*-caffeoylquinic acid, and presumably related caffeoylquinic acids, is from *p*-coumaroyl-CoA via 5-*O*-*p*-coumaroylquinic acid (Figure 1.15) (Hoffman *et al.* 2004). *p*-Coumaroyl-CoA is also a pivotal intermediate leading to the synthesis of flavonoids and stilbenes (Figures 1.15 and 1.16).

1.3.2 Flavonoids and stilbenes

1.3.2.1 The pathways to flavonoid formation

The C_6–C_3–C_6 flavonoid structure is the product of two separate biosynthesis pathways (Figure 1.16). The bridge and the aromatic B-ring constitute a phenylpropanoid unit synthesized from *p*-coumaroyl-CoA. The six carbons of ring-A originate from the condensation of three acetate units via the malonic acid pathway (Figure 1.17). The fusion of these two parts involves the stepwise condensation of *p*-coumaroyl-CoA with three malonyl CoA residues, each of which donates two carbon atoms, in a reaction catalysed by chalcone synthase (CHS). The product of this reaction is naringenin-chalcone. A slight modification in the pathway is involved in the production of isoflavones, such as daidzein, that are derived from isoliquiritigenin which, unlike naringenin-chalcone, lacks a 2′-hydroxyl group (Dixon 2004). The formation of isoliquiritigenin is catalysed by chalcone reductase, an NADPH-dependent enzyme that presumably interacts with CHS (Welle and Grisebach 1988) (Figure 1.16).

The next step in the flavonoid biosynthesis pathway is the stereospecific conversion on naringenin-chalcone to naringenin by chalcone isomerase (CHI). In legumes, CHI also catalyses the conversion of isoliquiritigenin to liquiritigenin (Forkmann and Heller 1999). The isomerization of naringenin-chalcone to naringenin is very rapid compared with the isomerization of isoliquiritigenin to liquiritigenin due to intramolecular hydrogen bonding in the substrate molecule. CHI enzymes isolated from non-leguminous plants are unable to catalyse the conversion of isoliquiritigenin to liquiritigenin. As a consequence, CHI has been classified into two groups. Type I CHI, which is found in both legumes and non-legumes, isomerizes only 2′-hydroxychalcones whereas Type II CHI, which is exclusive to legumes, accepts both 2′-deoxy and 2′-hydroxychalcones as a substrate (Shimada *et al.* 2003).

Naringenin is a central intermediate as from this point onwards the flavonoid biosynthetic pathway diverges into several side branches each resulting in the production of different classes of flavonoids including isoflavones, flavanones, flavones, flavonols, flavan-3-ols and anthocyanins (Figure 1.16).

1.3.2.2 Isoflavonoid biosynthesis

Isoflavonoids are found principally in leguminous plants and most of the enzymes involved in their biosynthesis have been identified and the genes cloned (Jung et al. 2003). The microsomal cytochrome P450 enzyme isoflavone synthase (IFS) catalyses the first step in this branch of the pathway converting naringenin and isoliquiritigenin into the isoflavones genistein and daidzein, respectively (Dixon and Ferreira 2002) (Figure 1.16). Further metabolism of the isoflavones, characterized by the fate of daidzein, results in a 4'-O-methylation catalysed by isoflavone-O-methyltransferase yielding formononetin and a 7-methylation to produce isoformononetin (Figure 1.18). Formononetin, in turn, undergoes a series of reactions including hydroxylation, reduction and dehydration to form the phytoalexin medicarpin. The enzymes involved in this three-step reaction are isoflavone reductase, vestotone reductase and dihydro-4'-methoxy-isoflavonol dehydratase (López-Meyer and Paiva 2002). Other less, well-characterized steps result in the conversion that yields a range of isoflavonoids including coumestans, rotenoids and pterocarpins (Figure 1.18).

1.3.2.3 Flavone biosynthesis

Flavone synthase is the enzyme responsible for the oxidation of flavanones to flavones. The conversion of naringenin to apigenin, illustrated in Figure 1.16 involves the introduction of a $\Delta^{2,3}$ double bond and this reaction requires NADPH and oxygen.

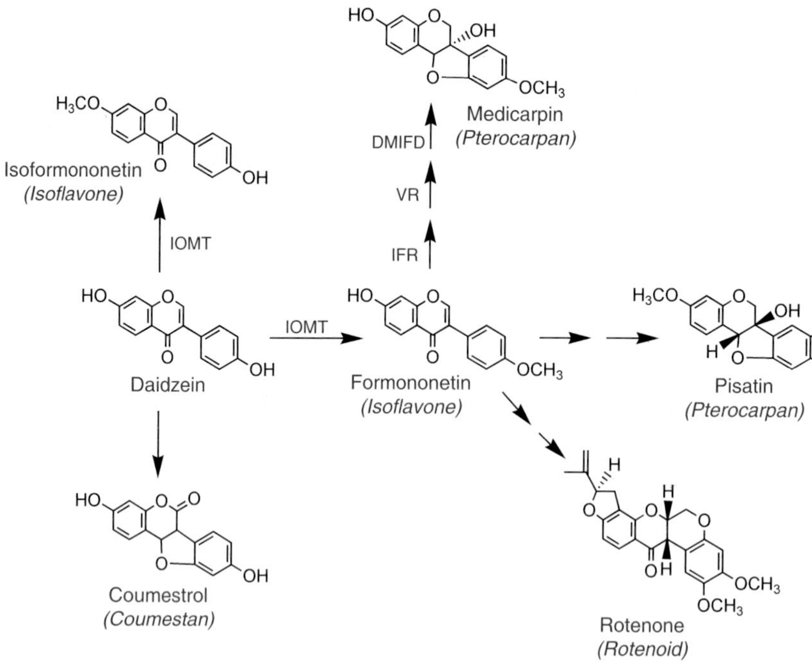

Figure 1.18 The isoflavone daidzein and biosynthetically related compounds. IOMT, isoflavone-O-methyltransferase; IFR, isoflavone reductase; VR, vestotone reductase; DMIFD, dihydro-4'-isoflavanol dehydratase.

1.3.2.4 Formation of intermediates in the biosynthesis of flavonols, flavan-3-ols, anthocyanins and proanthocyanidins

As a result of hydroxylation at C3, which is catalysed by flavanone 3-hydroxylase, flavanones are converted to dihydroflavonols as illustrated with the conversion of naringenin to dihydrokaempferol (Figure 1.16). The following steps are branch points in the pathway and involve flavonol synthase, which catalyses the introduction of a $\Delta^{2,3}$ double bond to convert dihydrokaempferol to the flavonol kaempferol, and flavonol 3′-hydroxylase, which is responsible for the synthesis of dihydroquercetin (Figure 1.16). Dihydroflavonol reductase then converts dihydroflavonols to leucoanthocyanidins, illustrated in Figure 1.16 with the synthesis of leucocyanidin from dihydroquercetin. Leucoanthocyanidins are key intermediates in the formation of flavan-3-ols proanthocyanidins and anthocyanidins. The enzyme leucocyanidin 4-reductase is responsible for the conversion of leucocyanidin to the flavan-3-ol (+)-catechin, whereas leucocyanidin deoxygenase catalyses the synthesis of cyanidin which anthocyanidin reductase converts to (−)-epicatechin. With regard to the synthesis of proanthocyanidins, the available evidence indicates that the conversion of leucocyanidin to the monomers (+)-catechin and (−)-epicatechin represents a minor fraction of the total flux and provides the terminal unit whereas the remainder, derived directly from leucocyanidin, provides the reactive extension units (Figure 1.16) (Tanner *et al.* 2003).

1.3.2.5 Stilbene biosynthesis

The stilbene *trans*-resveratrol is also synthesized by condensation of *p*-coumaroyl CoA with three units of malonyl CoA, each of which donates two carbon atoms, in a reaction catalysed by stilbene synthase. The same substrate yields naringenin-chalcone, the immediate precursor of flavonoids, when catalysed by CHS (Figure 1.16). Stilbene synthase and CHS have been shown to be structurally very similar and it is believed that both are members of a family of polyketide enzymes (Soleas *et al.* 1997). CHS is constitutively present in tissues whereas stilbene synthase is induced by a range of stresses including UV radiation, trauma and infection.

1.4 Genetic engineering of the flavonoid biosynthetic pathway

Using biochemical and molecular approaches, not only have many of the enzymes involved in the flavonoid pathway (Figure 1.16) been identified, cloned and characterized (Winkel-Shirley 2001) but a vast amount of data has also been generated on the mechanism and the regulation of these pathways. Enzymes are now believed to be compartmentalized into macromolecular complexes, which facilitate the rapid transfer of biosynthetic intermediates between catalytic sites without them diffusing into the cytoplasm of the cell. This compartmentalization of enzymes and metabolic channelling is also thought to be responsible for regulation and coordination of activities involved in the complex flavonoid pathway where several routes and side branches are active within the cell (Winkel 2004). With recent elucidation of 3-D structures of specific enzymes, information on their molecular mechanism, enzyme–substrate specificity and their involvement in metabolic channelling is also now available. New information is also emerging with regard to the regulation of phenylpropanoid and flavonoid pathways through transcriptional factors, which are regulatory proteins that activate metabolic pathways (Winkel-Shirley 2001). The vast amount of data generated on this topic together with the diverse functions of phenolic compounds have opened up possibilities to genetically engineer plants with tailor-made optimized flavonoid levels. In the longer term similar possibilities also exist for hydroxycinnamates.

1.4.1 Manipulating flavonoid biosynthesis

Early attempts to manipulate flavonoid biosynthesis were made for different reasons. During the last decade a number of genetic engineering studies on the flavonoid pathway were carried out to generate novel flower colours, in particular, blue and yellow flowering cultivars of ornamental plants such as *Dianthus* and *Petunia* (Forkmann and Martens 2001; Martens *et al.* 2003). Since defence against pathogens is one of the functions of flavonoids in planta, improved flavonoid production in this respect has also been attempted (Jeandet *et al.* 2002; Yu *et al.* 2003). With increasing awareness of the health benefits of flavonoids, other research has focused on increasing flavonoid levels in food crops as exemplified by studies using tomatoes (Muir *et al.* 2001) and potatoes (de Vos *et al.* 2000).

Two classes of genes can be distinguished within the flavonoid pathway. The structural genes encoding enzymes that directly participate in the formation of flavonoids and the regulatory genes that control the expression of the structural genes. Manipulation of the flavonoid pathway can be carried out by either down-regulating or over-expressing these genes. To date most of the structural and several regulatory genes have been cloned, characterized and used in gene transformation experiments to modify flavonoid synthesis in planta. The use of structural genes in metabolic engineering becomes more important when attempting to direct flavonoid synthesis towards branches that normally are absent in the host plant. This approach was used by Jung *et al.* (2000) who introduced the *IFS* gene into the non-legume Arabidopsis in order to convert naringenin, which is ubiquitous in higher plants, to the isoflavone genistein.

Regulatory genes control the expression of structural genes though the production of proteins called transcriptional factors. Transcriptional factors are believed to play an important role in regulating flavonoid production and other secondary metabolism pathways. Since transcriptional factors are able to control multiple steps within a pathway, they are potentially more powerful than structural genes, which control only a single step, when attempting to manipulate metabolic pathways in plants (Broun 2004). At least three main classes of transcriptional factors, which act in pairs or in triplicate, regulate the flavonoid pathway. For example, to successfully express the *Banyuls* gene, which encodes the anthocyanin reductase enzyme, the combinatorial control of all the three classes of transcriptional factors are required (Baudry *et al.* 2004).

The potential benefits of using transcription factors to modify fluxes through the flavonoid pathway have been highlighted by a number of studies using the maize transcriptional regulators LC and C1. The LC and C1 transcriptional factors are capable of activating different sets of structural genes in the flavonoid biosynthetic pathway (Quattrocchio *et al.* 1998). Their introduction into Arabidopsis and tobacco resulted in the production of anthocyanins in tissues where they are not normally synthesized (Schijlen *et al.* 2004). In another study, over-expression of LC and C1 in tomato resulted in an increase in flavonols in the flesh of tomato fruit. Total flavonol content of ripe transgenic tomatoes over-expressing LC and C1 was about 20-fold higher than that of the controls where flavonol production occurred only in the skin (Bovy *et al.* 2002; Le Gall *et al.* 2003). Similarly, expressing LC and C1 in potatoes resulted in enhanced accumulation of kaempferol and anthocyanins in the tubers (de Vos *et al.* 2000).

1.4.2 Constraints in metabolic engineering

Although there has been some success in metabolic engineering of the flavonoid content of crop plants, due to the complexity of the pathways involved, it nonetheless remains a challenging task to generate the desired flavonoids. In practice, the final result is dependent on a number of factors, including the approach used, the encoded function of the introduced gene and the type of promoter used as well as the regulation of the endogenous pathway (Lessard *et al.* 2002; Broun 2004). The desired results may not always be achieved. This was the case in a study by Yu *et al.* (2000) in which the C1 and R transcriptional factors were introduced into maize cell cultures expressing the soybean *IFS* gene. This yielded higher flavonol levels but did not lead to increased production of isoflavones. There are similar reports of an inability to induce the anticipated synthesis of anthocyanins in tomato (Bovy *et al.* 2002) and alfalfa (Ray *et al.* 2003).

The introduction of a new branch point into an existing pathway may interfere with endogenous flavonoid biosynthesis, and/or the transgenic enzyme may fail to compete with the native enzyme for the common substrate. This could, in part, be due to compartmentalization and metabolic channelling of substrates which may further complicate metabolic engineering strategies by limiting the access of substrates to introduced enzymes. This occurred when soybean-derived *IFS* was introduced into Arabidopsis and tomato (Jaganath 2005). The non-leguminous species did not synthesize genistein despite expression of the IFS protein. In soybean, IFS is a membrane-bound enzyme whereas in the transgenic plants expressing the *IFS* gene, IFS was located in the cytoplasm where presumably it was unable to access the substrate naringenin, which was tightly channelled towards flavonol production in the pre-existing compartmentalized multi-enzyme complex between CHS, CHI, FLS and F3H. As a consequence, despite the presence of both the appropriate enzyme and the substrate, genistein was not synthesized.

1.5 Databases

Information on the occurrence and levels of various flavonoids in fruits, vegetables, beverages and foods can be found in online databases prepared by the US Department of Agriculture, Agricultural Research Service (2002, 2003, 2004).

Acknowledgements

IBJ was supported by a fellowship from the Malaysian Agricultural Research and Development Institute.

References

Auger, C., Al Awwadi, N., Bornet, A. *et al.* (2004) Catechins and procyanidins in Mediterranean diets. *Food Res. Int.*, **37**, 233–245.

Balentine, D.A., Wiseman, S.A. and Bouwnes, L.C. (1997) The chemistry of tea flavonoids. *Crit. Rev. Food Sci. Nutr.*, **37**, 693–704.

Baudry, A., Heim, M.A. and Dubreucq, B. (2004) TT2, TT8, and TTG1 synergistically specify the expression of *BANYULS* and proanthocyanidin biosynthesis in *Arabidopsis thaliana*. *Plant J.*, **39**, 366–380.

Bovy, A., de Vos, R. and Kemper, M. (2002) High-flavonol tomatoes resulting from the heterologous expression of the maize transcription factor genes *LC* and *C1*. *Plant Cell*, **14**, 2509–2526.

Broun, P. (2004) Transcription factors as tools for metabolic engineering in plants. *Curr. Opin. Plant Biol.*, **7**, 202–209.

Burns, J., Yokota, T., Ashihara, H. *et al.* (2002) Plant foods and herbal sources of resveratrol. *J. Agric. Food Chem.*, **50**, 3337–3340.

Butler, L.G. (1989) Effects of condensed tannins on animal nutrition. In R.W. Hemmingway and J.J. Karchesy (eds), *Chemistry and Significance of Condensed Tannins*. Plenum Press, New York, pp. 391–402.

Clifford, M.N. (1986) Phenol–protein interactions and their possible significance for astringency. In G.G. Birch and M.G. Lindley (eds), *Interaction of Food Components*. Elsevier Applied Science, London, pp. 143–164.

Clifford, M.N. (1997) Astringency. In F.A. Tomás-Barberán and R.J. Robins (eds), *Phytochemistry of Fruits and Vegetables*. Clarendon Press, Oxford, pp. 87–107.

Croteau, R., Kutchan, T.M. and Lewis, N.G. (2000) Natural products (secondary metabolites. In B.B. Buchannan, W. Gruissem and R.L. Jones (eds), *Biochemistry and Molecular Biology of Plant*. American Society of Plant Physiologists, Rockville, MD, pp. 1250–1318.

Crozier, A., Lean, M.E.J., McDonald, M.S. *et al.* (1997) Quantitative analysis of the flavonoid content of commercial tomatoes, onions, lettuce and celery. *J. Agric. Food Chem.*, **45**, 590–595.

Crozier, A., Kamiya, Y., Bishop, G. *et al.* (2000) Biosynthesis of hormones and elicitor molecules. In B.B. Buchannan, W. Gruissem and R.L. Jones (eds), *Biochemistry and Molecular Biology of Plant*. American Society of Plant Physiologists, Rockville, MD, pp. 850–929.

Del Rio, D., Stewart, A.J., Mullen, W. *et al.* (2004) HPLC-MS^n analysis of phenolic compounds and purine alkaloids in green and black tea. *J. Agric. Food Chem.*, **52**, 2807–2815.

Dewick, P.M. (2002) *Medicinal Natural Products: A Biosynthetic Approach*, 2nd edn., John Wiley and Sons, Chichester.

de Vos, R., Bovy, A., Busink, H. *et al.* (2000) Improving health potential of crop plants by means of flavonoid pathway engineering. *Polyphenol Commun.* 25–26.

Dixon, R.A. (2004) Phytoestrogens. *Ann. Rev. Plant Biol.*, **55**, 225–261.

Dixon, R.A. and Ferreira, D. (2002) Genistein. *Phytochemistry*, **60**, 205–211.

Dixon, R.A., Xie, D.-Y. and Sharma, S.B. (2005) Proanthocyanidins – a final frontier in flavonoid research. *New Phytologist*, **165**, 9–28.

Forkmann, G. and Heller, W. (1999) Polyketides and other secondary metabolites including fatty acids and their derivatives. In D. Barton, K. Nakanishi and O. Meth-Cohn (eds), *Comprehensive Natural Products Biochemistry*. Sankawa, Elsevier, Amsterdam, pp. 713–747.

Forkmann, G. and Martens, S. (2001) Metabolic engineering and applications of flavonoids. *Curr. Opin. Biotechnol.*, **12**, 155–160.

Gaitan, E., Lindsay, R.H., Reichert, R.D. *et al.* (1989) Antithyriodic and gioterogenic effects of millet: role of C-glycosylflavones. *J. Clin. Endocrinol. Metab.*, **68**, 707–714.

Gross, G.G. (1992) Enzymes in the biosynthesis of hydrolysable tannins. In R.W. Heminway, P.E. Laks and S.J. Branham (eds), *Plant Polyphenols*. Plenum Press, New York, pp. 43–60.

Gu, L., Kelm, M.A., Hammerstone, J.F. *et al.* (2004) Concentrations of proanthocyanidins in common foods and estimates of consumption. *J. Nutr.*, **134**, 613–617.

Harborne, J.B. (1993) *The Flavonoids: Advances in Research Since 1986*. Chapman & Hall, London.

Haslam, E. (1998) *Practical Polyphenols – From Structure to Molecular Recognition and Physiological Action*. Cambridge University Press, Cambridge.

Hertog, M.G.L., Hollman, P.C.H. and Katan, M.B. (1992) Content of potentially anticarcinogenic flavonoids of 28 vegetables and 9 fruits commonly consumed in The Netherlands. *J. Agric. Food Chem.*, **40**, 2379–2383.

Hertog, M.G.L., Hollman, P.C.H. and van de Putte, B. (1993) Content of potentially anticarcinogenic flavonoids of tea, infusions, wines and fruit juices. *J. Agric. Food Chem.*, **41**, 1242–1246.

Hoffman, L., Maury, S., Martz, F. *et al.* (2004) Silencing of hydroxycinnamoyl-coenzymeA shikimate/quinate hydroxycinnomyoltransferase affects phenylpropanoid biosynthesis. *Plant Cell*, **16**, 1446–1465.

Jaganath, I.B. (2005) Dietary flavonoids: bioavailability and biosynthesis. PhD thesis, University of Glasgow.

Jeandet, P., Douillet-Breuil, A.C. and Bessis, R. (2002) Phytoalexins from the Vitaceae: biosynthesis, phytoalexins gene expression in transgenic plants, antifungal activity and metabolism. *J. Agric. Food Chem.*, **50**, 2731–2741.

Jung, W., Yu, O., Lau, S.M. *et al.* (2000) Identification and expression of isoflavone synthase, the key enzyme for biosynthesis of isofavones in legumes. *Nat. Biotechnol.*, **18**, 208–212.

Jung, W.S., Chung, III-M. and Heo, H.Y. (2003) Manipulating isoflavone levels in plants. *J. Plant Biotechnol.*, **5**, 149–155.

Kimura, Y., Okuda, H. and Arichi, S. (1985) Effects of stilbenes on arachidonate metabolism in leukocytes. *Biochim. Biophys. Acta*, **834**, 275–278.

Koes, R.E., Quattrocchio, F. and Mol, J.N.M. (1994) The flavonoid biosynthetic pathway in plants: function and evolution. *BioEssays*, **16**, 123–132.

Le Gall, G., Colquhoun, I.J., Davis, A.L. *et al.* (2003) Metabolite profiling of tomato (*Lycopersicon esculentum*) using H^{-1} NMR spectroscopy as a tool to detect potential unintended effects following a genetic modification. *J. Agric. Food Chem.*, **51**, 2447–2456.

Lessard, P.A., Kulaveerasingam, H., York, G.M. *et al.* (2002). Manipulating gene expression for the metabolic engineering of plants. *Metab. Eng.*, **4**, 67–79.

López-Meyer, M. and Paiva, N.L. (2002) Immunolocalization of vestitone reductase and isoflavone reductase, two enzymes involved in the biosynthesis of the phytoalexin medicarpin. *Physiol. Mol. Plant Pathol.*, **61**, 15–30.

Martens, S., Forkmann, G. and Britsch, L. (2003) Divergent evolution of flavonoid 2-oxoglutarate-dependent dioxygenases in parsley. *FEBS Lett.* **544**, 93–98.

Muir, S.R., Collins, G.J., Robinson, S. *et al.* (2001) Over-expression of petunia chalcone isomerase in tomato results in fruit containing increased levels of flavonols. *Nat. Biotechol.*, **19**, 470–474.

Niemetz, R. and Gross, G.G. (2005) Enzymology of gallotannin and ellagitannin biosynthesis. *Phytochemistry*, **66**, 2001–2011.

Pierpoint, W.S. (2000) Why do plants make medicines. *Biochemist*, **22**, 37–40.

Quattrocchio, F., Wing, J.F. and van der Woude, K. (1998) Analysis of bHLH and MYB domain proteins: species-specific regulatory differences are caused by divergent evolution of target anthocyanin genes. *Plant J.*, **13**, 475–488.

Ray, H., Yu, M., Auser, P. *et al.* (2003) Expression of anthocyanins and proanthocyanidins after transformation of alfalfa with maize LC. *Plant Physiol.*, **132**, 1448–1463.

Schijlen, E.G.W.M., Ric de Vos, C.H., van Tunen, A.J. *et al.* (2004) Modification of flavonoid biosynthesis in crop plants. *Phytochemistry*, **65**, 2631–2648.

Seto, R., Nakamura, H., Nanjo, F. et al. (1997) Preparation of epimers of tea catechins by heat treatment. *Biosci. Biotechnol. Biochem.*, **61**, 1434–1439.

Shimada, N., Aoki, T., Sato, S. et al. (2003) A cluster of genes encodes the two types of chalcone isomerase involved in the biosynthesis of general flavonoids and legume-specific 5-deoxy(iso)flavonoids in *Lotus japonicus*. *Plant Physiol.*, **131**, 941–951.

Soleas, G.J., Diamandis, E.P. and Goldberg, D.M. (1997) Resveratrol: a molecule whose time has come and gone? *Clin. Chem.*, **30**, 91–113.

Strack, D. (1997) Phenolic metabolism. In P.M. Dey and J.B. Harborne (eds), *Plant Biochemistry*. Academic Press, London, pp. 387–416.

Strack, D. and Wray V. (1992) Anthocyanins. In J.B. Harborne (ed.), *The Flavonoids: Advances in Research Since 1986*. Chapman & Hall, London, pp. 1–22.

Tanner, G.J., Franki, K.T., Abrahams, S. et al. (2003) Proanthocyanidin biosynthesis in plants: purification of legume leucoanthocyanidin reductase and molecular cloning of its cDNA. *J. Biol. Chem.*, **278**, 31647–31656.

Unno, T., Sugimoto, A. and Kakuda, T. (2000) Scavenging effect of tea catechins and their epimers on superoxide anion radicals generated by hypoxanthine and xanthine oxidase. *J. Sci. Food Agric.*, **80**, 601–606.

Welle, R. and Grisebach, H. (1988) Isolation of a novel NADPH-dependent reductase which coacts with chalcone synthase in the biosynthesis of 6′-deoxychalcone. *FEBS Lett.*, **236**, 221–225.

US Department of Agriculture, Agricultural Research Service (2002) *USDA-Iowa State University Database on the Isoflavone Content of Foods*. Nutrient Data Laboratory Web Site (http://www.nal.usda.gov/fnic/foodcomp/Data/isoflav/isoflav.html).

US Department of Agriculture, Agricultural Research Service (2003) *USDA Database on the Flavonoid Content of Selected Foods*. Nutrient Data Laboratory Web Site (http://www.nal.usda.gov/fnic/foodcomp/Data/flav/flav.html).

US Department of Agriculture, Agricultural Research Service (2004) *USDA Database on the Procyanidin Content of Selected Foods*. Nutrient Data Laboratory Web Site (http://www.nal.usda.gov/fnic/foodcomp/Data/PA/PA.html).

Winkel, B.S.J. (2004) Metabolic channelling in plants. *Ann. Rev. Plant Biol.*, **55**, 85–107.

Winkel-Shirley, B. (2001) Flavonoid biosynthesis. A colorful model for genetics, biochemistry, cell biology, and biotechnology. *Plant Physiol.*, **126**, 485–493.

Xie, D.-Y. and Dixon, R.A. (2005) Proanthocyanidin biosynthesis – still more questions than answers? *Phytochemistry*, **66**, 2127–2144.

Yang, B., Arai, K. and Kusu, F. (2000) Determination of catechins in human urine subsequent to tea ingestion by high-performance liquid chromatography with electrochemical detection. *Anal. Biochem.*, **283**, 77–82.

Yu, O., Jung, W., Shi, J. et al. (2000) Production of the isoflavones genistein and daidzein in non-legume dicot and monocot tissues. *Plant Physiol.*, **124**, 781–794.

Yu, O., Shi, J., Hession, A.O. et al. (2003) Metabolic engineering to increase isoflavone biosynthesis in soybean seed. *Phytochemistry*, **63**, 753–763.

Chapter 2
Sulphur-Containing Compounds

Richard Mithen

2.1 Introduction

There are two major sources of sulphur-containing plant compounds in the diet; those derived from the glucosinolate–myrosinase (substrate–enzyme) system found in cruciferous crops (Figure 2.1), such as cabbages, broccoli (*Brassica oleracea*) and watercress (*Nasturtium officinale*), and those derived from the alliin-alliinase system found within *Allium* crops (see Figure 2.7), such as garlic (*A. sativum*), onions (*A. cepa*) and leeks (*A. porrum*). Although biochemically and evolutionarily distinct from each other, both systems have a non-volatile substrate spatially separated from a glycosylated enzyme that, following tissue disruption, catalyses substrate hydrolysis to an unstable product that spontaneously and/or enzymatically converts to an array of products, many of which are volatile and have important sensory properties. The myrosinase and alliinase enzymes are both associated with lectin-like proteins and have several isoforms of unknown biological significance. The evolution of both the glucosinolate-myrosinase and the alliin-alliinase systems are probably related to the evolution of herbivore defence in ancestors of these crop plants and may still serve a role in deterring herbivores. However, certain herbivorous insects have evolved to be able to detoxify these systems, either through preventing hydrolysis or altering product formation. No doubt the production of these compounds was also a major reason behind the domestication of these crops and their widespread historical and current use as vegetables, salads, spices and herbal remedies. Epidemiological studies with both cruciferous crops and *Allium* crops suggest that they provide health benefits, particularly with regard to a reduction in risk of cancer. Experimental approaches with animal and cell models suggest that the sulphur-containing compounds of these crops may be the major bioactive agent. Additionally, these compounds may also protect against atherosclerosis and other inflammatory diseases.

Despite their similarities, there are important differences between these two systems. Diversity of product formation in the glucosinolate-myrosinase system is mainly due to structural variation of the glucosinolate substrate. In contrast, within the alliin-aliinase system the great diversity of organosulphur compounds arises from the high biological reactivity of the products of enzymic hydrolysis of a small number of alk(en)yl-L-cysteine sulphoxides, of which alliin, 2-propenyl-L-cysteine sulphoxide, is the most prominent. In this chapter, the biochemistry of the glucosinolate-myrosinase and the alliin-alliinase systems is discussed separately, and then the potential health benefits of sulphur-containing compounds discussed together, due to similarities in their mechanisms of action. *Brassica*

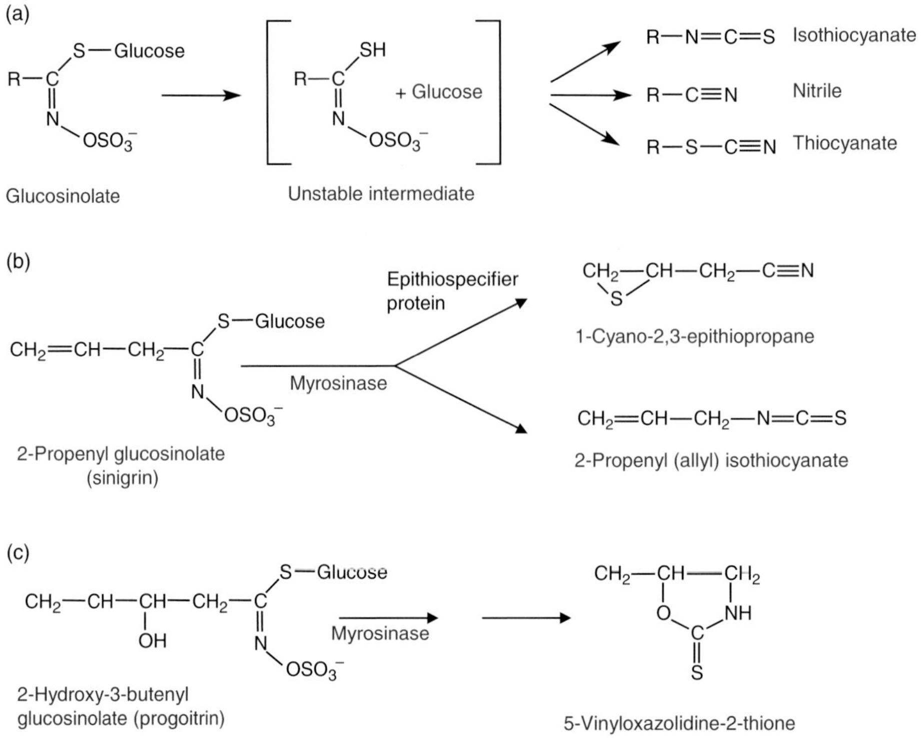

Figure 2.1 (a) Structure of glucosinolates (for details of side chain, R, see Figure 2.2), and its enzymic degradation to an unstable intermediate and subsequent conversion to isothiocyanate, nitrile and, more rarely, a thionitrile. (b) Glucosinolates with alkenyl side chains may degrade to either isothiocyanates or, in the presence of the ESP protein, epithionitriles. (c) Glucosinolates that have β-hydroxy group form unstable isothiocyanates that cyclize to produce oxazolidine-2-thiones that are the major antinutritional factor in rapeseed meal.

vegetables also contain S-methyl cysteine sulphoxide (Stoewsand 1995), which can degrade on heating to form a number of sulphur-containing volatile metabolites. These are not specifically discussed, but certain aspects of there degradation upon heating are similar to those of sulphur compounds within *Allium* crops.

2.2 The glucosinolate-myrosinase system

Glucosinolates are the main secondary metabolites found in cruciferous crops. Their presence is made known to us whenever we eat cruciferous vegetable and salad crops as they degrade immediately upon tissue damage to release a small number of products, of which isothiocyanates ('mustard oils') are the most well known. The chemical structure and concentration to which humans are exposed can be considered a consequence of three processes: First, the synthesis and accumulation of the glucosinolate molecule in the crop plants, which is dependent upon both genetic and environmental factors, second the hydrolysis of the glucosinolates to produce isothiocyanates and other products, which can

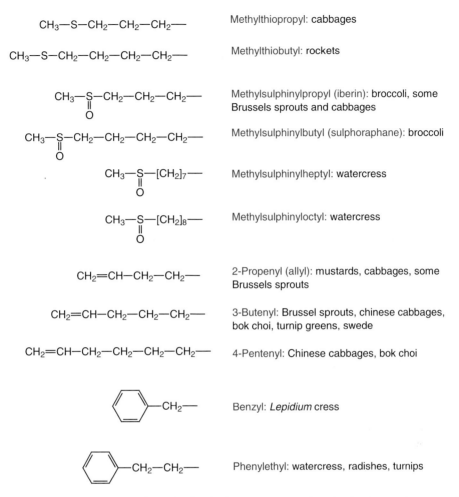

Figure 2.2 Glucosinolate side chains (for details of R, see Figure 2.1), found within frequently consumed cruciferous vegetable and salad crops. Most crops contain a mixture of a small number of these, combined with glucosinolates with indolyl side chains (Figure 2.3).

occur both within the plant and the gastrointestinal tract, and, third, the conjugation and subsequently metabolism of these degradation products, post-absorption.

2.3 Chemical diversity of glucosinolates in dietary crucifers

The glucosinolate molecule consists of a β-thioglucose unit, a sulphonated oxime unit and a variable side chain, derived from an amino acid. Glucosinolates with more than 120 side chain structures have been described (Fahey *et al.* 2001), although only about 16 of these are commonly found within crop plants (Figures 2.2 and 2.3). Seven of these 120 side chain structures correspond directly to a protein amino acid (alanine, valine, leucine, isoleucine, phenylalanine, tyrosine and tryptophan). The remaining glucosinolates have

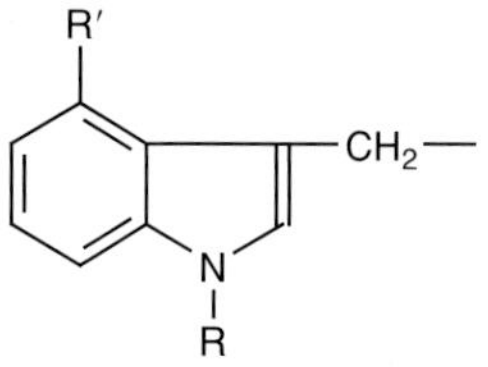

	R	R'
3-Indolylmethyl	H	H
4-Hydroxy-3-indolylmethyl	H	OH
4-Methoxy-3-indolylmethyl	H	CH$_3$O
1-Methoxy-3-indolylmethyl	CH$_3$O	H

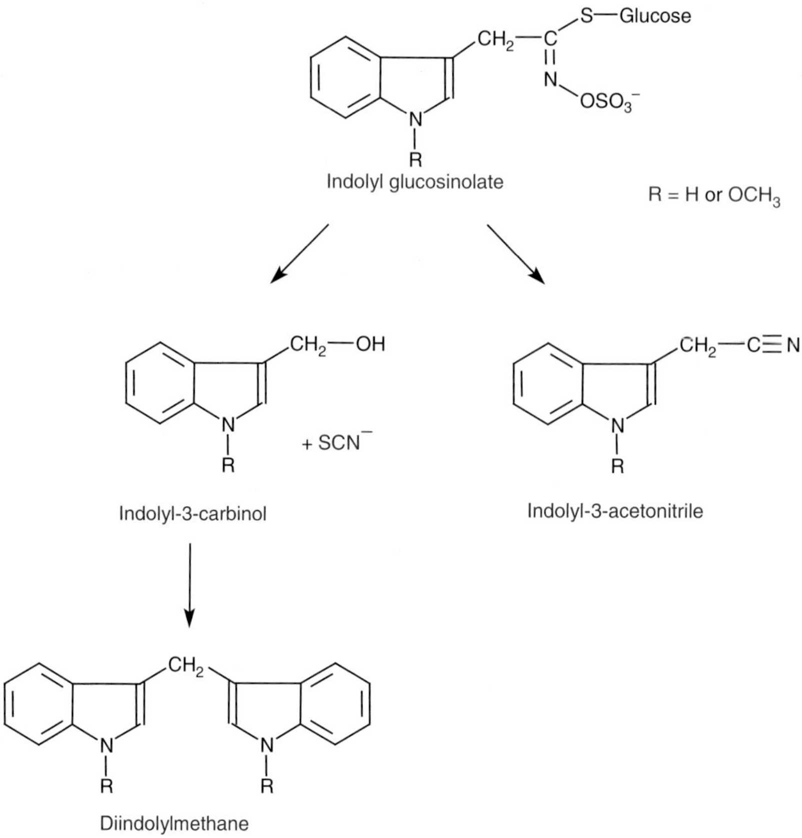

R = H or OCH$_3$

Indolyl glucosinolate

Indolyl-3-carbinol + SCN$^-$

Indolyl-3-acetonitrile

Diindolylmethane

Figure 2.3 Indolyl glucosinolates derived from tryptophan. These occur in most cruciferous vegetables, and degrade to a number of non-sulphur-containing degradation products. Indolyl-3-carbinol can conjugate with acetic acid to form ascorbigen.

side chain structures which arise in three ways. First, many glucosinolates are derived from chain-elongated forms of protein amino acids, notably from methionine but also from phenylalanine and branch chain amino acids. Second, the structure of the side chain may be modified after amino acid elongation and glucosinolate biosynthesis by, for example, the oxidation of the methionine sulphur to sulphinyl and sulphonyl, and by the subsequent loss of the ω-methylsulphinyl group to produce a terminal double bond.

Subsequent modifications may also involve hydroxylation and methoxylation of the side chain. Chain elongation and modification interact to result in several homologous series of glucosinolates, such as those with methylthioalkyl side chains ranging from $CH_3S(CH_2)_3-$ to $CH_3S(CH_2)_8-$, and methylsulphinylalkyl side chains ranging from $CH_3SO(CH_2)_3-$ to $CH_3SO(CH_2)_{11}-$. Third, some glucosinolates occur which contain relatively complex side chains such as *o*-(α-L rhamnopylransoyloxy)-benzyl glucosinolate in mignonette (*Reseda odorata*) and glucosinolates containing a sinapoyl moiety in raddish (*Raphanus sativus*). Comprehensive reviews of glucosinolate structure and biosynthesis are provided by Fahey *et al.* (2001) and Mithen (2001), respectively.

Despite the potentially large number of glucosinolates, the major cruciferous crops have a restricted range of glucosinolates. Fenwick *et al.* (1983) and Rosa *et al.* (1997) provide a useful summary of glucosinolate content in the major crop species. All of these have a mixture of indolylmethyl and *N*-methoxyindolylmethyl glucosinolates, derived from tryptophan (Figure 2.3), and either a small number of methionine-derived or phenylalanine-derived glucosinolate (Figure 2.2). The greatest diversity within a species is found within *B. oleracea*, such as broccoli, cabbages, Brussels sprouts and kales. These contain indolyl glucosinolates combined with a small number of methionine-derived glucosinolates. For example, broccoli (*B. oleracea* var. *italica*) accumulates 3-methylsuphinylpropyl and 4-methylsulphinylbutyl glucosinolates, while other botanical forms of *B. oleracea* have mixtures of 2-propenyl, 3-butenyl and 2-hydroxy-3-butenyl derivatives. Some cultivars of cabbage and Brussels sprouts also contain significant amounts of methylthiopropyl and methylthiobutyl glucosinolates. *B. rapa* (Chinese cabbage, bok-choi, turnip etc.) and *B. napus* (swede) contain 3-butenyl- and sometimes 4-pentenyl glucosinolates and often their hydroxylated homologues. In addition to methionine-derived glucosinolates, phenylethyl glucosinolate usually occurs in low levels in many vegetables. Several surveys of glucosinolate variation between cultivars of *Brassica* species have been reported, for example *B. rapa* (Carlson *et al.* 1987a; Hill *et al.* 1987) and *B. oleracea* (Carlson *et al.* 1987b; Kushad *et al.* 1999).

The distinctive taste of many minor horticultural cruciferous crops is due to their glucosinolate content. For example, watercress accumulates large amounts of phenylethyl glucosinolate, combined with low levels of 7-methylsulphinylheptyl and 8-methylsulphinyloctyl glucosinolates; rockets (*Eruca* and *Diplotaxis* species) possess 4-methylthiobutyl glucosinolate, and cress (*Lepidium sativum*) contains benzyl glucosinolate. Glucosinolates are also the precursors of flavour compounds in condiments derived from crucifers. For example, *p*-hydroxybenzyl glucosinolate accumulates in seeds of white (or English) mustard (*Sinapis alba*), 2-propenyl and 3-butenyl glucosinolates in seeds of brown mustard (*B. juncea*). 2-Propenyl and other glucosinolates occur in roots of horseradish (*Armoracia rusticana*) and wasabi (*Wasabia japonica*).

2.4 Biosynthesis

There have been major advances in the understanding of glucosinolate biosynthesis over the last decade through the use of the 'model' plant species *Arabidopsis thaliana* that has enabled a molecular genetic dissection of the biosynthetic pathway. This has resolved several aspects of the biochemistry of glucosinolates that had proved intractable via a

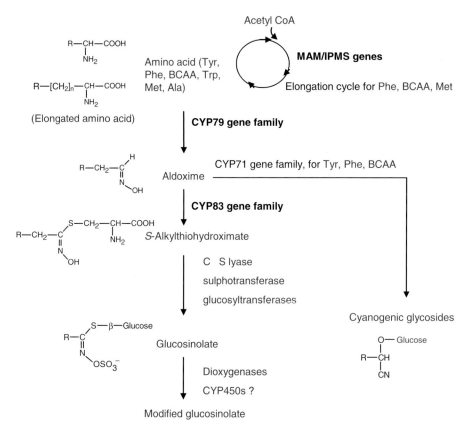

Figure 2.4 Summary of glucosinolate biosynthesis. Amino acids or elongated amino acids are converted to aldoximes by the action of enzymes encoded by the CYP79 gene family. The aldoxime conjugates with cysteine, which acts as the sulphur donor, which is then lysed and 'detoxified' by glycosylation and sulphation. The MAM and CYP79 enzymes have a high degree of specificity to the amino acid structure, whereas other enzymes of the pathway have lower specificity.

biochemical approach. The initial step is the conversion of either a primary amino acid or a chain elongated amino acid (see below) to an aldoxime (Figure 2.4). The work of Barbara Halkier and colleagues has elegantly shown that this conversion is due to gene products of the *CYP79* gene family, each of which has substrate specificity for different amino acid precursors. For example, within *Arabidopsis*, the products of *CYP79F1* and *F2* catalyse the conversion of elongated homologues of methionine to the corresponding aldoximes (Chen *et al.* 2003), CYP79B2 and CYP79B3 convert tryptophan to their aldoximes (Hull *et al.* 2000), and CYP79A2 converts phenylalanine to its aldoxime (Wittstock and Halkier 2000). The aldoxime conjugates with cysteine which acts as the sulphur donor and is then cleaved by a C–S lyase (Mikkelsen *et al.* 2004). The resultant potential toxic thiohydroximates are 'detoxified' by glycosylation by a soluble uridine diphosphate glucose (UDPG):thiohydroximate glucosyltransferase (S-GT) to produce a desulphoglucosinolate (Matsuo and Underhill 1969; Reed *et al.* 1993; Guo and Poulton 1994) and sulphation by a soluble 3′-phosphoadenosine 5′-phosphosulphate (PAPS):desulphoglucosinolate sulphotransferase (Jain *et al.* 1990). In contrast to the CYP79 enzymes, these latter steps exhibit no specificity towards the nature of the amino acid precursor.

Many of the glucosinolates found within crop plants are derived from chain elongated forms of methionine or phenylalanine (Figure 2.4). Biochemical studies, involving the administering of ^{14}C-labelled acetate and ^{14}C-labelled amino acids and subsequent analysis of the labelled glucosinolates (Graser *et al.* 2000), suggests that amino acid elongation is similar to that which occurs in the synthesis of leucine from 2-keto-3-methylbutanoic acid and acetyl CoA (Strassman and Ceci 1963). The amino acid is transaminated to produce an α-keto acid, followed by condensation with acetyl CoA, isomerization involving a shift in the hydroxyl group and oxidative decarboxylation to result in an elongated keto acid which is transaminated to form the elongated amino acid. The elongated keto acid can undergo further condensations with acetyl CoA to result in multiple chain elongations. Studies with *Arabidopsis* have identified genes encoding enzymes similar to isopropylmalate synthase as being particularly important in determining the extent of chain elongation of methionine prior to glucosinolate synthesis. The products of these genes catalyse the condensation of acetyl CoA, as the methyl donor, with a α-keto acid derived by amino acid transamination. Different members of this family can catalyse different numbers of rounds of elongation (Field *et al.* 2004).

Following glucosinolate synthesis, the side chains can be modified by hydroxylation, methoxylation, oxidation, desaturation, conjugation with benzoic acid and glycosylation. Within *Arabidopsis*, some of these modification genes have been characterized as 2-oxogluturate-dependent dioxygenases (Hall *et al.* 2001; Kliebenstein *et al.* 2001), but other types of genes are also likely to be involved.

2.5 Genetic factors affecting glucosinolate content

Two processes act independently to determine the glucosinolate content in cruciferous crops. First, that determining the types of glucosinolates, and, second, that determining the overall amount of these glucosinolates. The first of these processes is under strict genetic control. Thus, a particular genotype will express the same ratio of glucosinolate side chains when grown in different environments, although the overall level may vary considerably. In *B. oleacea*, a series of Mendelian genes have been identified and located on linkage maps that determine the length and chemical structure of the side chain (Magrath *et al.* 1994; Giamoustaris and Mithen 1996; Hall *et al.* 2001; Li and Quiros 2003). The underlying genes are likely to correspond directly to cloned genes that have been functionally analysed in *Arabidopsis*, described above. In addition, quantitative trait loci (QTLs) have been identified that determine the overall level of glucosinolates in both *B. napus* (Toroser *et al.* 1995; Howell *et al.* 2003) and *B. oleracea* (Mithen *et al.* 2003). Introgression of alleles at these QTL alleles derived from wild *B. species* has enabled broccoli to be developed with enhanced levels of specific glucosinolates (Mithen *et al.* 2003).

2.6 Environmental factors affecting glucosinolate content

The total level of glucosinolates is determined by alleles at several QTLs, discussed above, and by several environmental factors. As would be expected, nitrogen and sulphur supply affects the amounts of glucosinolates. Zhao *et al.* (1994) showed that sulphur and nitrogen supply affected the glucosinolate content of rapeseed, and described minor alterations in the ratios of methionine-derived glucosinolates and larger ones in the ratio of indolyl to methionine-derived glucosinolates. Within vegetable Brassicas, sulphur and nitrogen fertilization has been shown to affect glucosinolate expression in some studies (Kim *et al.* 2002; Sultana *et al.* 2002) but not others (Vallejo *et al.* 2003). It has been suggested that glucosinolates can act as a sulphur store so that in times of sulphur deficiency sulphur can be mobilized from glucosinolates into primary metabolism, although good evidence is lacking. The induction of glucosinolates following abiotic or biotic stresses has frequently been described. For example, pathogens (Doughty *et al.* 1995a), insect herbviores (Griffiths *et al.* 1994), salicylic acid (Kiddle *et al.* 1994), jasmonates (Bodnaryk and Rymeson 1994; Doughty *et al.* 1995b) all induce glucosinolates. In general, indolyl glucosinolates seem to be induced to a greater extent and for a longer time compared to methionine-derived glucosinolates. The majority of these experiments have been undertaken on glasshouse grown plants in which glucosinolate expression is usually considerably less than in plants grown in field experiments.

2.7 Myrosinases and glucosinolate hydrolysis

In the intact plant, glucosinolates are probably located in the vacuole of many cells and are also concentrated within specialized cells. Following tissue disruption, glucosinolates are hydrolysed by thioglucosidases, known as myrosinases, which are probably located in the cytoplasm, and also in specialized myrosin cells, to result in the generation of a small number of products (Figure 2.1). Bones and Rossiter (1996) and Rask *et al.* (2000) provide reviews of myrosinases, and only a brief summary is provided here. Many myrosinase isozymes have been detected in glucosinolate-containing plants, and myrosinase (i.e. thioglucosidase) activity has also been detected in insects, fungi and bacteria. In plants, the expression of different isozymes varies both between species and between organs of the same individual (Lenman *et al.* 1993). As yet no correlation with activity or substrate specifity towards glucosinolate chain structure (if any occurs) has been described. Molecular studies in *A. thaliana*, *Brassica* and *Sinapis* have shown that myrosinases comprise a gene family (Xue *et al.* 1992) within which there are three subclasses, denoted MA, MB and MC. Members of each of these subfamilies occur in *A. thaliana* (Xue *et al.* 1995). All myrosinases are glycosylated, and the extent of glycosylation varies between the subclasses. It is likely that the subdivision of myrosinases will be revised as new sequence data become available. Associated with myrosinases are myrosinase-binding proteins (MBP) (Lenman *et al.* 1990). This is a large class of proteins with masses ranging from 30 to 110 kDa, and often contains sequences of amino acid repeats. At least 17 MBPs have so far been found in *A. thaliana*. Some of the MBPs have lectin-like activity (Taipelensuu *et al.* 1997). The role of these proteins is far from understood. They are induced upon wounding or by application of jasmonates (Geshi and Brandt 1998). It has been suggested that they may stabilize the myrosinase enzymes themselves, or possibly interact and bind to carbohydrates on the surface of invading pathogens. In addition, another class of glycoproteins, designated myrosinase-associated proteins (MyAP) has also been identified in complexes with myrosinase and MBPs. As with MBPs, the role of these is not understood. They have sequence similarities to lipases. Isoforms which occur in leaves are inducible upon wounding or treatment with methyl jasmonate (Taipalensuu *et al.* 1996).

2.8 Hydrolytic products

When tissue disruption occurs, myrosinase activity results in the cleavage of the thioglucose bond to give rise to unstable thiohydroximate-O-sulphonate. This aglycone spontaneously rearranges to produce several products (Figure 2.1). Most frequently, it undergoes a Lossen rearrangement to produce an isothiocyanate. If the isothiocyanate contains a double bond, and in the presence of an epithiospecifier protein (see below), the isothiocyanate may rearrange to produce an epithionitrile (Figure 2.1b). At lower pH, the unstable intermediate may be converted directly to a nitrile with the loss of sulphur. Conversion to nitriles is also enhanced in the presence of ferrous ions (Uda *et al.* 1986). A small number of glucosinolates have been shown to produce thiocyanates, although the mechanism by which this occurs is not known. Aglycones from glucosinolates which contain β-hydroxylated side chains, such 2-hydroxy-3-butenyl (progoitrin) found in the seeds of oilseed rape and some horticultural Brassicas, such as Brussels sprouts and Chinese cabbage, spontaneously cyclize to form the corresponding oxazolidine-2-thiones (Figure 2.1c). The epithiospecifier protein (ESP) was first described by Tookey (1973) and purified from *B. napus* (Bernardi *et al.* 2000; Foo *et al.* 2000). This protein appears to have no enzymatic activity of its own, and does not interact with glucosinolates, but only the unstable thiohydroximate-O-sulphonate following myrosinase activity. Foo *et al.* (2000) suggest that ESP has a mode of action similar to a cytochrome P450, such as in iron-dependent epoxidation reactions. While this protein has only been considered in the context of the generation of epithionitriles, it is likely to be involved in the production of nitriles from glucosinolates such as methylsulphinylalkyls. In this case, the sulphur from the glucone is lost as it cannot be reincorporated into the degradation product due the lack of a terminal double bond. Indolyl glucosinolates also form unstable isothiocyanates, which degrade to the corresponding alcohol and may condense to form diindolylmethane, and conjugate with ascorbic acid to form ascorbigen. At more acidic pH, indolyl glucosinolates can form indolyl-3-acetonitrile and elemental sulphur (Figure 2.3).

Cooking has important consequences for glucosinolate degradation. When raw crucifers such as radishes and watercress (*Nasturtium* spp.), are eaten, degradation of glucosinolates is dependent upon the endogenous myrosinase enzymes and, if present, ESP which may result in significant nitrile production. Cooking has two major effects. First, relatively mild heat treatment denatures ESP, whereas more prolonged cooking results in the denaturation of myrosinase. Thus, mild cooking, such as steaming or microwaving for approximately two minutes, can result in an enhancement of isothiocyanates as it prevents conversion of the thiohydroximate-O-sulphonate to nitriles. More prolonged cooking prevents glucosinolate degradation as myrosinase is inactivated and intact glucosinolates, which are heat stable, are ingested. The degradation of glucosinolates to isothiocyanates still occurs, but due to microbial activity in the gastrointestinal tract.

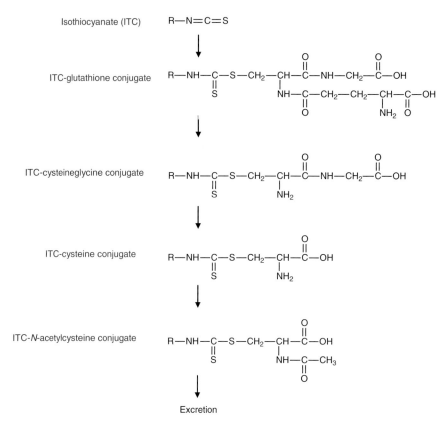

Figure 2.5 Human metabolism of isothiocyanate. Initial conjugation with glutathione occurs rapidly in an epithelial cell. Subsequent metabolism via the mercapturic acid pathway lead to N-acetylcysteine conjugates that are excreted in the urine.

2.9 Metabolism and detoxification of isothiocyanates

Humans metabolize isothiocyanates by conjugation with glutathione, and then subsequent metabolism is via the mercapturic acid pathway, and excretion, as mainly the N-acetylcysteine-isothiocyanate conjugates (Mennicke *et al.* 1983, 1988; Figure 2.5). The presence of N-acetylcysteine-isothiocyanate conjugates in urine provides a useful marker for isothiocyanate and glucosinolate consumption (Duncan *et al.* 1997; Chung *et al.* 1998; Sharpio *et al.* 1998; Rose *et al.* 2000). Epithionitriles are also metabolized in a similar manner (Brocker *et al.* 1984).

2.10 The alliin-alliinase system

Allium species, such as onions, garlic and leeks are a rich source of sulphur-containing compounds, many of which are volatile and give rise to the characteristic flavour and aroma of these species. The volatile compounds are formed by the hydrolysis of non-volatile alkyl-

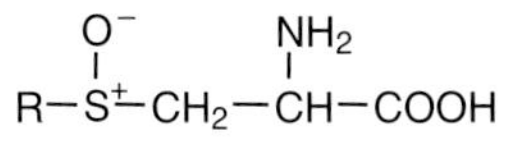

$$\underset{R}{R-\overset{O^-}{\underset{|}{S^+}}-CH_2-\overset{NH_2}{\underset{|}{CH}}-COOH}$$

R	
CH_3-	Methyl
$CH_3-CH_2-CH_2-$	Propyl
$CH_3-CH=CH-$	1-Propenyl- (Onions)
$CH_2=CH-CH_2-$	2-Propenyl- (Garlic)

Figure 2.6 Alkyl and alkenyl substituted L-cysteine sulphoxides found in *Allium* crops. All of these compounds are found in most *Allium* crops, but 1-propenyl-L-cysteine sulphoxide predominates in onions, and 2-propenyl-L-cysteine sulphoxide predominate in garlic.

and alkenyl-substituted L-cysteine sulphoxides (ACSOs, Figures 2.6 and 2.7) due to the action of the enzyme alliinase following tissue disruption. Of these precursors, 2-propenyl-L-cysteine sulphoxide (alliin) is characteristic of leeks and garlic, and 1-propenyl-L-cysteine sulphoxide (isoalliin) is characteristic of onions. Methyl and propyl-L-cysteine sulphoxides also occur in these and other *Allium* species. The ACSOs are located in the cytoplasm, while alliinase is localized within the vacuole (Lancaster *et al.* 2000). A number of forms of this enzyme have been biochemically characterized and genes cloned from a variety of *Allium* species. Several isoforms may exist within one species, some preferentially expressed in roots or leaves and with different activities towards different ACSOs (Manabe *et al.* 1998; Lancaster *et al.* 2000). Within garlic, alliinase has been shown to aggregate with low molecular mass lectins into stable active complexes (Peumans *et al.* 1997; Smeets *et al.* 1997).

In onions, the action of alliinase on 1-propenyl-L-cysteine sulphoxide results in the formation of the unstable 1-propenylsulphenic acid. This compound is either converted into propanethial S-oxide, which is the major lachrymatory factor generated when onions are cut by another enzyme termed Lachrymatory Factor (LF) synthase (Imai *et al.* 2002), or condenses with itself to form a thiosulphinate or with other thiosulphinates (Figure 2.7). The lacrimatory factor, propanethial S-oxide, is highly reactive and hydrolyses to propionaldehyde, sulphuric acid and hydrogen sulphide, and other sulphur-containing derivatives. Garlic lacks LF synthase, and all of the 2-propenyl-L-cysteine sulphoxide is converted to the diallyl thiosulphinate, allicin (Imai *et al.* 2002). The thiosulphinates are unstable and through both spontaneous and enzymic action a large array of volatile organosulphur compounds are generated (Figure 2.8). In addition to the action of alliinase, 2-propenyl-L-cysteine sulphoxide can cyclize to form a stable compound, particularly during cooking (Yanagita *et al.* 2003). This cyclic compound (cycloalliin) can account for up to 50% of all sulphur-containing compounds. In garlic, thiosulphinates are converted to predominantly mono- di- and tri-sulphides and other compounds such as ajoene (Figure 2.8). Similar compounds are generated from onions. However, due to the isomeric nature of alliin in garlic and onions, garlic derivatives have a thioallyl moiety, while onion derivatives have a thiopropyl group, and, as a consequence, have somewhat different biological activities.

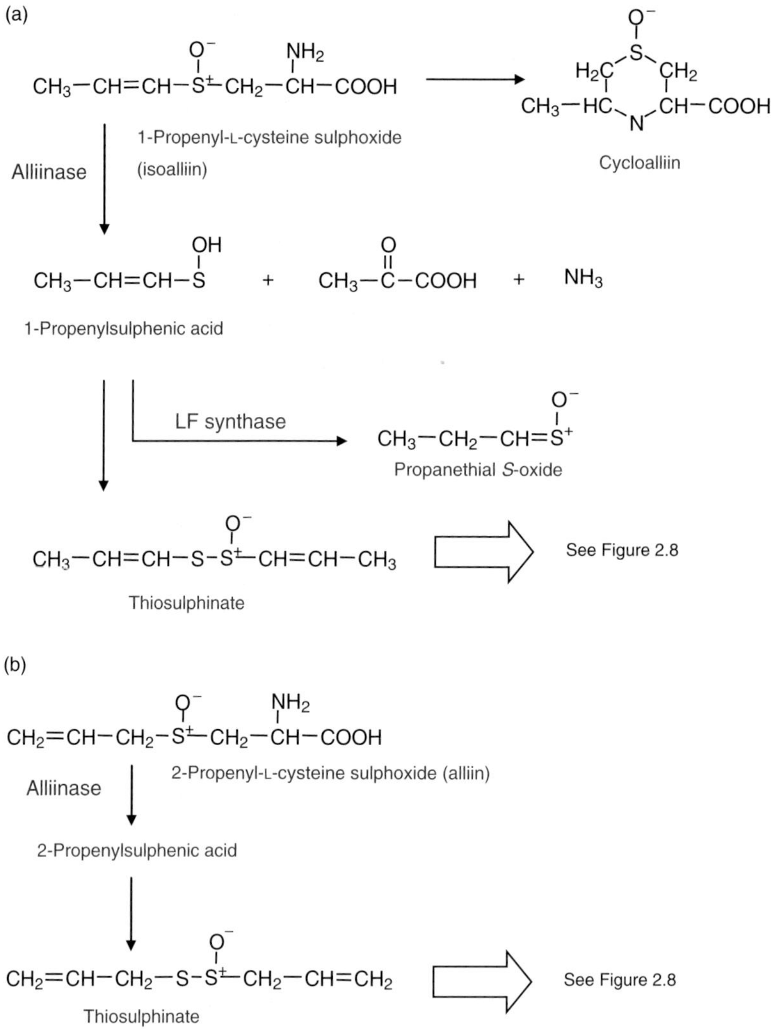

Figure 2.7 (a) Degradation of 1-propenyl-L-cysteine in onions to cycloalliin, propanethial S-oxide (the major lacrimatory compound derived from onions), and to an unstable thiosulphinate which further degrades (see Figure 2.8). (b) Degradation of 2-propenyl-L-cysteine in garlic, which lacks LF synthase.

The human metabolism of these sulphur compounds is complex and far from understood. Moreover, due to the complexity and number of potential products, it is difficult to generalize. Several compounds probably conjugate with glutathione post-absorption, and are metabolized via the mercapturic acid pathways, in a similar manner to isothiocyanates. N-acetyl-S-allyl-L-cysteine (allylmercapturic acid), derived from diallyl disulphide, has been detected in urine from humans who have consumed garlic (de Rooji et al. 1996). The characteristic breath and perspiration odours found following garlic consumption are

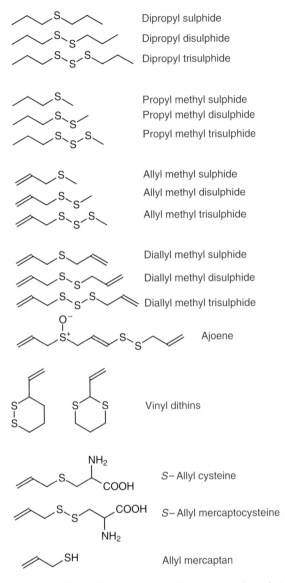

Figure 2.8 Sulphur compounds derived from enzymic and spontaneous degradation of thiosulphinates.

due to several sulphides, including allyl methyl sulphide, allyl methyl disulphide, diallyl disulphide, that reach the lungs or sweat glands via the blood stream.

2.11 Biological activity of sulphur-containing compounds

The importance of sulphur-containing compounds as flavour compounds is well known. Isothiocyanates provide the characteristic hot and pungent flavours of many of our cruciferous salad crops and condiments and contribute to flavour components of cooked cruciferous vegetables. The contribution of these to flavour is however complex. Bitterness in some cultivars of Brussel sprouts and other crucifers is possibly due to the levels of 2-hydroxy-3-butenyl and 3-butenyl glucosinolates and indole glucosinolate products, especially ascorbigen, but the direct cause is difficult to elucidate. Likewise, organosulphur compounds from *Allium* are of fundamental importance to the taste of these crops.

2.12 Anti-nutritional effects in livestock and humans

The presence of glucosinolates in the seeds of oilseed cruciferous crops significantly reduces the livestock feeding quality of the meal left following oil extraction from seeds (Bell 1984; Griffiths *et al.* 1998). This is largely due to the presence of 2-hydroxy-3-butenyl glucosinolate which degrades to 5-vinyloxazolidine-2-thione and causes thyroid dysfunction by acting as an inhibitor of thyroxine synthesis (Elfving 1980). With the expression of ESP, this glucosinolate can also degrade to 1-cyano-2-hydroxy-3-butene which can result in the enlargement of the liver and kidneys (Gould *et al.* 1980). Thiocyanates, which can also be derived from glucosinolates, can also have goitrogenic activity by acting as iodine competitors. There is no evidence for any goitrogenic effect of *Brassica* consumption in humans; inclusion of 150 g of Brussel sprouts in the diet of adult volunteers had no effect on thyroid hormones (McMillan *et al.* 1986).

Some glucosinolate products have been implicated as possible carcinogens. Foremost amongst these are indolyl-3-carbinol and indole-3-acetonitrile from hydrolysis of indolyl glucosinolates. These can be metabolized in the presence of nitrite to form mutagenic *N*-nitroso compounds (Wakabayashi *et al.* 1985; Tiedink *et al.*, 1990). 2-Propenyl and phenylethyl isothiocyanates have also been shown to induce clastogenic changes in mammalian cell lines (Musk *et al.* 1995). However, considerable amounts of epidemiological data and experimental studies suggest that glucosinolate and their hydrolytic products act as anti-carcinogenic agents in the diet.

2.13 Beneficial effects of sulphur-containing compounds in the human diet

2.13.1 *Epidemiological evidence*

Epidemiological evidence for health benefits of sulphur-containing compounds in the diet is mainly concerned with the correlation between crucifer or *Allium* vegetable consumption and reduction in the risk of cancer. While sulphur compounds are the predominant secondary metabolites in these vegetables, they also contain significant quantities of water and lipid soluble vitamins, minerals and other secondary metabolites, notably flavonoids and some cinnamates. For example, onions are a particularly rich source of quercetin glucosides (see Chapter 7). Thus, care must be taken in interpreting epidemiological studies; biological activity may be due not to sulphur compounds, but to other plant metabolites, or, more likely, to complex interactions between these metabolites.

Despite these qualifications, the epidemiological evidence does suggest that sulphur-containing compounds from these two crop types may have significant protective effects against cancer at a variety of sites. Crucifer consumption has been inversely correlated with risk of breast, lung, colon and prostate cancer, particularly, although not exclusively, with individuals who have a null allele at the glutathione S-transferase M1 (GSTM1) and/or glutathione S-transferase T1 (GSTT1) locus, either on the basis of food frequency questionnaires to assess consumption or by the quantification of N-acetylcysteine-isothiocyanate conjugates in the urine (Lin *et al.* 1998; van Poppel *et al.* 1998; London *et al.* 2000; Zhao *et al.* 2001; Ambrosone *et al.* 2004). Likewise, consumption of garlic, leeks, chives and onions has been reported to have a protective effect against esophageal, stomach and colorectal cancer (Bianchini and Vainio 2001).

2.13.2 Experimental studies and mechanisms of action

Feeding extracts of cruciferous and *Allium* vegetables, or administering individual sulphur-containing compounds, to rats and mice treated with a carcinogen has provided further evidence of the protective effect of these compounds. The results of these studies are largely consistent with those having model cell culture systems. A major type of biological activity for both glucosinolate degradation products and organosulphur compounds from *Allium* is the modulation of Phase I and Phase II genes, the former involved with the activation of carcinogens, and the latter with their detoxification, mainly through conjugation with sulphate, glucuronic acid and glutathione (see below). In addition, both classes of compounds have antiproliferative and anti-inflammatory effects. There is extensive literature on the mechanisms of the potential health benefits of these compounds, and only a brief summary is provided below.

2.13.2.1 Inhibition of Phase I CYP450

Several CYP450s act as Phase I genes that can activate carcinogens. Isothiocyanates have been shown to inhibit some of these enzymes. For example, CYP1A1 and CYP2B1/2 are inhibited in isothiocyanate-treated rat hepatocyte, whereas the expression of CYP3A4, the major CYP in human liver, was decreased in human hepatocytes (Conaway *et al.* 1996; Maheo *et al.* 1997). In contrast, treatment of cells with degradation products of indolyl glucosinolates, such as 3,3′-diindolylmethane, ascorbigen, indole-3-carbinol and indolo [3,2-β] carbazole, results in induction of CYP1A1 (Bonnesen *et al.* 2001). Allyl sulphide compounds have been shown both to reduce expression and to enhance expression of CYP2E1, important for nitrosamine activation (Siess *et al.* 1997).

2.13.2.2 Induction of Phase II enzymes

Isothiocyanates are potent inducers of Phase II detoxification enzymes, such as quinone reductase and glutathione transferase (Faulkner *et al.* 1998; Fahey and Talalay 1999; Rose *et al.* 2000; Basten *et al.* 2002; Munday and Munday 2004). The induction of these enzymes enables the excretion of potential carcinogens prior to harmful effects and is thought to be an effective mechanism to reduce the risk of carcinogenesis. The mechanism by which this occurs is likely to be by the activation of the transcription factors Nrf2. Upon exposure to

ITCs, Nrf2 dissociates from the cytoplasmic Keap1 protein and translocates to the nucleus where it binds to the antioxidant or electrophilic response element (ARE/EpRE) in the 5′ flanking region of Phase II gene, and initiates transcription. In this manner a suite of detoxification enzymes that all possess ARE domains are co-regulated (Dinkova-Kostova *et al.* 2002). The induction of these enzymes has been demonstrated in a variety of cells and tissues, such as rat and murine and human hepatocytes, CaCo-2 and LS-174 cells (originating from human colon adenocarcinoma), LNCaP (human prostrate cell lines) and lung, colon and liver tissue from rats and mice. Induction of Phase II enzymes via broccoli consumption has also been shown to reduce the risk of developing cardiovascular problems of hypertension and atherosclerosis within rats (Wu and Juurlink 2001; Wu *et al.* 2004). In a similar manner several studies have demonstrated that allyl sulphides from garlic and onions can enhance glutathione transferase activity (Takada *et al.* 1994; Guyonnet *et al.* 1999; Andorfer *et al.* 2004).

2.13.2.3 Antiproliferative activity

Both isothiocyanates and allylsulphides have antiproliferative effects on the growth of cell cultures. This may be due to inhibition of cell cycle and/or induction of apoptosis. Isothiocyanates induce apoptosis in a variety (but not all) of cell cultures (Yu *et al.* 1998; Chen *et al.* 2002; Fimognari *et al.* 2002; Misiewicz *et al.* 2003). Treatment of cell cultures results in caspase-3-like activity and proteolytic cleavage of poly (ADP-ribose) polymerase. Phenylethyl isothiocyanate has been shown to induce apoptosis by induction of p53 protein expression and p53-dependent transactivation. In contrast, p53 induction was not observed in ITC-treated cells, and other pathways of apoptosis have been suggested (Gamet-Payrastre *et al.* 2000). In addition to induction of apoptosis, ITCs have been shown to inhibit cell cycle via a variety of mechanism, including the inhibition of the cyclin-dependent kinase cdk4 and reduced cyclin D1 (Witschi *et al.* 2002), along with an induction of the cell cycle inhibitor p21WAF-1/Cip-1 (Wang *et al.* 2004), and induction of mitotic block via disruption of α-tubulin (Jackson and Singletary 2004; Smith *et al.* 2004). Likewise, diallyl thiosulphinate (allicin) and several derivatives can also induce apoptosis (Shirin *et al.* 2001; Xiao *et al.* 2003; Oommen *et al.* 2004) and inhibit cell cycle (Hirsch *et al.* 2000).

2.13.2.4 Anti-inflammatory activity

Isothiocyanate has also been shown to have anti-inflammatory activity mediated by its interactions with the transcription factor NF-κB. Treatment of Raw 264.7 macrophages resulted in a potent decrease in lipopolysaccharide-induced secretion of pro-inflammatory and pro-carcinogenic signalling factors, such as nitric oxide, prostaglandin E-2 and tumour necrosis A. Further studies revealed that ITC treatment down regulate the transcription of several enzymes via its direct, or indirect, interaction with NF-κB (Heiss *et al.* 2001). Likewise, it has been shown that sulphur compounds derived form garlic can have anti-inflammatory activity, associated with inhibition of NF-κB (Keiss *et al.* 2003) and antioxidant activity (Higuchi *et al.* 2003).

2.13.2.5 Reduction in Helicobacter pylori

Colonization of the stomach by the bacteria *Helicobacter pylori* is associated with increased risk of stomach ulcers and cancer. Isothiocyanates and garlic derivatives have been shown to be able to inhibit the growth and kill off *H. pylori* in both in vitro studies and in experimental animals (O'Gara *et al.* 2000; Fahey *et al.* 2002; Canizares *et al.* 2004).

References

Ambrose, C.B., McCann, S.E., Freudenheim, J.L. *et al.* (2004) Breast cancer risk in premenopausal women is inversely associated with consumption of broccoli, source of isothiocyanates, but is not modified by GST genotype. *J. Nutr.*, **134**, 1134–1138.

Andorfer, J.H., Tchaikovskaya, T. and Listowsky, I. (2004) Selective expression of glutathione *S*-transferase in the murine gastrointestinal tract in response to dietary organosulphur compounds. *Carcinogenesis*, **25**, 359–367.

Basten, G.P., Bao, Y. and Williamson, G. (2002) Sulforaphane and its glutathione conjugate but not sulforaphane nitrile induce UDP-glucuronosyl transferase (UGT1A1) and glutathione transferase (GSTA1) in cultured cells. *Carcinogenesis*, **23**, 1399–1404.

Bell, J.M. (1984) Nutrients and toxicants in rapeseed meal: a review. *J. Animal Sci.*, **58**, 996–1010.

Bernardi, R., Negri, A., Ronchi, S. *et al.* (2000) Isolation of the epithiospecifier protein from oil-rape (*Brassica napus* ssp. oleifera) seed and its characterization. *FEBS Lett.*, **467**, 296–298.

Bianchini, F. and Vainio, H. (2001) *Allium* vegetables and organosulphur compounds: do they help to prevent cancer? *Environ. Health Perspect.*, **109**, 893–902.

Bodnaryk, R.P. and Rymerson, R.T. (1994) Effect of wounding and jasmonates on the physio-chemical properties and flea beetle defence responses of canola seedlings, *Brassica napus* L. *Can. J. Plant Sci.*, **74**, 899–907.

Bones, A.M. and Rossiter, J.T. (1996) The myrosinase-glucosinolate system, its organisation and biochemistry. *Physiol. Plant.*, **97**, 194–208.

Bonnesen, C., Eggleston, I.M. and Hayes, J.D. (2001) Dietary indoles and isothiocyanates that are generated from cruciferous vegetables can both stimulate apoptosis and confer protection against DNA damage in human colon cell lines. *Cancer Res.*, **61**, 6120–6130.

Brocker, E.R., Benn, Lüthy, J. and von Däniken A. (1984). Metabolism and distribution of 3-4-epithiobutanenitrile in the rat. *Food Chem. Toxicol.*, **22**, 227–232.

Canizares, P., Gracia, I., Gomex, L.A. *et al.* (2004). Allyl thiosulfinates, the bacteriostatic compounds of garlic against *Helicobacter pylori*. *Biotechnol. Prog.*, **20**, 397–401.

Carlson, D.G., Daxenbichler, M.E., Tookey, H.L. *et al.* (1987a) Glucosinolates in turnip tops and roots: cultivars grown for greens and/or roots. *J. Am. Soc. Hortic. Sci.*, **112**, 179–183.

Carlson, D.G., Daxenbichler, M.E., Van Etten, C.H. *et al.* (1987b) Glucosinolates in crucifer vegetables: broccoli, Brussels sprouts, cauliflowers, collard, kale, mustard greens and kohlrabi. *J. Am. Soc. Hortic. Sci.*, **112**, 173–179.

Chen, S., Glawischnig, E., Jorgensen, K. *et al.* (2003) CYP79F1 and CYP79F2 have distinct functions in the biosynthesis of aliphatic glucosinolates in *Arabidopsis*. *Plant J.*, **33**, 923–937.

Chen, Y.R., Han, J., Kori, R. *et al.* (2002) Phenylethyl isothiocyanate induce apoptotic signaling via suppressing phosphatase activity against c-Jun N terminal kinase. *J. Biol. Chem.*, **277**, 39334–39342.

Chung, F.-L., Jiao, D., Getahun, S.M. *et al.* (1998) A urinary biomarker for uptake of dietary isothiocyanates in humans. *Cancer Epidemiol. Biomarkers Prev.*, **7**, 103–108.

Conaway, C.C., Jiao, D. and Chung, F.-L. (1996) Inhibition of rat liver cytochrome P450 isoenzymes by isothiocyanates and their conjugates, a structure–activity relationship study. *Carcinogenesis*, **17**, 2423–2427.

de Rooij, B.M., Boogaard, P.J., Rijksen, D.A. *et al.* (1996) Urinary excretion of *N*-acetyl-*S*-allyl-cysteine upon garlic consumption by human volunteers. *Arch. Toxicol.*, **70**, 635–639.

Dinkova-Kostova, A.T., Holtzclaw, W.D., Cole, R.N. *et al.* (2002) Direct evidence that sulfhydryl groups of Kep1 are the sensors regulating induction of Phase II enzymes that protect against carcinogens and oxidants. *Proc. Natl. Acad. Sci. USA*, **99**, 11908–11913.

Doughty, K.J., Porter, A.J.R., Morton, A.M. *et al.* (1995a) Variation in glucosinolate content of oilseed rape (*Brassica napus* L.) leaves. II. Response to infection by *Alternaria brassicae* (Berk.). Sacc. *Ann. Appl. Biol.*, **118**, 469–477.

Doughty, K.J., Kiddle, G.A., Morton, A.M. *et al.* (1995b) Selective induction of glucosinolates in oilseed rape leaves by methyl jasmonate. *Phytochemistry*, **38**, 347–350.

Duncan, J., Rabot, S. and Nugon-Baudon, L. (1997) Urinary mercapturic acids as markers for the determination of isothiocyanate release from glucosinolates in rats fed a cauliflower diet. *J. Sci. Food Agric.*, **73**, 214–220.

Elfving, S. (1980) Studies on the naturally occurring goitrogen 5-vinyl-2-thiooxazolidone. *Ann. Clin. Res.*, **28**, 1–47.

Fahey, J.W. and Talalay, P. (1999) Antioxidant functions of sulforaphane: a potent inducer of Phase II detoxification enzymes. *Food Chem. Toxicol.*, **37**, 973–979.

Fahey, J.W., Zalcmann, A.T. and Talalay, P. (2001) The chemical diversity and distribution of glucosinolates and isothiocyanates among plants. *Phytochemistry*, **56**, 5–51.

Fahey, J.W., Haristoy, X., Dolan, P.M. *et al.* (2002) Sulforaphane inhibits extracellular, intracellular and antibiotic-resistant strains of *Helicobacter pylori* and prevents benzo α pyrene-induced stomach tumors. *Proc. Natl Acad. Sci. USA*, **99**, 7610–7615.

Faulkner, K., Mithen, R.F. and Williamson, G. (1998) Selective increase of the potential anticarcinogen 4-methylsulphinylbutyl glucosinolate in broccoli. *Carcinogenesis*, **19**, 605–609.

Fenwick, G.R., Heaney, R.K. and Mullin, W.J. (1983). Glucosinolates and their breakdown products in food and food plants. *CRC Crit. Rev. Food Sci. Nutr.*, **18**, 123–20.

Field, B., Cardon, G., Traka, M. *et al.* (2004) Glucosinolate and amino acid biosynthesis in *Arabidopsis thaliana*. *Plant Physiol.*, **135**, 828–839.

Fimognari, C., Nusse, M., Cesari, R. *et al.* (2002) Growth inhibition, cell cycle arrest and apoptosis in human T-cell leukemia by the isothiocyanate sulforaphane. *Carcinogensis*, **23**, 581–586.

Foo, H.L., Grønning, L.M., Goodenough, L. *et al.* (2000) Purification and characterisation of epithiospecifer protein from *Brassica napus*, enzymic intramolecular sulphur addition within alkenyl thiohydroximates derived from alkenyl glucosinolate hydrolysis. *FEBS Lett.*, **468**, 243–246.

Gamet-Payrastre L., Li, P., Lumeau S. *et al.* (2000) Sulforaphane, a naturally occurring isothiocyanate, induces cell cycle arrest and apoptosis in HT29 human colon cancer cells. *Cancer Res.*, **60**, 1426–1433.

Geshi, N. and Brandt, A. (1998) Two jasmonate-inducible myrosinase-binding proteins from *Brassica napus* L. seedlings with homology to jacalin. *Planta*, **204**, 295–304.

Giamoustaris, A. and Mithen, R.F. (1996) Genetics of aliphatic glucosinolates IV. Side chain modifications in *Brassica oleracea*. *Theor. Appl. Genet.*, **93**, 1006–1010.

Gould, D.H., Gumbmann, M.R. and Daxenbichler, M.E. (1980) Pathological changes in rats fed the Crambe meal – glucosinolate hydrolytic products 2S-1-cyano-2-hydroxy-3,4-epithiobutanes – for 90 days. *Food Cosmet. Toxicol.*, **18**, 619–625.

Graser, G., Schneider B., ldham N.J. *et al.* (2000) The methionine chain elongation pathway in the biosynthesis of glucosinolates in *Eruca sativa* (Brassicaceae). *Arch. Biochem. Biophys.*, **378**, 411–419.

Griffiths, D.W., Birch, A.N.E. and Macfarlane-Smith, W.H. (1994) Induced changes in the indole glucosinolate content of oilseed rape and forage rape (*Brassica napus*) plants in response to either turnip root fly or artificial root damage. *J. Sci. Food Agric.*, **65**, 171–178.

Griffiths, D.W., Birch, A.N.E. and Hillman, J.R. (1998) Antinutritional compounds in the Brassicaceae,. Analysis, biosynthesis, chemistry and dietary effects. *J. Hortic. Sci. Biotechnol.*, **73**, 1–18.

Guo, L. and Poulton, J.E. (1994) Partial purification and characterisation of *Arabidopsis thaliana* UDPG: thiohydroximate glucosyltransferase. *Phytochemistry*, **36**, 1133–1138.

Guyonnet, D., Siess, M.H., Le Bon, A.M. *et al.* (1999) Modulation of phase II enzymes by organo-sulphur compounds from *Allium* vevegtables in rat tissues. *Toxicol. Appl. Pharmacol.*, **154**, 50–58.

Hall, C., McCallum, D., Prescott, A. *et al.* (2001) Biochemical genetics of glucosinolate chain modification in *Arabidopsis* and *Brassica. Theor. Appl. Gen.*, **102**, 369–374.

Heiss, E., Herhaus, C., Klimo, K. *et al.* (2001) Nuclear factor kappa B is a molecular target for sulforaphane-mediated anti-infammatory mechanisms. *J. Biol. Chem.*, **276**, 32008–32015.

Higuchi, O., Tateshita, K. and Nishimura, H. (2003) Antioxidant activity of sulphur-containing cmpounds in *Allium* species for human low density lipoprotein oxidation in vitro. *J. Agric. Food Chem.*, **51**, 7208–7214.

Hill, C.B., Williams, P.H., Carlson, D.G. *et al.* (1987). Variation in glucosinolates in oriental brassica vegetables. *J. Am. Soc. Hortic. Sci.*, **122**, 309–313.

Hirsch, K., Danilenko, M., Giat, J. *et al.* (2000) Effect of purified allicin, the major ingredient of freshly crushed garlic, on cancer cell proliferation. *Nutr. Cancer*, **38**, 245–254.

Howell, P.M., Sharper, A.G. and Lydiate, D.J. (2003) Homologous loci control the accumulation of seed glucosinolates in oilseed rape (*Brassica napus*). *Genome*, **46**, 454–460.

Hull, A.K., Vij, R. and Celenza, J.L. (2000) Arabidopsis cytochrome P450s that catalyse the first step of tryptophan-dependent indole-3-acetic acid biosynthesis. *Proc. Natl. Acad. Sci. USA*, **97**, 2379–2384.

Imai, S., Tsuge, N., Tomotake, M. *et al.* (2002) Plant biochemistry: an onion enzyme that makes the eyes water. *Nature*, **419**, 685.

Jackson, S.J. and Singletary, K.W. (2004) Sulforaphane: a naturally occurring mammary carcinoma mitotic inhibitor, which disrupts tubulin polymerization. *Carcinogenisis*, **25**, 219–227.

Jain, J.C., GrootWassink, J.W.D., Kolenovsky, A.D. *et al.* (1990). Purification and properties of 3′-phosphoadenosine-5′-phophosulphate desulphoglucosinolate sulphotransferase from *Brassica juncea* cell cultures. *Phytochemistry*, **29**, 1425–1428.

Keiss, H.P., Dirsch, V.M., Hartung, T. *et al.* (2003) Galic (*Allium sativum*) modulates cytokine expression in lipopolysaccharide-activated human blood there by inhibiting NF-kappaB activity. *J. sNutr.*, **133**, 2171–2175.

Kiddle, G.A., Kevin, J.D. and Wallsgrove, R.M. (1994) Salicylic acid-induced accumulation of glucosinolates in oilseed rape (*Brassica napus*) leaves. *J. Exp. Bot.*, **45**, 1343–1346.

Kim, S.J., Matsuo, T. and Watanabe, M. (2002) Effect of nitrogen and sulphur application on the glucosinolate conent in vegetable turnip rape (*Brassica rapa* L.). *Soil Sci. Plant Nutr.*, **48**, 43–49.

Kliebenstein, D.J., Lambrix, V.M., Reichelt, M. *et al.* (2001) Gene duplication in the diversification of secondary metabolism: tandem 2-oxogluturate-dependent dioxygenases control glucosinolate biosynthesis in Arabidopsis. *Plant Cell*, 681–693.

Kushad, M.M., Brown, A.F., Kurilich, A.C. *et al.* (1999) Variation of glucosinolates in vegetable crops of *Brassica oleracea. J. Agric. Food Chem.*, **47**, 1541–1548.

Lancaster, J.E., Shaw, M.L., Joyce, M.D. *et al.* (2000) A novel alliinase from onion roots. Biochemical characterization and cDNA cloning. *Plant Physiol.*, **122**, 1269–1279.

Lenman, M., Rödin, J., Josefssn, L.-G. *et al.* (1990) Immunological characterisation of rapeseed myrosinase. *Eur. J. Biochem.*, **194**, 747–753.

Lenman, M., Falk, A., Rödin, J. *et al.* (1993) Differential expression of myrosinase gene families. *Plant Physiol.*, **103**, 703–711.

Li, G. and Quiros, C.F. (2003) In planta side-chain glucosinolate modification in Arabidopsis by introduction of dioxygenase Brassica homolog BoGSL-ALK. *Theor. Appl. Gen.*, **106**, 1116–1121.

Lin, H.J., Probst-Hensch, N.M., Louie, A.D. *et al.*, (1998) Glutathione transferase null genotype, broccoli, and lower prevalence of colorectal adenomas. *Cancer Epidemiol., Biomarkers Prev.*, **7**, 647–652.

London, S.J., Yuan, J.M., Chung, F.L. *et al.* (2000) isothiocyanates, glutathione S-transferase M1 and T1 polymorphisms, and lung-cancer risk: a prospective study of men in Shanghai, China. *Lancet*, **356**, 724–729.

Magrath, R., Bano, F., Morgner, M. *et al.* (1994) Genetics of aliphatic glucosinolates. 1. Side-chain elongation in *Brassica napus* and *Arabidopsis thaliana.*, *Heredity* **72**, 290–299.

Maheo, K., Morel, F., Langouet, S. *et al.* (1997) Inhibition of cytochrome P-450 and induction of S-transferase by sulforaphane in primary human and rat hepatocytes. *Cancer Res.*, **57**, 3649–3652.

Manabe, T., Jasumi, A., Sugiyama, M. *et al.* (1998) Alliinase [S-alk(en)yl-L-cysteine sulphoxide lyase] from *Allium tuberosum* (Chinese chive) – purification, localization, cDNA cloning and heterologous functional expression. *Eur. J. Biochem.*, **257**, 21–30.

Matsuo, M. and Underhill, E.W. (1969) Biosynthesis of mustard oil glucosides XII. A UDP-glucose thiohydroxamate glycosyltransferase from *Tropaeolum majus. Biochem., Biophys. Res. Commun.*, **36**, 18–23.

McMillan, M., Spinks, E.A. and Fenwick, G.R. (1986) Preliminary observations on the effect of dietary brussels sprouts on thyroid function. *Human Toxicol.*, **5**, 5–19.

Mennicke, W.H., Görler, K. and Krumbiegler, G. (1983) Metabolism of naturally occuring isothiocyanates in the rat. *Xenobiotica*, **13**, 203–227.

Mennicke, W.H., Görler, K., Krumbiegler, G. *et al.* (1988) Studies on the metabolism and excretion of benzyl isothiocyanate in man. *Xenbiotica*, **4**, 441–447 .

Mikkelsen, M.D., Naur, P. and Halkier, B.A. (2004) Arabidopsis mutants in the $C-S$ lyase of glucosinolate biosynthesis establish a critical role for indole-3-acetaldoxime in auxin homeostasis. *Plant J.*, **37**, 770–777.

Misiewicz, I., Skupinska, K. and Kasprzycka-Guttman, T. (2003) Sulforaphane and 2-oxohexyl isothiocyanate induce cell growth arrest and apoptosis in L-1210 leukenia and ME-18 melanoma cells. *Oncol. Rep.*, **10**, 2045–2050.

Mithen, R. (2001) Glucosinolates and their degradation products. *Adv. Bot. Res.* **35**, 214–262.

Mithen. R., Faulkner, K., Magrath, R. *et al.* (2003) Development of isothiocyanate-enriched broccoli, and its enhanced ability to induce Phase II detoxification enzymes in mammalian cells. *Theor. Appl. Gen.*, **106**, 727–734.

Munday, R. and Munday, C.M. (2004) Induction of Phase II detoxification enzymes in rats by plant-derived isothiocyanates: comparison of allyl isothiocyanate with sulforaphane and related compounds. *J. Agric. Food Chem.*, **52**, 1867–1871.

Musk, S.R., Stephenson, P., Smith, T.K. *et al.* (1995) Selective toxicity of compounds naturally present in food towards the transformed phenotype of human colorectal cell line HT29. *Nutr. Cancer*, **24**, 289–298.

O'Gara, E.A., Hill, D.J. and Maslin, D.J. (2000) Activities of garlic oil, garlic powder and the diallyl constituents against *Helicobacter pylori*. *Appl. Environ. Microbiol.*, **66**, 2269–2273.

Oommen, S., Anto, R.J., Srinivas, G. *et al.* (2004) Allicin (from garlic) induces caspase-mediated apoptosis in cancer cells. *Eur. J. Pharmacol.*, **485**, 97–103.

Peumans, W.J., Smeets, K., Van Nerum, K. *et al.* (1997) Lectin and alliinase are the predominant proteins in nectar from leek (*Allium porrum* L) flowers. *Planta*, **201**, 298–302.

Rask, L., Andréasson, E., Ekbom, B. *et al.* (2000) Myrosinase, gene family evolution and herbivore defense in Brassicaceae. *Plant Mol. Biol.*, **42**, 92–113.

Reed, D.W., Davin, L., Jain, J.C. *et al.* (1993) Purification and properties of UDP-glucose, thiohydroximate glycosyltransfrase from *Brassica napus* L. seedlings. *Arch. Biochem. Biophys.*, **305**, 526–532.

Rosa, E.A.S., Heaney, R.K., Fenwick, G.R. *et al.* (1997) Glucosinoates in crop plants. *Hortic. Rev.*, **19**, 99–215.

Rose, P., Faulkner, K., Williamson, G. *et al.* (2000) 7-Methylsulfinylheptyl and 8-methylsulfinyloctyl isothiocyanates from watercress are potent inducers of Phase II enzymes. *Carcinogenesis*, **21**, 1983–1988.

Sharpio, T.A., Fahey, J.W., Wade, K.L. *et al.* (1998) Human metabolism and excretion of cancer chemoprotective glucosinolates and isothiocyanates of cruciferous vegetables. *Cancer Epidemiol. Biomarkers. Prev.*, **7**, 1091–1100.

Shirin, H., Pinto, J.T., Kawabata, Y. *et al.* (2001) Antiproliferative effects of *S*-allylmercaptocysteine on colon cancer cells when tested alone or in combination with sulindac sulfide. *Cancer Res.*, **61**, 725–731.

Siess, M.H., Le Bon, A.M., Canivenc-Lavier, M.C. *et al.* (1997) Modification of hepatic drug-metabolising enzymes in rats treated with alkyl sulfides. *Cancer Lett.*, **120**, 195–201.

Smeets, K., Van Damme, E.J., Van Leuven. *et al.* (1997) Isolation and characterization of lectins and lectin-alliinase complexes from bulbs of garlic (*Allium sativum*) and ramsons (*Allium ursinum*). *Glycoconj. J.*, **14**, 331–343.

Smith, T.K., Lund, E.K., Parker, M.L. *et al.* (2004) Allyl isothiocyanate causes mitotic block, loss of cell adhesion and disrupted cytoskeletal structure in HT29 cells. *Carcinogenesis*, **102**, 369–374.

Stoewsand, G.S. (1995) Bioactive organosulphur phytochemicals in *Brassica oleracea* vegetables – a review. *Food Chem. Toxicol.*, **33**, 537–543.

Strassman, M. and Ceci, L.N. (1963) Enzymic formation of α-isopropylmalic acid, an intermediate in leucine biosynthesis. *J. Biol. Chem.*, **238**, 2445–2452.

Sultana, T., Savage, G.P., McNeil, D.L. *et al.* (2002) Effects of fertilization on the allyl isothiocyanate profile of above-ground tissues of New Zealand-grown wasabi. *J. Sci. Food Agric.*, **82**, 1477–1482.

Taipalensuu, J., Falk, A. and Rask, L. (1996) A wound- and methyl jasmonate-inducible transcript coding for a myrosinase-associated protein with similarities to an early nodulin. *Plant Physiol.*, **110**, 483–491.

Taipalensuu, J., Eriksson, S. and Rask, L. (1997) The myrosinase-binding protein from *Brassica napus* seeds possesses lectin activity and has a highly similar vegetatively expressed wound-inducible counterpart. *Eur. J. Biochem.*, **250**, 680–688.

Takada, N., Kitano, M., Chen, T. *et al.* (1994) Enhancing effects of organosulphur compounds from garlic and onions on hepatocarcinogenesis in rats: association with increased cell proliferation and elevated ornithine decarboxylase activity. *Jpn. J. Can. Res.*, **85**, 1067–1072.

Tookey, H.L. (1973). Separation of a protein required for epithiobutane formation. *Can. J. Biochem.*, **51**, 1654–1660.

Toroser, D., Thormann, C.E., Osborn, T.C. *et al.* (1995) RFLP mapping of quantitative trait loci controlling seed aliphatic-glucosinolate content in oilseed rape (*Brassica napus* L.). *Theor. Appl. Gen.*, **91**, 802–808.

Uda, Y., Kurata, T. and Arakawa, N. (1986) Effects of pH and ferrous ions on the degradation of glucosinolates by myrosinase. *Agric. Biol. Chem.*, **50**, 2735–2740.

Vallejo, F., Tomas-Barberan, F.A., Benavente-Garcia, A.G. *et al.* (2003) Total and individual glucoinolate contents in inflorescences of eight broccoli cultivars grown under various climatic and fertilization conditions. *J. Sci. Food Agric.*, **83**, 307–313.

Van Poppel, G., Verhoeven, D.T., Verhagen, H. *et al.* (1999) *Brassica* vegetables and cancer prevention. Epidemiology and mechanisms. *Adv. Exp. Med. Biol.*, **472**, 159–168.

Wakabayashi, K., Nagao, M., Ochiai, M. *et al.* (1985) A mutagen precursor in Chinese cabbage, indole acetonitrile, which becomes mutagenic on nitrite treatment. *Mutat. Res.*, **143**, 17–21.

Wang, L., Liu, D., Ahmed, T. *et al.* (2004) Targeting cell cycle machinery as a molecular mechanism of sulforaphane in prostate cancer prevention. *Int. J. Oncol.*, **24**, 187–192.

Witschi, H., Espiritu, I., Suffia, M. *et al.* (2002) Expressions of cyclin D1/2 in the lungs of strain A/J mice fed chemopreventive agents. *Carcinogenesis*, **23**, 289–294.

Wittstock, U. and Halkier, B.A. (2000) Cytochrome P450CYP79A2 from *Arabidopsis thaliana* L. catalyzes the conversion of L-phenylalanine to phenylacetaldoxime in the biosynthesis of benzylglucosinolate. *J. Biol. Chem.*, **275**, 14659–14666.

Wu, L. and Juurlink, B.H. (2001) The impaired glutathione system and it regulation by sulforaphane in vascular smooth muscle cells from spontaneously hypertensive rats. *J. Hypertens.*, **19**, 1819–1825.

Wu, L., Ashraf, M.H., Facci, M. *et al.* (2004) Dietary approach to attenuate oxidative stress, hypertension and inflammation in the cardiovascular system. *Proc. Natl. Acad. Sci. USA*, **1001**, 7094–7099.

Xiao, D., Pinto, J.T., Soh, J.W. *et al.* (2003) Induction of apoptosis by the garlic-derived compound S-allylmercaptocysteine (SAMC) is associated with microtubule depolymerisation and c-Jun NH (2)-terminal kinase 1 activation. *Cancer Res.*, **63**, 6825–6837.

Xue, J., Lenman, M., Falk, A. *et al.* (1992) The glucosinolate-degrading enzyme myrosinase in Brassicaceae is encoded by a gene family. *Plant Mol. Biol.*, **18**, 387–398.

Xue, J., Jorgenson, M., Pihlgren, U. *et al.* (1995) The myrosinase gene family in *Arabidopsis thaliana*: gene organsiation, expression and evolution. *Plant Mol. Biol.*, **27**, 911–922.

Yanagita, T., Han, S.Y., Wang, Y.M. *et al.* (2003) Cycloalliin, a cyclic sulphur imino acid, reduces serum triacylglycerol in rats. *Nutrition*, **19**, 140–143.

Yu, R., Mandlekar, S., Harvey, K.J. *et al.* (1998) Chemopreventive isothiocyanates induce apoptosis and caspase-3-like protease activity. *Cancer Res.*, **58**, 402–408.

Zhao, B., Seow, A., Lee, E.J.D. *et al.* (2001) Dietary isothiocyanates, glutathione S-transferase-M1,-T1 polymorphisms and lung cancer risk among Chinese women in Singapore. *Cancer Epidemiol. Biomarkers Prev.*, **10**, 1063–1067.

Zhao, F.J., Evans, E.J., Bilsborrow, P.E. *et al.* (1994) Influence of nitrogen and sulphur on the glucosinolate profile of rapeseed (*Brassica napus* L.). *J. Sci. Food Agric.*, **64**, 295–304.

Chapter 3
Terpenes

Andrew J. Humphrey and Michael H. Beale

3.1 Introduction

The terpenes, or isoprenoids, are one of the most diverse classes of metabolites. The *Dictionary of Natural Products* (Buckingham 2004) lists over 30 000, mainly of plant origin, encompassing flavours and fragrances, antibiotics, plant and animal hormones, membrane lipids, insect attractants and antifeedants, and mediators of the essential electron-transport processes which are the energy-generating stages of respiration and photosynthesis. Some of these key molecules are illustrated in Figure 3.1. In addition, terpenoid motifs are recurring structural and functional features of a host of bioactive natural products, from the phytol side chain of the chlorophylls and the diterpenoid skeleton of the anti-cancer drug Taxol to the core structural unit of tetrahydrocannabinol, the major bioactive component of marijuana.

This chapter aims to give an overview of terpene biosynthesis from the perspective of the plant scientist and a reflection on its significance to the environment and human health. The interested reader seeking more information on the health and environmental properties of particular terpenes is directed to Chapter 5 of Paul Dewick's *Medicinal Natural Products: A Biosynthetic Approach* (2001). The same author has produced regular review articles in *Natural Product Reports* (Dewick 1997, 1999, 2002) covering in some depth the biochemical and mechanistic aspects of terpene biosynthesis.

Although the final chemical structures of the terpenes are as diverse as their functions, all terpenes are derived from a sequential assembly of molecular building blocks as shown in Figure 3.2, each of which consists of a branched chain of five carbon atoms (Dewick 2001). Classically it was thought that the terpenes were assembled from isoprene (**1**) (Ruzicka 1953), hence their alternative name of isoprenoids. It is now known that the actual five-carbon building blocks in vivo are the interconvertible isomers isopentenyl pyrophosphate (IPP, **2**) and dimethylallyl pyrophosphate (DMAPP, **3**). These two building blocks are condensed together in a sequential fashion by the action of enzymes called prenyltransferases. The products include geranyl (**4**), farnesyl (**5**) and geranylgeranyl (**6**) pyrophosphates, squalene (**7**) and phytoene (**8**), which are the direct precursors of the major families of terpenes (Figure 3.3). Subsequent modifications to the carbon backbone (typically by enzyme-catalysed cyclization, oxidation and skeletal rearrangement steps) give rise to the multitude of isoprenoid structures illustrated in Figure 3.1 and throughout this review.

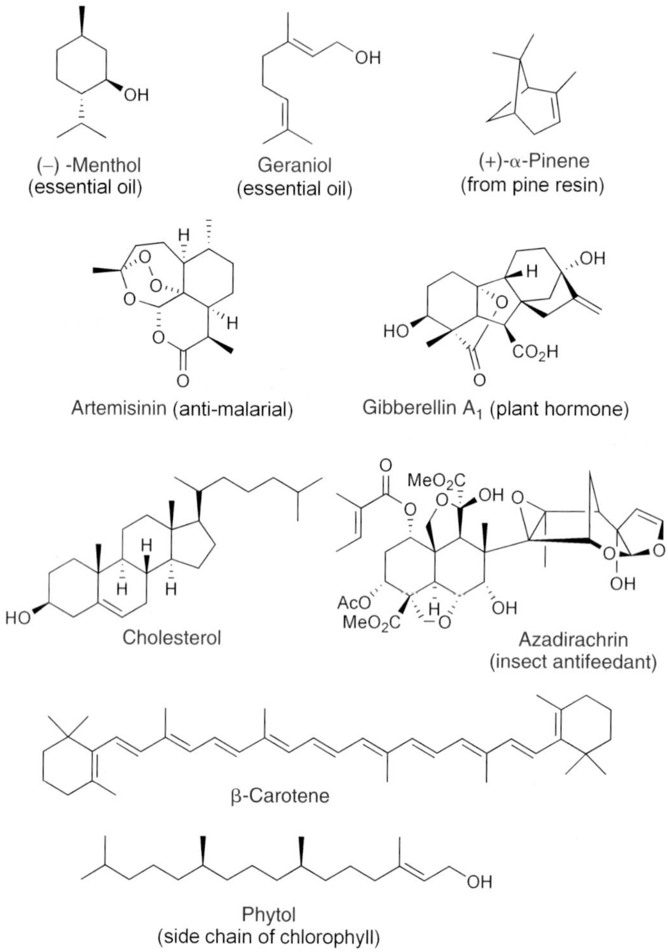

Figure 3.1 Examples of terpenes.

The biosynthesis of these molecular building blocks is mediated by a host of different enzymes. However, in the eukaryotes it is not the case that any one cell will possess a general pool of precursor molecules. The biosynthesis of terpenes in fact shows a striking segregation, with different subcellular structures possessing their own machinery for terpene biosynthesis and quite often generating their own entirely separate pools of biosynthetic intermediates. This segregation is at its most dramatic in green plants, where two entirely separate enzymatic systems are responsible for terpene biosynthesis. One system, in the cytosol, generates most of the sesquiterpenes, the triterpenes and sterols; the other, in the plastids, generates the essential oil monoterpenes, the diterpenes and carotenoids (Figure 3.2). The understanding of this segregation, and of the 'cross-talk' between plastidic and cytosolic biosynthetic machinery, is still developing, and presents one of the most significant challenges in plant molecular biology today (Lichtenthaler 1999; Eisenreich *et al.* 2004).

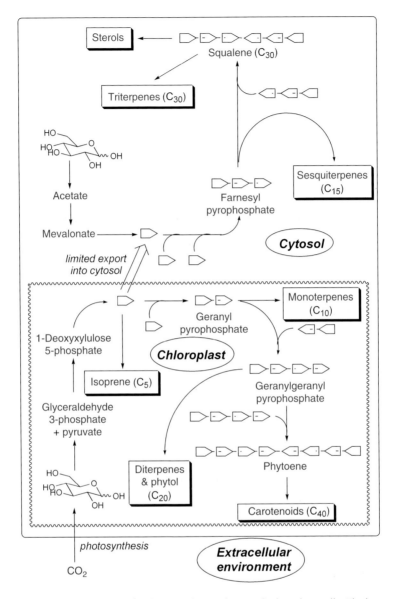

Figure 3.2 Schematic representation of terpene biosynthesis in higher plant cells. Blocks represent 5-carbon isoprenyl units (IPP or DMAPP).

3.2 The biosynthesis of IPP and DMAPP

3.2.1 *The mevalonic acid pathway*

The identification of the chemical pathway by which living organisms synthesize IPP and DMAPP, and hence the whole terpene family, is a remarkable example of scientific detective

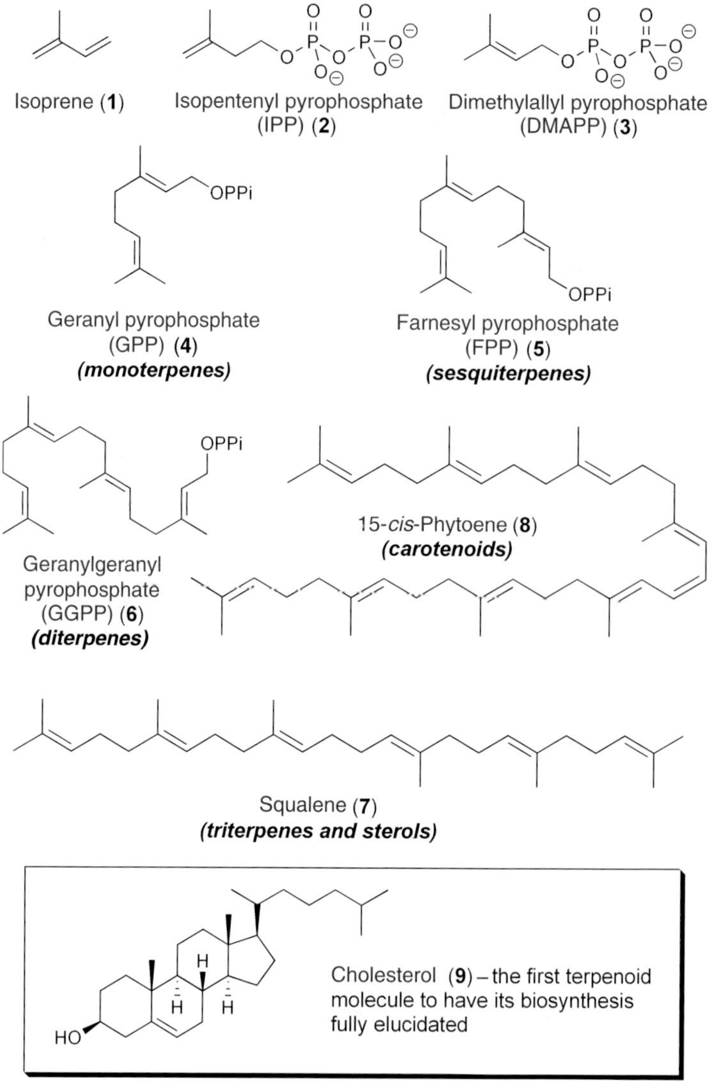

Figure 3.3 Chemical structures of the terpene precursors. OPPi represents the pyrophosphate group, $OP_2O_6^{3-}$.

work spanning more than 50 years. The first chapter of the story unfolded with the elucidation of the biosynthesis of cholesterol (9), as reviewed by two of the founding fathers of biological chemistry, John Cornforth (1959) and Konrad Bloch (1965), whose contributions to this field earned them separate Nobel Prizes. Their discoveries were made possible by the emerging technology of radioisotope tracer studies, in which small molecules containing a radioactive label at a defined position in the molecule were fed to cell cultures, and the location of the radiolabel in the ensuing metabolites was traced and precisely identified.

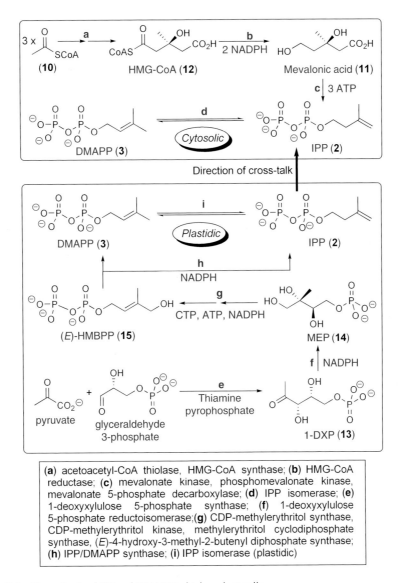

Figure 3.4 Biosynthesis of IPP and DMAPP in higher plant cells.

The first key discovery was that the entire cholesterol molecule was assembled by a sequential condensation of acetate molecules in their biologically activated form as coenzyme A (CoA) thioesters **10** (Bloch 1965). The key intermediate in the process was (3*R*)-mevalonic acid (MVA, **11**), a six-carbon compound originally discovered as a byproduct in 'distillers' solubles'. Mevalonic acid is formed by the enzymatic reduction of 3-hydroxy-3-methylglutaryl coenzyme A (HMG-CoA, **12**) which in turn is formed by the head-to-tail condensation of three molecules of acetate (Figure 3.4) (Cornforth 1959). Mevalonic acid is enzymatically converted to IPP with loss of carbon dioxide, and it was subsequently possible to show that IPP and its isomer, DMAPP, were incorporated directly into cholesterol by yeasts and human liver cells (Bloch 1965). The chemical and enzymatic details of the mevalonate pathway and its regulation are now well understood (see Brown (1998) for a recent review). The use of statin drugs which inhibit the key reduction of HMG-CoA to mevalonate is a cornerstone of cholesterol-lowering therapy in the treatment of heart disease (Endo 1992).

3.2.2 The 1-deoxyxylulose 5-phosphate (or methylerythritol 4-phosphate) pathway

For a long time the ubiquity of the mevalonate pathway in terpene biosynthesis was accepted as a biochemical dogma. However, over the years a trickle of evidence began to amass that in green plants (Treharne *et al.* 1966; Shah and Rogers 1969), algae (Schwender *et al.* 1996) and certain bacteria (Pandian *et al.* 1981; Zhou and White 1991), it was not the mevalonate pathway which gave rise to the majority of the terpenes after all. The failure of green plant chloroplasts and the bacterium *E. coli* to incorporate radioactively labelled mevalonate or acetate into β-carotene, phytol and bacterial ubiquinone (which contains a long isoprenyl sidechain) was at first explained as a failure in the uptake mechanism which prevented the labelled probe from crossing the membrane boundaries efficiently. A similar theory was later put forward to explain the observation that upregulation of mevalonate pathway gene expression in tobacco and potato cells had no effect on carotenoid or phytol biosynthesis, even though levels of sterols were considerably enhanced (Chappell 1995).

It was more difficult to explain away the unusual pattern of incorporation of ^{13}C, a magnetically active nucleus whose chemical environment could be probed by Nuclear Magnetic Resonance (NMR) spectroscopy, which was provided in the form of labelled glucose. Glucose labelled at carbon 1 (see Figure 3.5) was expected to be carried through the glycolytic pathway to [2-^{13}C]-acetyl CoA, and hence *via* the mevalonate pathway to IPP and DMAPP labelled at positions 2, 4 and 5. What was actually observed was an incorporation of ^{13}C at positions 1 and 5 (Rohmer *et al.* 1993; Lichtenthaler *et al.* 1997). Furthermore, mevinolin, an inhibitor of HMG-CoA reductase, was found to have no effect on carotenoid biosynthesis in green algae (Schwender *et al.* 1996) and in green tissues of higher plants (Bach and Lichtenthaler 1983).

The painstaking work of Michel Rohmer, Duilio Arigoni and their respective co-workers led to the elucidation of an entirely separate pathway for terpene biosynthesis operating in the plastids of green tissues and oil gland cells in plants, and in most eubacteria. In these organisms IPP is derived, not from mevalonic acid, but from 1-deoxyxylulose 5-phosphate (1-DXP, **13**) (Rohmer *et al.* 1996; Arigoni *et al.* 1997; Sprenger *et al.* 1997), formed from the glycolytic intermediates glyceraldehyde 3-phosphate and pyruvate as shown in Figure 3.4. The key step in the biosynthesis is the skeletal rearrangement and reduction of 1-DXP to form 2C-methylerythritol 4-phosphate (MEP, **14**) (Takahashi *et al.* 1998) using the biological reducing agent NADPH as cofactor. The enzyme responsible for the rearrangement, 1-deoxyxylulose 5-phosphate reductoisomerase (DXR), has been cloned from numerous microbial and plant sources (see Eisenreich *et al.* (2004) for a summary) and from a protozoan, the malaria parasite *Plasmodium falciparum* (Jomaa *et al.* 1999). DXR from peppermint (Lange and Croteau 1999a) differs from the bacterial enzyme

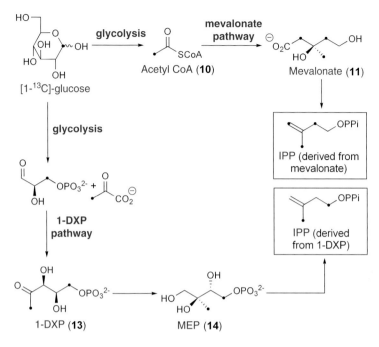

Figure 3.5 The metabolic fate of a carbon-13 label traced through the two possible pathways for IPP biosynthesis when administered in the form of [1-^{13}C]-glucose. ^{13}C label is indicated with a bold dot.

(Takahashi *et al.* 1998) in possessing an *N*-terminal amino acid sequence which is not part of the mature protein but functions as a 'transit peptide' which is used by the plant cell to target the nascent enzyme to the organelle where terpene biosynthesis takes place. Once inside the plastid, the transit peptide is cleaved by proteolysis and the remainder folds into the active, soluble protein. Transit peptides are a ubiquitous feature of the plastidic terpene biosynthesis pathway in plants (for other examples see Williams *et al.* 1998; Bohlmann *et al.* 2000; Okada *et al.* 2000).

MEP is converted to IPP *via* a chemical sequence involving the reductive removal of three molecules of water. The chemistry of these final steps (the details of which have been comprehensively reviewed by Eisenreich *et al.* (2004)) was unravelled in a dramatic early demonstration of the power of genomics. Genes implicated in the conversion of MEP to IPP were found to cluster together in the genomes of organisms possessing the 1-DXP pathway, whilst being absent from organisms without this pathway. Cloning of these putative genes and overexpression of the gene products in *E. coli* rapidly elucidated the nature of the biosynthetic precursors of IPP, which are shown in Figure 3.6. The speed with which the genes were successively cloned and the corresponding enzyme activities identified was such that the entire pathway had been fully described within five years of the initial identification of 1-DXP as the key intermediate. Plant homologues of the *E. coli* genes were also rapidly identified by database searching, and their functions confirmed by overexpression or by virus-induced gene silencing (Page *et al.* 2004).

The absence of 1-DXP pathway genes in humans makes this pathway particularly attractive as a potential target in the treatment of bacterial or parasitic infections. Drugs which

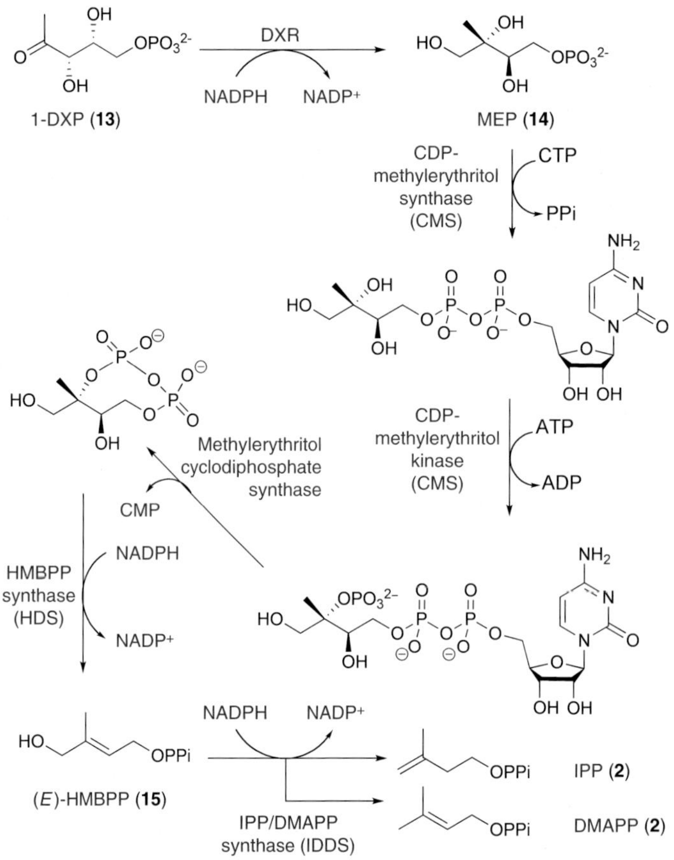

Abbreviations: ADP, adenosine 5'-diphosphate; ATP, adenosine 5'-triphosphate; CDP, cytidine 5'-diphosphate; CMP, cytidine 5'-monophosphate; CTP, cytidine 5'-triphosphate; $NADP^+$, nicotinamide adenine dinucleotide phosphate (oxidised form); NADPH, nicotinamide adenine dinucleotide phosphate (reduced form); OPPi, pyrophosphate

Figure 3.6 The biosynthesis of IPP and DMAPP from 1-deoxyxylulose 5-phosphate.

inhibit key steps of the pathway, such as the reduction and rearrangement of 1-DXP to MEP, could have a lethal effect on the target organism without comparable toxicity in human patients. In particular, the development of 1-DXP pathway inhibitors to target the malaria parasite *Plasmodium falciparum*, an organism which over time has grown resistant to many of the more well-established drug treatments, shows considerable potential as a therapeutic strategy (Jomaa *et al.* 1999).

3.2.3 Interconversion of IPP and DMAPP

An unusual feature of the 1-DXP pathway is that not all organisms which possess this pathway also possess *isoprenyl diphosphate isomerase* (IDI), the enzyme which interconverts

IPP and DMAPP (Ershov *et al.* 2000). This enzyme, whilst a key feature of the mevalonate pathway, is not vital to the 1-DXP pathway because the final step in the pathway (the reductive dehydration of (*E*)-4-hydroxy-3-methylbut-2-enyl pyrophosphate or HMBPP, 15) was found to produce IPP *and* DMAPP in an approximately 6:1 ratio in green plant plastids (Adam *et al.* 2002). Despite this, IDI *is* present in plastids. When viral infection was used to silence expression of this enzyme in the tobacco *Nicotiana benthamiana*, the result was an 80% reduction of essential chlorophyll and carotenoid levels in infected leaves. This suggests that IDI, whilst not strictly necessary for viable plants, is highly beneficial to green tissue growth (Page *et al.* 2004).

3.2.4 Biosynthesis of IPP and DMAPP in green plants

The 1-DXP pathway is absent from higher animals, yeasts and fungi, but in green plants the 1-DXP and mevalonate pathways co-exist in separate cellular compartments. The 1-DXP pathway, operating in the plastids, is responsible for the formation of essential oil monoterpenes (such as menthol in peppermint (*Mentha* × *piperita*) (Eisenreich *et al.* 1997) and linalyl acetate in *M. citrata* (Fowler *et al.* 1999)), some sesquiterpenes (Maier *et al.* 1998; Adam *et al.* 1999), diterpenes (Eisenreich *et al.* 1996), and carotenoids and phytol (Lichtenthaler *et al.* 1997). The mevalonate pathway, operating in the cytosol, gives rise to triterpenes, sterols and most sesquiterpenes (Brown 1998; Lichtenthaler 1999). Recent evidence in *Arabidopsis thaliana* (Kasahara *et al.* 2002; Laule *et al.* 2003) and lima bean (Piel *et al.* 1998) suggests that there is a limited degree of 'cross-talk' between the pathways (see Figure 3.2), in particular that IPP from the 1-DXP pathway can cross through the plastid membranes to the cytosol and be incorporated to some degree into terpenes normally formed *via* the mevalonate pathway. It has been suggested that this unidirectional traffic might be designed to give the plant an advantage in times of nutrient stress, prenyl pyrophosphates formed in the plastids from photosynthetically derived glucose being exported to assist the biosynthesis of sterols and other essential terpenes in the cytosol (Piel *et al.* 1998; Page *et al.* 2004). Metabolites from the cytosolic MVA pathway are unable to compensate fully for deficiencies in plastidic terpene biosynthesis (as shown by the dwarf albino *cla1* mutants of *A. thaliana*, which lack an active 1-DXP synthase), although there is recent evidence to suggest that limited import of MVA-derived intermediates into *Arabidopsis* plastids may occur under certain developmental and environmental conditions (Kasahara *et al.* 2002).

Cross-talk is also evident in the biosynthesis of certain sesquiterpenes and diterpenes. In chamomile flowers (see Section 3.5.1) it is thought that export of isoprenoid intermediates formed by the 1-DXP pathway takes place into a cellular space which has access to IPP derived from both 1-DXP *and* mevalonate pathways (Adam *et al.* 1999).

3.3 Enzymes of terpene biosynthesis

3.3.1 Prenyltransferases

There are three main classes of enzymes involved in the processing of isoprenoid building blocks to form the vast range of structures present in the terpene family (Liang *et al.* 2002).

The isoprenyl pyrophosphate synthases (IPPS) catalyse the formation of long-chain prenyl pyrophosphates by the addition of an allylic cation generated from DMAPP, GPP or FPP to one or more molecules of IPP. Terpene synthases are responsible for the conversion of prenyl pyrophosphates into the classic terpene structures; a large number of these are cyclic structures so these synthases are also known as terpene cyclases. Protein prenyltransferases catalyse the addition of isoprenoid chains to a protein or peptide. Here we consider mainly the first two of these.

The compartmentation of terpene biosynthesis in plants is reflected in the fact that prenyltransferases which may be closely related at the genetic level can occur in completely separate subcellular compartments. This is illustrated by the geranylgeranyl pyrophosphate (GGPP) synthase family in *Arabidopsis* (Okada *et al.* 2000). Out of five GGPP synthases identified by gene sequence similarity searches, two were plastid-localized and implicated in gibberellin (GA) and carotenoid biosynthesis; two were localized to the endoplasmic reticulum, where they would be exposed to the cytosolic (mevalonate-derived) IPP pool and are most likely involved with protein prenylation; and one was localized in the mitochondria where it is implicated in the biosynthesis of prenylquinones involved in energy transduction. The five identified GGPP synthases were also expressed differently in different tissues; one of the plastidic GGPP synthases was mainly expressed in aerial tissues and one was expressed predominantly in roots.

3.3.2 Mechanism of chain elongation

Terpene biosynthesis occurs *via* the combination of isoprenyl pyrophosphate units in a sequential fashion, beginning with the formation of geranyl pyrophosphate (GPP, **4**) from one molecule of IPP and one of DMAPP, under the action of isoprenyl pyrophosphate synthases (IPPS). In the language of organic chemistry, the reaction is said to proceed *via* an $S_N 1$ mechanism; elimination of the pyrophosphate moiety, an excellent leaving group, from DMAPP leaves an allyl cation which then carries out an electrophilic attack on the electron-rich double bond of IPP (Poulter and Rilling 1976). The two building blocks join together in a 'head-to-tail' fashion as shown in Figure 3.7. The dissociation of DMAPP is favoured by chelation of the pyrophosphate moiety to a divalent magnesium ion in the enzyme active site; other non-covalent interactions in the active site stabilize the charged intermediates (Tarshis *et al.* 1996), reducing the activation energy for the reaction. The newly formed double bond can be of *cis* (Z) or *trans* (E) geometry. In plants, IPPS enzymes produce almost entirely *trans* products. A few *cis*-IPPS enzymes have been isolated, but almost all are from bacterial sources and are involved with the synthesis of very long-chain isoprenoids (C_{55} or more).

This mode of action is common to nearly all IPPS enzymes. In GPP synthase, the reaction stops at this point, and GPP is released from its binding site. In farnesyl pyrophosphate (FPP) synthase, the reaction proceeds a step further: GPP itself dissociates in the active site and the resulting allylic carbocation attacks a further molecule of IPP (Figure 3.7). The product, FPP (**5**), is also a substrate for a synthase which produces the 20-carbon geranylgeranyl pyrophosphate (GGPP, **6**) and so on. Longer-chain IPPS enzymes from many organisms will generally accept a range of these allylic pyrophosphates as prenyl donors, although the exact specificity varies from one organism to another (Dewick 2002).

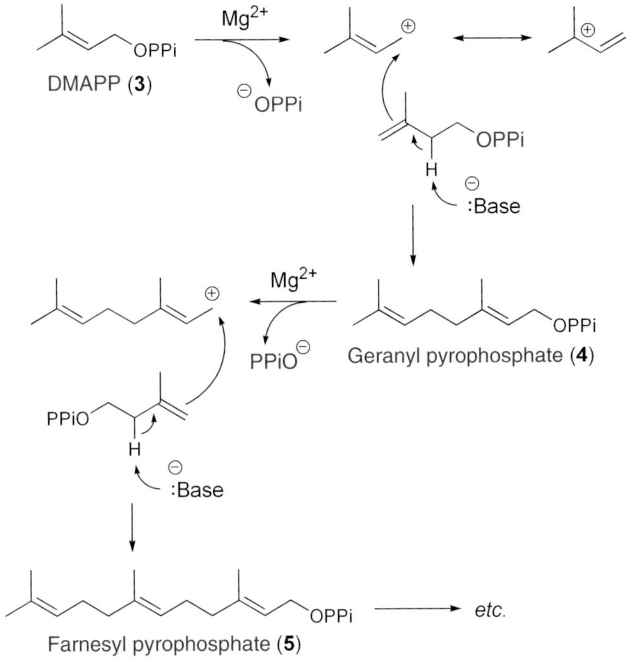

Figure 3.7 Electrophilic mechanism for the prenyltransferase-catalysed extension of an isoprenoid chain.

IPP is the near-universal prenyl acceptor in the IPPS enzymes ('irregular monoterpene' synthases are exceptions (vide infra)). In the protein prenyltransferases, IPP is absent, and a cysteine sulphur atom from the target protein acts as the electron-rich prenyl acceptor.

The mechanism of the *trans*-IPPS enzymes has been studied mainly using cloned enzymes from bacterial, yeast or vertebrate sources. Crystal structures of several of these enzymes are also available, a number of which show the isoprenoid substrates bound in the active sites (for a review see Liang *et al.* (2002)). These synthases, and their homologues in plants, show a high degree of amino acid sequence homology and a number of key conserved regions. Crucial to the enzymes' activity are two separate aspartate-rich motifs, DDxxD. These motifs serve as coordination sites for the Mg^{2+} ions which bind the pyrophosphate moieties of the prenyl donor and IPP, respectively (Tarshis *et al.* 1996) (Figure 3.8). Mutagenesis studies have shown that substitution of these key aspartate residues can reduce prenyltransferase activity or even eliminate it altogether.

Mutagenesis and crystal structure studies have also cast light on another important aspect of prenyltransferase chemistry: how the enzymes know when to stop. The active sites of *trans*-IPPSs are located in a deep cleft in the heart of the enzyme, large enough to accommodate the final product of the enzymic reaction, with a hydrophobic 'pocket' around the region of the donor substrate furthest from the pyrophosphate group. In the avian FPP synthase, the conserved DDxxD residues are located opposite one another in the centre of the cleft (Tarshis *et al.* 1994). In FPP and GGPP synthases, a large amino acid (leucine, tyrosine or phenylalanine), located upstream of the donor-binding DDxxD motif, protrudes into the active site cleft (Figure 3.8). The result of this is that when

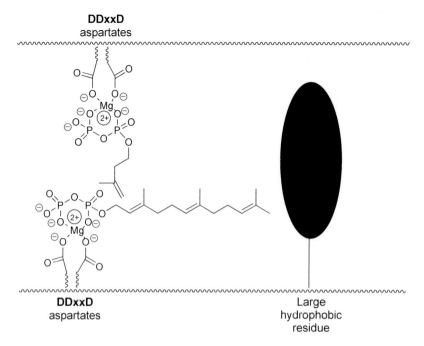

Figure 3.8 Schematic representation of the active site of FPP synthase. The large hydrophobic residue provides a block to further chain elongation.

the incipient polyprenyl chain reaches a certain size, it is unable to undergo any further condensation reactions because the intermediate will no longer bind properly in the active site (Ohnuma *et al.* 1996). In longer-chain IPPSs, this large residue is typically replaced by a much smaller amino acid (often alanine), and large residues further along the amino acid sequence impose a 'cut-off' point beyond which chain elongation will not continue. Site-directed mutagenesis of these key residues has enabled a number of groups to create engineered prenyltransferases which produce longer-or shorter-chain isoprenoids in vitro than the wild-type enzymes, according to the nature of the change (Ohnuma *et al.*1998).

3.3.3 Terpene synthases (including cyclases)

The terpene synthases are structurally and mechanistically very closely related to the IPPS. A number of these enzymes have been cloned and overexpressed from plant sources (Chappell 1995); thale cress (*Arabidopsis thaliana*), whose genome is now fully sequenced, has 32 functional terpene synthase DNA sequences in its genome (Aubourg *et al.* 2002). Conserved amino acid sequences, particularly DDxxD motifs for binding of Mg^{2+} and pyrophosphate in the substrates, are common and allow terpene synthases from a wide variety of sources to be identified by sequence similarity searches (Bohlmann *et al.* 1998). Sequence analysis of the genes which encode known terpene synthases in plants has enabled them to be grouped into six subfamilies whose members each share >40% sequence identity.

The major difference between the IPPSs and the terpene cyclases lies in the fact that terpene cyclases do not bind IPP. The first step in the terpene cyclase-catalysed reaction is still in most cases the loss of the pyrophosphate group with the formation of a polyprenyl cation in the active site. This time, however, it is an electron-rich double bond elsewhere in the molecule which serves as an internal nucleophile, with the result being formation of a cyclic structure. The active sites of terpene cyclase enzymes are tailored to fold the polyprenyl pyrophosphates into the optimum conformation for intramolecular attack to take place, with hydrophobic residues to force the prenyl chain into the desired conformation, and aromatic residues such as tyrosine to stabilize the positive charge on the intermediate carbocation (Starks *et al.*1997). Initial isomerization, for example to linalyl or nerolidyl pyrophosphate (vide infra), is often important to present the correct geometry at the electron-accepting end of the molecule to allow cyclization to occur (Bohlmann *et al.* 1998).

As with the IPPSs, site-directed mutagenesis of key active site residues can alter the folding of the substrate, with the result that alternative cyclization paths can be followed, and novel products obtained (Rising *et al.* 2000).

In many terpene cyclases, cyclizations are also accompanied by Wagner-Meerwein rearrangements – migrations of carbon–hydrogen or carbon–carbon σ-bonds towards sites of positive charge. Substantial rearrangements of the carbon skeletons of terpenes can result: hence the plethora of structural types associated with the sesquiterpenes, diterpenes and sterols. The most famous examples of this behaviour are the methyl group migrations which occur during the biosynthesis of sterols (vide infra), but spectacular rearrangements of the whole carbon skeleton are known, for example, in the biosynthesis of patchoulol (Croteau *et al.* 1987).

3.4 Isoprenoid biosynthesis in the plastids

3.4.1 Biosynthesis of monoterpenes

The C_{10} isoprenoids constitute 1000 or so metabolites, most of which are colourless, volatile oils with highly distinctive aromas and flavours (Sangwan *et al.* 2001; Mahmoud and Croteau 2002). They are best known as components of the essential oils of flowers and herbs, and of turpentine, the volatile component of the oleoresins of coniferous plants. In vivo many of the essential oil monoterpenes function as pollinator attractants, while the components of oleoresins serve as defence compounds by virtue of their toxicity to invading organisms. Their distinctive flavour and olfactory properties mean that many of these species are of considerable commercial value (Lange and Croteau 1999b). A great many essential oils have long found application as herbal remedies, and the recent resurgence of interest in this field (Castleman 2003; Heinrich 2003) has added to the scientific importance of the constituent monoterpenes and to their economic significance.

Monoterpenes are synthesized and stored in specialized structures (Sangwan *et al.* 2001) whose cells express all the genes necessary for monoterpene biosynthesis. In conifers, specialized epithelial cells form resin ducts with large central storage cavities. In lemongrass, essential oil accumulates in cells inside the leaf blades, while in mints, the essential oil is generated in the secretory cells of hair-like structures called glandular trichomes, which cover the surface of the leaves and stems, and the oil is stored in cavities beneath the cuticles. The major organelle in which monoterpene biosynthesis occurs is the leucoplast, a non-pigmented plastid compartment in the cells of the oil glands (Turner et al. 1999). Environmental factors (light quality, rainfall and nutrient availability) all have an influence on the terpene composition and the amount of plant essential oils, as does the growth stage of the oil-producing tissue (for a review see Sangwan et al. 2001).

The most well-known producers of essential oil monoterpenes are herbs such as the mint (*Mentha*) and sage (*Salvia*) families. *Arabidopsis thaliana* does not form glandular trichomes and has only recently been found to produce monoterpenes (Chen et al. 2003). Genes coding for monoterpene synthases in *Arabidopsis* have been cloned and overexpressed (Bohlmann et al. 2000; Bouvier et al. 2000; Chen et al. 2003), but the biological role of the monoterpene products is presently unclear.

The key prenyltransferase in the biosynthesis of monoterpenes is GPP synthase. To date, only a few GPP synthases have been fully characterized (Burke et al. 1999; Bouvier et al. 2000). Enzyme-catalysed isomerization of GPP *via* loss of pyrophosphate as a leaving group produces the allylic linalyl cation (**16**). Attack by an extraneous nucleophile can take place at either end of the allylic system and ultimately gives rise to linear monoterpenes such as geraniol (**17**), linalool (**18**) and linalyl acetate (**19**), all common components of a number of essential oils (Figure 3.9).

Linalyl cations are conformationally rigid owing to the delocalized π-electrons of the allylic system. The (E)-linalyl cation initially formed by dissociation of GPP is unreactive to terpene cyclases, since the remaining double bond in the molecule is geometrically and electronically constrained from acting as an internal nucleophile. However linalyl pyrophosphate (**20**), the isomer of GPP formed by the recombination of the linalyl cation and the pyrophosphate ion, is able to rotate freely into a *cisoid* conformation. From here, ionization to the (Z)-linalyl cation allows **20** to act as the precursor of nerol (**21**) and the cyclic monoterpenes, of which there are several families (Bohlmann et al. 1998). Figure 3.10 shows the synthesis of (−)-limonene (**22**), the precursor of the menthane family of monocyclic monoterpenes in mints. A limonene synthase from spearmint (*Mentha spicata*) has been overexpressed and found to catalyse both the dissociation of GPP and the cyclization of the resultant linalyl cation. **20** is a genuine intermediate in the reaction; mutation of key arginine residues produced an enzyme which would no longer accept GPP but would happily cyclize (3S)-**20** to limonene (Williams et al. 1998). The other members of the menthane series, including the commercially valuable products (−)-menthol (**23**) and (−)-carvone (**24**) (Figure 3.10), are derived *via* hydroxylation of limonene by cytochrome P$_{450}$ monooxygenase enzymes (Schalk and Croteau 2000). The regioselectivity of the hydroxylases varies from species to species – in peppermint, hydroxylation is predominantly at C-3, whereas in spearmint it occurs mainly at C-6 – although the responsible hydroxylases are closely related (the limonene 3-hydroxylase from peppermint and the 6-hydroxylase from spearmint share 70% amino acid sequence identity). Limonene synthase in mints is localized in the leucoplasts, but the hydroxylases responsible for downstream conversion of limonene are non-plastidic enzymes residing in the smooth endoplasmic reticulum (Turner et al. 1999).

The high commercial value of menthol has led to considerable interest in the production of transgenic mints with enhanced oil qualities (Mahmoud and Croteau 2002). The essential oil of peppermint comprises numerous menthanes (Figure 3.10) shows a representative

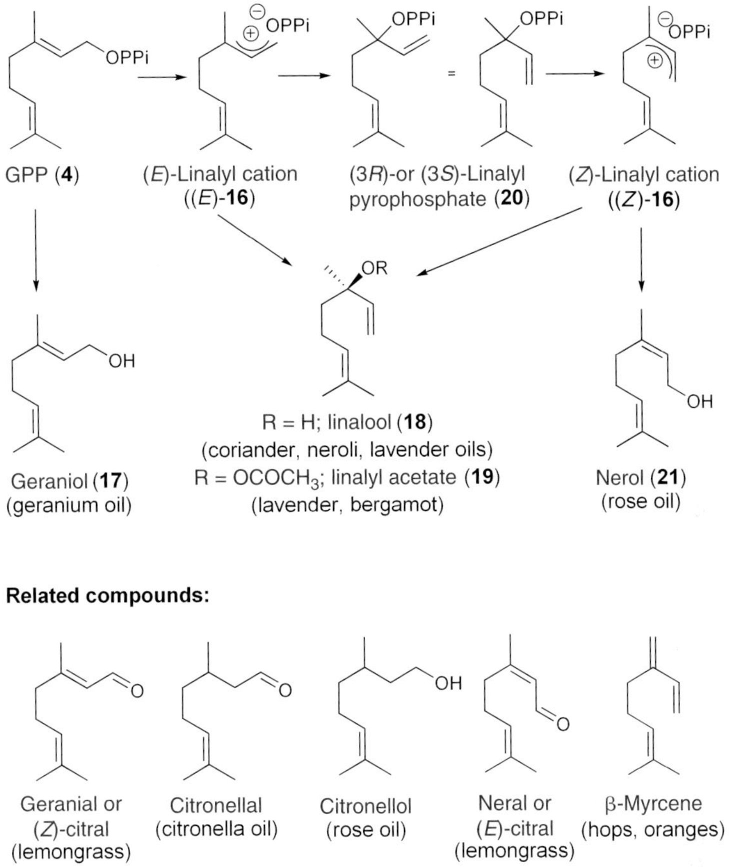

Figure 3.9 Monoterpenes: the geraniol/linalool family.

sample), all of which contribute in some way to the flavour profile of the oil. Poor light conditions and high temperatures can lead to the accumulation of unwanted components such as menthofuran (**25**). Suppression of menthofuran synthase at the genetic level has allowed improved yields of menthol in the oil (Mahmoud and Croteau 2001). The various hydroxylases involved in the downstream processing of limonene are also targets for genetic engineering; mutation of the limonene 6-hydroxylase in spearmint converted it to a 3-hydroxylase, allowing the enzyme to produce (−)-isopiperitenol instead of (−)-carveol (Schalk and Croteau 2000).

Thymol (**26**) and **carvacrol** (**27**), isolated from herbs such as thyme and savory, are members of the menthane family in which the cyclohexane structure has been oxidized to an aromatic (phenolic) ring. Oils containing these phenolic terpenes have been shown to be particularly effective as antibacterial agents (Kalemba and Kunicka 2003).

The more complicated cyclic carbon skeletons of the bornanes, thujanes and pinanes are also derived from intramolecular cyclization of the linalyl cation. Members of all three families are present in the essential oil of common sage (*Salvia officinalis*) which contains some 75 separate volatile compounds (Santos-Gomes and Fernandes-Ferreira

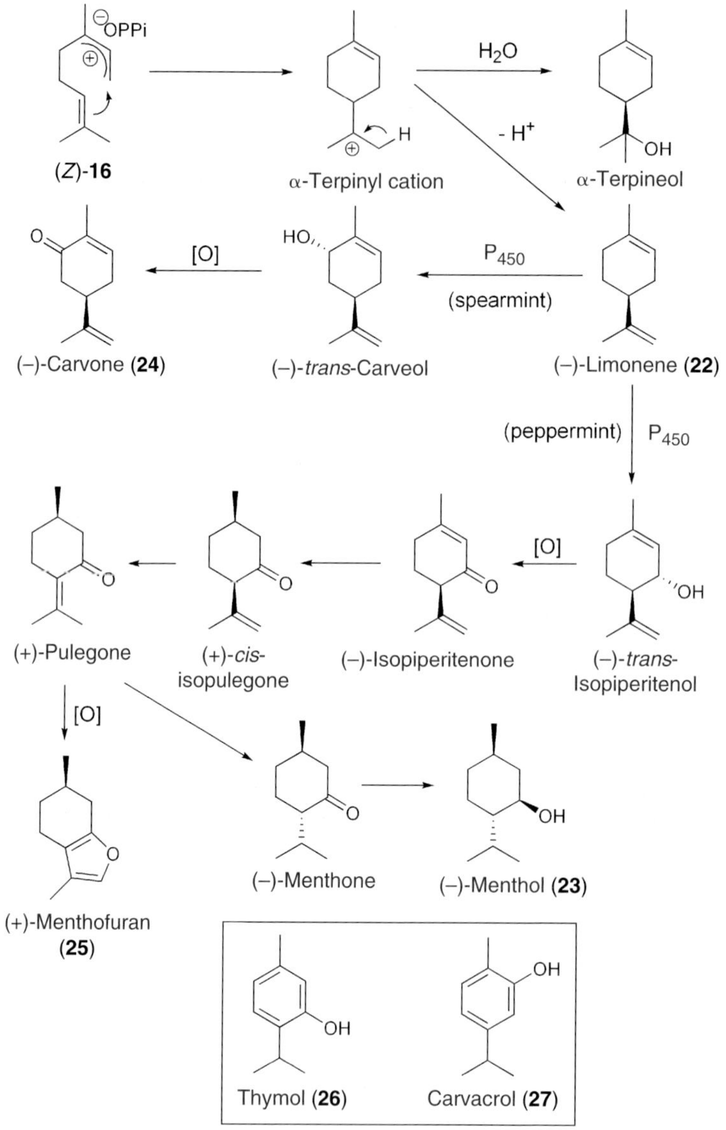

Figure 3.10 Monoterpenes: the menthane family.

2003). Figure 3.11 shows the biosynthesis of (+)-bornyl pyrophosphate (the precursor of (+)-borneol (**28**) and (+)-camphor (**29**)) and of (+)-sabinene (**30**) (the precursor of the thujones) from (3R)-**20** cyclized in an *anti,endo* conformation (Wise et al. 1998). Other related products include camphene (**31**) and 1,8-cineole (**32**). The chemistry involved in the formation of the final products includes Wagner-Meerwein rearrangements of hydride and (in the case of camphene) a skeletal carbon-carbon bond as well as simple cyclizations. The biosynthesis of (−)-α-and β-pinene (**33** and **34**) proceeds along similar lines from (3S)-**20** and is shown in Figure 3.12.

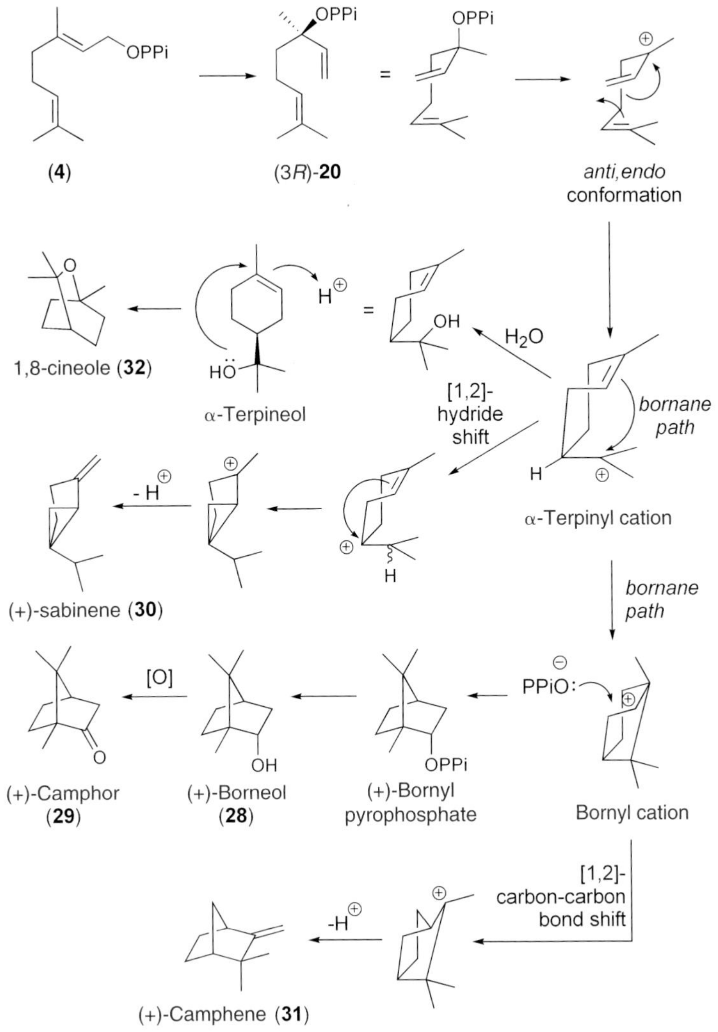

Figure 3.11 Monoterpenes: the (+)-bornane family.

The terpene cyclases responsible for these transformations allow a degree of flexibility in the folding of the intermediate cations so that they produce not just one product but several, in varying proportions (Gambliel and Croteau 1984). For instance, it appears that in sage, a single enzyme is responsible for the synthesis of both (+)-bornyl pyrophosphate and (+)-α-pinene (Wise et al. 1998). Sage oil contains both (+)- and (−)-enantiomers of α-pinene, the two enantiomers being formed by different sets of cyclase enzymes, one acting on (3R)-20 and one on (3S)-20 (Gambliel and Croteau 1984). It is presumed that initial enzyme-catalysed isomerisation of GPP, the natural substrate of the synthases responsible for these transformations, generates linalyl pyrophosphate of the required stereochemistry before subsequent ionization and cyclization of the enzyme-bound intermediate.

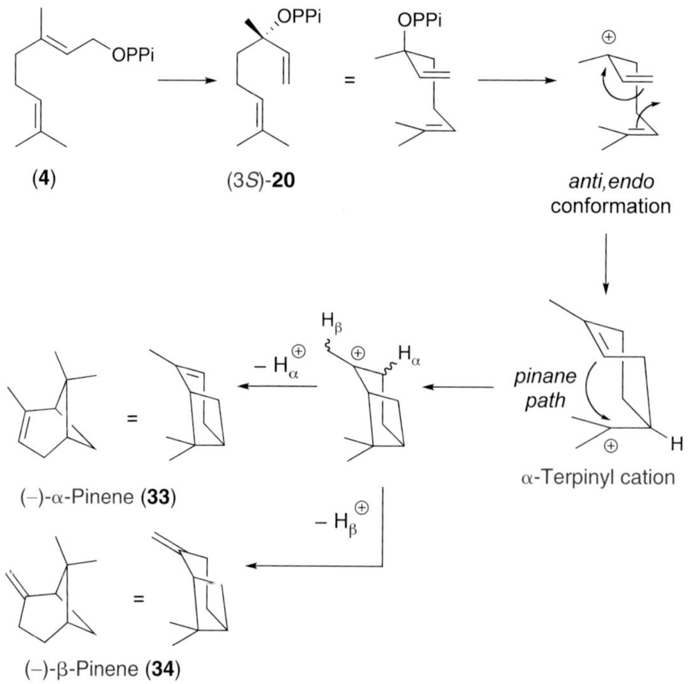

Figure 3.12 Monoterpenes: the (−)-pinane family.

(−)-α-Thujone (**35**, Figure 3.13) is one of the most notorious monoterpenes. It is the major bioactive ingredient of the hallucinogenic liquor absinthe, a favourite of artists and writers in the nineteenth and early twentieth centuries. It is widely held to have been responsible for psychoses and suicides, possibly including that of Vincent van Gogh (Arnold 1988). α-Thujone in absinthe derives from the oil of wormwood (*Artemisia absinthum*), which is widely used in herbal medicine as a relief for stomach, liver and gall bladder complaints. Thujones are also a component of many essential oils which have applications as folk medicines, including *Salvia* species (Santos-Gomes and Fernandes-Ferreira 2003). The mode of action of thujone in humans and the basis of its toxicity have only recently been established; it is now known as a modulator of the γ-aminobutyric acid (GABA) receptor (Höld *et al.* 2000) and as an inducer of porphyria (Bonkovsky *et al.* 1992). The dual behaviour of thujone-rich oils as both toxins *and* medicines highlights an important issue in the exploitation of plant natural products. Regulation of herbal medicines is generally much less rigorous than that of synthetically derived pharmaceuticals, and the possible risks associated with 'natural' treatments are often much less well understood than those of synthetic drugs (Höld *et al.* 2000). The need to mitigate against unwanted physiological effects in therapeutic preparations is consequently one of the major challenges in the successful development of traditional medicines as modern therapeutic products (Walker 2004; see also Section 3.6).

The iridoids constitute a large family of highly oxygenated monoterpenes, mixtures of which are present in many medicinal plants, including olive leaves (well known for

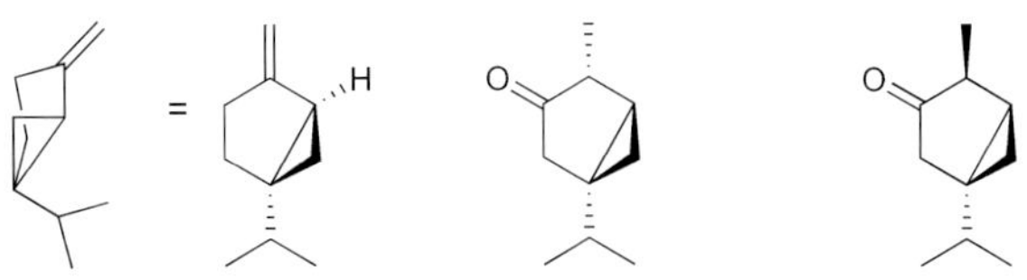

(+)-Sabinene (**30**)   (−)-α-Thujone (**35**)   (−)-β-Thujone
                         (1S, 4R, 5R)   or isothujone
                                                   (1S, 4S, 5R)

Figure 3.13 Monoterpenes: the thujones.

their hypotensive qualities), valerian (*Valeriana officinalis*, a herb which has long been used as a tranquillizer), and 'Picroliv' (an extract of the medicinal plant *Picrorhiza kurroa*, which has liver regenerative properties) (Ghisalberti 1998). They are derived from geraniol or nerol *via* oxidation of a terminal methyl group (Tietze 1983). Cyclizations and further cytochrome P_{450}-dependent oxidations produce a characteristic bicyclic skeleton, as found in valtrate (**36**) (the principal component of the essential oil of valerian) as well as in loganin (**37**) and its many analogues (Figure 3.14). The cyclopentane ring of loganin can itself be cleaved in a further P_{450}-dependent step, leading to secologanin (**38**), which provides the carbon skeleton for many powerfully bioactive secondary metabolites including the indole and *Cinchona* alkaloids (e.g. quinine (**39**), the original anti-malarial drug). Closely related to the iridoids are the various diastereomers of nepetalactone (**40**). Nepetalactones are pheromones emitted by sexual female aphids, but they are also a major constituent of the essential oil of several plants including catmint (*Nepeta racemosa*) (Birkett and Pickett 2003) – in fact they are the agents responsible for this plant's power as a cat attractant.

The 'irregular' monoterpenes, typified by chrysanthemic and pyrethric acids (**41** and **42**), form the carboxylic acid portions of the pyrethrin esters, powerful natural insecticides which are the basis for a whole family of economically valuable agrochemicals (the pyrethroids) (George et al. 2000). Irregular monoterpenes occur in flowering plants of the *Compositae* family and are unusual in that they are formed not from GPP but from the 'head-to-head' condensation of two molecules of DMAPP as shown in Figure 3.15. This results in the formation of the distinctive cyclopropane unit in this family of molecules (Rivera *et al.* 2001). Dephosphorylation and cytochrome P_{450}-dependent oxygenation then generate a terminal carboxylic acid.

3.4.2 Biosynthesis of diterpenes

The diterpenes are formed by cyclization of geranylgeranyl pyrophosphate (GGPP, **6**). The increased length of the linear isoprenoid diphosphate compared with GPP and the increased number of double bonds provide more possibilities for folding and cyclization reactions than observed for monoterpenes. The resulting increase in the number of possible cyclic carbon skeleta and the greater diversity of downstream oxidation products makes the diterpene family relatively large. Diterpenes are usually considered as secondary metabolites but in plants many diterpenes perform essential physiological or ecological functions, particularly as growth hormones (the gibberellin family) or defence compounds (phytoalexins, molecules whose production is triggered by infection or predation

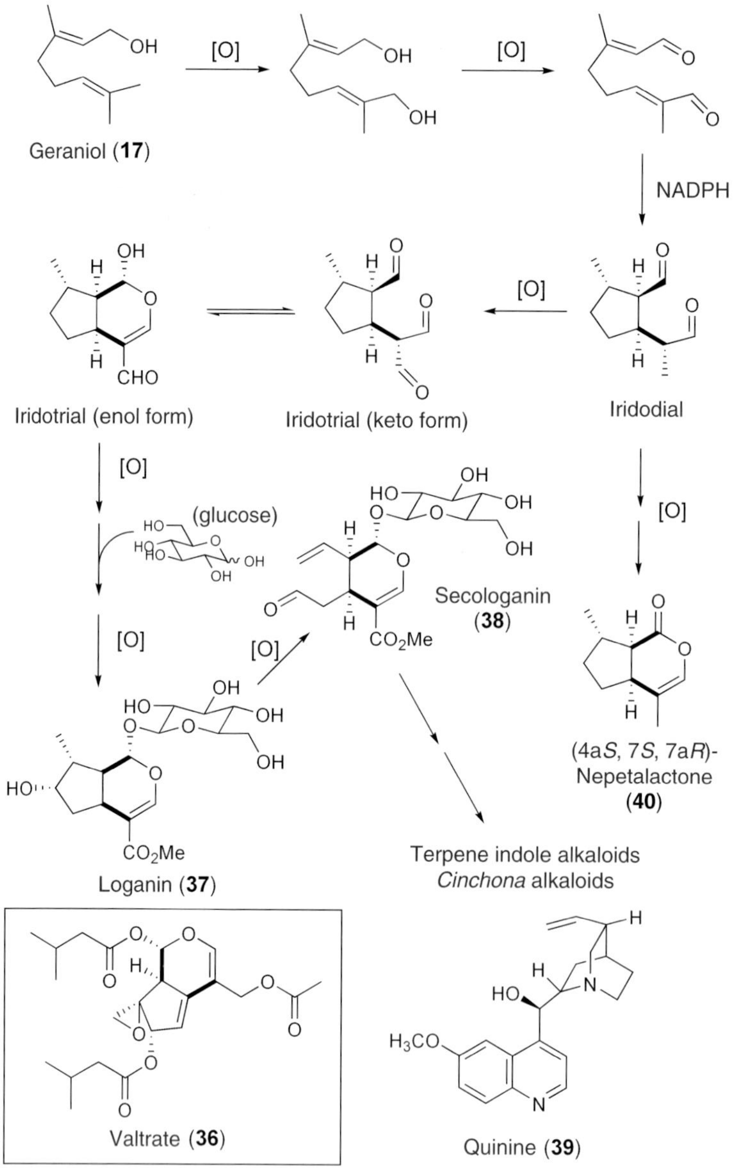

Figure 3.14 Outline biosynthesis of the iridoid family of oxygenated monoterpenoids.

and which usually possess antibacterial or antifungal activity). Numerous diterpenes also possess medicinal properties.

The biosynthetic cyclization reactions leading to the various diterpene skeleta are summarized in Figure 3.16 (for a review see MacMillan and Beale (1999)). In addition to the well-established mode of cyclization driven by pyrophosphate ionization (sometimes referred to as 'Type I'), a second mode ('Type II'), driven by protonation of double bonds,

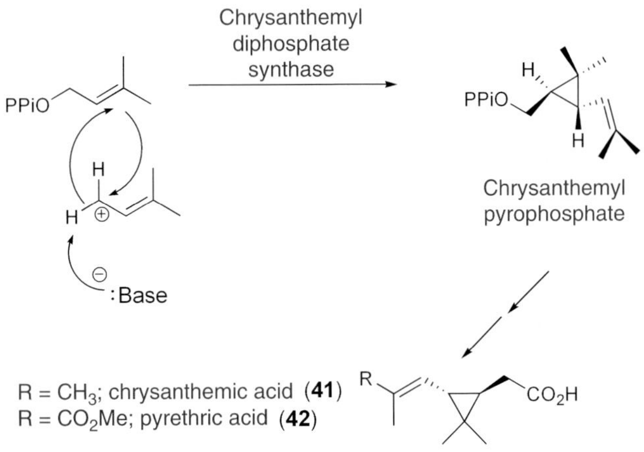

Figure 3.15 Proposed mechanism for the biosynthesis of chrysanthemic and pyrethric acids.

R = CH$_3$; chrysanthemic acid (**41**)
R = CO$_2$Me; pyrethric acid (**42**)

is also apparent. The structural diversity of the diterpene family is largely due to the fact that the products of diterpene synthase enzymes may be formed by either 'Type I' or 'Type II' chemistry, or by both processes acting in tandem – sometimes as a result of cooperative activity by two different synthases, in other cases by the action of a single enzyme with bifunctional properties.

The biosynthesis of casbene (**43**), a phytoalexin of castor bean, is an example of direct pyrophosphate ionization-driven cyclization of GGPP. Casbene synthase has been cloned and overexpressed in *E. coli* (Huang et al. 1998); the enzyme catalyses formation of a 14-membered ring with the distal double bond acting as nucleophile, followed by a cyclopropanation as shown in Figure 3.16.

In the context of human health, one of the most important diterpenes of the modern era is **Taxol** (**44**), a potent anticancer drug that is used in the treatment of a variety of cancers. This highly oxygenated and substituted compound is formed from GGPP in the bark of the Pacific yew tree, *Taxus brevifolia*, in 19 enzymatic steps. Difficulties in obtaining enough taxol for clinical use from its natural source have led to a massive effort to delineate its biosynthesis, in preparation for possible biotechnological solutions to this problem (Jennewein and Croteau 2001). Many of the genes involved have now been cloned (Jennewein and Croteau 2001; Jennewein et al. 2004). A key step in Taxol biosynthesis is the formation of taxa-4(5),11(12)-diene (**45**) by a bifunctional diterpene synthase (Figure 3.17) (Lin et al. 1996). The cyclization of GGPP by taxadiene synthase begins with diphosphate ionization but the initial formation of a 14-membered ring (cf. casbene synthase) is followed by a second internal cyclization. A neutral bicyclic olefin, verticillene (**46**), is implicated as an enzyme-bound intermediate in the reaction; protonation of **46** provides the driving force for a third intramolecular nucleophilic attack on the resulting positive centre, with taxadiene **45** released as the final product. Over 350 taxane diterpenoids are known to occur naturally (Jennewein and Croteau 2001), although most are found in such low quantities that their biological activity is presently unknown.

More commonly, however, diterpene biosynthesis begins not with dissociation of GGPP but with protonation at the distal double bond, producing the bicyclic structures

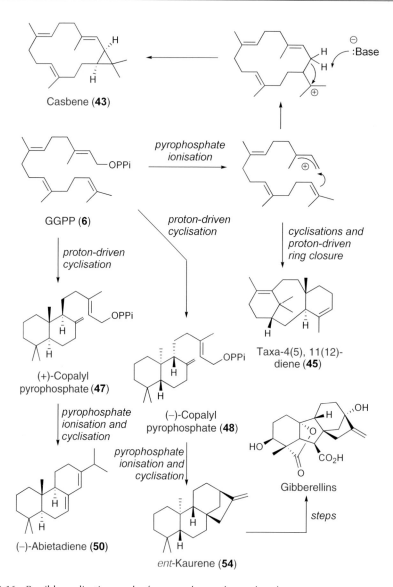

Figure 3.16 Possible cyclization modes for geranylgeranyl pyrophosphate.

(+)-copalyl (or labdadienyl) pyrophosphate (CPP, **47**), and its enantiomer *ent-* or (−)-copalyl diphosphate (*ent*-CPP, **48**) (Figure 3.16). Over 5000 compounds based on the labdane diterpene skeleton are known (Prisic *et al.* 2004). The versatility of the cyclization chemistry employed by these enzymes is well illustrated by abietadiene synthase, a bifunctional diterpene cyclase which has been cloned from grand fir (Peters *et al.* 2001). Here, proton-induced cyclization of GGPP to **47** is followed by loss of pyrophosphate and cyclization to a sandaracopimarenyl cation (**49**) (Figure 3.18). The reaction sequence can be terminated by deprotonation at any of several sites, with a Wagner-Meerwein shift of the

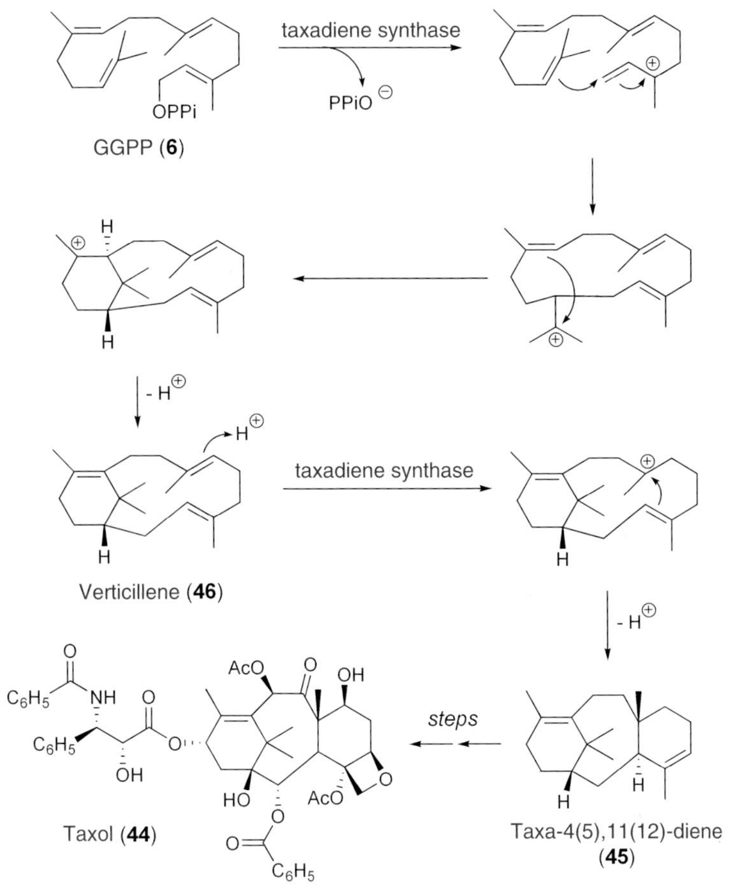

Figure 3.17 Diterpenes: the mechanism of taxadiene synthase.

C13 methyl group as an accompanying reaction, so that several possible products may be generated, the major ones being abietadiene (**50**) and levopimaradiene (**51**). Intriguingly, this bifunctional enzyme contains *two* completely separate active sites, one accepting GGPP and the other accepting (+)-copalyl pyrophosphate, with free diffusion of intermediates between them (Peters *et al.* 2001). Resin diterpenes such as abietic acid (**52**) are formed by the oxidation of the terpene cyclase products; they are secreted in oleoresin along with volatile monoterpenes in response to wounding, and their polymerization after the monoterpenes have evaporated forms a hard protective coating over the damaged area.

Levopimaradiene **51** is also the biosynthetic precursor of the gingkolides, a series of highly oxygenated diterpenoids from the medicinal tree *Gingko biloba* (Figure 3.18) (Schepmann *et al.* 2001). *G. biloba* evolved some 150 million years ago and is the last remaining member of an order of small, primitive trees now otherwise known only as fossils. Extracts of *Gingko* have long been used in the treatment of cardio- and cerebrovascular diseases and dementia; their effectiveness is believed to be a result of the stimulatory activity of the gingkolides on blood circulation, especially in the brain.

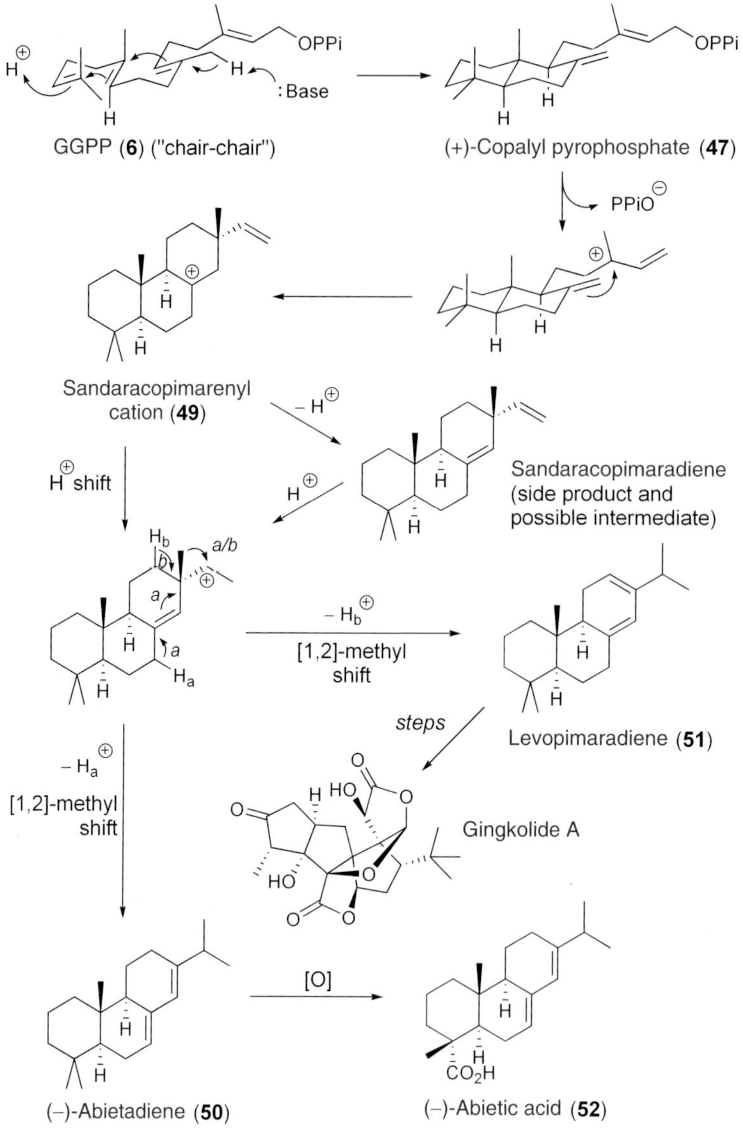

Figure 3.18 Diterpenes: the biosynthesis of abietic acid and the gingkolides.

The project to sequence the genome of rice (*Oryza sativa*) has provided a great deal of insight at the genetic level into the production of a host of labdane diterpene phytoalexins. An outline biosynthesis is shown in Figure 3.19 (Mohan et al. 1996). The 'Type II' cyclization of GGPP and the subsequent 'Type I' ionization of the intermediate labdane diphosphates are catalysed by separate enzymes. The oryzalexin and phytocassane families arise from cyclization of *ent*-CPP **48** formed by inducible *ent*-CPP synthases which are distinct from the constitutive *ent*-CPP synthase involved in GA biosynthesis

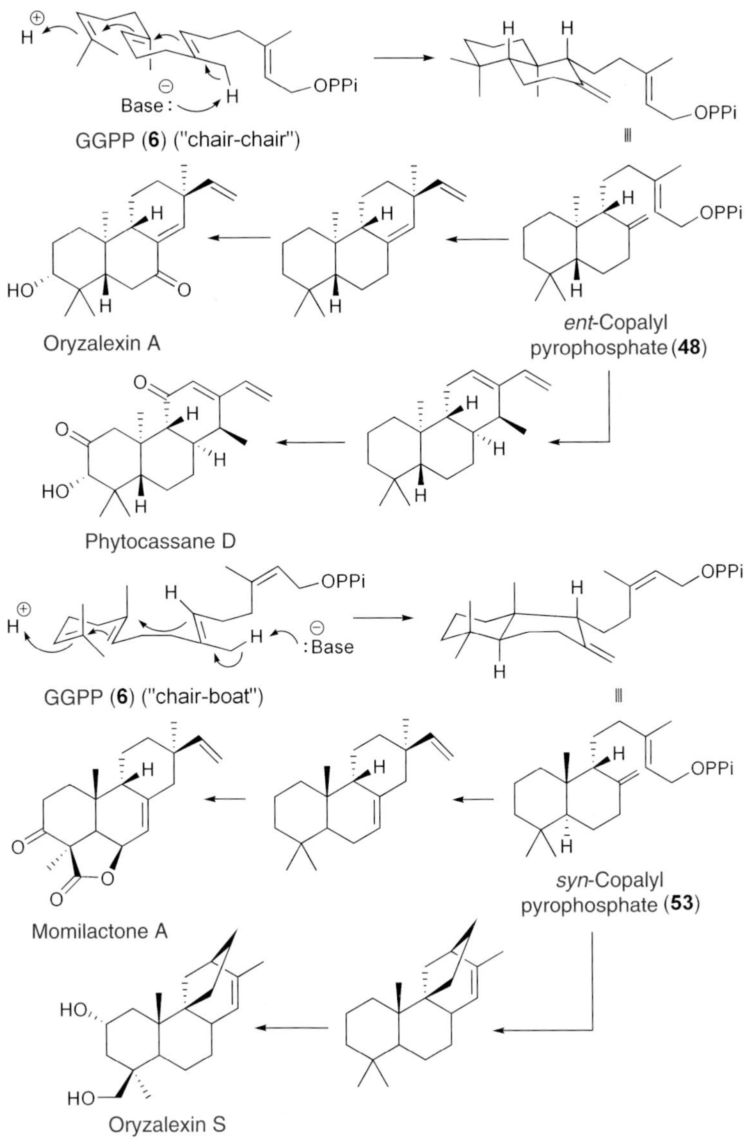

Figure 3.19 Diterpenes: two cyclization modes of GGPP in the biosynthesis of rice phytoalexins.

(Prisic et al. 2004). The momilactones and oryzalexin S, however, derive from an unusual diastereomeric labdane diphosphate: 9,10-*syn*-copalyl pyrophosphate 53 (Mohan et al. 1996). The normal 'chair-chair' conformational folding of GGPP disfavours the formation of *syn*-CPP compared with (+)-or (−)-CPP as its formation would require an unfavourable axial placing of the bulky sidechain, so it is assumed that the formation of *syn*-CPP must proceed via a 'chair-boat' conformational arrangement of GGPP as shown in Figure 3.19.

Over 100 members of the gibberellin (GA) diterpenoid family are now known. GAs are phytohormones responsible for regulating a host of key physiological processes in the life cycle of the plant: these include seed germination, elongation of stems, flowering and seed development, as well as the plant's responses to environmental stimuli such as light (Richards *et al.* 2001). The importance of these compounds in plant development has led to an extensive study of their biosynthesis and its regulation (for reviews see Hedden and Kamiya 1997; Crozier *et al.* 2000). Plants with mutations in GA biosynthesis genes tend to be characterized by a dwarf phenotype, delayed flowering and sterile male flowers, although application of exogenous GA can often restore wild-type growth patterns. By contrast, mutants with unregulated GA biosynthesis tend to be pale and elongated. *ent*-CPP **48**, the immediate precursor of the GA family, is cyclized and rearranged to *ent*-kaurene **54** by the action of *ent*-kaurene synthase (Figure 3.20). *ent*-CPP synthases and *ent*-kaurene synthases (sometimes referred to by their older names of *ent*-kaurene synthases A and B respectively) have been purified, and their genes overexpressed, from a variety of sources, including pumpkin, pea and *Arabidopsis* (Hedden and Kamiya 1997); they are separable proteins but often form closely associated complexes in vivo. The conversion of *ent*-kaurene to the gibberellins is mediated by an arsenal of oxygenating enzymes: first a membrane-bound cytochrome P_{450} mono-oxygenase which converts *ent*-kaurene to gibberellin A_{12} (**55**), establishing the characteristic ring structure of the GA family, then a series of closely related, soluble, 2-oxoglutarate-dependent dioxygenases which produce the various intermediates leading to the key physiologically active products shown in Figure 3.20. Several of these dioxygenases have wide substrate specificity, and may act on several different GA intermediates. The result is a network of intersecting biosynthetic pathways which operate concurrently in planta (MacMillan and Beale 1999). An intricate regulatory mechanism exists *in planta* to control the expression of key GA biosynthetic genes and the metabolism of the gibberellins themselves (Richards *et al.* 2001). At the metabolite level, physiologically active GAs such as gibberellins A_1 (**56**), A_3 (**57**, the original 'gibberellic acid') and A_7 (**58**) are distinguished by hydroxylation at C3. These bioactive gibberellins are themselves deactivated by further hydroxylations.

The coffee diterpenes cafestol (**59**) and kahweol (**60**) (Figure 3.21) also possess an *ent*-kaurene skeleton. Unfiltered coffee brews contain 1–2 g/L of lipid, of which 10–15% is made up of fatty acid esters of these diterpenes (Urgert and Katan 1997). Filtration through paper, percolation through a packed solid layer of coffee particles or the processing involved in the preparation of instant coffee removes this lipid component; however it remains present in Scandinavian boiled, cafetière or Middle Eastern coffee and to a lesser extent in mocha and espresso. A link has now been established between intake of coffee diterpenes (particularly cafestol) and elevated levels of serum cholesterol. A link has also been observed between consumption of coffee diterpenes and elevated serum levels of liver enzymes such as alanine aminotransferase, so it is possible that the mode of action of these compounds involves some disruptive effect on liver cells, affecting lipid metabolism; this has not, however, been clinically proven. It is noteworthy that in Scandinavia between 1970 and 1990 a widespread shift from boiled to filtered coffee coincided with a 10% decrease in serum cholesterol levels in the general public and a dramatic reduction in mortality from coronary heart disease (Urgert & Katan 1997).

Enzymes catalysing the initial steps in diterpene biosynthesis are localized in the plastids and depend on the 1-DXP pathway for their supply of GGPP. As with the monoterpenes,

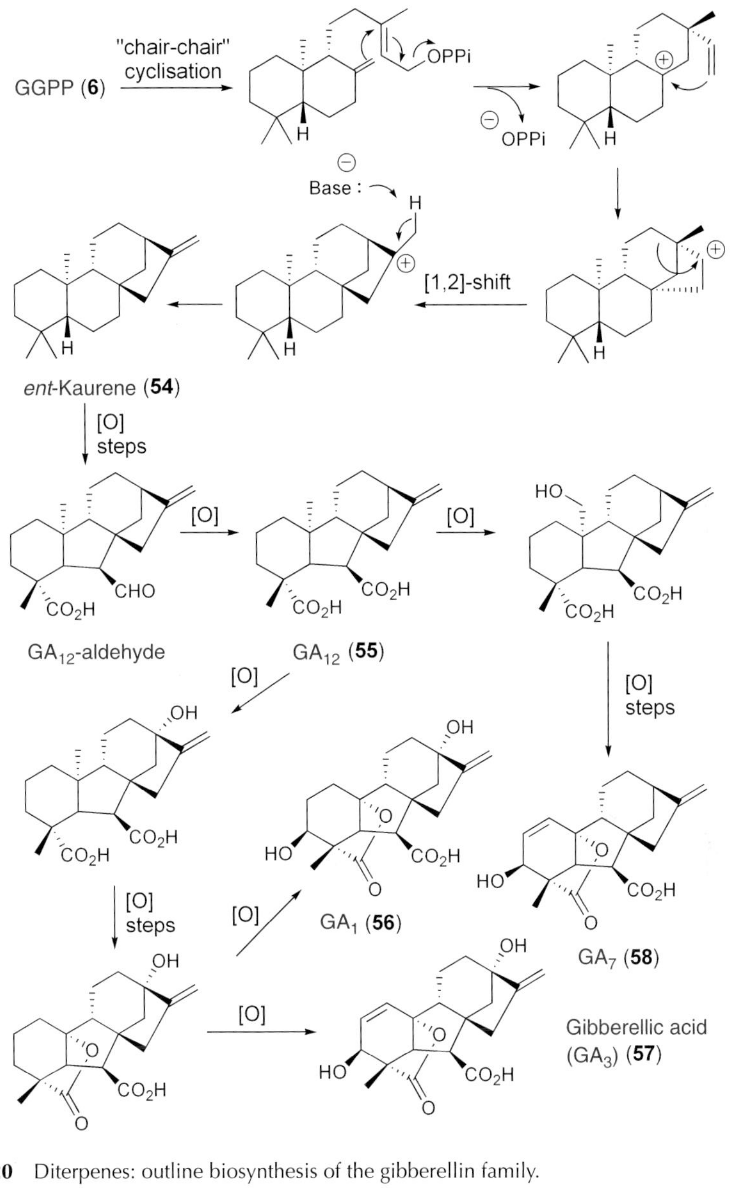

Figure 3.20 Diterpenes: outline biosynthesis of the gibberellin family.

however, the oxygenating enzymes controlling later biosynthetic steps are often located outside the plastid. The interrelationship between diterpene synthases from a wide range of plants has been studied at the amino acid sequence level (Bohlmann et al. 1998; MacMillan & Beale 1999). As well as the DDxxD motif associated with Mg^{2+}-pyrophosphate binding, another conserved motif, DxDD (where x is usually valine or isoleucine) is also found in 'Type I' diterpene synthases and is associated with protonation of double bonds.

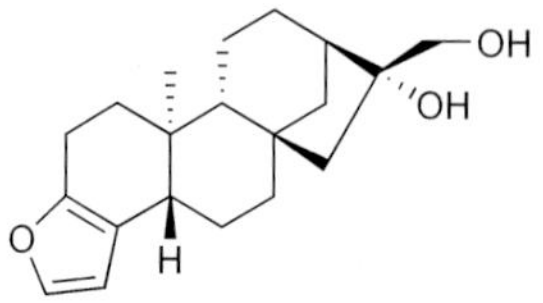

Cafestol (**59**)

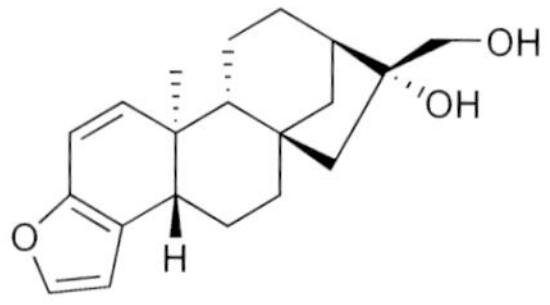

Kahweol (**60**)

Figure 3.21 Coffee diterpenes.

3.4.3 Biosynthesis of carotenoids

GGPP is also the biosynthetic precursor of the carotenoids, a family of C_{40} isoprenoids (tetraterpenes) which carry out essential functions in the life cycle of green plants (for a review see Fraser & Bramley (2004)). The highly extended, conjugated double bonds of carotenoids make them ideal as accessory pigments in the light-harvesting steps of photosynthesis. The same extended systems of conjugated double bonds are responsible for the intense colouration of the carotenoids, a property which plants harness to good effect by making use of them as pigments to attract insects, birds and animals to their flowers and fruits. Many of these natural pigments are also of considerable economic value, both in the ornamental garden and as food additives.

Carotenoids play a key role in human diet by virtue of their metabolism to vitamin A (retinol, **61**). As the prosthetic group associated with the light-harvesting pigment rhodopsin, retinol is critically involved in visual processes. It is also a participant in growth and development processes, so vitamin A deficiency, which is estimated to inflict 124 million children worldwide, can have serious consequences for long-term health (Beyer *et al.* 2002). The antioxidant properties of carotenoids have also been implicated in the protection against heart disease and cancer (Fraser and Bramley 2004).

The enzymes of carotenoid biosynthesis are nuclear encoded but the sites of biosynthesis are the chloroplasts or (in the case of the pigments) the chromoplasts, which are non-chlorophyll-containing plastids. Biosynthesis is initiated by the condensation of two molecules of GGPP (**6**) which, unusually, occur in a 'head-to-head' fashion under the action of the enzyme phytoene synthase (PSY) (Figure 3.22) (Dogbo *et al.*1988). A cyclopropane ring-containing intermediate, pre-phytoene pyrophosphate (**62**), has been implicated in the reaction; this reaction closely parallels that which occurs in the synthesis of chrysanthemic acid (see Section 3.4.1) and squalene (vide infra). PSY purified from *Capsicum* chromoplasts was found to have an absolute requirement for Mn^{2+} for activity (most terpene synthases use Mg^{2+} to bind pyrophosphate).

The predominant product of PSY in plants is the counter-intuitively numbered 15-*cis*-phytoene (**8**). Phytoene is dehydrogenated in a series of enzymatic steps to yield lycopene (**63**), the pigment responsible for the familiar orange colour of ripening tomato fruits. The final product is the all-*trans* geometric isomer (Fraser and Bramley 2004). From lycopene the pathway branches as shown in Figure 3.23: one branch leads via α-carotene (**64**) to lutein (**65**), the predominant carotenoid in the light-harvesting complex of the chloroplasts, the other, to β-carotene (**66**) and the oxygenated xanthophyll series. Branching occurs by the operation of two different pathways for cyclization of the terminal

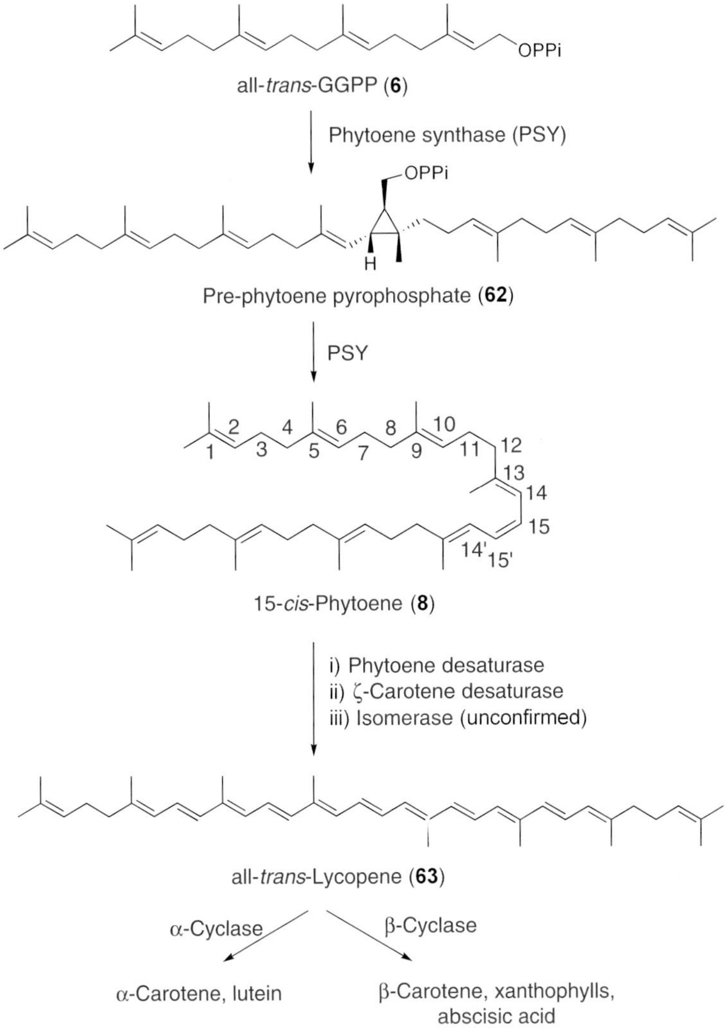

Figure 3.22 The biogenesis of the carotenoids. The numbers shown represent the convention in carotenoid chain labelling and do not correspond to the systematic numbering used in other terpene systems.

unit of the carotene chains, catalysed by different, but related, cyclase enzymes. In the β-series, the major products are symmetrical though reactions tend to occur in two-step processes, one at either end of the carotenoid chain; the asymmetrical intermediates are isolable and are well characterized. The conversion of β-carotene through the xanthophyll series is mediated by a series of ferredoxin-dependent hydroxylases. Xanthophylls serve as accessory pigments in the light-harvesting steps of photosynthesis; the interconversion of zeaxanthin (**67**) and violaxanthin (**68**) plays an important role in protecting green tissues from radiation damage through the dissipation of excess energy (Niyogi *et al.* 1998).

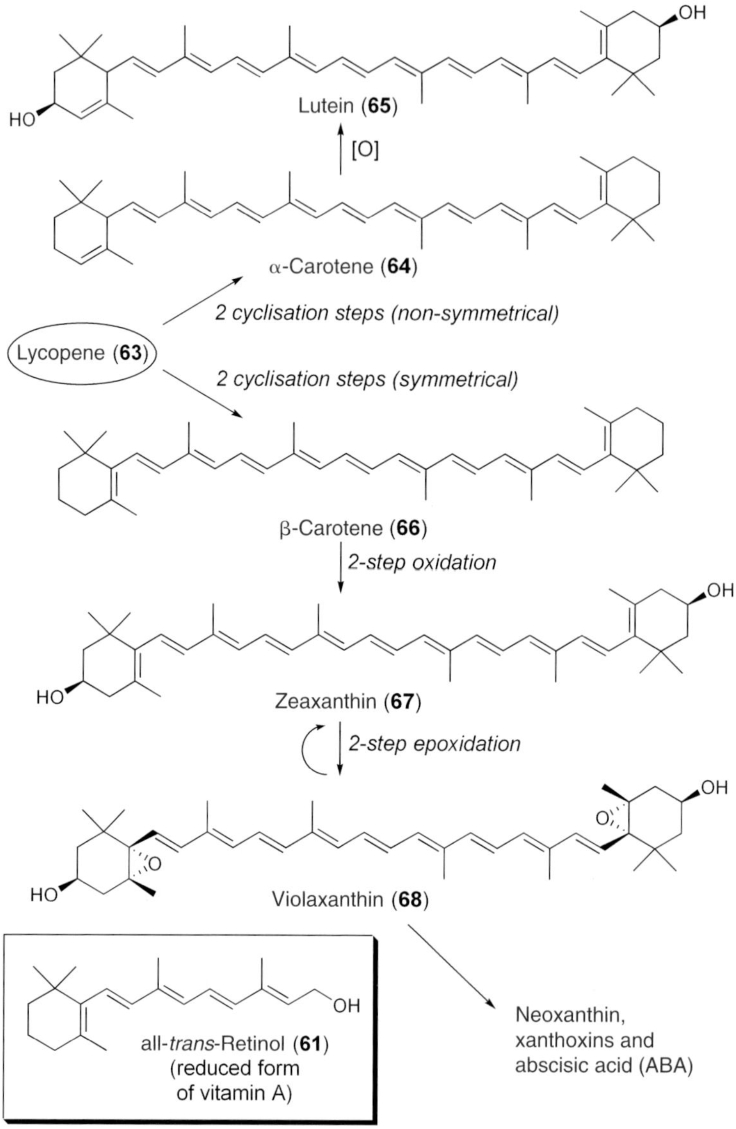

Figure 3.23 Carotenoids: the biosynthesis of β-carotene and the xanthophylls from lycopene.

The xanthophylls are also precursors in the biosynthesis of the important plant hormone (+)-abscisic acid (ABA, **69**). ABA plays a key role in the regulation of plant growth, often acting in opposition to the stimulating effect of the gibberellins and is also a modulator of plant responses to drought stress (Wilkinson and Davies 2002). The biosynthesis of ABA from carotenoids in green plants has been reviewed by Oritani and Kiyota (2003) and arises from the asymmetrical oxidative cleavage of the unusual allenic xanthophyll 9-*cis*-neoxanthin (**70**) which is formed from violaxanthin **68** as shown in Figure 3.24.

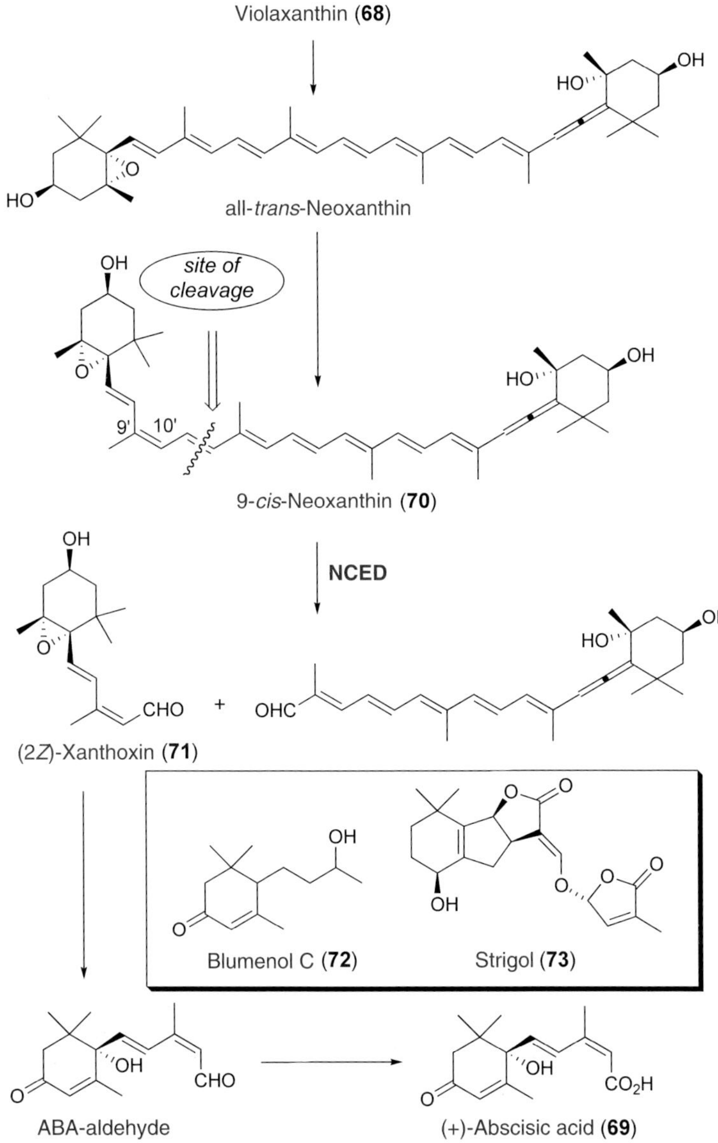

Figure 3.24 Major pathway for the biosynthesis of (+)-abscisic acid in plants.

Xanthoxin (**71**), the key metabolite resulting from the cleavage reaction, is exported to the cytosol where it is further converted to ABA mainly via a route involving dehydrogenation, epoxide ring opening and P_{450}-mediated oxidation of the terminal aldehyde group to a carboxylic acid. ABA synthesized in the roots is carried upwards in the xylem but ABA is also biosynthesized in green tissues above ground, so both rhizosphere and aerial stress conditions can influence the production of ABA and its site of action (Wilkinson and Davies 2002).

The enzyme which mediates the key event in the ABA biosynthetic pathway, the cleavage of 9-*cis*-neoxanthin **70**, is 9-*cis*-epoxycarotenoid dioxygenase (NCED). The enzyme has been cloned from several plant sources including maize and tomato. Homologues of NCED have been identified in many plants and in humans and represent a large family of closely related enzymes mediating a host of carotenoid cleavage reactions, including those generating vitamin A from the carotenes (Giuliano *et al.* 2003). Other plant metabolites which may be derived from NCED-type cleavage of carotenoids include the blumenol (**72**) family of C13 anti-fungal compounds produced by mycorrhizal barley roots, traditionally classed as sesquiterpenes though derived from the 1-DXP pathway (Maier *et al.* 1998), and the strigol (**73**) family of semiochemicals secreted by the roots of maize, sorghum and other crops (Bouwmeester *et al.* 2003).

3.5 Isoprenoid biosynthesis in the cytosol

3.5.1 Biosynthesis of sesquiterpenes

The tens of thousands of known C_{15} terpenes derive from farnesyl pyrophosphate (**5**), which can be cyclized to produce about 300 different skeletal structures. The majority of known sesquiterpenes have been isolated from fungi, marine organisms and *Streptomyces* species, but a large number are also produced by flowering plants, where they serve a variety of functions. Like monoterpenes, many are found as components of essential oils. Other sesquiterpenes function as insect attractants, antifeedants or phytoalexins.

Sesquiterpene biosynthesis begins with loss of pyrophosphate from FPP under the action of sesquiterpene synthase enzymes, generating an allylic cation which is highly susceptible to intramolecular attack. Cyclization of the farnesyl cation may take place onto either of the remaining double bonds, with the result that 6-, 10- or 11-membered rings may be formed (Figure 3.25). Stereoelectronic considerations dictate the favoured cyclization path, with the enzyme providing the desired cyclization geometry (Cane 1990). In many cyclizations, isomerization of the C2–C3 *trans*-double bond to produce a *cis*-geometry in the final products is observed, which is electronically impossible in a single step. This takes place *via* isomerization of all-*trans*-FPP to nerolidyl pyrophosphate (**74**), in which free rotation about the C2–C3 bond is possible. Dissociation of nerolidyl pyrophosphate can then produce either *cis*- or *trans*-allylic cations. This process is analogous to the isomerization of geranyl and linalyl pyrophosphates under the influence of monoterpene synthases (Bohlmann *et al.* 1998). The resulting cyclic carbocations can deprotonate, with the formation of a free hydrocarbon product, or be trapped by water or another nucleophile, resulting in the wide range of functionalized carbon skeletons which are typical of the sesquiterpenes.

The germacrene skeleton is a 10-membered ring containing all *trans* double bonds. Germacrenes are volatile sesquiterpenes found in many plant extracts and are precursors for a whole family of commercially valuable molecules. The parent germacryl cation (**75**) can deprotonate in a variety of ways, and may also undergo Wagner-Meerwein hydride shifts, as shown in Figure 3.26. Germacrene A and D synthases from goldenrod (*Solidago canadensis*) have been cloned and overexpressed in *E. coli* (Prosser *et al.* 2002, 2004). The two enantiomers of germacrene D (**76**) are produced in goldenrod by separate, but very closely related, enzymes (85% amino acid sequence identity); each accepts FPP **5** as

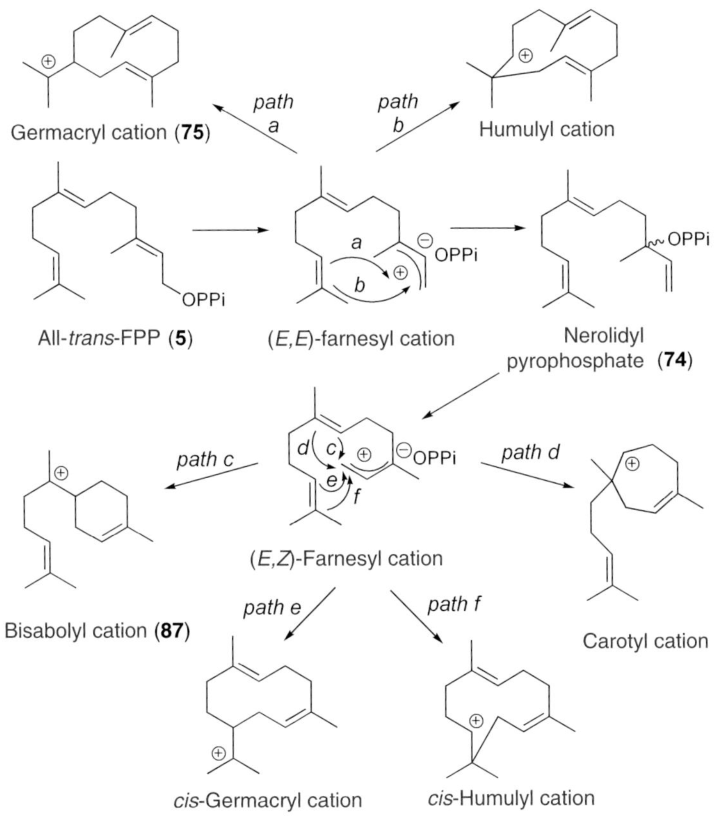

Figure 3.25 The major pathways of sesquiterpene biosynthesis.

substrate and produces the same cationic intermediate (**75**), but subtly different hydride shift mechanisms are responsible for the two enantiomeric products (Prosser *et al.* 2004). Derivatives of germacrene A (**77**) include parthenolide (**78**), the pain-relieving ingredient of feverfew (*Tanacetum parthenium*, a well-established traditional remedy for migraine, digestive and gynaecological problems (Castleman 2003)). Not all germacrene derivatives are benign to human consumption, however; many have allergenic or cytotoxic properties believed to originate in the highly reactive α,β-unsaturated lactone moieties which are common structural features of these sesquiterpenes (Robles *et al.* 1995). Even fresh feverfew can induce mouth ulcers and allergic contact dermatitis.

Protonation-induced cyclization of germacrene A generates the *cis*-decalin skeleton of the eudesmyl cation, an intermediate in the biosynthesis of the phytoalexin capsidiol (**79**). The eudesmane skeleton is converted to the 5-epi-aristolochene (**80**) skeleton by Wagner-Meerwein rearrangements of a hydride and a methyl group (Cane 1990). Hydroxylation of **80** then produces capsidiol. The entire process of formation of *epi*-aristolochene from FPP is mediated by a single synthase which has been purified from tobacco (Starks *et al.* 1997). The intermediacy of germacrene A in the biosynthesis was demonstrated by mutation of

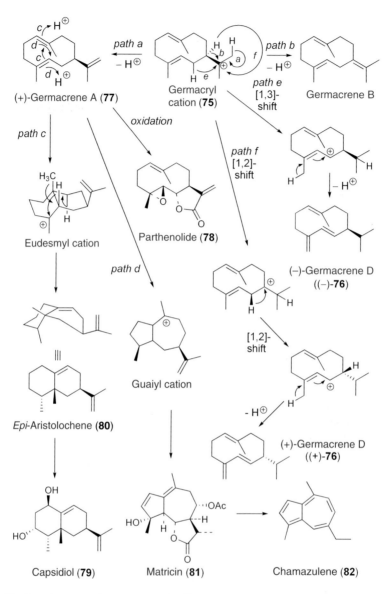

Figure 3.26 Sesquiterpenes: the germacrene family.

a tyrosine residue which mediates the key protonation step; in the mutant enzyme, free germacrene A was released as the product (Rising *et al.* 2000).

The guaianes, like the eudesmanes, are derived from a further internal cyclization of germacrene A. The family includes matricin (**81**) and chamazulene (**82**), major components of the extract of German chamomile (*Matricaria chamomilla*), which has anti-inflammatory properties (Castleman 2003). Related to the guaianes is the tricyclic sesquiterpene patchoulol (**83**), a perfumery raw material which is the major component

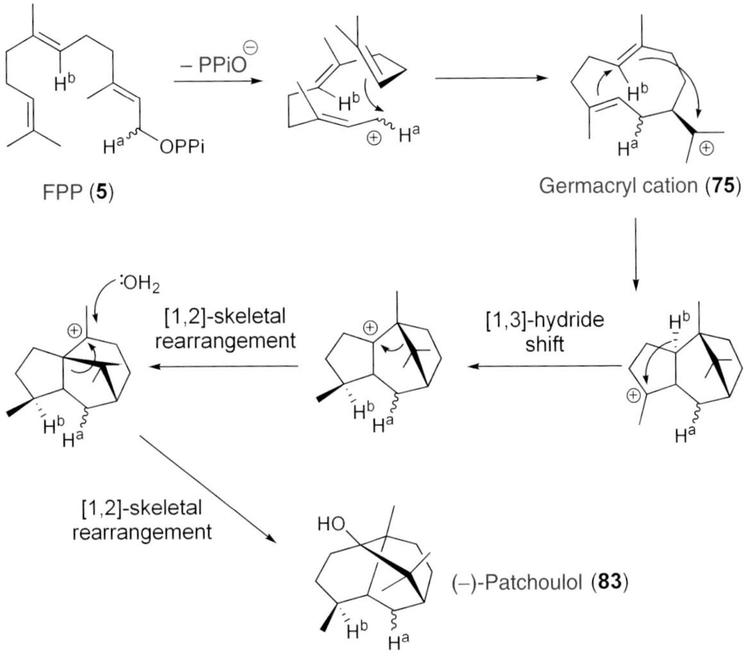

Figure 3.27 Sesquiterpenes: the biosynthesis of (−)-patchoulol from FPP.

of patchouli (*Pogestomon cablin*) oil. Key protons have been traced through the biosynthetic sequence by radiolabelling FPP incubated with cell-free extracts of *P. cablin*; the results suggested that patchoulol formation proceeded entirely *via* an elaborate sequence of Wagner-Meerwein skeletal rearrangements of cationic intermediates (Croteau *et al.* 1987) (Figure 3.27). The entire sequence of cyclizations and rearrangements starting from FPP is catalysed by a single synthase which has been purified from *P. cablin* leaves (Munck & Croteau 1990). A number of olefinic side products are also generated by the same enzyme, *via* deprotonation of the various cationic intermediates.

α-Bisabolol (**84**) and its oxides **85** and **86** are components of *Matricaria* oil and also have anti-inflammatory properties. They are derived from the bisabolyl cation (**87**), a cyclization product of the (E,Z)-farnesyl cation (Figure 3.28). The bisabolyl cation is also the precursor to the *cis*-decalin skeleton of amorpha-4,11-diene **88** (Bouwmeester *et al.* 1999), one of whose derivatives is the novel anti-malarial compound artemisinin (qinghaosu) (**89**). The anti-malarial activity of this compound, which is extracted from sweet wormwood (*Artemisia annua*), appears to be associated with the unusual peroxide linkage in the molecule (Haynes & Vonwiller 1997); its mode of action differs from that of all previous anti-malarial drugs.

The remaining possible cyclization modes of FPP give rise to the *cis*-germacryl (10-membered), and the humulyl and *cis*-humulyl (11-membered), cations as shown in Figure 3.29. α-Humulene **90** is found in the oil of hops (*Humulus lupulus*) and occurs as a minor component of many essential oils. β-Caryophyllene (**91**) is a humulyl derivative present in clove and cinnamon oil. The *cis*-germacryl cation is the precursor of

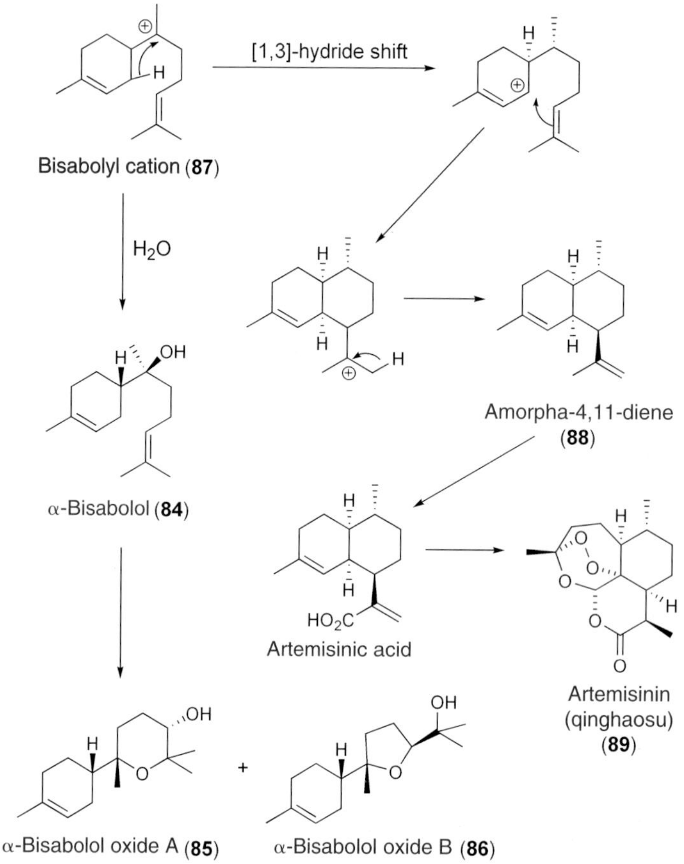

Figure 3.28 Sesquiterpenes: the bisabolene family.

the *trans*-decalin structures typified by α-cadinene (**92**), a component of juniper oil; a Wagner-Meerwein [1,3]-hydride shift is necessary to effect the key cyclization. The related δ-cadinene (**93**) is a precursor of gossypol (**94**), an unusual pigment containing two aromatic sesquiterpene units linked by a phenolic coupling mechanism (Bell 1967). Gossypol is a phytoalexin isolated from the seeds of cotton (*Gossypium* spp.) (Heinstein *et al.* 1962) and the (−)-enantiomer has potent activity as a male contraceptive.

The biochemical origins of the sesquiterpenes are generally considered to lie with the mevalonate pathway. Classical incorporation experiments with radiolabelled acetate or mevalonate have confirmed the biogenesis of phytoalexins such as capsidiol (Vögeli & Chappell 1988) and gossypol (Heinstein *et al.* 1962) from mevalonate; comparative studies of sesquiterpene and sterol biosynthesis in these systems have also shown that plant cells are capable of regulating the flux of FPP through the competing biosynthetic paths when challenged, for example, by fungal infection (Vögeli and Chappell 1988). The lack of plastid targeting sequences in the genes of enzymes such as *epi*-aristolochene (Rising *et al.* 2000) and germacrene A (Prosser *et al.* 2002) synthases suggests a cytosolic location for these enzymes, where they will be exposed to the mevalonate-derived pool of IPP and DMAPP.

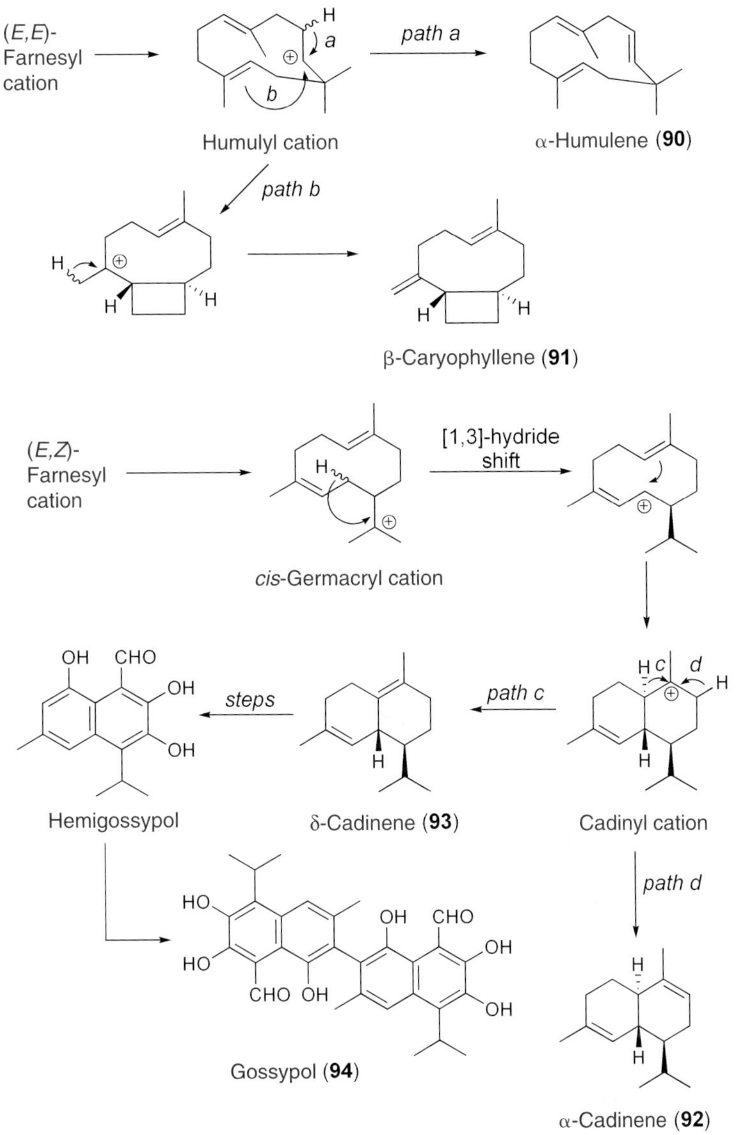

Figure 3.29 Sesquiterpenes: the humulene and cadinene families.

This does not mean that all sesquiterpene biosynthesis is independent of the 1-DXP pathway, however. In one plant system, quite the opposite has been found to be true. Bisabolol oxides (**85** and **86**), matricin (**81**) and chamazulene (**82**) from chamomile flowers have been shown to contain two isoprene units derived from the 1-DXP pathway and one derived at least partially from the mevalonate pathway (Adam *et al.* 1999). A proposed explanation for this labelling pattern is shown in Figure 3.30. Here, sesquiterpene biosynthesis is thought to begin, unusually, in the plastids with the formation of IPP and GPP *via*

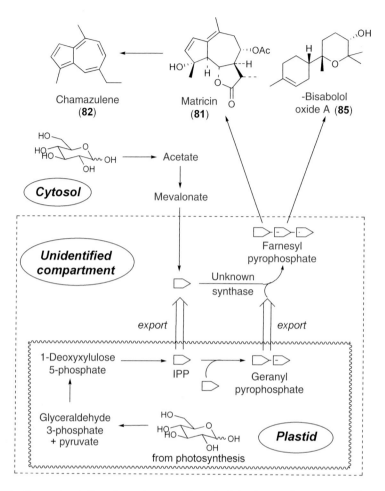

Figure 3.30 Compartmentation of sesquiterpene biosynthesis in chamomile as proposed by Adam et al. (1999). Blocks represent isoprenoid precursor units (IPP or DMAPP).

the 1-DXP pathway. These intermediates are then exported to a separate cellular region which also has access to the cytosolic (mevalonate-derived) pool of IPP. In this region, the final isoprene unit is added by an FPP synthase. The fact that the mevalonate pathway does not contribute at all to the biosynthesis of the first two isoprene units (Adam et al. 1999) suggests that the affinity of the participating FPP synthase for GPP as a substrate must be very much greater than its affinity for DMAPP, which would mainly be present from the mevalonate-derived precursor pool.

The full details of sesquiterpene biosynthesis in chamomile are not yet known. Isolation of the prenyltransferase(s) responsible for the final steps in the biosynthesis and identification of the site of biosynthesis are likely to shed further light on the conundrum of biosynthetic compartmentation in flowering plants. These studies may also offer a challenge to the accepted wisdom that sesquiterpene biosynthesis in other flowering plants is always simply a matter of the processing of mevalonate-derived FPP in the cytosol.

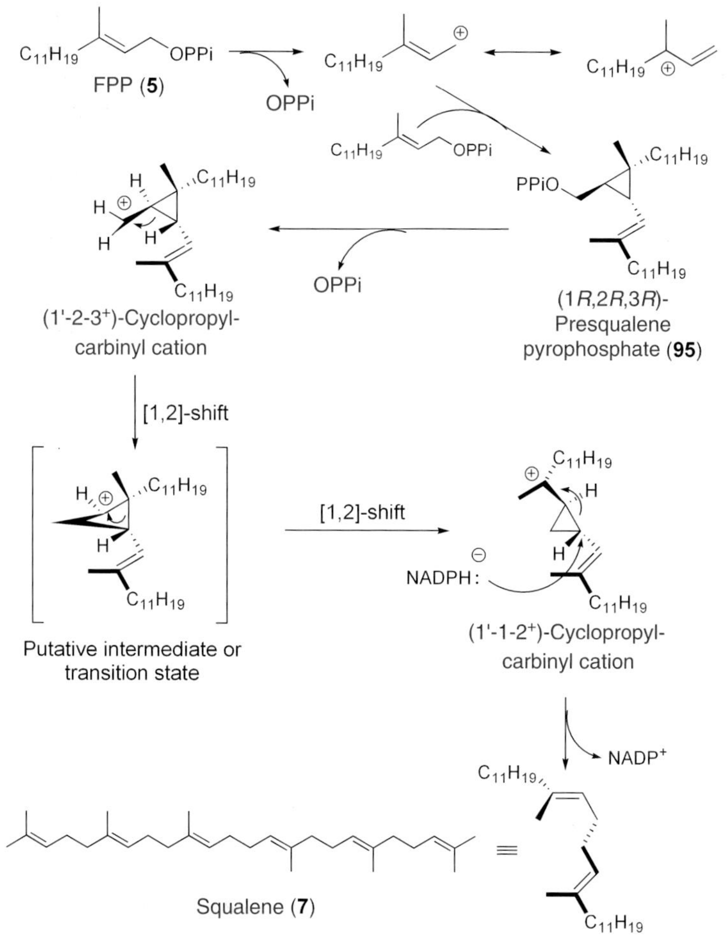

Figure 3.31 Biosynthesis of presqualene pyrophosphate and squalene from FPP according to Blagg et al. (2002).

3.5.2 Biosynthesis of triterpenes

The triterpenes encompass several families of polycyclic isoprenoids. The parent C_{30} carbon chain is derived from the 'head-to-head' condensation of two molecules of FPP to form *squalene* (7). The reaction parallels those involved in the formation of chrysanthemyl pyrophosphate and phytoene (vide supra) and proceeds, as they do, via a cyclopropane intermediate, presqualene pyrophosphate (95); a series of Wagner-Meerwein rearrangements are involved before the final product is released (Figure 3.31) (Blagg et al. 2002).

Cyclization of squalene in eukaryotes is initiated by protonation of the epoxide derivative, squalene oxide 96, formed by flavin-mediated oxidation of squalene with molecular O_2. A wide variety of polycyclic structures with different ring sizes and stereochemistries can result, controlled by the precise substrate folding imposed by the enzyme catalysing the cyclization. Each known mode of cyclization gives rise to a different subgroup of triterpenes (Abe *et al.* 1993). The best-known and most extensively studied of these subgroups is the steroid family. Steroids are triterpenoids containing a nearly flat cyclopenta[*a*]phenanthrene skeleton and an aliphatic sidechain. The most familiar example is cholesterol (**9**), the primary steroid in human membranes. In plants, cholesterol is widely distributed but often present in minute quantities; other steroids not found in animals (the phytosterols) tend to predominate. Steroid biosynthesis in plants proceeds via mevalonate and FPP in the cytosol but differs from the pathway in mammals and fungi because of the intermediacy of the unusual cyclopropane ring-containing sterol cycloartenol (**97**, Figure 3.32). A series of intricately co-ordinated Wagner-Meerwein rearrangements accompanies the cascade of cyclization events which produce the cycloartenol skeleton (Abe *et al.* 1993). The available evidence suggests that these occur in a stepwise fashion through rigid enzyme-bound intermediates, rather than as a concerted process, but it is convenient to illustrate all these migrations together as in Figure 3.32. The complete conversion of squalene oxide to cycloartenol is catalysed by a single synthase enzyme (Brown 1998). Wagner-Meerwein shifts of hydride and methyl groups are also invoked in the rearrangement or ring opening reactions of cycloartenol which produce the different phytosterol structures (Figure 3.33). Hydroxylases, dehydrogenases and methylating enzymes are all involved in the subsequent functionalization of the steroid skeleton; several of these enzymes (particularly the methyltransferases) also provide control points in regulating the flow of material through the steroid biosynthetic pathway (Brown 1998; Piironen *et al.* 2000).

In plants, as in mammals, steroids play a wide variety of different roles. The nearly planar aliphatic core and 3-hydroxyl head group of the phytosterols enable them to be incorporated into phospholipid membranes in cells, where they play a critical role in controlling the membrane fluidity or rigidity. The chemical structures of phytosterols are finely tuned to their physiological function; the flexible aliphatic side chains of sitosterol (**98**) and campesterol (**99**), two of the most abundant phytosterols, can stack in an ordered fashion in the membranes, increasing their rigidity, but the *trans*-double bond in the side chain of stigmasterol (**100**) interferes with membrane ordering and promotes greater fluidity (Piironen *et al.* 2000). Phytosterols also have roles in cell growth and proliferation, in controlling cell permeability and regulating ion transport, and as precursors of the *brassinosteroid* hormones.

The chair-boat-chair-boat conformation show in Figure 3.32 is not the only mode of cyclization possible for the key triterpene precursor, squalene oxide. A chair-chair-chair-boat arrangement (Figure 3.34), for instance, generates the dammarane skeleton. This is the core structural unit of the widely distributed β-amyrin (**101**) and related pentacyclic triterpenes, and of the ginsenoside family of triterpene glycosides which are the major bioactive components of the restorative herb ginseng (Castleman 2003; Sparg *et al.* 2004). The dammarane skeleton is also found in euphol (**102**) and tirucallol (**103**), the precursors of the limonoid series (Barton *et al.* 1961). The limonoids are highly oxygenated, truncated terpenoids produced as defensive agents by plants of the Rutaceae (citrus) and Meliaceae (mahogany) families. Limonin (**104**) and its glucoside (which forms after the opening of the D-ring lactone) are obtained in large quantities as waste products from the processing of citrus juices and have recently been found to have surprisingly high levels of anti-cancer

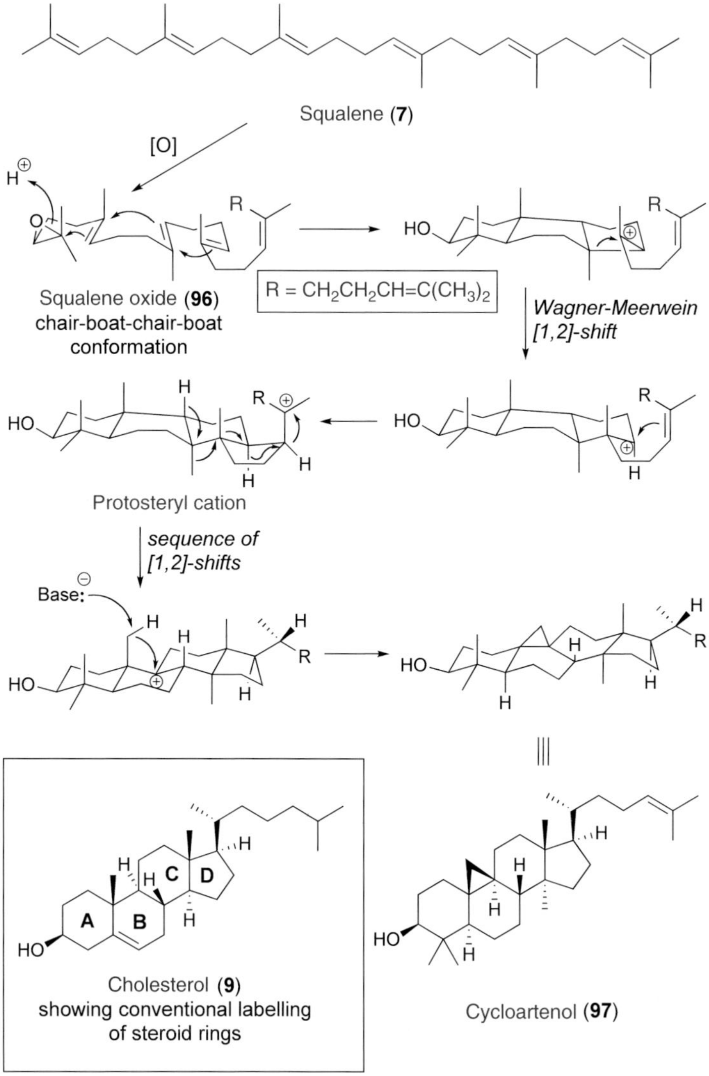

Figure 3.32 Cyclization of squalene oxide to cycloartenol.

activity in animal studies (Manners & Hasegawa 1999). Studies on the bioavailability of limonoids in human diet are now in progress since these compounds could represent a cheap, non-toxic cancer preventative, either in their natural form in citrus juice and peel or as food additives.

The neem (margosa or Indian lilac) tree (*Azadirachta indica*) is particularly well known for producing a complex mixture of limonoids which can be extracted from the bark, leaves and seed oil. Extracts of neem have been employed for many years in the Indian subcontinent as insect repellents and folk medicines and can be used to preserve stored grains and pulses (Boeke et al. 2004). The best known of the neem limonoids is azadirachtin (**105**),

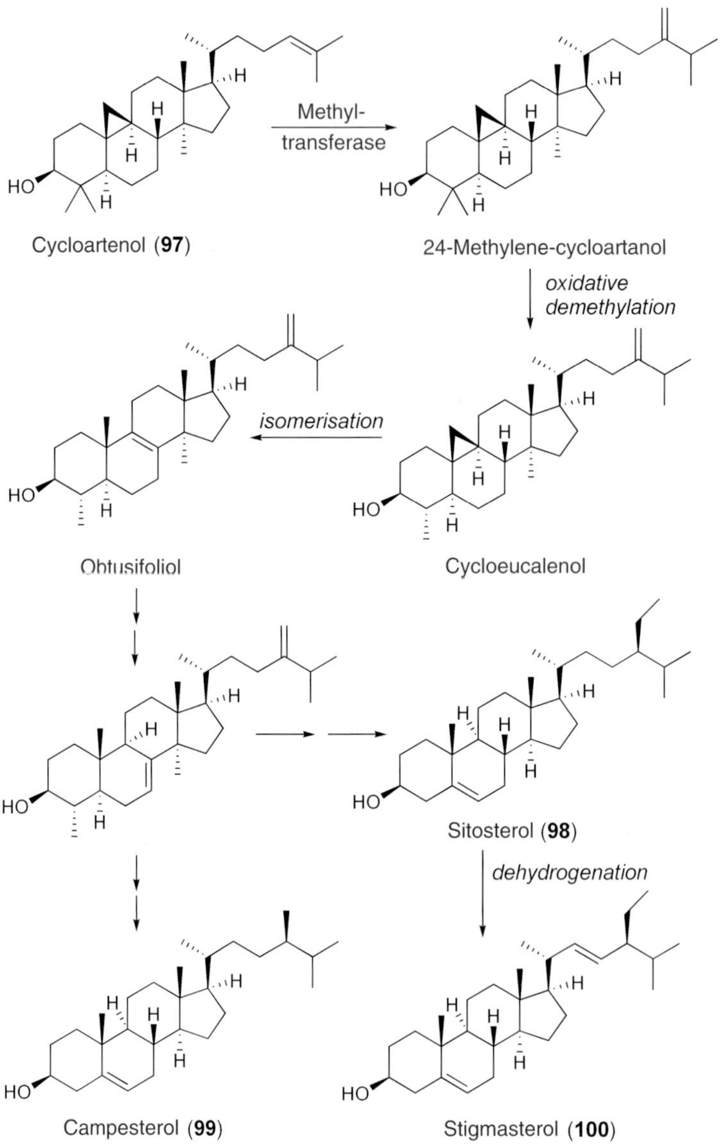

Figure 3.33 Outline biosynthesis of the major phytosterols from cycloartenol.

the most potent known insect antifeedant. Pure azadirachtin is remarkably non-toxic to plants and higher animals (Boeke et al. 2004). Other components present in crude extracts of neem used medicinally or as agricultural pesticides can have toxic effects at high dosages, however; so there is considerable current interest in developing azadirachtin-enriched extracts, for instance from plant cell culture (George et al. 2000).

As well as the free triterpenes and sterols, plant cells often accumulate significant quantities of triterpene derivatives such as glycosides (as shown above by the ginsenosides and

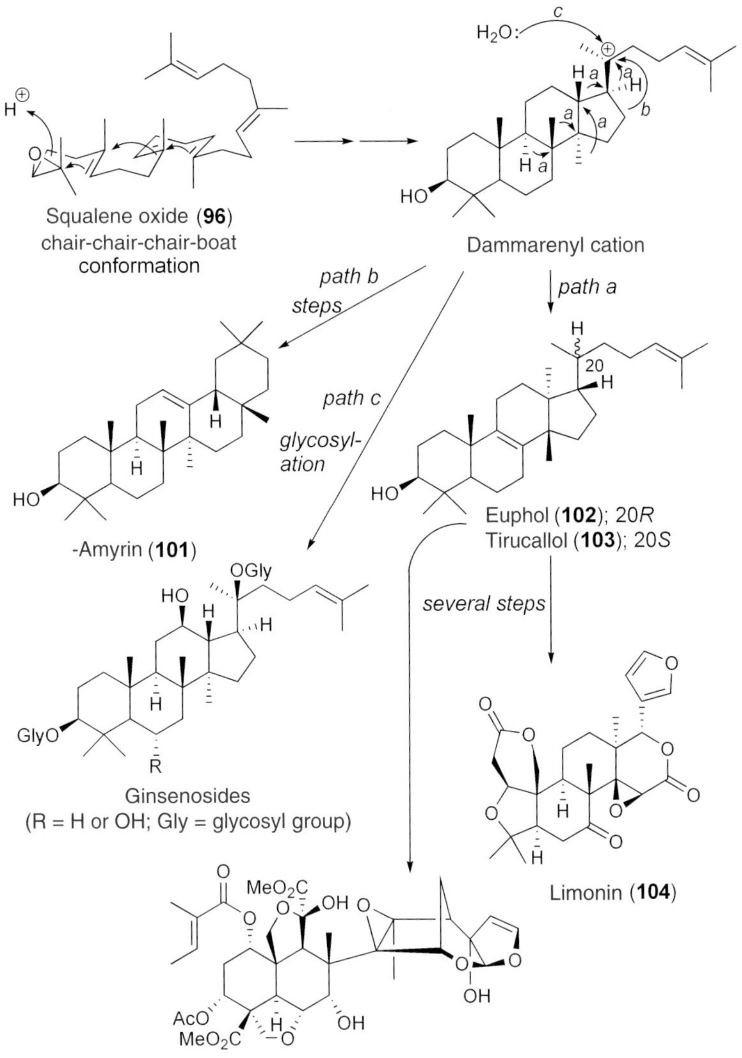

Figure 3.34 Outline biosynthesis of the dammarane triterpenoids including ginsenosides and limonoids.

limonoid glucosides) or fatty acid esters. Saponins (Sparg et al. 2004) are bitter-tasting triterpene glycosides found in many dicotyledonous plants, especially legumes (Shi et al. 2004). The majority of saponins are derived from dammarane-type triterpenes. They have soap-like properties, causing foaming in aqueous solution. They can also disrupt membranes e.g. membranes of red blood cells in the human body, so many are highly toxic if injected (hence their traditional use as arrow poisons). Oral toxicity in humans is usually low, however. The therapeutic effects of a large number of folk medicines are thought to be associated with their saponin content (Sparg et al. 2004); in several of the more familiar

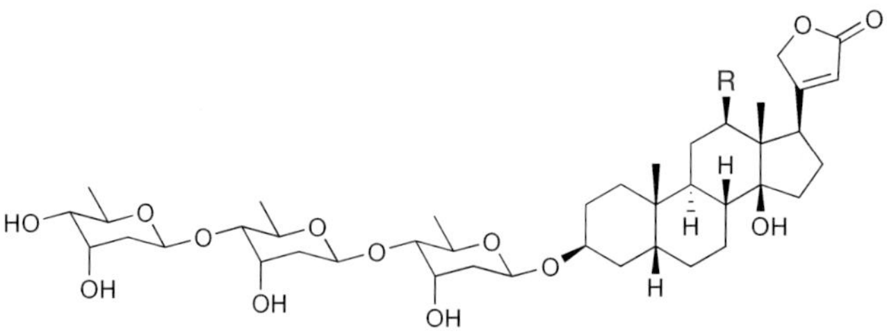

R = H; digitoxin (**106**) isolated from foxglove (*Digitalis purpurea*)
R = OH; digoxin from *D. lanata*; widely used treatment for heart failure.

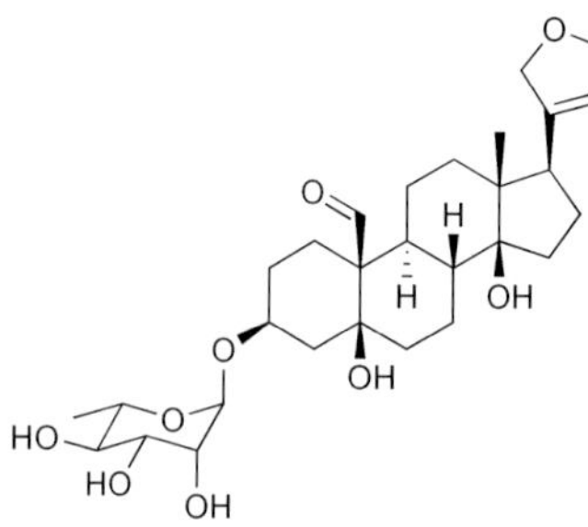

Convallatoxin, from lily of the valley (*Convallaria majalis*)

Figure 3.35 Examples of cardioactive triterpene glycosides.

examples (e.g. liquorice extract, used in the treatment of stomach ulcers (Castleman 2003)) the therapeutic effect is probably analogous to the effect of anti-inflammatory steroids. More recent studies in human nutrition have noted the effectiveness of legume saponins as cholesterol-lowering agents (Shi *et al.* 2004) and of ginseng and soybean saponins as anti-cancer agents (Shi *et al.* 2004; Sparg *et al.* 2004). Steroidal saponins are also known; tubers of the *Dioscorea* (yam) family are a particularly rich source and are a major source of raw material for the semi-synthesis of steroid drugs.

Another family of triterpenoids is represented by the cardenolides and bufadienolides from plants such as foxglove (*Digitalis purpurea*), lily of the valley (*Convallaria majalis*) and hellebore (*Helleborus niger*) (Melero *et al.* 2000). Cardenolide glycosides such as digitoxin (**106**) have potent cardiac stimulatory activity and are used in emergency heart surgery. This is also the reason for the high toxicity of the parent plant material; even a slight overdose of these compounds can be fatal. They act as inhibitors of the Na^+, K^+-ATPase or 'sodium pump' which is critical to the correct functioning of cardiac muscles. Some examples of cardioactive glycosides are shown in Figure 3.35. The *cis*-fused A/B-ring junction in these compounds, not often seen in triterpenes, is thought to arise from an unusual hydrogenation of a cholesterol derivative and is critical to their biological activity.

3.6 Terpenes in the environment and human health: future prospects

It is unquestionably the case that both the economic importance of plant terpenes and their significance in human health will continue to grow (Sangwan *et al.* 2001). Recent advances in mass spectroscopy, chromatographic and NMR technologies have made the metabolic profiling of complex plant extracts possible (Wang et al. 2004), enabling the active components of these mixtures to be identified and fully characterized. The cultivation of plant material specifically for its terpene content is now a major economic activity (Lange & Croteau 1999b). New uses for familiar material such as sage leaves (Santos-Gomes & Fernandes-Ferreira 2003) and citrus peelings (Manners & Hasegawa 1999) have been found as understanding of the properties of their constituent terpenes has grown. Increasing numbers of terpenes are being found to have anti-bacterial, anti-malarial and anti-cancer properties; the relatively low toxicity and high bioavailability of dietary monoterpenes (Crowell 1999) and carotenoids (Fraser & Bramley 2004) in particular make these compounds attractive as potential therapeutic agents.

The popularity of 'natural' herbal remedies continues to soar (Castleman 2003). In America alone, spending on herbal medicines rose fourfold between 1990 and 1997, to an estimated US$5.1 billion (Wang et al. 2004). The biochemical and physiological basis for the efficacy of these remedies is now much easier to establish, although their usage is not without its difficulties. The chemical compositions of herbal extracts can vary widely with the plant variety and the growth conditions used, as well as the mode of preparation; standardization of extracts from different growing regions can be lax or even non-existent (Wang et al. 2004). The regulatory regimes governing the use of botanical supplements fall somewhere between those of food additives and those of pharmaceuticals (Walker 2004), and the standards required for the risk assessment of food additives, for instance, are not necessarily appropriate for herbal medicines, where components of the botanical preparation can be expected to have appreciable physiological effects, including possible unwanted side effects (Höld et al. 2000). Risk assessment of herbal remedies is conducted on a case-by-case basis, the establishment of a 'decision tree' approach to the risk assessment enabling existing knowledge to be used wherever available; this approach also enables manufacturers to address gaps in the existing knowledge before products are brought to market (Walker 2004).

Extensive studies are now under way around the world on the effectiveness of essential oils and their constituent terpenes as antibacterial and antifungal agents, both medicinally and as natural preservatives which can be added to foodstuffs. Although in vitro studies have routinely shown many monoterpenes (Kalemba & Kunicka 2003) and diterpenes (Ulubelen 2003) to have significant bactericidal effects, formulations which are effective in transmitting the potential health benefits to humans are not always easy to obtain. Nevertheless, applications such as the use of tea tree oil (containing terpineols) in acne treatment and thymol in toothpaste continue to enjoy considerable success. The main disadvantage in the use of essential oil terpenes as food preservatives lies in the strong flavours often associated with these compounds. Nevertheless, the consumer reaction against artificial food additives certainly favours the development of essential oil products as more benign alternatives.

It has long been known that diets rich in phytosterols can offset the build up of cholesterol levels in the human bloodstream (Piironen et al. 2000). Sterols generally enter the digestive system as fatty acid esters and pass through the human gut in two forms: an oily, esterified phase, and a micellar phase, in which the fatty acid esters are hydrolysed by lipases, releasing free sterols which can pass through the gut lining and enter the bloodstream (Figure 3.36). Phytosterols are more hydrophobic than cholesterol and

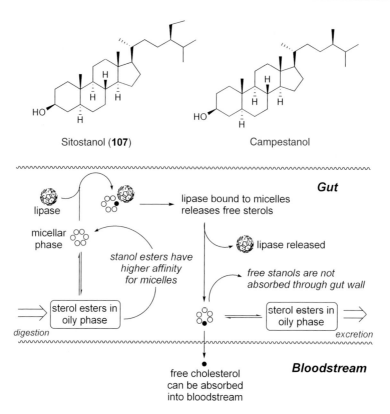

Figure 3.36 Stanol esters and their part in the lowering of cholesterol levels.

have a much higher affinity for the micelles which are associated with sterol digestion. The free phytosterols are, however, less well absorbed than cholesterol and persist in the micellar phase. Cholesterol is forced to remain in the esterified, oily phase and is eventually excreted without being absorbed. The resulting reduction in intake of dietary cholesterol also increases the rate of metabolism of cholesterol synthesized naturally by the human body, leading to its more rapid excretion in bile acid. Even more effective than the phytosterols in this regard are their reduced forms, the stanols, which are virtually unabsorbable by the human gut. The margarine Benecol™ contains esters of sitostanol (**107**) obtained by the hydrogenation of oils from wood pulp. It has been shown that an intake of 2 g of stanols (the amount obtained from an average daily intake of 25 g margarine) can lower serum levels of dangerous LDL-cholesterol by 14% (Law 2000). This is estimated to correspond to a ~25% reduction in the risk of life-threatening heart disease – more than can otherwise be achieved by a reduction in saturated fat intake alone. When used in combination with a careful dietary regime and statin drugs, the effects in severely hypercholesterolemic patients can be even more dramatic. The success of Benecol™ margarine has led its parent company, Raisio, to launch an entire range of 'nutraceuticals' including yoghurts and cheese spreads containing stanol esters. Its success has also prompted a resurgence of interest in the dietary use of stanol and phytosterol esters from other sources, for example soybean.

The rapid pace at which understanding of the genetic basis of terpene biosynthesis has grown leads to the prospect of genetically engineering plants for improved production of commercially valuable terpenes. Since the discovery of the 1-DXP pathway, considerable interest has been focussed on overexpression of key enzymes in this pathway (particularly 1-DXP synthase) in planta (Lichtenthaler 1999). Since the plastidic 1-DXP pathway is largely independent of the pathway for terpene biosynthesis in the cytosol, the hope is that engineering this pathway could enhance the production of essential oil monoterpenes without adversely affecting the biosynthesis of other key metabolites such as phytosterols. Overexpression of GPP synthase has been proposed as a possible technique for boosting monoterpene levels in *Mentha*, although this technique runs the risk of depleting the stock of biosynthetic intermediates available for the synthesis of gibberellins and carotenoids, which are essential for the plants' well-being (Mahmoud & Croteau 2002). Essential oil monoterpenes are not the only compounds whose yields might be increased through genetic engineering of the parent plants. Insecticides (George *et al.* 2000) and medicinal drugs are also candidates – the recent delineation of almost all of the 19 genes implicated in the biosynthesis of the diterpene skeleton of Taxol (Jennewein *et al.* 2004) may well be the prelude to a completely new strategy for production of this hugely valuable anti-cancer drug.

Genetic engineering of carotenoid production in primary food crops has been intensely investigated in research programmes aimed at reducing vitamin A deficiency or boosting levels of beneficial dietary antioxidants such as lycopene or β-carotene. The achievements and pitfalls of this research programme to date have been reviewed by Fraser & Bramley (2004). Overexpression of phytoene synthase, which controls the flux of GGPP into carotenoid biosynthesis, has been used to boost lycopene levels in GM tomatoes (Fraser *et al.* 2001); intervention at this point in the carotenoid biosynthesis pathway can however deplete the pool of GGPP available for gibberellin biosynthesis with consequent adverse effects on plant health. An even more dramatic (if controversial) result has been achieved with the production of the so-called 'Golden Rice' (Beyer *et al.* 2002). Rice endosperm contains negligible levels of carotenoids, and vitamin A deficiency among communities dependent on the rice harvest is common. By transforming rice with the phytoene synthase and lycopene β-cyclase genes from daffodil (*Narcissus pseudonarcissus*) along with a bacterial phytoene desaturase, carotenoids (β-carotene, zeaxanthin and lutein) were found to accumulate in sufficient quantities to allow for 10–20% of the recommended daily intake (RDA) of provitamin A in a single rice meal. The technology to produce 'Golden Rice' has been donated by its developers to the International Rice Research Institute in the Philippines. However, questions still remain about the bioavailability of vitamin A from 'Golden Rice', and general wariness about the large-scale use of GM crops in developing countries has meant that wide uptake of this technology has yet to be realized.

Degenhardt *et al.* (2003) have proposed another use for genetic engineering of terpene production: to create crop plants which can defend themselves against insect pests without the need for pesticide treatment. Since the discovery that herbivory-stimulated emission of certain terpenes such as linalool can serve as a signal to attract predators of the herbivores (Kessler & Baldwin 2001), there has been considerable interest in enhancing the levels of chemical signals produced by the plants or equipping plants with the genetic machinery to enable them to produce non-natural chemical signals in response to herbivory. The area has met with only limited success mainly due to the plants' propensity for inactivating overproduced terpenes by converting them to non-volatile forms such as the β-glucosides. It remains to be seen whether success in this field will lead to such genetically engineered plants being welcomed as an environmentally benign alternative to existing cultivation regimes. A more immediately useful application of the chemical signals used by plants to defend themselves may be in the development of intercropping regimes to help protect valuable food crops such as maize (Khan et al. 2000).

A more successful application of terpenes in crop protection has been found with the development of (4aS,7S,7aR)-nepetalactone **40** from catmint as an insect attractant. As well as luring male aphids, nepetalactone serves as an attractant to numerous aphid predators such as lacewings (Birkett & Pickett 2003). Formulations of catmint oil distillate, impregnated into a polymer which releases nepetalactone over a period of months, are now commercially available. This represents the first example of plant cultivation on a large scale for insect pheromone production – less than 35 tonnes of plant material yield over 30 kg of nepetalactone-rich oil at a production cost of approximately £1/g.

There are certain circumstances where it might be advantageous to suppress, rather than enhance, the biosynthesis of particular terpenes. One terpenoid family where this may bring unprecedented benefit is the strigolactone family, typified by strigol (**73**) and its analogues. The precise biosynthetic origin of these semiochemicals, traditionally classed as sesquiterpenoids, is still unclear; they are secreted, for an unknown purpose, by the roots of crops such as maize and sorghum (Siame et al. 1993) and act as growth triggers for parasitic weeds of the genera *Striga* (witchweed) and *Orobanche* (broomrape). The seeds of these parasites lie dormant until strigolactone secretion from the host plants triggers them to germinate and attach themselves to the root systems of the host. The resultant crop damage affects two-thirds of the 70 million hectares of African agricultural land, making these weeds the most significant biological cause of crop damage throughout the continent (Khan et al. 2000; Bouwmeester et al. 2003). The development of crop strains with suppressed capacity to secrete strigolactones – either through selective breeding or mutation – could offer a means to reduce the economic damage to countries ill suited to invest in high-technology solutions. At the same time, there is no doubt that a greater understanding of the biosynthetic machinery which gives rise to the strigolactones will provide new opportunities for controlling infestation by *Striga* and *Orobanche*, perhaps suggesting routes towards non-aggressive chemical treatments or genetic modification in order to produce resistant crops.

In conclusion: the state of knowledge in the field of terpene biosynthesis has advanced enormously in the past 20 years, creating unprecedented opportunities for science to play its part in bringing the health and nutritional benefits of plant terpenes out of the research lab and into the public domain. The pace of this development may even accelerate in the twenty-first century as detailed genomic and metabolomic studies lead to the discovery of valuable new roles for these highly versatile natural products.

References

Abe, I., Rohmer, M. and Prestwich, G.D. (1993) Enzymatic cyclization of squalene and oxidosqualene to sterols and triterpenes. *Chem. Rev.*, **93**, 2189–2206.

Adam, K.P., Thiel, R. and Zapp, J. (1999) Incorporation of 1-[1-^{13}C]-deoxy-D-xylulose in chamomile sesquiterpenes. *Arch. Biochem. Biophys.*, **369**, 127–132.

Adam, P., Hecht, S., Eisenreich, W.G. *et al.* (2002) Biosynthesis of terpenes: studies on 1-hydroxy-2-methyl-2-(*E*)-butenyl 4-diphosphate reductase. *Proc. Natl. Acad. Sci. USA*, **99**, 12108–12113.

Arigoni, D., Sagner, S., Latzel, C., Eisenreich, W., Bacher, A. and Zenk, M.H. (1997) Terpenoid biosynthesis from 1-deoxy-D-xylulose in higher plants by intramolecular skeletal rearrangement. *Proc. Natl. Acad. Sci. USA*, **94**, 10600–10605.

Arnold, W.N. (1988) Vincent van Gogh and the thujone connection. *J. Am. Med. Assoc.*, **260**, 3042–3044.

Aubourg, S., Lecharny, A. and Bohlmann, J. (2002) Genomic analysis of the terpenoid synthase (AtTPS) gene family of *Arabidopsis thaliana*. *Mol. Genet. Genom.*, **267**, 730–745.

Bach, T.J. and Lichtenthaler, H.K. (1983) Inhibition by mevinolin of plant growth, sterol formation and pigment accumulation. *Physiol. Plant.*, **59**, 50–60.

Barton, D.H.R., Pradhan, S.K., Sternhell, S. and Templeton, J.F. (1961) Triterpenoids. Part 25. The constitutions of limonin and related bitter principles. *J. Chem. Soc.*, 255–275.

Bell, A.A. (1967) Formation of gossypol in infected or chemically irritated tissues of *Gossypium* species. *Phytopathology*, **57**, 759–764.

Beyer, P., Al-Babili, S., Ye, X. *et al.* (2002) Golden rice: introducing the β-carotene biosynthesis pathway into rice endosperm by genetic engineering to defeat vitamin A deficiency. *J. Nutr.*, **132**, 506S–510S.

Birkett, M.A. and Pickett, J.A. (2003) Aphid sex pheromones: from discovery to commercial production. *Phytochemistry*, **62**, 651–656.

Blagg, B.S.J., Jarstfer, M.B., Rogers, D.H. and Poulter, C.D. (2002) Recombinant squalene synthase. A mechanism for the rearrangement of presqualene diphosphate to squalene. *J. Am. Chem. Soc.*, **124**, 8846–8853.

Bloch, K. (1965) The biological synthesis of cholesterol. *Science*, **150**, 19–28.

Boeke, S.J., Boersma, M.G., Alink, G.M. *et al.* (2004) Safety evaluation of neem (*Azadirachta indica*) derived pesticides. *J. Ethnopharmacol.*, **94**, 25–41.

Bohlmann, J., Meyer-Gauen, G. and Croteau, R. (1998) Plant terpenoid synthases: molecular biology and phylogenetic analysis. *Proc. Natl. Acad. Sci. USA*, 1998, **95**, 4126–4133.

Bohlmann, J., Martin, D., Oldham, N.J. and Gershenzon, J. (2000) Terpenoid secondary metabolism in *Arabidopsis thaliana*: cDNA cloning, characterization, and functional expression of a myrcene/(*E*)-β-ocimene synthase. *Arch. Biochem. Biophys.*, **375**, 261–269.

Bonkovsky, H.L., Cable, E.E., Cable, J.W. *et al.* (1992) Porphyrogenic properties of the terpenes camphor, pinene, and thujone (with a note on historic implications for Absinthe and the illness of Vincent van Gogh). *Biochem. Pharmacol.*, **43**, 2359–2368.

Bouvier, F., Suire, C., d'Harlingue, A., Backhaus, R.A. and Camara, B. (2000) Molecular cloning of geranyl diphosphate synthase and compartmentation of monoterpene synthesis in plant cells. *Plant J.*, **24**, 241–252.

Bouwmeester, H.J., Wallaart, T.E., Janssen, M.H.A. *et al.* (1999) Amorpha-4,11-diene synthase catalyses the first probable step in artemisinin biosynthesis. *Phytochemistry*, **52**, 843–854.

Bouwmeester, H.J., Matusova, R., Sun, Z.K. and Beale, M. H. (2003) Secondary metabolite signalling in host-parasitic plant interactions. *Curr. Opin. Plant Biol.*, **6**, 358–364.

Brown, G.D. (1998) The biosynthesis of steroids and triterpenoids. *Nat. Prod. Rep.*, **15**, 653–696.

Buckingham, J. (ed.) (2004) *Dictionary of Natural Products*. Version 9.2 on CD-ROM. Chapman & Hall/CRC Press, London, New York.

Burke, C.C., Wildung, M.R. and Croteau, R. (1999) Geranyl diphosphate synthase: cloning, expression, and characterization of this prenyltransferase as a heterodimer. *Proc. Natl. Acad. Sci. USA*, **96**, 13062–13067.

Cane, D.E. (1990) Enzymatic formation of sesquiterpenes. *Chem. Rev.*, **90**, 1089–1103.

Castleman, M. (2003) *The New Healing Herbs*. Hinkler Books Pty Ltd., Dingley, Australia.

Chappell, J. (1995) Biochemistry and molecular biology of the isoprenoid biosynthetic pathway in plants. *Ann. Rev. Plant Physiol. Plant Mol. Biol.*, **46**, 521–547.

Chen, F., Tholl, D., D'Auria, J.C., Farooq, A., Pichersky, E. and Gershenzon, J. (2003) Biosynthesis and emission of terpenoid volatiles from *Arabidopsis* flowers. *Plant Cell*, **15**, 481–494.

Cornforth, J.W. (1959) Biosynthesis of fatty acids and cholesterol considered as chemical processes. *J. Lipid Res.*, **1**, 3–28.

Croteau, R., Munck, S.L., Akoh, C.C., Fisk, H.J. and Satterwhite, D.M. (1987) Biosynthesis of the sesquiterpene patchoulol from farnesyl pyrophosphate in leaf extracts of *Pogostemon cablin* (Patchouli): mechanistic considerations. *Arch. Biochem. Biophys.*, **256**, 56–68.

Crowell, P.L. (1999) Prevention and therapy of cancer by dietary monoterpenes. *J. Nutrition* **129**, 775S–778S.

Crozier, A., Kamiya, Y., Bishop, G. *et al.* (2000) Biosynthesis of harmones and elicitor molecules. In B.B. Buchannan, W. Gruissem and R.L. Jones (eds), Biochemistry and Molecular Biology of Plant. American Society of Plant Physiologists, Rockville, MD, pp. 850–929.

Degenhardt, J., Gershenzon, J., Baldwin, I.T. and Kessler, A. (2003) Attracting friends to feast on foes: engineering terpene emission to make crop plants more attractive to herbivore enemies. *Curr. Opin. Biotechnol.*, **14**, 169–176.

Dewick, P.M. (1997) The biosynthesis of C5-C25 terpenoid compounds. *Nat. Prod. Rep.*, **14**, 111–144.

Dewick, P.M. (1999) The biosynthesis of C5-C25 terpenoid compounds. *Nat. Prod. Rep.* **16**, 97–130.

Dewick, P.M. (2001) The mevalonate and deoxyxylulose phosphate pathways: terpenoids and steroids. In P.M. Dewick (ed.), *Medicinal Natural Products: A Biosynthetic Approach*, 2nd edn. John Wiley & Sons Ltd., Chichester, pp. 167–289.

Dewick, P.M. (2002) The biosynthesis of C5-C25 terpenoid compounds. *Nat. Prod. Rep.*, **19**, 181–222.

Dogbo, O., Laferrière, A., D'Harlingue, A. and Camara, B. (1988) Carotenoid biosynthesis: isolation and characterization of a bifunctional enzyme catalyzing the synthesis of phytoene. *Proc. Natl. Acad. Sci. USA*, **85**, 7054–7058.

Eisenreich, W., Menhard, B., Hylands, P.J., Zenk, M.H. and Bacher, A. (1996) Studies on the biosynthesis of taxol: the taxane carbon skeleton is not of mevalonoid origin. *Proc. Natl. Acad. Sci. USA*, **93**, 6431–6436.

Eisenreich, W., Sagner, S., Zenk, M.H. and Bacher, A. (1997) Monoterpenoid essential oils are not of mevalonoid origin. *Tetrahedron Lett.*, **38**, 3889–3892.

Eisenreich, W., Bacher, A., Arigoni, D. and Rohdich, F. (2004) Biosynthesis of isoprenoids *via* the non-mevalonate pathway. *Cell. Mol. Life Sci.*, **61**, 1401–1426.

Endo, A. (1992) The discovery and development of HMG-CoA reductase inhibitors. *J. Lipid Res.*, **33**, 1569–1582.

Ershov, Y., Gantt, R.R., Cunningham, F.X. and Gantt, E. (2000) Isopentenyl diphosphate isomerase deficiency in *Synechocystis* sp. strain PCC6803. *FEBS Lett.*, **473**, 337–340.

Fowler, D.J., Hamilton, J.T.G., Humphrey, A.J. and O'Hagan, D. (1999) Plant terpene biosynthesis. The biosynthesis of linalyl acetate in *Mentha citrata*. *Tetrahedron Lett.*, **40**, 3803–3806.

Fraser, P.D. and Bramley, P.M. (2004) The biosynthesis and nutritional uses of carotenoids. *Prog. Lipid Res.*, **43**, 228–265.

Fraser, P.D., Römer, S., Kiano, J.W. *et al.* (2001) Elevation of carotenoids in tomato by genetic manipulation. *J. Sci. Food Agric.*, **81**, 822–827.

Gambliel, H. and Croteau, R. (1984) Pinene Cyclase-I and Cyclase-II. Two enzymes from sage (*Salvia officinalis*) which catalyze stereospecific cyclizations of geranyl pyrophosphate to monoterpene olefins of opposite configuration. *J. Biol. Chem.*, **259**, 740–748.

George, J., Bais, H.P. and Ravishankar, G.A. (2000) Biotechnological production of plant-based insecticides. *Crit. Rev. Biotechnol.*, **20**, 49–77.

Ghisalberti, E.L. (1998) Biological and pharmacological activity of naturally occurring iridoids and secoiridoids. *Phytomedicine*, **5**, 147–163.

Giuliano, G., Al-Babili, S. and von Lintig, J. (2003) Carotenoid oxygenases: cleave it or leave it. *Trends Plant Sci.*, **8**, 145–149.

Haynes, R.K. and Vonwiller, S.C. (1997) From Qinghao, marvelous herb of antiquity, to the antimalarial trioxane Qinghaosu – and some remarkable new chemistry. *Acc. Chem. Res.*, **30**, 73–79.

Hedden, P. and Kamiya, Y. (1997) Gibberellin biosynthesis: enzymes, genes and their regulation. *Ann. Rev. Plant Physiol. Plant Mol. Biol.*, **48**, 431–460.

Heinrich, M. (2003) Ethnobotany and natural products: the search for new molecules, new treatments of old diseases or a better understanding of indigenous cultures? *Curr. Top. Med. Chem.*, **3**, 141–154.

Heinstein, P.F., Smith, F.H. and Tove, S.B. (1962) Biosynthesis of ^{14}C-labeled gossypol. *J. Biol. Chem.*, **237**, 2643–2646.

Höld, K.M., Sirisoma, N.S., Ikeda, T., Narahashi, T. and Casida, J.E. (2000) α-Thujone (the active component of absinthe): γ-aminobutyric acid type A receptor modulation and metabolic detoxification. *Proc. Natl. Acad. Sci. USA*, **97**, 3826–3831.

Huang, K.-X., Huang, Q.-L. and Scott, A.I. (1998) Overexpression, single-step purification, and site-directed mutagenetic analysis of casbene synthase. *Arch. Biochem. Biophys.*, **352**, 144–152.

Jennewein, S. and Croteau, R. (2001) Taxol: biosynthesis, molecular genetics, and biotechnological applications. *Appl. Microbiol. Biotechnol.*, **57**, 13–19.

Jennewein, S., Wildung, M.R., Chau, M., Walker, K. and Croteau, R. (2004) Random sequencing of an induced *Taxus* cell cDNA library for identification of clones involved in Taxol biosynthesis. *Proc. Natl. Acad. Sci. USA*, **101**, 9149–9154.

Jomaa, H., Wiesner, J., Sanderbrand, S. *et al.* (1999) Inhibitors of the nonmevalonate pathway of isoprenoid biosynthesis as antimalarial drugs. *Science*, **285**, 1573–1576.

Kalemba, D. and Kunicka, A. (2003) Antibacterial and antifungal properties of essential oils. *Curr. Med. Chem.*, **10**, 813–829.

Kasahara, H., Hanada, A., Kuzuyama, T., Takagi, M., Kamiya, Y. and Yamaguchi, S. (2002) Contribution of the mevalonate and methylerythritol phosphate pathways to the biosynthesis of gibberellins in *Arabidopsis*. *J. Biol. Chem.*, **277**, 45188–45194.

Kessler, A. and Baldwin, I.T. (2001) Defensive function of herbivore-induced plant volatile emissions in nature. *Science* **291**, 2141–2144.

Khan, Z.R., Pickett, J.A., van den Berg, J., Wadhams, L.J. and Woodcock, C.M. (2000) Exploiting chemical ecology and species diversity: stem borer and *Striga* control for maize and sorghum in Africa. *Pest Management Sci.*, **56**, 957–962.

Lange, B.M. and Croteau, R. (1999a) Isoprenoid biosynthesis via a mevalonate-independent pathway in plants: cloning and heterologous expression of 1-deoxy-D-xylulose 5-phosphate reductoisomerase from peppermint. *Arch. Biochem. Biophys.*, **365**, 170–174.

Lange, B.M. and Croteau, R. (1999b) Genetic engineering of essential oil production in mint. *Curr. Opin. Plant Biol.*, **2**, 139–144.

Laule, O., Fürholz, A., Chang, H-S. *et al.* (2003) Crosstalk between cytosolic and plastidial pathways of isoprenoid biosynthesis in *Arabidopsis thaliana*. *Proc. Natl. Acad. Sci. USA*, **100**, 6866–6871.

Law, M. (2000) Plant sterol and stanol margarines and health. *Brit. Med. J.*, **320**, 861–864.

Liang, P.-H., Ko, T-P. and Wang, A.H-J. (2002) Structure, mechanism and function of prenyltransferases. *Eur. J. Biochem.*, **269**, 3339–3354.

Lichtenthaler, H.K. (1999) The 1-deoxy-D-xylulose 5-phosphate pathway of isoprenoid biosynthesis in plants. *Ann. Rev. Plant Physiol. Plant Mol. Biol.*, **50**, 47–65.

Lichtenthaler, H.K., Schwender, J., Disch, A. and Rohmer, M. (1997) Biosynthesis of isoprenoids in higher plant chloroplasts proceeds *via* a mevalonate-independent pathway. *FEBS Lett.*, **400**, 271–274.

Lin, X., Hezari, M., Koepp, A.E., Floss, H.G. and Croteau, R. (1996) Mechanism of taxadiene synthase, a diterpene cyclase that catalyzes the first step of taxol biosynthesis in Pacific yew. *Biochemistry*, **35**, 2968–2977.

MacMillan, J. and Beale, M.H. (1999) Diterpene biosynthesis. In D.E. Cane (ed.), *Comprehensive Natural Products Chemistry, Vol. 2: Isoprenoids including Carotenoids and Steroids* Elsevier, Oxford, pp. 217–243.

Mahmoud, S.S. and Croteau, R.B. (2001) Metabolic engineering of essential oil yield and composition in mint by altering expression of deoxyxylulose phosphate reductoisomerase and menthofuran synthase. *Proc. Natl. Acad. Sci. USA*, **98**, 8915–8920.

Mahmoud, S.S. and Croteau, R.B. (2002) Strategies for transgenic manipulation of monoterpene biosynthesis in plants. *Trends Plant Sci.*, **7**, 366–373.

Maier, W., Schneider, B. and Strack, D. (1998) Biosynthesis of sesquiterpenoid cyclohexenone derivatives in mycorrhizal barley roots proceeds via the glyceraldehyde 3-phosphate/pyruvate pathway. *Tetrahedron Lett.*, **39**, 521–524.

Manners, G.D. and Hasegawa, S. (1999) Squeezing more from citrus fruits. *Chem. Ind.*, 542–545.

Melero, C.P., Medarde, M. and San Feliciano, A. (2000) A short review on cardiotonic steroids and their aminoguanidine analogues. *Molecules*, **5**, 51–81.

Mohan, R.S., Yee, N.K.N., Coates, R.M. *et al.*, (1996) Biosynthesis of cyclic diterpene hydrocarbons in rice cell suspensions: conversion of 9,10-*syn*-labda-8(17),13-dienyl diphosphate to 9β-pimara-7,15-diene and stemar-13-ene. *Arch. Biochem. Biophys.*, **330**, 33–47.

Munck, S.L. and Croteau, R. (1990) Purification and characterization of the sesquiterpene cyclase patchoulol synthase from *Pogostemon cablin*. *Arch. Biochem. Biophys.* **282**, 58–64.

Niyogi, K.K., Grossman, A.R. and Björkman, O. (1998) *Arabidopsis* mutants define a central role for the xanthophyll cycle in the regulation of photosynthetic energy conversion. *Plant Cell*, **10**, 1121–1134.

Ohnuma, S.-I., Narita, K., Nakazawa, T. *et al.* (1996) A role of the amino acid residue located on the fifth position before the first aspartate-rich motif of farnesyl diphosphate synthase on determination of the final product. *J. Biol. Chem.*, **271**, 30748–30754.

Ohnuma, S.-I., Hirooka, K., Tsuruoka, N. *et al.* (1998) A pathway where polyprenyl diphosphate elongates in prenyltransferase. *J. Biol. Chem.*, **273**, 26705–26713.

Okada, K., Saito, T., Nakagawa, T., Kawamukai, M. and Kamiya, Y. (2000) Five geranylgeranyl diphosphate synthases expressed in different organs are localized into three subcellular compartments in *Arabidopsis*. *Plant Physiol.*, **122**, 1045–1056.

Oritani, T. and Kiyota, H. (2003) Biosynthesis and metabolism of abscisic acid and related compounds. *Nat. Prod. Rep.*, **20**, 414–425.

Page, J.E., Hause, G., Raschke, M. *et al.* (2004) Functional analysis of the final steps of the 1-deoxy-D-xylulose 5-phosphate (DXP) pathway to isoprenoids in plants using virus-induced gene silencing. *Plant Physiol.*, **134**, 1401–1413.

Pandian, S., Saengchjan, S. and Raman, T.S. (1981) An alternative pathway for the biosynthesis of isoprenoid compounds in bacteria. *Biochem. J.*, **196**, 675–681.

Peters, R.J., Ravn, M.M., Coates, R.M. and Croteau, R.B. (2001) Bifunctional abietadiene synthase: Free diffusive transfer of the (+)-copalyl diphosphate intermediate between two distinct active sites. *J. Am. Chem. Soc.*, **123**, 8974–8978.

Piel, J., Donath, J., Bandemer, K. and Boland, W. (1998) Mevalonate-independent biosynthesis of terpenoid volatiles in plants: induced and constitutive emission of volatiles. *Angew. Chem. Intl. Edn.*, **37**, 2478–2481.

Piironen, V., Lindsay, D.G., Miettinen, T.A., Toivo, J. and Lampi, A.M. (2000) Plant sterols: biosynthesis, biological function and their importance to human nutrition. *J. Sci. Food Agric.*, **80**, 939–966.

Poulter, C.D. and Rilling, H.C. (1976) Prenyltransferase: the mechanism of the reaction. *Biochemistry*, **15**, 1079–1083.

Prisic, S., Xu, M.M., Wilderman, P.R. and Peters, R.J. (2004) Rice contains two disparate *ent*-copalyl diphosphate synthases with distinct metabolic functions. *Plant Physiol.*, **136**, 4228–4236.

Prosser, I., Phillips, A.L., Gittings, S. *et al.* (2002) (+)-(10*R*)-germacrene A synthase from goldenrod (*Solidago canadensis*): cDNA isolation, bacterial expression and functional analysis. *Phytochemistry*, **60**, 691–702.

Prosser, I., Altug, I.G., Phillips, A.L., Konig, W.A., Bouwmeester, H.J. and Beale, M.H. (2004) Enantiospecific (+)- and (−)-germacrene D synthases, cloned from goldenrod, reveal a functionally active variant of the universal isoprenoid-biosynthesis aspartate-rich motif. *Arch. Biochem. Biophys.*, **432**, 136–144.

Richards, D.E., King, K.E., Ait-ali, T. and Harberd, N.P. (2001) How gibberellin regulates plant growth and development: a molecular genetic analysis of gibberellin signaling. *Ann. Rev. Plant Physiol. Plant Mol. Biol.*, **52**, 67–88.

Rising, K.A., Starks, C.M., Noel, J.P. and Chappell, J. (2000) Demonstration of germacrene A as an intermediate in 5-*epi*-aristolochene synthase catalysis. *J. Am. Chem. Soc.*, **122**, 1861–1866.

Rivera, S.B., Swedlund, B.D., King, G.J. *et al.* (2001) Chrysanthemyl diphosphate synthase: isolation of the gene and characterization of the recombinant non-head-to-tail monoterpene synthase from *Chrysanthemum cinerariaefolium*. *Proc. Natl. Acad. Sci. USA*, **98**, 4373–4378.

Robles, M., Aregullin, M., West, J. and Rodriguez, E. (1995) Recent studies on the zoopharmacognosy, pharmacology and neurotoxicity of sesquiterpene lactones. *Planta Med.*, **61**, 199–203.

Rohmer, M., Knani, M., Simonin, P., Sutter, B. and Sahm, H. (1993) Isoprenoid biosynthesis in bacteria – a novel pathway for the early steps leading to isopentenyl diphosphate. *Biochem. J.*, **295**, 517–524.

Rohmer, M., Seemann, M., Horbach, S., Bringer-Meyer, S. and Sahm, H. (1996) Glyceraldehyde 3-phosphate and pyruvate as precursors of isoprenic units in an alternative non-mevalonate pathway for terpenoid biosynthesis. *J. Am. Chem. Soc.*, **118**, 2564–2566.

Ruzicka, L. (1953) The isoprene rule and the biogenesis of terpenic compounds. *Experientia*, **9**, 357–367.

Sangwan, N.S., Farooqi, A.H.A., Shabih, F. and Sangwan, R.S. (2001) Regulation of essential oil production in plants. *Plant Growth Regul.*, **34**, 3–21.

Santos-Gomes, P.C. and Fernandes-Ferreira, M. (2003) Essential oils produced by *in vitro* shoots of sage (*Salvia officinalis* L.). *J. Agric. Food Chem.*, **51**, 2260–2266.

Schepmann, H.G., Pang, J.H. and Matsuda, S.P.T. (2001) Cloning and characterization of *Ginkgo biloba* levopimaradiene synthase, which catalyzes the first committed step in ginkgolide biosynthesis. *Arch. Biochem. Biophys.*, **392**, 263–269.

Schalk, M. and Croteau, R. (2000) A single amino acid substitution (F363I) converts the regiochemistry of the spearmint (−)-limonene hydroxylase from a C6- to a C3-hydroxylase. *Proc. Natl. Acad. Sci. USA*, **97**, 11948–11953.

Schwender, J., Seemann, M., Lichtenthaler, H.K. and Rohmer, M. (1996) Biosynthesis of isoprenoids (carotenoids, sterols, prenyl side-chains of chlorophylls and plastoquinone) *via* a novel pyruvate/glyceraldehyde 3-phosphate non-mevalonate pathway in the green alga *Scenedesmus obliquus*. *Biochem. J.*, **316**, 73–80.

Shah, S.P.J. and Rogers, L.J. (1969) Compartmentation of terpenoid biosynthesis in green plants. *Biochem. J.*, **114**, 395–405.

Shi, J., Arunasalam, K., Yeung, D., Kakuda, Y., Mittal, G. and Jiang, Y.M. (2004) Saponins from edible legumes: chemistry, processing, and health benefits. *J. Med. Food*, **7**, 67–78.

Siame, B.A., Weerasuriya, Y., Wood, K., Ejeta, G. and Butler, L.G. (1993) Isolation of strigol, a germination stimulant for *Striga asiatica*, from host plants. *J. Agric. Food Chem.*, **41**, 1486–1491.

Sparg, S.G., Light, M.E. and van Staden, J. (2004) Biological activities and distribution of plant saponins. *J. Ethnopharmacol.*, **94**, 219–243.

Sprenger, G.A., Schörken, U., Wiegert, T. *et al.* (1997) Identification of a thiamin-dependent synthase in *Escherichia coli* required for the formation of the 1-deoxy-D-xylulose 5-phosphate precursor to isoprenoids, thiamin, and pyridoxol. *Proc. Natl. Acad. Sci. USA*, **94**, 12857–12862.

Starks, C.M., Back, K., Chappell, J. and Noel, J.P. (1997) Structural basis for cyclic terpene biosynthesis by tobacco 5-*epi*-aristolochene synthase. *Science*, **277**, 1815–1820.

Takahashi, S., Kuzuyama, T., Watanabe, H. and Seto, H. (1998) A 1-deoxy-D-xylulose 5-phosphate reductoisomerase catalyzing the formation of 2C-methyl-D-erythritol 4-phosphate in an alternative non-mevalonate pathway for terpenoid biosynthesis. *Proc. Natl. Acad. Sci. USA*, **95**, 9879–9884.

Tarshis, L.C., Yan, M.J., Poulter, C.D. and Sacchettini, J.C. (1994) Crystal structure of recombinant farnesyl diphosphate synthase at 2.6-Å resolution. *Biochemistry*, **33**, 10871–10877.

Tarshis, L.C., Proteau, P.J., Kellogg, B.A., Sacchettini, J.C. and Poulter, C.D. (1996) Regulation of product chain length by isoprenyl diphosphate synthases. *Proc. Natl. Acad. Sci. USA*, **93**, 15018–15023.

Tietze, L.-F. (1983) Secologanin, a biogenetic key compound – Synthesis and biogenesis of the iridoid and secoiridoid glycosides. *Angew. Chem. Intl. Edn. Engl.*, **22**, 828–841.

Treharne, K.J., Mercer, E.I. and Goodwin, T.W. (1966) Incorporation of [^{14}C]-carbon dioxide and [2-^{14}C]-mevalonic acid into terpenoids of higher plants during chloroplast development. *Biochem. J.*, **99**, 239–245.

Turner, G., Gershenzon, J., Nielson, E.E., Froehlich, J.E. and Croteau, R. (1999) Limonene synthase, the enzyme responsible for monoterpene biosynthesis in peppermint, is localized to leucoplasts of oil gland secretory cells. *Plant Physiol.*, **120**, 879–886.

Ulubelen, A. (2003) Cardioactive and antibacterial terpenoids from some *Salvia* species. *Phytochemistry*, **64**, 395–399.

Urgert, R. and Katan, M.B. (1997) The cholesterol-raising factor from coffee beans. *Ann. Rev. Nutr.*, **17**, 305–324.

Vögeli, U. and Chappell, J. (1988) Induction of sesquiterpene cyclase and suppression of squalene synthetase activities in plant cell cultures treated with fungal elicitor. *Plant Physiol.*, **88**, 1291–1296.

Walker, R. (2004) Criteria for risk assessment of botanical food supplements. *Toxicol. Lett.*, **149**, 187–195.

Wang, Y., Tang, H., Nicholson, J.K. *et al.* (2004) Metabolomic strategy for the classification and quality control of phytomedicine: a case study of chamomile flower (*Matricaria recutita* L.). *Planta Med.*, **70**, 250–255.

Wilkinson, S. and Davies, W.J. (2002) ABA-based chemical signalling: the co-ordination of responses to stress in plants. *Plant Cell Environ.*, **25**, 195–210.

Williams, D.C., McGarvey, D.J., Katahira, E.J. and Croteau, R. (1998) Truncation of limonene synthase preprotein provides a fully active 'pseudomature' form of this monoterpene cyclase and reveals the function of the amino-terminal arginine pair. *Biochemistry*, **37**, 12213–12220.

Wise, M.L., Savage, T.J., Katahira, E. and Croteau, R. (1998) Monoterpene synthases from common sage (*Salvia officinalis*) – cDNA isolation, characterization, and functional expression of (+)-sabinene synthase, 1,8-cineole synthase, and (+)-bornyl diphosphate synthase. *J. Biol. Chem.*, **273**, 14891–14899.

Zhou, D. and White, R.H. (1991) Early steps of isoprenoid biosynthesis in *Escherichia coli*. *Biochem. J.*, **273**, 627–634.

Chapter 4
Alkaloids

Katherine G. Zulak, David K. Liscombe, Hiroshi Ashihara and Peter J. Facchini

4.1 Introduction

Alkaloids are a diverse group of low molecular weight, nitrogen-containing compounds mostly derived from amino acids and found in about 20% of plant species. As secondary metabolites, alkaloids are thought to play a defensive role in the plant against herbivores and pathogens. Due to their potent biological activity, many of the approximately 12 000 known alkaloids have been exploited as pharmaceuticals, stimulants, narcotics and poisons (Wink 1998). Plant-derived alkaloids currently in clinical use include the analgesics morphine and codeine, the anti-neoplastic agent vinblastine, the gout suppressant colchicine, the muscle relaxants (+)-tubocurarine and papaverine, the anti-arrhythmic ajmaline, and the sedative scopolamine. Other well known alkaloids of plant origin include caffeine, nicotine, and cocaine, and the synthetic O-diacetylated morphine derivative heroin.

Research in the field of plant alkaloid biochemistry began with the isolation of morphine in 1806. Remarkably, the structure of morphine was not elucidated until 1952 due to the stereochemical complexity of the molecule. Since then, extensive chemical, biochemical and molecular research over the last half-century have led to an unprecedented understanding of alkaloid biosynthesis in plants (Facchini 2001). In this chapter, we review the biosynthesis, ethnobotany and pharmacology of the major groups of alkaloids produced in plants with a focus on the role of important alkaloids in diet and human health.

4.2 Benzylisoquinoline alkaloids

Many benzylisoquinoline alkaloids are used as pharmaceuticals due to their potent pharmacological activity, which is often an indication of the biological function of the approximately 2500 known members of this group. For example, the effectiveness of morphine as an analgesic, colchicine as a microtubule disrupter and (+)-tubocurarine as a neuromuscular blocker suggests that these alkaloids function as herbivore deterrents. The anti-microbial properties of sanguinarine and berberine suggest that they confer protection against pathogens. Benzylisoquinoline alkaloids occur mainly in basal angiosperms including the Ranunculaceae, Papaveraceae, Berberidaceae, Fumariaceae, Menispermaceae and Magnoliaceae.

The importance of alkaloids since the birth of human civilization is well illustrated by the drug opium, which is obtained from opium poppy (*Papaver somniferum*) and contains

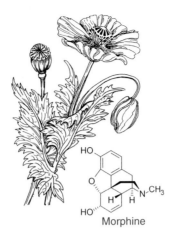

Figure 4.1 *Papaver somniferum* and morphine.

the analgesic morphine and numerous related alkaloids (Figure 4.1). Morphine is named after the Greek god Morpheus, the creator of sleep and dreams in *Ovid*. In the epic story the *Odyssey*, opium is used as an ingredient in a wine-based drink called *Nepenthes* (Greek *ne*: not, *penthos*: sorrow) that was consumed by soldiers before combat to dull the horrors of battle (Hesse 2002). The Sumarians incorporated the euphoria induced by opium into religious rituals at the end of the third millennium BC The Roman Emperor Nero murdered his stepbrother Britannicus with a mix of hemlock and opium in 50 AD. In most cultures, opium use was restricted to pain relief until the seventeenth century when recreational use of the drug began in China. The Opium Wars were fought between the British and Chinese to maintain free trade of the drug between the two countries. Opium remains the only commercial source for morphine and codeine.

Benzylisoquinoline alkaloid biosynthesis begins with a lattice of decarboxylations, *ortho*-hydroxylations and deaminations that convert tyrosine to both dopamine and 4-hydroxyphenylacetaldehyde (4-HPAA) (Figure 4.2). Molecular clones for the aromatic L-amino acid decarboxylase (TYDC) that converts tyrosine and dopa to tyrosine and dopamine, respectively, have been isolated. Norcoclaurine synthase (NCS) condenses dopamine and 4-HPAA to yield (S)-norcoclaurine, the central precursor to all benzylisoquinoline alkaloids in plants. (S)-Norcoclaurine is converted to (S)-reticuline by a 6-O-methyltransferase (6OMT), an N-methyltransferase (CNMT), a P450 hydroxylase (CYP80B) and a 4′-O-methyltransferase (4′OMT). Molecular clones have been isolated for all of the enzymes involved in the conversion of (S)-norcoclaurine to (S)-reticuline, which is a branch-point intermediate in the biosynthesis of many different types of benzylisoquinoline alkaloids (Facchini 2001). Intermediates of the (S)-reticuline pathway also serve as precursors to more than 270 dimeric bisbenzylisoquinoline alkaloids such as (+)-tubocurarine. The molecular clone for a P450-dependent oxidase (CYP80A) that couples (R)-N-methylcoclaurine to (R)- or (S)-N-methylcoclaurine to yield bisbenzylisoquinoline alkaloids, respectively, has been isolated from barberry (*Berberis stolonifera*).

Much work has focused on the branch pathways leading to benzophenanthridine alkaloids, such as sanguinarine, protoberberine alkaloids, such as berberine, and morphinan

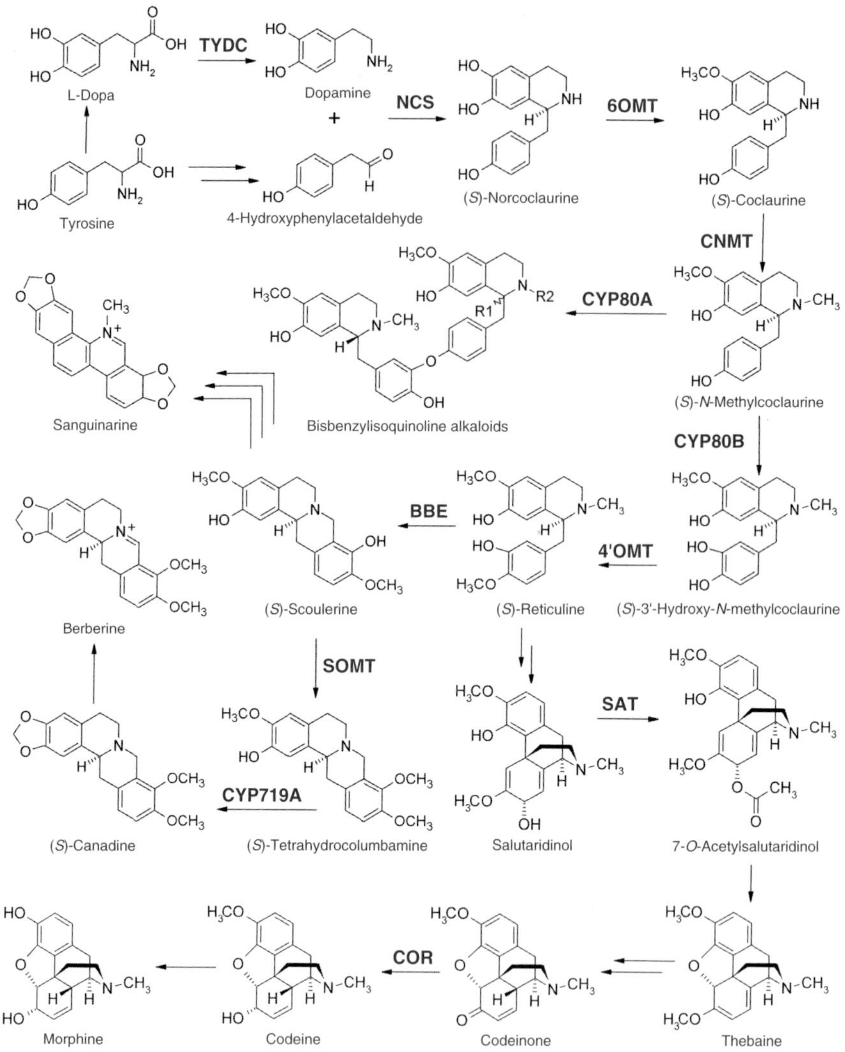

Figure 4.2 Biosynthesis of benzylisoquinoline alkaloids. Enzyme abbreviations: TYDC, tyrosine/dopa decarboxylase; NCS, norcoclaurine synthase; 6OMT, norcoclaurine 6-O-methyltransferase; 4′OMT, 3′-hydroxy-N-methylcoclaurine 4′-O-methyltransferase; CYP80A, berbamunine synthase; CYP80B, N-methylcoclaurine 3′-hydroxylase; BBE, berberine bridge enzyme; SOMT, scoulerine 9-O-methyltransferase; CYP719A, (S)-canadine synthase; SAT, salutaridinol 7-O-acetyltransferase; COR, codeinone reductase.

alkaloids, such as morphine. Most of the enzymes involved in these pathways, and five corresponding molecular clones, have been isolated (Figure 4.2). The first committed step in benzophenanthridine and protoberberine alkaloid biosynthesis involves the conversion of (S)-reticuline to (S)-scoulerine by the berberine bridge enzyme (BBE). (S)-Scoulerine can be converted to (S)-stylopine by two P450-dependent oxidases. Following the N-methylation of (S)-stylopine by a specific methyltransferase, two additional P450-dependent enzymes convert (S)-cis-N-methylstylopine to dihydrosanguinarine, which is oxidized to yield sanguinarine. Root exudates from many Papaveraceae species, such as bloodroot (*Sanguinaria canadensis*) and (*Eschsholzia californica*), are red due to the accumulation of sanguinarine and other benzophenanthridine alkaloids.

In some plants, especially the Berberidaceae and Ranunculaceae, (S)-scoulerine is methylated by the SAM-dependent scoulerine-9-O-methyltransferase (SOMT) to yield (S)-tetrahydrocolumbamine (Figure 4.2). Molecular clones for SOMT and the P450-dependent (S)-canadine synthase (CYP719A), which converts (S)-tetrahydrocolumbamine to (S)-canadine, have been isolated. (S)-Canadine is oxidized to yield berberine. In some *Papaver* species, (S)-reticuline is epimerized to (R)-reticuline as the first committed step in morphinan alkaloid biosynthesis. Subsequently, (R)-reticuline is converted in two steps to (7S)-salutaridinol by a P450-dependent enzyme and an NADPH-dependent oxidoreductase. The morphinan alkaloid thebaine is produced from (7S)-salutaridinol via acetyl coenzyme A:salutaridinol-7-O-acetyltransferase (SAT). Thebaine is converted to codeinone, which is reduced to codeine by the NADPH-dependent enzyme codeinone reductase (COR). Molecular clones for SAT and COR have been isolated from opium poppy. Finally, codeine is demethylated to yield morphine.

The cell type-specific localization of benzylisoquinoline alkaloid biosynthesis has been determined in two plant species. In *P. somniferum* (Papaveraceae), CYP80B, BBE and COR enzymes are localized to sieve elements in the vascular system of the plant, whereas the corresponding gene transcripts are restricted to adjacent companion cells. No secondary product pathway has previously been localized to these cell types. The biosynthesis of morphine in the phloem breaks the long-standing paradigm that sieve element functions are restricted to the translocation of solutes and information macromolecules in plants. Specialized cells accompanying the phloem, known as laticifers, are now known to serve as the site of benzylisoquinoline alkaloid accumulation, and not biosynthesis, in opium poppy. Remarkably, the overall process from gene expression through product accumulation requires three unusual cell types and predicts the intercellular translocation of biosynthetic enzymes and products. In meadow rue (*Thalictrum flavum*) (Ranunculaceae), protoberberine alkaloids accumulate in the endodermis of the root at the onset of secondary growth and in the pith and cortex of the rhizome (Samanani *et al.* 2002). Gene transcripts encoding nine enzymes involved in protoberberine alkaloid biosynthesis were co-localized to the immature root endodermis and the protoderm of leaf primordia in the rhizome; thus, benzylisoquinoline alkaloid biosynthesis and accumulation are temporally and spatially related in *T. flavum* roots and rhizomes, respectively. Benzylisoquinoline alkaloid biosynthetic enzymes are also compartmentalized at the subcellular level due to the toxicity of the pathway intermediates and products. The non-cytosolic enzymes involved in the conversion of the benzophenanthridine and morphinan branch pathways are localized to the endoplasmic reticulum (ER), or ER-derived endomembranes.

Morphine is perhaps the most extensively used analgesic for the management of acute pain associated with injury, neuropathic conditions and cancer. Morphine acts on the central nervous system by activating membrane opioid receptors. The pharmacological effects of morphine vary enormously with dosage. Small doses induce euphoria and sedation, whereas high doses cause pupil dilation, irregular respiration, pale skin, a deep sleep and eventual death within 6–8 h due to respiratory paralysis (Hesse 2002). Even at moderate doses, morphine causes constipation, loss of appetite, hypothermia, a slow heart rate and urine retention. Despite these side effects, morphine is the most effective method of pain relief in modern medicine. Heroin, a diacetylated derivative of morphine, is tenfold more powerful than morphine but has limited clinical application. Nevertheless, it is estimated that 90% of morphine production is used for the illicit synthesis of heroin as a recreational drug.

Although codeine is present in opium poppy at a much lower concentration than morphine, 95% of all licit morphine is chemically methylated to produce codeine due to its greater versatility and demand. Pharmacologically, codeine displays similar but less potent properties compared with morphine. Codeine is often combined with other analgesics, such as acetaminophen or acetylsalicylic acid, which leads to its misuse as a mood-altering drug. The affinity of codeine for opioid receptors is low; thus, the euphoric effects of codeine are thought to result from the O-demethylation of codeine to morphine by a genetically polymorphic P450-dependent enzyme (CYP2D6) that controls morphine levels in the body. Approximately 7% of Caucasians lack this enzyme and cannot metabolize codeine. Codeine-dependency can be treated with inhibitors of CYP2D6, such as quinidine and fluoxetine. Codeine is also commonly used as a cough-suppressant and can be found in most prescription cough syrups. In this regard, codeine reduces bronchial secretion and suppresses the cough centre of the medulla oblongata (Schmeller and Wink 1998).

Sanguinarine, which is common in the Papaveraceae and Fumariaceae, has been used in oral hygiene products to treat gingivitis and plaque formation due to its anti-microbial and anti-inflammatory properties. Bloodroot accumulates sanguinarine and was used by Native Americans to purify the blood, relieve pain, heal wounds and reduce fevers. Sanguinarine has been shown to control inflammation by regulating a relevant transcription factor, to modulate apoptosis as a potential chemotherapy drug, and to inhibit angiogenesis.

Berberine is the major protoberberine alkaloid in goldenseal (*Hydrastis canadensis*), which is native to Canada and the northeastern United States (Figure 4.3). Native Americans used the plant to dye cloth, reduce inflammation, stimulate digestion, treat infections, regulate menstrual abnormalities, and induce abortions. Possible pharmacological effects of ingesting goldenseal extracts include an inhibition of phosphodiesterase activity, modulation of potassium and calcium channels, and a release of nitric oxide (Abdel-Haq 2000). Goldenseal is often combined with dried roots of *Echinacea spp.* to prevent colds and flu. Sales of goldenseal surpassed $44 million in 1999, making it the third most popular herbal in the United States.

Many other plants produce isoquinoline alkaloids with pharmaceutical and herbal applications or potential. Psychotrine, emetine, and cephaline in the dried rhizome and root of ipecac (*Cephaelis ipecacuanha*) show anti-amoebic, anti-tumour, and anti-viral activities. Psychotrine derivatives are among the most potent inhibitors of the human immunodeficiency virus reverse transcriptase. The most common use of ipecac extracts is to induce vomiting as a treatment for poisoning. Celandine (*Chelidonium majus*), which is widely used in Chinese herbal medicine, accumulates chelidonine, berberine, and coptisine in aerial organs, and sanguinarine and chelerythrine in the roots. Extracts of *C. majus* shoots display anti-microbial and anti-inflammatory properties. Chelidonine has been investigated as an anti-tumour drug, but exhibits toxic effects at therapeutic doses. (+)-Tubocurarine, which is obtained from tubo curare (*Chondronendron tomentosum*), was used as an arrowhead poison with effects similar to strychnine. The drug is currently

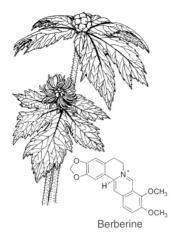

Figure 4.3 *Hydrastis canadensis* and berberine.

used as a muscle relaxant during surgery, and to control muscle spasms and convulsions. (+)-Tubocurarine acts as a competitive antagonist of nicotinic acetycholine receptors at neuromuscular junctions, effectively blocking nerve impulses to muscle fibers (Schmeller and Wink 1998).

4.3 Tropane alkaloids

The tropane alkaloids possess an 8-azabicyclo[3.2.1]octane nucleus and are found in plants of three families, the Solanaceae, Erythroxylaceae, and Convolvulaceae. Although several tropane alkaloids are used as legitimate pharmaceuticals most are known for their toxicity, which can be a major problem due to the attractive berries produced by several solanaceous plants. Fewer than three berries of henbane (*Hyoscyamus niger*) or deadly nightshade (*Atropa belladonna*), both of which contain (−)-hyoscine (scopolamine) and hyoscyamine, can cause death in infants. Plants that produce tropane alkaloids have been used as both medicines and poisons throughout history. Cleopatra purportedly tested the effects of henbane and deadly nightshade on her slaves to identify the best poison for suicide, although she found the toxic effects of tropane alkaloids too painful and selected asp venom instead. The wives of the Roman emperors, Augustus and Claudius, used deadly nightshade for murder. The highly toxic thorn-apple (*Datura stramonium*), which is a rich source of scopolamine and hyoscyamine, is widely distributed throughout the world and has long been used as a sedative (Figure 4.4). Extracts of Hindu Datura (*Datura metel*) were used to sedate and lure virgins into prostitution and by the prostitutes themselves to sedate their clients. Colombian Indians used *Datura* species for infanticide by smearing extracts on the nipples of the mother. Mandrake (*Mandragora officinarum*) was considered a potent and prized aphrodisiac, but the high scopolamine content makes the plant toxic.

Renaissance women enlarged the pupils of their eyes to appear more attractive using atropine-containing extracts of *A. belladonna*. Atropine, the racemic form of hyoscyamine, is a muscarinic acetylcholine receptor antagonist used to dilate the pupil during

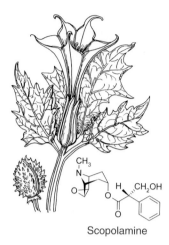

Figure 4.4 *Datura stramonium* and scopolamine.

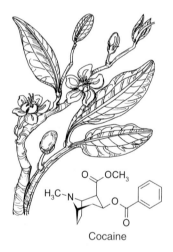

Figure 4.5 *Erythroxylum coca* and cocaine.

opthalmological examinations and to treat cases of poisoning, particularly by organophosphorous insecticides, nerve gas and the toxic principles of the red fly agaric mushroom (*Amanita muscaria*). Scopolamine continues to be used to prevent motion sickness and, in combination with morphine, as a sedative during labour.

Outside the Solanaceae, tropane alkaloids occur in two other plant families. Within the Erythroxylaceae, the genus *Erythroxylum* comprises about 200 widely distributed, tropical species found mainly in South America and Madagascar. Peruvian coca (*Erythroxylum coca*) is the only plant currently cultivated for cocaine production, which occurs at concentrations between 0.2% and 1% (w/w) in the leaves (Griffin and Lin 2000) (Figure 4.5). A few other *Erythroxylum* species also produce cocaine, including Trujillo coca (*E. novogranatense* var. *truxillense*) and Amazonian coca (*E. coca* var. *ipadu*). Cocaine is used licitly and illicitly as a local anesthetic in ophthamology, a central nervous system stimulant, and to improve physical endurance. Peruvian Indians used coca for at least 1000 years before the arrival of Europeans. In the late 1800s, several coca-based beverages were used as mild stimulants, including coca wines. In 1886, John Smyth Pemberton produced a nonalcoholic beverage he called Coca-Cola, which included extracts from South American coca that contained little cocaine. Originally, the cocaine content of Coca-Cola was less than 5 mg per 100 mL, which was not enough to cause stimulation or addiction. Ironically, the stimulatory effect of Coca-Cola still comes from the caffeine content. By 1906, cocaine was eliminated from Coca-Cola (Hobhouse 1985).

Calystegines are polyhydroxynortropanes found primarily in the Convolvulaceae, which includes sweet potato (*Ipomoea batatas*), but also occur in potato (*Solanum tuberosum*) and other members of the Solanaceae (Schimming *et al.* 1998). The calystegines from the roots of field bindweed (*Convolvulus arvensis*) and hedge bindweed (*Calystegia sepium*) are potent inhibitors of β-glucosidase and possess strong therapeutic potential (Scholl *et al.* 2001). Ingestion of calystegine-producing plants causes neurological dysfunction (e.g. staggering and incoordination) associated with glycosidase inhibition and lysosomal storage disorders (Watson *et al.* 2001). However, calystegines do not pose an acute toxic threat to humans or livestock.

Present commercial sources of scopolamine and hyoscyamine include the genera *Datura*, *Brugmansia* and *Duboisia*. *Datura stramonium* is a noxious weed of cereal crops throughout the world and is widely cultivated. Plants of the genus *Brugmansia* are native to South America and are most often used by indigenous tribes of southern Columbia. Since 1968, Red angel's trumpet (*Brugmansia sanguinea*) has been cultivated in Ecuador and yields approximately 400 tonnes of dried leaf annually containing approximately 0.8% (w/w) scopolamine. The pituri bush (*Duboisia hopwoodii*), a small western and central Australian shrub, is one of the few scopolamine- and hyoscyamine-producing plants used by aboriginals. Australian *Duboisia* cultivation has been a multimillion-dollar industry since 1970. Although *Duboisia* cultivation has declined, the scopolamine content of new hybrid varieties has more than doubled to between 1.5% and 2.5% (w/w) (Griffin and Lin 2000).

Tropane alkaloid biosynthesis begins with the SAM-dependent N-methylation of putrescine by the enzyme putrescine N-methyltransferase (PMT), cDNAs of which have been isolated from *A. belladonna* and *H. niger* (Suzuki *et al.* 1999) (Figure 4.6). N-methylputrescine is then oxidatively deaminated by a diamine oxidase to 4-methylaminobutanal, which undergoes spontaneous cyclization to form the reactive N-methyl-Δ^1-pyrrolinium cation. The N-methyl-Δ^1-pyrrolinium cation is thought to condense with acetoacetic acid to yield hygrine as a precursor of the tropane ring, although the enzymology of this reaction is not known. A key branch point in tropane alkaloid biosynthesis is the conversion of tropinone to either tropine or pseudotropine (or ψ-tropine), which possess opposite stereochemistry at the 3-hydroxyl position (Nakajima and Hashimoto 1999). Two different NADPH-dependent enzymes catalyse the stereospecific reduction of tropinone: the 3-carbonyl of tropinone is reduced to the 3α-hydroxy group of tropine by tropinone reductase I (TR-I) and to the 3β-hydroxy group of ψ-tropine by tropinone reductase II (TR-II). Genes encoding both TR-I and TR-II have been identified in other tropinone alkaloid-producing species, but not in tobacco, which accumulates nicotine rather than tropane alkaloids. Hyoscyamine and scopolamine are

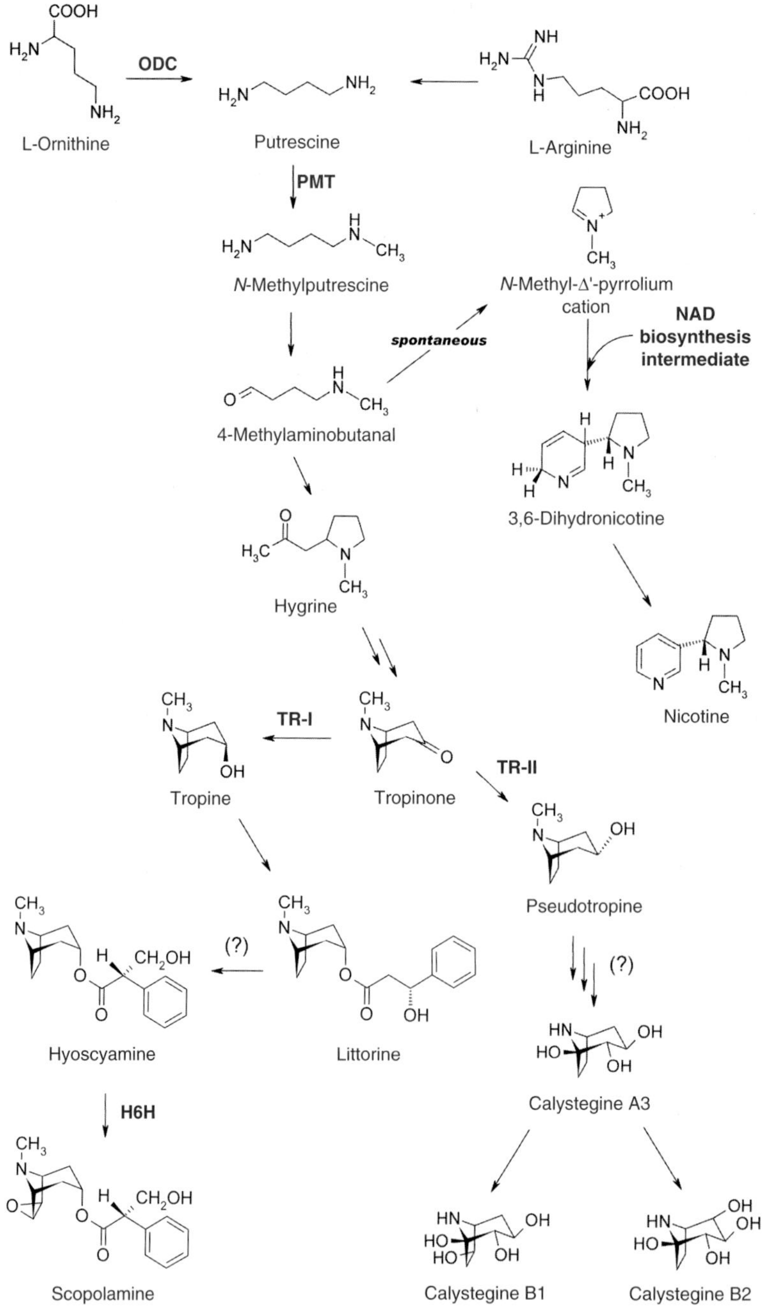

Figure 4.6 Biosynthesis of tropane alkaloids and nicotine. Enzyme abbreviations: ODC, ornithine decarboxylase; PMT, putrescine N-methyltransferase; TR-I, tropinone reductase-I; TR-II, tropinone reductase-II; H6H, hyoscyamine 6β-hydroxylase.

derived from tropine via TR-I, whereas TR-II yields pseudotropine, which is converted to calystegines (Figure 4.6) and other nortropane alkaloids.

Hyoscyamine is produced by the condensation of tropine and the phenylalanine-derived intermediate tropic acid, although the pathway is not fully understood (Duran-Patron *et al.* 2000). Hyoscyamine can be further converted to its epoxide scopolamine by 6β-hydroxylation of the tropane ring followed by intramolecular epoxide formation via removal of the 7β-hydrogen (Figure 4.6). Both reactions are catalysed by a 2-oxoglutarate-dependent dioxygenase, hyoscyamine 6β-hydroxylase (H6H). Only scopolamine-producing plants exhibit H6H activity (Hashimoto *et al.* 1991). The distribution of hyoscyamine in the Solanaceae is wider than that of scopolamine suggesting that only certain phylogenetic lineages acquired the *h6h* gene (Kanegae *et al.* 1994).

Pathway intermediates appear to undergo intercellular translocation during the biosynthesis of tropane alkaloids. Genes encoding PMT and H6H are expressed only in the root pericycle. In contrast, TR-I is localized in the endodermis and outer root cortex, whereas TR-II is found in the pericycle, endodermis and outer root cortex. The differential localization of TR-I and TR-II to a unique cell type compared with PMT and H6H shows that an intermediate between PMT and TR-I (or TR-II) moves from the pericycle to the endodermis and that an intermediate between TR-I and H6H moves back to the pericycle. Localized in the pericycle PMT has efficient access to putrescine, ornithine and arginine precursors unloaded from the phloem. Similarly, scopolamine produced in the pericycle can be readily translocated to aerial organs via the adjacent xylem. Grafts between shoots from tropane alkaloid-producing species and roots from non-producing species result in plants that do not accumulate tropane alkaloids. However, reciprocal grafts produce plants that accumulate tropane alkaloids in aerial organs; thus, tropane alkaloids are synthesized in roots and translocated to aerial organs via the xylem.

The physiological effects of atropine poisoning, which are usually suffered by young children who consume *Atropa belladonna* berries, include tachycardia, mydriasis, inhibited glandular secretions and smooth muscle relaxation. Excitatory effects of atropine on the central nervous system cause irritability and hyperactivity, accompanied by a considerable rise in body temperature potentiated by the inability to sweat (Rang *et al.* 1999). Anticholinergic poisoning of adolescents following deliberate ingestion of the common Angel's trumpet (*Datura suaveolens*) to induce hallucinations has also been documented (Francis and Clarke 1999). Scopolamine causes sedation and possesses an anti-emetic effect, which makes it useful in the treatment of motion sickness (Rang *et al.* 1999). Autumn mandrake (*Mandragora autumnalis*) is often mistaken for the edible borage (*Borago officinalis*) or intentionally used as a hallucinogen. A number of *M. autumnalis* poisonings have been reported in Italy, the native habitat for this plant (Piccillo *et al.* 2002).

4.4 Nicotine

Tobacco (*Nicotiana tabacum*), a native plant of the Americas, has been cultivated since 5000–3000 BC and was in widespread use when Christopher Columbus arrived in the New World in 1492 (Musk and de Klerk 2003) (Figure 4.7). Tobacco was sniffed, chewed, eaten, drunk, applied topically to kill parasites and used in eye drops and enemas. The act of smoking tobacco appears to have evolved from snuffing and is currently the most

Figure 4.7 *Nicotiana tabacum* and nicotine.

common means of administration. Tobacco was used ceremonially, medicinally and for social activities. Ironically, one of the first medicinal uses of tobacco was based on its purported anti-cancer properties.

The active principle in most *Nicotiana* species is the simple alkaloid nicotine, which is composed of a pyridine ring joined to a N-methylpyrrolidine ring. The pyridine ring is derived from quinolinic acid and the pyrrolidine ring, as with the tropane alkaloids, comes from putrescine. The nicotine biosynthetic pathway consists of at least eight enzymatic steps. Nicotine biosynthesis involves the formation of an N-methyl-Δ^1-pyrrolinium cation, which is also a precursor of tropane alkaloids (Figure 4.6). In the case of nicotine biosynthesis the N-methyl-Δ^1-pyrrolinium cation is thought to condense with an intermediate from the nicotinamide adenine dinucleotide (NAD) pathway (either 1,2-dihydropyridine or nicotinic acid) to form 3,6-dihydronicotine. This reaction might be catalysed by a poorly defined enzyme known as nicotine synthase (Katoh and Hashimoto 2004). Genes encoding putresine N-methyltransferase (PMT), the first committed step in the biosynthesis of the N-methyl-Δ^1-pyrrolinium cation, and quinolinate phosphoribosyltransferase (QPRtase), the enzyme responsible for the conversion of quinolinic acid to nicotinic acid, have been cloned and characterized (Hibi *et al.* 1994; Sinclair *et al.* 2000). The final step in nicotine biosynthesis is the removal of a proton from the pyridine ring by a predicted 1,2-dihydronicotine dehydrogenase enzyme.

Nicotine biosynthesis is induced by the wounding of tobacco leaves or by insect damage. Nicotine is produced in the roots of *Nicotiana* species and transported to aerial organs through the xylem (Shoji *et al.* 2000). Genes encoding known nicotine biosynthetic enzymes are expressed only in tobacco roots. The perception of tissue damage in the leaves stimulates the release of a signal molecule (most likely jasmonate), which induces nicotine biosynthesis in the roots. Nicotine biosynthetic gene transcript levels increase in the roots after mechanical damage to aerial organs or by treatment with methyljasmonate (MeJa) (Shoji *et al.* 2000; Sinclair *et al.* 2000). Nicotine biosynthesis is controlled by two regulatory loci, *NIC1* and *NIC2*, but the functions of the gene products have not been identified (Hibi *et al.* 1994). Two genes encoding enzymes involved in NAD biosynthesis,

L-aspartate oxidase and quinolate synthase, are also partially regulated by *NIC* loci. These genes are abundantly expressed and induced by MeJa only in tobacco roots; thus, transcription of genes involved in the aspartate/NAD and nicotine pathways are controlled by common regulatory mechanisms (Katoh and Hashimoto 2004).

Cigarette smoke is more acidic than pipe or cigar smoke and requires inhalation into the lungs for effective uptake of nicotine. In contrast, nicotine administered via pipes or cigars is more readily absorbed through the oral mucosa (Musk and de Klerk 2003). In the lungs, the large surface area of the respiratory epithelium is exposed to the smoke, which promotes the absorption of nicotine. This provides a more immediate sense of satisfaction for the smoker and potentiates a rapid addiction to nicotine.

Smoking tobacco is a major cause of heart disease, stroke, peripheral vascular disease, chronic obstructive pulmonary disease, lung and other cancers, and various gastrointestinal disorders (Musk and de Klerk 2003). Smoking can cause many other health problems including osteoporosis, impaired fertility, inflammatory bowel disease, diabetes and hypertension. Tobacco smoke contains a multitude of chemicals including polycyclic aromatic hydrocarbons; thus, nicotine is not solely responsible for these disorders. However, nicotine is one of the most biologically active chemicals in nature, binding to several different receptors and activating a number of key signal transduction pathways. Many of the physiological effects of nicotine, including addiction, are exerted by its action on nicotinic acetylcholine receptors. Nicotine modulates the phosphatidylinositol pathway and increases intracellular calcium levels, which are both universal signalling components in physiological processes (Campain 2004). Although the cancer-causing properties of tobacco smoke were once solely associated with carcinogenic tars, nicotine exhibits genotoxic effects by inhibiting programmed cell death (apoptosis) and promoting oxidative damage by reactive oxygen species (Campain 2004).

4.5 Terpenoid indole alkaloids

Terpenoid indole alkaloids are a large group of about 3000 compounds found mainly in the Apocynaceae, Loganiaceae and Rubiaceae. Indole alkaloids consist of an indole moiety provided by tryptamine and a terpenoid component from the iridoid glucoside secologanin. Many have attracted pharmacological interest including the tranquillizing alkaloids of the passion flower (*Passiflora incarnata*), the ophthalmic alkaloids related to physostigmine from the calabar bean (*Physostigma venenosum*) and the anti-neoplastic agents from Chinese 'happy' tree (*Camptotheca accuminata*). The well-known central nervous stimulants strychnine and yohimbine are also indole alkaloids. Perhaps the most important from a health perspective are the anti-neoplastic agents vincristine and vinblastine from the Madagascar periwinkle (*Catharanthus roseus*) (Figure 4.8). The importance of *C. roseus* as a source of anti-cancer medicines has prompted intensive research into the biology of alkaloid biosynthesis in this plant.

Hindus have used the Indian snakeroot (*Rauwolfia serpentaria*) for centuries as a febrifuge, an antidote to poisonous snakebites, and a treatment for dysentery and other intestinal afflictions. The plant is a perennial, evergreen shrub that grows in India, Pakistan, Sri Lanka, Burma and Thailand. Reserpine, the major indole alkaloid present in roots, stems and leaves of *R. serpentina* at levels of 1.7–3.0% (w/w), is an effective hypotensive.

Figure 4.8 *Catharanthus roseus* and vinblastine.

More than 90% of the alkaloids in roots accumulate in the bark. The poisonous indole alkaloids strychnine and bruicine are produced in tropical plants of the genus *Strychnos* (Loganiaceae), including nux-vomica (*Strychnos nux vomica*) and snakewood (*Strychnos colubrina*). Nux-vomica is described as a lethal poison and a cure for demonic possession in the *Kitab al-sumum*, or Book of Poisons, which dates back to the ninth century. Use of nux-vomica spread rapidly from Asia to North Africa and, subsequently, to the Western world. In Europe, *S. nux vomica* was mainly used as a poison to kill dogs, cats and rodents. Members of the *Strychnos* genus were also used by Amazonian Indians as arrowhead poisons, or 'curares', which are highly toxic in the bloodstream.

The anti-malarial drug quinine comes from the stem and root bark (often called Peruvian, fever or Jesuits' bark) of *Cinchona* species. In fact, *C. calisaya* accumulates up to 12% (w/w) quinine, representing one of the highest alkaloid contents in any plant. The earliest documented use of *Cinchona* was in Ecuador and involved a Spanish nobleman who was cured of malaria using 'fever bark'. The bark was brought to Europe in 1693 and gained popularity as an anti-malarial drug. Demand for *Cinchona* bark increased in the nineteenth century, which resulted in the death of many trees due to massive bark-stripping projects (Hesse 2002).

The conversion of tryptophan to tryptamine by tryptophan decarboxylase (TDC) represents one of the first steps in terpenoid indole alkaloid biosynthesis (Figure 4.9). The P450-dependent enzymes geraniol 10-hydroxylase (G10H) and secologanin synthase (SS) catalyse the first and last steps, respectively, in the secologanin pathway. The condensation of tryptophan and secologanin to form strictosidine is catalysed by strictosidine synthase (STR). Strictosidine is deglucosylated by strictosidine β-D-glucosidase (SGD), and the resulting strictosidine aglycoside is converted to 4,21-dehydrogeissoschizine. Many important terpenoid indole alkaloids, such as catharanthine, are produced via 4,21-dehydrogeissoschizine, but the enzymology has not been established. In contrast, the formation of vindoline from tabersonine is well characterized (Meijer *et al.* 1993). Tabersonine is converted to 16-hydroxytabersonine by tabersonine 16-hydroxylase (T16H), and, subsequently, 16-hydroxytabersonine is 16-*O*-methylated, undergoes hydration of

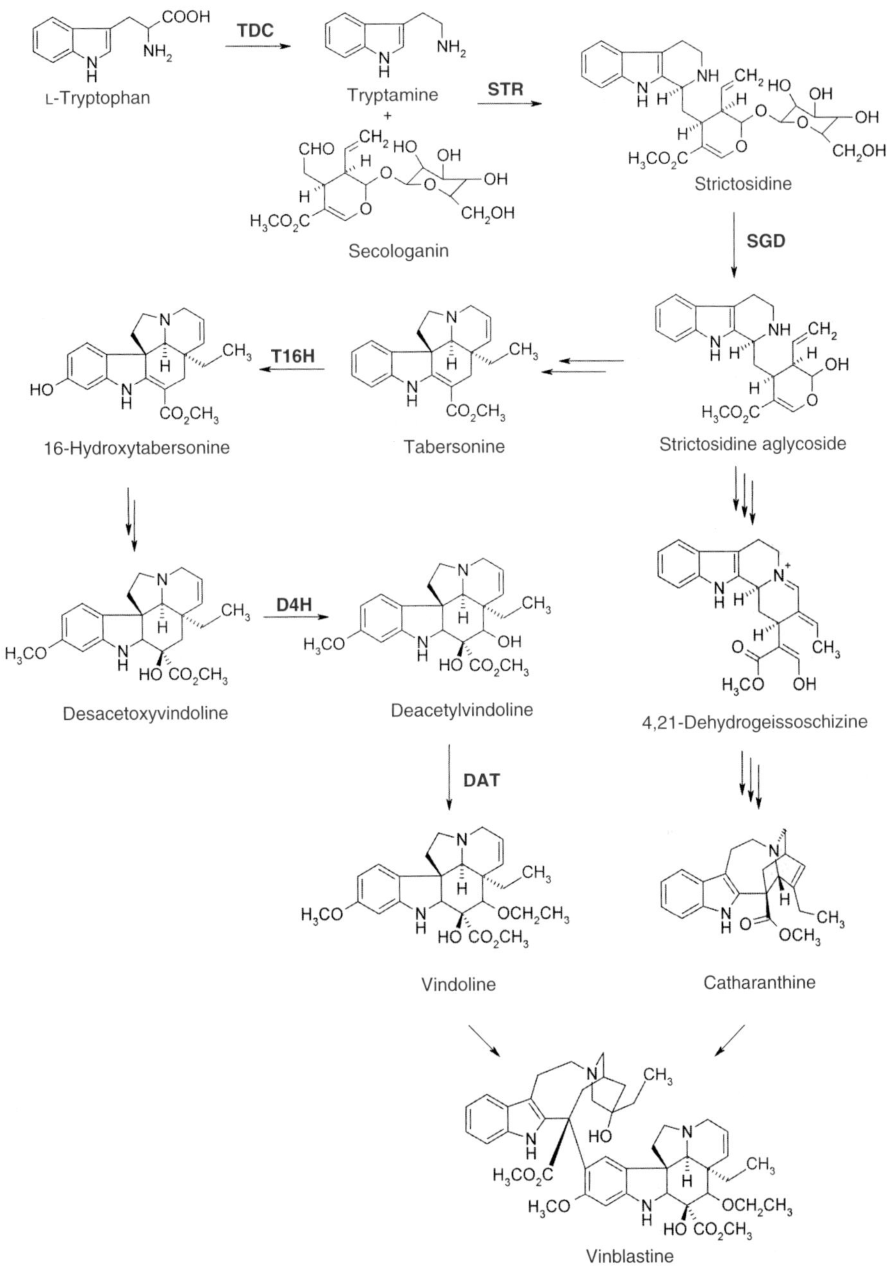

Figure 4.9 Biosynthesis of monoterpenoid indole alkaloids. Enzyme abbreviations: TDC, tryptophan decarboxylase; STR, strictosidine synthase; SGD, strictosidine β-d-glucosidase; T16H, tabersonine 16-hydroxylase; D4H, desacetoxyvindoline 4-hydroxylase; DAT, deacetylvindoline 4-O-acetyltransferase.

the 2,3-double bond and N-methylated at the indole-ring nitrogen to yield desacetoxyvindoline. The O- and N-methyltransferases involved in the formation of desacetoxyvindoline have been isolated, but the enzyme responsible for the hydration reaction is not known. The penultimate step in vindoline biosynthesis involves the hydrolysis of desacetoxyvindoline by a 2-oxoglutarate-dependent dioxygenase (D4H), whereas the final step is catalysed by an acetylcoenzyme A-dependent 4-O-acetyltransferase (DAT). Vindoline is coupled to catharanthine by a non-specific peroxidase to yield vinblastine. Enzymes implicated in a limited number of other terpenoid indole alkaloid pathways have been isolated. These include a P450-dependent monooxygenase and a reductase catalysing the synthesis and reduction, respectively, of vomilenine, an intermediate of ajmalicine biosynthesis (Falkenhagen and Stockigt 1995), and a P450-dependent monooxygenase that converts tabersonine to lochnericine.

Terpenoid indole alkaloid biosynthetic enzymes are associated with at least three different cell types in *C. roseus*: TDC and STR are localized to the epidermis of aerial organs and the apical meristem of roots, D4H and DAT are restricted to the laticifers and idioblasts of leaves and stems, and G10H is found in internal parenchyma of aerial organs (St-Pierre *et al.* 1999; Burlat *et al.* 2004); thus, vindoline pathway intermediates must be translocated between cell types. Moreover, enzymes involved in terpenoid indole alkaloid biosynthesis in *C. roseus* are also localized to at least five subcellular compartments: TDC, D4H and DAT are in the cytosol, STR and the peroxidase that couples catharanthine to vinblastine are localized to the vacuole indicating transport of tryptamine across the tonoplast, SGD is a soluble enzyme associated with the cytoplasmic face of the endoplasmic reticulum, the P450-dependent monooxygenases are integral endomembrane proteins, and the N-methyltransferase involved in vindoline biosynthesis is localized to thylakoid membranes (De Luca and St-Pierre 2000).

The molecular regulation of terpenoid indole alkaloid biosynthesis in *C. roseus* is the best characterized among plant alkaloid pathways. Two transcription factors, GT-1 and 3AF1, that interact with the TDC promoter have been identified (Ouwerkerk *et al.* 1999), and GT-1 also binds to STR1 and CPR promoters. Two inducible G-box binding factors, CrGFB1 and CrGFB2 (Ouwerkerk and Memelink 1999), and three APETALA2-domain (ORCA) factors (Menke *et al.* 1999) also interact with the STR1 promoter. ORCA3 was isolated using T-DNA activation tagging and shown to activate expression of TDC, STR, SGD, D4H and cytochrome P450 reductase when overexpressed in *C. roseus* cells (van der Fits and Memelink 2000). A MYB-like regulator protein was also isolated using a yeast-one hybrid screen with the STR1 promoter as bait (van der Fits *et al.* 2000).

Early interest in *C. roseus* was focused on a purported hypoglycemic activity, but plant extracts did not affect serum glucose levels. A decrease in bone marrow content and the onset of leucopenia, however, led to the isolation of the anti-neoplastic agent vinblastine in 1958 (McCormack 1990). Extracts of *C. roseus* and isolated vinblastine were also active against murine leukaemia. Vinblastine and vincristine are commonly used in cancer therapy due to their ability to bind microtubules and inhibit hydrolysis of GTP, thus arresting cell division at metaphase (Lobert *et al.* 1996). The binding of vinblastine and vincristine to tubulin occurs at protein domains different from other drugs, such as colchicines, that disrupt microtubules (McCormack 1990). Other biological effects associated with vinblastine and vincristine treatment include inhibition of protein, nucleic acid, and lipid biosynthesis, and reduced protein kinase C, which modulates cell growth and differentiation. Vinblastine is a component of chemotherapy for metastatic testicular cancer, Hodgkin's disease and other lymphomas, and a variety of solid neoplasms. Vincristine is the preferred treatment for acute lymphocytic leukaemia in children, management of Hodgkin's disease and other lymphomas, and pediatric tumours (McCormack 1990). Vinblastine has also been used in patients with autoimmune blood disorders due to its immunosuppressant properties (Schmeller and Wink 1998). However, vinblastine is toxic to bone marrow and white blood cells, causes nausea and sometimes results in neuropathic effects. Both drugs are expensive since *C. roseus* plants are the only source and the compounds are present at low levels.

Other terpenoid indole alkaloids possess important biological activities. Ajmaline is an anti-arrhythmic drug that prolongs the refractory period of the heart by blocking Na^+ channels, prolonging action potentials, and increasing depolarization thresholds (Schmeller and Wink 1998). Strychnine, or 'rat poison', is a competitive antagonist that blocks postsynaptic receptors of glycine in the spinal cord and motor neurons, resulting in hyperexcitability, convulsions and spinal paralysis (Neuwinger 1998). Aqueous extracts of *Strychnos* bark also cause convulsive and contradictory relaxation of the neuromuscular junction and cardiac muscle due to the presence of bisnordihydrotoxiferine, a sedative that affects the central nervous system (Neuwinger 1998). *S. nux-vomica* seeds with reduced strychnine levels are used to treat rheumatism, musculoskeletal injuries and limb paralysis (Chan 2002).

Camptothecin is currently one of the most important compounds in cancer research due to its activity against leukaemias and other cancers resistant to vincristine. Camptothecin inhibits nucleic acid biosynthesis and topoisomerase I, which is necessary for the relaxation of DNA during vital cellular processes. However, camptothecin is relatively unstable under physiological conditions prompting the preparation of more stable derivatives. Camptothecin also inhibits the replication of DNA viruses by disrupting the normal function of DNA in cellular ontogenesis. Side effects of camptothecin include haematopoietic depression, diarrhoea, alopecia, haematuria, and other urinary tract irritations.

Quinine is used as an anti-pyretic to combat malaria, to induce uterine contractions during labour, and to treat infectious diseases (Schmeller and Wink 1998). Synthetic substitutes have also been developed, but quinine has regained popularity due to the increased resistance of malaria against synthetic drugs. Quinine antagonizes muscarinic acetylcholine receptors and α-adrenoceptors, and inhibits nucleic acid synthesis through DNA intercalation and reduced carbohydrate metabolism, in the malarial parasites *Plasmodium falciparum*, *P. ovale*, *P. vivax* and *P. malariae*. Quinine also binds to sarcoplasmic reticulum vesicles, diminishes binding and uptake of calcium, and inhibits Na^+/K^+-ATPases. Quinine is also used as a bittering agent in soft drinks, such as quinine water or Indian tonic water that contain 0.007% quinine hydrosulfate. Concerns have been raised about the effects of the excessive consumption of quinine-containing beverages on the eye, apparently due to an immune reaction, and on the male reproductive system.

Reserpine is reported to influence the concentration of glycogen, acetylcholine, γ-amino butyric acid, nucleic acids and anti-diuretic hormones. The effects of reserpine include respiratory inhibition, stimulation of peristalsis, myosis, relaxation of nictitating membranes and influence on the temperature-regulating centre. The drug is primarily used to treat young patients suffering from mild hypertension but is also marketed as an aphrodisiac, an energy booster and a treatment for male impotence. Yohimbine acts as an $α_2$-adrenoceptor antagonist, potentiating the release of norepinephrine (noradrenaline) into the synaptic cleft, and down-regulates tyrosinase activity in human melanocytes (Fuller *et al.* 2000). However, yohimbine can cause anxiety, panic attacks and flashbacks in patients with post-traumatic stress disorder.

4.6 Purine alkaloids

The purine alkaloid caffeine (1,3,7-trimethylxanthine) was discovered in coffee (*Coffea arabica*) and tea (*Camellia sinensis*) in the 1820s (Ashihara and Crozier 2001) (Figures 4.10 and 4.11). Theobromine is also a commercially important purine alkaloid found in the seeds of cacao (*Theobroma cacao*). Three groups of plants that accumulate purine alkaloids

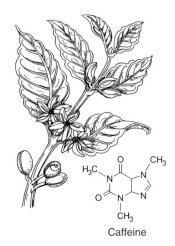

Figure 4.10 *Coffea arabica* and caffeine.

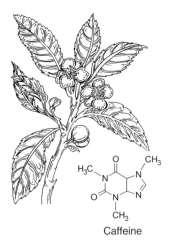

Figure 4.11 *Camellia sinensis* and caffeine.

can be identified based on the types of alkaloids they produce: caffeine-producing plants include coffee, tea and maté (*Ilex paraguariensis*); theobromine-producing plants are represented by cacao, cocoa tea (*Camellia ptilophylla*) and *Camellia irrawadiensis*; and methyluric acid-producing plants consist of *Coffea dewevrei, Coffea liberica, C. excelsa* and kucha tea (Camellia assamica). Coffee and tea are among the most popular non-alcoholic beverages in the world. Although coffee is preferred in Western, industrialized nations, tea consumption is highest in Asian societies (Hertog *et al.* 1997).

Chocolate was introduced to Europe when Christopher Columbus captured a Mayan trading canoe containing cocoa beans in 1502 (Jamieson 2004). Cocoa beans were used as currency in New Spain and were valued as a trading commodity. It was not until twenty years later when the Spanish conquered the Aztec empire that the true value of the beans was realized. Cocoa bean crops date back to the fifth century in the Americas and were maintained on a large scale by the Aztecs. After the Spanish conquered the Aztecs, cocoa drinking became a common practice among the Spanish elite, especially with women, but did not become popular in London, Paris or Rome until the mid-seventeenth century.

Coffee cultivation began in Yemen in the mid-fifteenth century (Jamieson 2004). Initially, coffee was restricted to the Arab world and was used in religious practices requiring wakefulness and trance-like states. Coffee rapidly spread throughout the Middle East, and then to Istanbul, and by the sixteenth century coffee houses began to appear in Europe. Northern European countries quickly established a monopoly over the coffee trade, and coffee plantations were soon established worldwide. Tea has been the beverage of choice in China since the Tang Dynasty (6816–906 BC) and was introduced into Holland, and then England, about 10–20 years after the introduction of coffee (Jamieson 2004). However, tea was difficult to grow in Europe and efforts to cultivate the crop were unsuccessful, allowing China to remain the sole producer of tea until the mid-nineteenth century. Tea quickly took its place in northern Europe as a drink associated with women, domestics, and gentility, and became a ritual of European bourgeois social life. Presently, over 2000 varieties of tea are grown in more than 25 countries. The three main categories of tea – black, green and oolong – come from the same plant species, but the leaves are processed differently. Black tea is allowed to oxidize for several hours before the leaves are dried, whereas green tea is steamed immediately after picking to prevent oxidation. Oolong tea is partially fermented.

Several *Ilex* spp. are used for the preparation of maté (Filipi and Ferraro 2003). Concern about maté being carcinogenic *per se* (Morton 1989) have been discounted (IARC 1991) and can be explained by the traditional method of preparation and consumption that allows scalding hot beverage to damage the oesophagus facilitating induction of carcinogenesis during the error prone stage of tissue repair by some other carcinogen. Drinking the beverage without scalding the oesophagus has a reduced statistical association with increased cancer risk (Sewram *et al.* 2003). Maté also has a significant theobromine content (Clifford and Ramìrez-Martìnez 1990).

Purine alkaloid biosynthesis begins with xanthosine, a purine nucleoside produced from the degradation of purine nucleotides found in free pools (Figure 4.12). The first committed step in caffeine biosynthesis involves the N-methylation of xanthosine to form 7-methylxanthosine, using S-adenosylmethionine (SAM) as the methyl donor. The cDNA encoding the enzyme responsible for this reaction, S-adenosylmethionine: 7-methylxanthosine synthase (MXS) has recently been cloned (Mizuno *et al.* 2003; Uefuji *et al.* 2003). The recombinant enzyme is specific for xanthosine, ruling out the possibility of

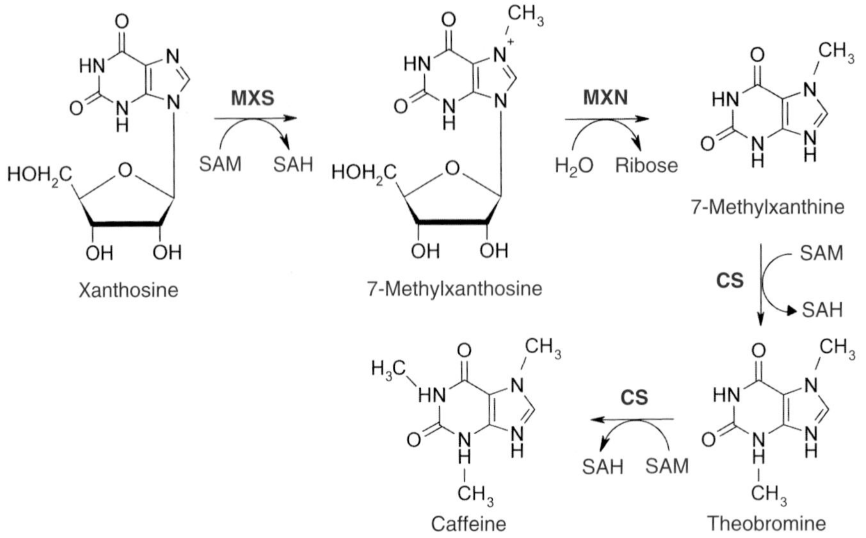

Figure 4.12 Biosynthesis of purine alkaloids. Enzyme abbreviations: CS, caffeine synthase; MXN, 7-methylxanthosine nucleosidase; MXS, 7-methylxanthosine synthase.

xanthosine monophosphate as a substrate, as originally hypothesized. 7-Methylxanthosine is converted to 7-methylxanthine by a specific N-methyl nucleosidase (MXN) that has been purified from tea. Conversion of 7-methylxanthine to caffeine involves two successive N-methyltransferases first detected in tea but also found in immature fruits and cell cultures of coffee. In tea, the same N-methyltransferase catalyses the conversions of 7-methylxanthine to theobromine and theobromine to caffeine. A molecular clone encoding this enzyme, known as caffeine synthase (CS), was isolated from tea (Kato et al. 2000). A caffeine synthase gene was also isolated from coffee exhibiting 40% identity to caffeine synthase from tea and the same dual methylation activity (Mizuno et al. 2003). Theobromine synthase (SAM: 7-methylxanthine N-methyltransferase), also isolated from coffee, is specific for the conversion of 7-methylxanthine to theobromine.

Caffeine synthase, the majority of SAH hydrolase activity, and parts of the adenine-salvage pathway are localized to chloroplasts. In coffee SAM synthase is confined to the cytosol and SAM synthase genes from tobacco and parsley lack a transit peptide. However, SAM synthase from tea is a chloroplastic enzyme, encoded by a nuclear gene (Koshiishi et al. 2001). The proposed model for the subcellular localization of caffeine biosynthesis begins with the production of homocysteine and its conversion to methionine in the chloroplasts. Methionine is then converted to SAM in the cytosol and transported back into the chloroplast to serve as the methyl donor in caffeine biosynthesis. Purine alkaloids are stored in vacuoles where they are thought to form complexes with chlorogenic acids (Mosli-Waldhauser and Baumann 1996).

Caffeine levels vary in developing coffee fruits from 0.2% to 2% (w/w) in the pericarp, and remain above 1% (w/w) in seeds. Caffeine is also detectable in leaves and cotyledons but absent in roots and older shoots. Young leaves also contain theobromine albeit at lower levels than caffeine. In tea, most of the caffeine is localized to the leaves, with small amounts in the stem and root. In developing fruits of *Theobroma cacao*, purine alkaloid levels also change over time (Zheng *et al.* 2004) with theobromine as the major alkaloid, followed by caffeine. As the fruits age, theobromine levels decrease sharply in the pericarp, with large amounts of theobromine accumulating in the seeds of mature fruits. In young leaves of cacao, theobromine is initially dominant, but caffeine levels increase as the leaves age. However, purine alkaloid levels decrease substantially in mature leaves of *T. cacao* (Koyama *et al.* 2003).

Caffeine is mostly ingested in coffee, tea and cola soft drinks, and remains the world's most widely used pharmacologically active substance. However, the acute and chronic risks and benefits of caffeine use are not fully understood. As an antagonist of endogenous adenosine receptors, caffeine causes vasoconstriction and increases blood pressure. Other short-term, unpleasant side effects include palpitations, gastrointestinal disturbances, anxiety, tremors and insomnia. In rare cases, caffeine ingestion can lead to cardiac arrhythmias. These effects are not associated with chronic caffeine consumption but rather with acute increases in plasma concentrations of the drug. Possible mechanisms controlling the cardiac effects of caffeine include antagonism of adenosine receptors, inhibition of phosphodiesterases, activation of the sympathetic nervous system, stimulation of the adrenal cortex and renal effects. Boiled, unfiltered coffee also contains specific lipids (particularly, cafestol and kahweol) that can increase cholesterol levels. The half-life of caffeine in the human body is less than 5 h, with overnight abstinence resulting in almost complete depletion of caffeine (James 1997). Withdrawal symptoms caused by missing a morning cup of coffee can be severe and include weariness, apathy, weakness, drowsiness, headaches, anxiety, decreased motor behaviour, increased heart rate and muscle tension, and less commonly, nausea, vomiting and flu-like symptoms (Jamieson 2004).

Although the effects of caffeine are not dependent on age, newborns and premature infants have a greatly extended half-life for caffeine (>100 h) due to incomplete development of the liver, in which caffeine-metabolizing enzymes are produced. Exercise increases the rate of elimination, whereas alcohol, obesity, liver disease and the use of oral contraceptives can reduce caffeine excretion. Smoking enhances caffeine metabolism in the liver, which might explain the strong statistical association between smoking and the consumption of caffeinated beverages.

Caffeine might also be potentially beneficial to human health since it increases extracellular levels of acetylcholine and serotonin by binding to adenosine receptors in the human brain suggesting that caffeine usage can reduce age-related cognitive decline. Caffeine also improves performance of tasks requiring verbal memory and information processing speed and has a positive effect on mood at low doses (Griffiths *et al.* 1990). The drug is added to many over-the-counter and prescription medicines including anti-inflammatory drugs (included in analgesic formulations), ephedrine (to promote weight loss) and ergotamine (to treat migraines). Caffeine might have analgesic properties of its own for specific types of pain, such as headaches (Ward *et al.* 1991). Caffeine also induces apoptosis by activating p53-mediated Bax and caspase-3 pathways suggesting possible chemoprotective properties (He *et al.* 2003). It is important to note that the pharmacological effects of caffeine are species-dependent. The extrapolation of rodent data to humans must take such differences into consideration (Arnaud 1993).

Trigonelline, or N-methylnicotinic acid, is a secondary metabolite derived from pyridine nucleotides. Trigonelline was first isolated from fenugreek (*Trigonella foenum-graecum*) and has since been found in many plant species. Coffee beans contain a large amount of trigonelline, which is thermally converted to nicotinic acid and some flavour compounds during roasting (Mazzafera 1991). The direct precursor of trigonelline is nicotinic acid, which appears to be produced as a degradation product of nicotinic acid adenine dinucleotide (NAD). Trigonelline is synthesized by SAM-dependent nicotinate N-methyltransferase, which has been found in coffee, pea and cultured soybean cells (Upmeier et al. 1988).

4.7 Pyrrolizidine alkaloids

Pyrrolizidine alkaloids are the leading plant toxins that have deleterious effects on the health of humans and animals. Over 360 different pyrrolizidine alkaloids are found in approximately 3% of the world's flowering plants. These noxious natural products are primarily restricted to the Boraginaceae (many genera), Asteraceae (tribes *Senecionae* and *Eupatoriae*), Fabaceae (mainly the genus *Crotalaria*) and Orchidaceae (nine genera). Additional pyrrolizidine alkaloid-producing plants are scattered throughout six other unrelated families (Ober et al. 2003).

Most pyrrolizidine alkaloids are esters of basic alcohols known as necine bases. The most frequently studied pyrrolizidine alkaloids are formed from the polyamines putrescine and spermidine and possess one of three common necine bases: retronecine, heliotridine and otonecine. Putrescine is utilized exclusively as a substrate in secondary metabolism, whereas spermidine is a universal cell-growth factor involved in many physiological processes in eukaryotes. Spermidine biosynthesis begins with the decarboxylation of SAM by SAM decarboxylase (Graser and Hartmann 2000) (Figure 4.13). The aminopropyl group

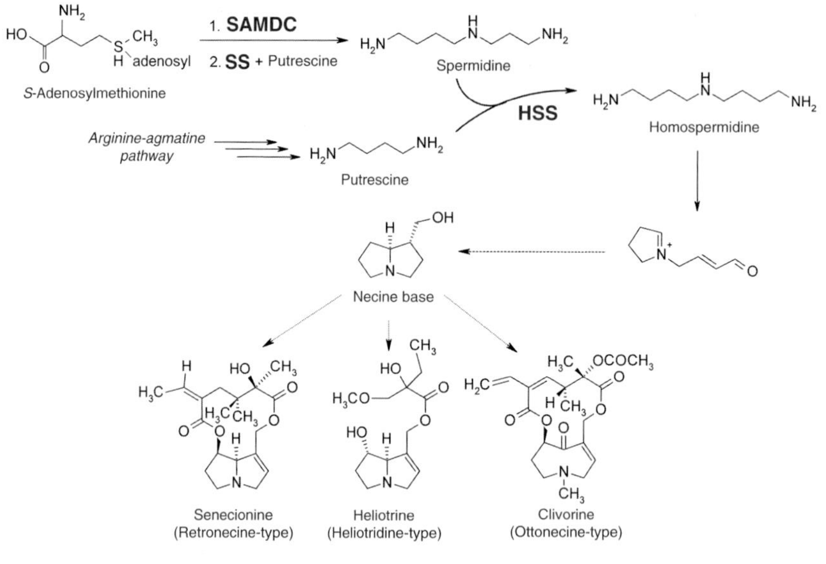

Figure 4.13 Biosynthesis of pyrrolizidine alkaloids showing the sites of action of enzymes for which corresponding genes have been isolated. Enzyme abbreviations: SAMDC, S-adenosylmethionine decarboxylase; HSS, homospermidine synthase; SS, spermidine synthase.

is then transferred from decarboxylated SAM to putrescine by spermidine synthase to form spermidine. Putrescine can be produced from ornithine by ornithine decarboxylase (ODC). However, putrescine is derived from the arginine-agmatine pathway in pyrrolizidine alkaloid-producing plants due to the absence of ODC activity (Hartmann et al. 1988). Homospermidine, the first pathway-specific intermediate in pyrrolizidine alkaloid biosynthesis, is formed from putrescine and spermidine by homospermidine synthase (HSS) (Ober and Hartmann 1999). Homospermidine formation is the only known example of a functional moiety of spermidine used in a secondary metabolic pathway. The necine base moiety is formed from homospermidine via consecutive oxidative deaminations (Ober and Hartmann 1999) and subsequently converted to senecionine (Figure 4.13). A cDNA encoding HSS has been cloned from spring groundsel (*Senecio vernalis*) (Ober and Hartmann 1999). RNA blot and RT-PCR experiments have shown that HSS transcripts are restricted to roots. Immunolocalization of HSS demonstrated that the enzyme is restricted to distinct groups of endodermal and adjacent parenchyma cells in the root cortex located directly opposite the phloem. Immunogold labelling showed that HSS is found exclusively in the cytoplasm of the endodermis and adjoining cortical cells, supporting the characterization of the enzyme as a soluble cytosolic protein (Moll et al. 2002).

HSS exhibits strong sequence similarity to deoxyhypusine synthase (DHS), a ubiquitous enzyme responsible for the activation of eukaryotic initiation factor 5A (eIF5A) (Ober and Hartmann 1999). DHS catalyses the transfer of an aminobutyl moiety from spermidine to a lysine residue of eIF5A. In contrast, HSS does not accept eIF5A as a substrate. However, HSS and DHS both catalyse the formation of homospermidine by the aminobutylation of putrescine, although this reaction is rarely catalysed by DHS in vivo. HSS is thought to have evolved from DHS after duplication of the single-copy *dhs* gene. The product lost the ability to bind and react with eIF5A but retained HSS activity. HSS is a well-documented example of the evolutionary recruitment of a primary metabolic enzyme into a secondary metabolic pathway.

In *Senecio* species, pyrrolizidine alkaloids are produced in actively growing roots as senecionine *N*-oxides, which are transported via the phloem to above-ground organs (Hartmann and Dierich 1998). Senecionine *N*-oxides are subsequently modified by one or two species-specific reactions (i.e. hydroxylation, dehydrogenation, epoxidation or *O*-acetylation) that result in the unique pyrrolizidine alkaloid profile of different plants. Specific carriers are thought to participate in the phloem loading and unloading of senecionine *N*-oxides, because plants that do not produce pyrrolizidine alkaloids are unable to translocate these polar intermediates (Moll et al. 2002). Pyrrolizidine alkaloids are spatially mobile but do not show any turnover or degradation (Hartmann and Dierich 1998). Inflorescences appear to be the major sites of pyrrolizidine alkaloid accumulation in the common ragwort (*Senecio jacobaea*) (Figure 4.14) and *S. vernalis*, with jacobine occurring in flowers. The quantitative and qualitative accumulation of pyrrolizidine alkaloids are both genetically determined and vary substantially in *S. jacobaea* populations (Macel et al. 2004). The average pyrrolizidine alkaloid content of *S. jacobaea* inflorescences is 1.3 mg/g dry weight.

Pyrrolizidine alkaloids are poisonous to humans and cause losses in livestock, especially grazing animals. Humans are exposed to pyrrolizidine alkaloids through consumption of plants containing these toxins, contaminated staple products, herbal teas or medicines, and dietary supplements (Fu et al. 2004). Poisoning by pyrrolizidine alkaloids is endemic in India, Jamaica and parts of Africa. Native American and Hispanic populations in western

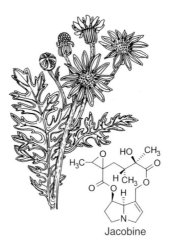

Figure 4.14 *Senecio jacobaea* and jacobine.

and southwestern United States are also at risk due to use of traditional herbal medicines. A significant herbal source, widely available around the world, is comfrey (*Symphytum officinale*). The well-demonstrated toxicity and carcinogenicity of comfrey has led the governments of Australia, Canada, Germany and the United Kingdom to restrict or even ban its sale entirely (Helferich and Winter 2001).

The most noxious pyrrolizidine alkaloids display acute and chronic toxicity, and genotoxicity, and are produced by species of the genera *Senecio*, *Crotalaria* and *Heliotropium*. The most potent genotoxic and tumorigenic pyrrolizidine alkaloids are the retronecine- and otonecine-type macrocyclic diesters (Figure 4.13). Although pyrrolizidine alkaloids exhibit differential potency, the functional groups at C-7 and C-9 of pyrrolic ester derivatives generally bind to and cross-link DNA and protein. Acute exposure causes considerable hepatotoxicity with haemorrhagic necrosis, whereas chronic poisoning primarily affects the liver, lungs and blood vessels (Fu et al. 2004).

Pyrrolizidine alkaloids require metabolic activation to become toxic. Three main pathways are involved in the metabolism of both retronecine- and heliotridine-type pyrrolizidine alkaloids (Fu et al. 2004). The first is the liver microsomal carboxylesterase-mediated hydrolysis of ester moieties linked to positions C-7 and C-9 resulting in the formation of necine bases and necic acids. Carboxylesterases exhibit species-specific substrate selectivity, which results in resistance to pyrrolizidine alkaloids in some animal species. The second pathway is the CYP3A-mediated N-oxidation of the necine bases to the corresponding pyrrolizidine alkaloid N-oxides. The third involves oxidation, followed by hydroxylation by CYP3A, of the necine base at positions C-3 or C-8 to yield the corresponding 3- or 8-hydroxynecine derivatives and spontaneous dehydration to produce dehydropyrrolizidine alkaloids. Otonecine-type pyrrolizidine alkaloids possess a structurally distinct necine base moiety metabolized by the hydrolysis of ester functional groups to form corresponding necine bases and acids. Retronecine-type pyrrolic esters are subsequently formed via oxidative N-demethylation by CYP3A, followed by ring closure and dehydration. Pyrrolic esters can effectively cross-link DNA and protein and appear to be largely responsible for the toxicity of pyrrolizidine alkaloids. Pyrrolic esters can also react

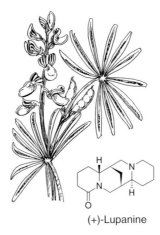

Figure 4.15 *Lupinus angustifolius* and (+)-lupanine.

with endogenous cellular constituents, such as glutathione, to create detoxified products. The detoxification of pyrrolic esters is mediated by glutathione-*S*-transferase or by non-enzymatic conjugation with glutathione (Fu *et al.* 2004). Although CYP3A is responsible for the metabolism of most pyrrolizidine alkaloids, CYP2B has also been implicated in some species (Huan *et al.* 1998).

4.8 Other alkaloids

4.8.1 Quinolizidine alkaloids

More than 200 known quinolizidine alkaloids are mainly distributed in the family Leguminoseae. Over 100 of these compounds are produced by members of the genus *Lupinus*, which includes more than 400 species mostly inhabiting the Americas (Ruiz and Sotelo 2001). The seeds of *Lupinus* spp. are rich in protein and have been used by humans for food and livestock feed for centuries. Lupin flour is often used as a source of protein enrichment in breads and pastas. In developing countries, cultivated or wild lupins are often the major source of protein in the human diet and are also used to feed domestic animals, because the plants can be grown in poor soils not suitable for the cultivation of traditional crops, such as soya (*Glycine max*). Quinolizidine alkaloids are generally toxic compounds that must be removed by soaking seeds in water for extended periods prior to consumption.

The quinolizidine alkaloid content of seeds is usually 1–2% (w/w) and is responsible for the characteristic bitter taste of lupin seeds. The development of commercial cultivars of 'sweet' varieties of narrow-leaved lupin (*Lupinus angustifolius*) with quinolizidine alkaloid contents of less than 0.015% (w/w) has expanded the potential use of lupin as a food source (Figure 4.15). However, quinolizidine alkaloids are involved in the defence response in plants, and consequently sweet lupin varieties are less resistant to pathogens and herbivores (Ruiz and Sotelo 2001).

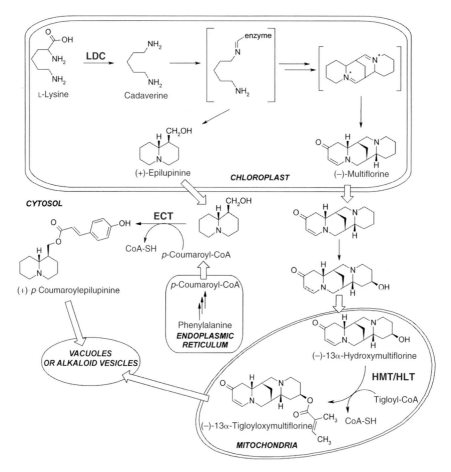

Figure 4.16 Biosynthesis of quinolizidine alkaloids. Enzyme abbreviations: ECT, p-coumaroyl-CoA: (+)-epilupinine O-p-coumaryltransferase; HMT/HLT, Tigloyl-CoA:(−)-13a-hydroxymultiflorine/(+)-13α-hydroxylupanine-O-tigloyltransferase; LDC, lysine decarboxylase.

The biochemistry and molecular biology of quinolizidine alkaloid biosynthesis have not been fully characterized. Quinolizidine alkaloids are formed from lysine via lysine decarboxylase (LDC) whereby cadaverine is the first detectable intermediate (Figure 4.16). Biosynthesis of the quinolizidine ring is thought to arise from the cyclization of cadaverine units via an enzyme-bound intermediate (Suzuki et al. 1996). Lysine decarboxylase and the quinolizidine skeleton-forming enzyme have been detected in chloroplasts of *Lupinus polyphyllus* (common lupin) (Wink and Hartmann 1982). Once the quinolizidine skeleton has been formed it is modified by dehydrogenation, hydroxylation or esterification to generate the diverse array of alkaloid products.

Lupins produce both tetracyclic (e.g. lupanine) and bicyclic quinolizidine alkaloids (i.e. lupinine) (Figure 4.15), which typically accumulate as esters of tiglic acid, p-coumaric acid, acetic acid and ferulic acid (Suzuki et al. 1996). The biological significance of these esters is not clear. Quinolizidine alkaloid esters are mainly distributed in the genera

Lupinus, Cytisus, Pearsonia, Calpurnia and *Rothia*. Some species lacking the acyltransferase required for ester formation accumulate quinolizidine alkaloid aglycones. The substrate specificity of these acyltransferases largely defines the profile of alkaloids of different lupin species. Two acyltransferases involved in quinolizidine alkaloid ester biosynthesis have been purified and characterized. Tigloyl-CoA:(−)-13α-hydroxymultiflorine/(+)-13α-hydroxylupanine O-tigloyltransferase (HMT/HLT) catalyses the transfer of a tigloyl group from tigloyl-CoA to the 13α-hydroxyl of 13α-hydroxymultiflorine or 13α-hydroxylupanine (Figure 4.16) (Suzuki *et al.* 1994). The tigloyl moiety is likely derived from isoleucine as in the biosynthesis of tigloyl esters of tropane alkaloids. HMT/HLT activity is localized to the mitochondrial matrix but not to chloroplasts where *de novo* quinolizidine alkaloid biosynthesis is thought to occur (Suzuki *et al.* 1996). A second acyltransferase, p-coumaroyl-CoA: (+)-epilupinine O-p-coumaroyltransferase (ECT), transfers p-coumaroyl to the hydroxyl moiety of (+)-epilupinine (Saito *et al.* 1992). ECT is present in an organelle distinct from the mitochondria and chloroplasts but has not been unambiguously localized. Although bitter and sweet lupins exhibit distinct alkaloid accumulation profiles, there are no discernable differences in acyltransferase activity between such varieties.

Pharmacological investigations have identified lupanine, 13-hydroxylupanine and sparteine as the primary toxic constituents of lupin seeds. Lupanine and 13-hydroxylupanine block ganglionic transmission, decrease cardiac contractility and contract uterine smooth muscle (Yovo *et al.* 1984). Acute oral LD_{50} value for lupanine is 410 mg/kg of body mass in mice. In seeds of wild Mexican lupin species (*L. angustifolius*), lupanine (11.5 mg/g dry weight) is generally the major alkaloid, whereas Canadian wild lupins (*L. reflexus*) contain more sparteine (26.6 mg/g dry weight) (Ruiz and Sotelo 2001). *L. reflexus* was identified as the most toxic lupin to mice due to its high sparteine content. A mortality rate of 33% occurred in animals fed with 6–12 g of seed/kg of body mass and increased to 100% when consumption was elevated to 15 g/kg of body mass. In contrast, *L. angustfolius* quinolizidine alkaloids were fed to rats at concentrations up to 5 g/kg of body mass for 90 days and no toxicological effects were detected (Robbins *et al.* 1996). However, inflammatory lesions were found on the livers of 8 out of 120 rats fed on similar levels of quinolizidine alkaloids (Butler *et al.* 1996). In comparison to rats, quinolizidine alkaloids are not metabolized as rapidly or extensively in humans and are, thus, not as toxic. Nevertheless, an acceptable daily intake (ADI) of 10 mg/kg of body mass has been recommended for humans, a level unlikely to be reached in Western diets.

4.8.2 Steroidal glycoalkaloids

Steroidal glycoalkaloids are found in many agricultural products obtained from members of the Solanaceae, including potatoes (*Solanum tuberosum*), tomatoes (*Lycopersicon esculentum*) and eggplants (*Solanum melongena*). Solanaceous steroidal alkaloids are of the spirosolane- or solanidane-type, which generally occur as glycosides. Surprisingly, little is known about the enzymes involved in steroidal glycoalkaloid biosynthesis. Cholesterol is considered a precursor to steroidal glycoalkaloids, but intermediates in the pathway have not been identified. The potato gene encoding solanidine-UDP-glucose glucosyltransferase has been cloned and was used to produce potato varieties with reduced steroidal glycoalkaloid levels.

Steroidal glycoalkaloids and aglycones have been linked to several health benefits, including reduced cholesterol levels, protection from *Salmonella typhimurium* infection and malaria, and cancer prevention (Cham 1994; Friedman *et al.* 2003). However, the properties that confer these health benefits also produce the toxicity of steroidal glycoalkaloids. The cultivated potato (*Solanum tuberosum*) contains two major steroidal glycoalkaloids: α-chaconine and α-solanine, both trisaccharides of the common aglycone solanidine. These natural products have been implicated in human poisoning caused by consumption of green potato tubers. α-Chaconine, the most toxic steroidal glycoalkaloid, is a potent teratogen, induces blood cell lysis, exhibits strong lytic properties and inhibits acetylcholinesterase (AchE) and butyrylcholinesterase (BchE). BchE and AchE break down anaesthetics and acetylcholine, respectively, and are important for normal nerve and muscle function. Inhibition of BchE and AchE also alters the response of patients to anaesthesia. The adverse effects of steroidal glycoalkaloids also include reduced respiratory activity and blood pressure, interference with sterol and steroid metabolism, and bradycardia or haemolysis resulting primarily from membrane disruption (Al Chami *et al.* 2003). Steroidal glycoalkaloid consumption has also been implicated as an environmental cause of schizophrenia (Christie 1999).

The concentrations of steroidal glycoalkaloids in unpeeled potatoes are typically between 2 and 0.15 mg/g of tuber. Steroidal glycoalkaloid content can increase postharvest by environmental factors such as light, mechanical injury and storage (Friedman *et al.* 2003). Exposure of potato tubers to light results in substantially higher chlorophyll accumulation and can induce a three-fold increase in steroidal glycoalkaloid content. For non-green tubers, a 60-kg person would have to consume at least 1.2 kg of potatoes to receive the estimated toxic dose (Phillips *et al.* 1996). This quantity is reduced to 400 g if the tubers are green. Traditional plant-breeding strategies have produced potato-blight-resistant potatoes, but the content of the protective steroidal glycoalkaloids was high enough to pose a substantial risk of human intoxication and commercial production was prohibited in the United Kingdom.

Potato leaves are sometimes consumed as a regular dietary component. However, leaves are more of a health concern than tubers because of their higher steroidal glycoalkaloid content, which is typically 0.6–0.8 mg/g. The toxic threshold for potato leaves might be reached by consuming only 150 g of tissue (Phillips *et al.* 1996). Direct exposure of human tissues to steroidal glycoalkaloids can cause lytic cell necrosis and inflammation, such that the primary effect of consuming potato leaves is typically damaging to the oral, oesophageal and stomach epithelia. The toxicity of these compounds can also produce severe symptoms including fever, rapid pulse, low blood pressure, rapid respiration and neurological disorders. Although direct damage to the alimentary tract is uncommon, steroidal glycoalkaloids can cause gastrointestinal disturbances such as vomiting, diarrhoea and abdominal pain (Korpan *et al.* 2004).

Western false hellebore (*Veratrum californicum*), a plant that grows in areas grazed by sheep throughout the western United States (James *et al.* 2004), produces the steroidal alkaloid 11-deoxojervine or cyclopamine. Although cyclopamine is relatively innocuous to adult mammals, it is a potent teratogen that produces serious craniofacial malformations in newborns ranging from upper jaw deformation to cyclopia (Keeler 1986). Cyclopamine has been demonstrated to affect developmental signalling pathways in humans.

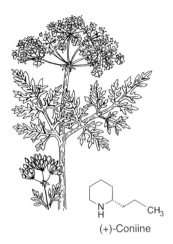

Figure 4.17 *Conium maculatum* and coniine.

4.8.3 Coniine

Poison hemlock (*Conium maculatum*) is native to Europe and western Asia, and contains eight structurally similar piperidine alkaloids including coniine and γ-coniceine (Figure 4.17). These two compounds are most abundant and are primarily responsible for the toxicity of poison hemlock, coniine being about eight times more toxic than γ-coniceine. Though no longer used for medicinal purposes, the dried fruits of *C. maculatum* were once used as an analgesic and for their sedative effects (Dewick 2002).

Unlike most piperidine alkaloids, the piperidine alkaloids of *C. maculatum* are synthesized from eight acetate units that produce a polyketoacid. An aminotransferase and an NADPH-dependent reductase participate in the cyclization of this polyketide to form γ-coniceine (Lopez *et al.* 1999). The alkaloid profile of poison hemlock changes substantially during the year due to the variations in light intensity and moisture. Coniine predominates during a sunny, dry summer, whereas coniine and γ-coniceine accumulate to similar levels during overcast, wet seasons. The highest alkaloid concentrations are typically found in ripe fruits.

The toxic effects of poison hemlock occur when plants or seeds are ingested. Symptoms of acute poisoning are similar in all animals and include muscular weakness, lack of coordination, trembling, pupil dilation, excess salivation and cold limbs. The toxicity of coniine, γ-coniceine and methylconiine occurs through blockage of spinal reflexes via the medulla. Higher doses stimulate skeletal muscles and induce neuromuscular blockage through antagonism of nicotinic acetylcholine receptors (Lopez *et al.* 1999). A lethal dose affects the phrenic nerve, paralysing the respiratory muscles. Chronic toxicity is only relevant in pregnant animals during critical periods of fetal development. Coniine and γ-coniceine also modulate the regulation of amniotic fluid levels, increasing the pressure in the amniotic sac. Offspring exposed to coniine during fetal development exhibit malformations, such as multiple congenital contractures. Structure-activity relationships have been established for the chronic toxicity of coniine. For example, the side chain must be at least as large as the propyl group of coniine and γ-coniceine.

Coniine persists in the meat and milk of animals that feed on poison hemlock (Frank and Reed 1990). Humans can be exposed through the food chain. For example, an individual was documented as suffering from acute renal failure after consuming wildfowl that had eaten hemlock (Scatizzi *et al.* 1993). Humans have also been poisoned after mistaking poison hemlock for other members of the Umbelliferae – the root for parsnip (*Pastinaca sativa*), the leaves for parsley (*Petroselinum crispum*) and the seeds for anise (*Pimpinella anisum*) (Lopez *et al.* 1999).

4.8.4 Betalains

Betalains represent two groups of water-soluble natural products that replace anthocyanins as the major pigments in flowers, fruits and vegetative tissues in most plants of the Caryophyllales (Strack *et al.* 2003). The red-violet betacyanins are immonium conjugates of betalamic acid and *cyclo*-dopa, whereas the yellow betaxanthins are immonium conjugates of betalamic acid and amino acids or amines. Various glycosides and acylglycosides of betaxanthins have also been identified.

Betalamic acid is produced by hydroxylation of tyrosine to dopa by tyrosinase, subsequent ring opening by dopa 4,5-dioxygenase, and a spontaneous rearrangement. Betalain derivatives result from the formation of aldamines between betalamic acid and *cyclo*-dopa (i.e. in the case of betacyanins) or amino acids (i.e. in the case of betaxanthins). Glycosides (e.g. betanin) and acylglycosides (e.g. lampranthin I and II) formation is catalysed by glucosyltransferases and acyltransferases, respectively (Strack *et al.* 2003).

The generation of free radicals and reactive oxygen species associated with cellular and metabolic injury inflicts oxidative stress and can contribute to the development of several chronic human diseases through the peroxidation of proteins and nucleic acids. Anti-oxidant enzymes, assisted by small molecule anti-oxidants, normally offer protection against oxygen free radicals. Along with vitamin C and carotenoids, betalains have been shown to be potent anti-oxidants in vitro (Kanner *et al.* 2001). However, despite the many natural sources of betalains only red beet (*Beta vulgaris*) and prickly pear are consumed regularly as foods. The dominant betalain in red beet is betanin, a glucoside containing both phenolic and cyclic amine moiety, which are excellent electron donors and confer anti-oxidant activity. Betacyanins are thought to prevent lipid peroxidation by interacting with peroxyl and alkoxyl free radicals. Consumption of red beets and beet products might help prevent oxidative stress-related disorders. Although reports of betaninuria show that some pigments are absorbed, there is no evidence for the benefits of betalains as anti-oxidants in the consumer. Although not as widespread as anthocyanins or carotenoids, betalains are also well-suited for colouring low-acid foods (Strack *et al.* 2003). The use of natural pigments for food colouring has increased as the health effects of synthetic dyes are questioned.

4.9 Metabolic engineering

Metabolic manipulation can be achieved either by over-expression or knockout of key enzymes to increase or decrease the accumulation of particular end-products, or to divert existing pathways to create new products. However, the rational engineering of alkaloid pathways is limited by our incomplete understanding of metabolic regulatory mechanisms. Nevertheless, the field offers great promise as illustrated by the examples discussed below.

The tropane alkaloid pathway was one of the first targets for the application of genetic engineering to alter the alkaloid profile of a plant. The introduction of H6H from *Atropa belladonna* into *Hyoscyamus niger*, a plant which does not normally possess this enzyme, resulted in the almost exclusive accumulation of scopolamine rather than hyoscyamine (Yun *et al.* 1992). Manipulation of a single enzymatic step resulted in a profound change in alkaloid profile. Over-expression of both the *PMT* and *H6H* genes in *H. niger* roots resulted in twice the accumulation of scopolamine compared with introducing *H6H* alone (Zhang *et al.* 2004).

SOMT has been the target of metabolic engineering in both Japanese gold thread (*Coptis japonica*) and *Eschscholzia californica*. Increased SOMT activity in transgenic *C. japonica* cells resulted in a marginally higher accumulation of berberine, rather than coptisine. When SOMT was expressed in *E. californica* cells, benzylisoquinoline alkaloid metabolism was diverted from sanguinarine biosynthesis and toward columbamine as a novel product (Sato *et al.* 2001). Sense constructs of the *Papaver somniferum BBE* gene were introduced into *E. californica* roots and resulted in the accumulation of more sanguinarine and related alkaloids (Park *et al.* 2003). Anti-sense-*BBE* constructs resulted in a reduced accumulation of alkaloids.

A different approach to modify the production of alkaloids in plants involves the use of transcriptional regulators to simultaneously alter the level of multiple enzymes in complex metabolic pathways, such as terpenoid indole alkaloid biosynthesis. Constitutive over-expression of *ORCA3* in *C. roseus* cells resulted in the activation of several genes including 1-deoxy-D-xylulose 5-phosphate synthase (involved in geraniol synthesis), anthranilate synthase (involved in tryptophan synthesis), TDC, STR, D4H and cytochrome P450 reductase (van der Fits and Memelink 2000). However, some genes including G10H, which is necessary for secologanin synthesis, showed no induction. The addition of exogenous loganin to cells with elevated *ORCA3* expression resulted in a threefold increase in terpenoid indole alkaloid accumulation compared with controls.

Recently, coffee plants were engineered using RNAi technology to reduce the caffeine content (Ogita *et al.* 2003). The repression of theobromine synthase transcript levels resulted in 50–70% less caffeine and 30–80% less theobromine compared with wild type plants. The biological suppression of caffeine biosynthesis offers a new opportunity to alter the quality of this popular beverage. Metabolic engineering offers a promising technology to improve the productivity and/or quality of important plants that accumulate desirable or undesirable alkaloids. In exploring solutions, genetic engineers must be cautious to recognize the inherent limitations of plant metabolism.

Acknowledgement

We are grateful to Joy MacLeod for the plant illustrations.

References

Abdel-Haq, H., Cometa, M.F., Palmery, M. *et al.* (2000) Relaxant effects of *Hydrastis canadensis* L. and its major alkaloids on guinea pig isolated trachea. *Pharmacol. Toxicol.*, 87, 218–222.

Al Chami, L., Mendez, R., Chataing, B. et al. (2003) Toxicological effects of α-solamargine in experimental animals. *Phytother. Res.*, **17**, 254–258.

Arnaud, M.J. (1993) Metabolism of caffeine and other components of coffee. In S. Garattini (ed.), *Caffeine, Coffee and Health*. Raven Press, New York, pp. 43–95.

Ashihara, H. and Crozier, A. (2001) Caffeine: a well known but little mentioned compound in plant science. *Trends Plant Sci.*, **6**, 407–413.

Burlat, V., Oudin, A., Courtois, M. et al. (2004) Co-expression of three MEP pathway genes and *geraniol 10-hydroxylase* in internal phloem parenchyma of *Catharanthus roseus* implicates multicellular translocation of intermediates during the biosynthesis of monoterpene indole alkaloids and isoprene-derived primary metabolites. *Plant J.*, **38**, 131–141.

Butler, W.H., Ford, G.P. and Creasy, D.M. (1996) A 90-day feeding study of lupin (*Lupinus angustifolius*) flour spiked with lupin alkaloids in the rat. *Food Chem. Toxicol.*, **34**, 531–536.

Campain, J.A. (2004) Nicotine: potentially a multifunctional carcinogen? *Toxicol. Sci.*, **79**, 1–3.

Cham, B.E. (1994) Solasodine glycosides as anti-cancer agents: pre-clinical and clinical studies. *Asia Pac. J. Pharmacol.*, **9**, 113–118.

Chan, T.Y.K. (2002) Herbal medicine causing likely strychnine poisoning. *Hum. Exp. Toxicol.*, **21**, 467–468.

Christie, A.C. (1999) Schizophrenia: is the potato the environmental culprit? *Med. Hypotheses*, **53**, 80–86.

Clifford, M.N. and Ramìrez-Martìnez, J.R. (1990) Chlorogenic acids and purine alkaloid content of Maté (*Ilex paraguariensis*) leaf and beverage. *Food Chem.* **35**, 13–21.

De Luca, V. and St-Pierre, B. (2000) The cell and developmental biology of alkaloid biosynthesis. *Trends Plant Sci.*, **5**, 168–173.

Dewick, P.M. (2002) *Medicinal Natural Products: A Biosynthetic Approach*, 2nd edn. John Wiley & Sons Ltd., West Sussex.

Duran-Patron, R., O'Hagan, D., Hamilton, J.T. et al. (2000) Biosynthetic studies on the tropane ring system of the tropane alkaloids from Datura stramonium. *Phytochemistry*, **53**, 777–784.

Facchini, P.J. (2001) Alkaloid biosynthesis in plants: Biochemistry, cell biology, molecular regulation, and metabolic engineering applications. *Annu. Rev. Plant Physiol. Plant Mol. Biol.*, **52**, 29–66.

Falkenhagen, H. and Stockigt, J. (1995) Enzymatic biosynthesis of vomilenine, a key intermediate of the ajamaline pathway, catalyzed by a novel cytochrome P450 dependent enzyme from plant cell cultures of *Rauwolfia serpentina*. *Z. Naturforsch.*, 1995, 45–53.

Filip, R. and Ferraro, G.E. (2003) Researching on new species of 'Maté': *Ilex brevicuspis*. Phytochemical and pharmacology study. *Eur. J. Nutr.*, **42**, 50–54.

Francis, P.D. and Clarke, C.F. (1999) Angel trumpet lily poisoning in five adolescents: clinical findings and management. *J. Paediatr. Child Health*, **35**, 93–95.

Frank, A.A. and Reed, W.M. (1990) Comparative toxicity of coniine, an alkaloid of *Conium maculatum* (poison hemlock), in chickens, quails, and turkeys. *Avian Dis.*, **34**, 433–437.

Friedman, M., Roitman, J.N. and Kozukue, N. (2003) Glycoalkaloid and calystegine contents of eight potato cultivars. *J. Agric. Food Chem.*, **51**, 2964–2973.

Fu, P.P., Xia, Q., Lin, G. et al. (2004) Pyrrolizidine alkaloids – genotoxicity, metabolism enzymes, metabolic activation, and mechanisms. *Drug Metab. Rev.*, **36**, 1–55.

Fuller, B.B., Drake, M.A. and Spaulding, D.T. (2000) Downregulation of tyrosine activity in human melanocyte cell cultures by yohimbine. *J. Invest. Dermatol.*, **114**, 268–276.

Graser, G. and Hartmann, T. (2000) Biosynthesis of spermidine, a direct precursor of pyrrolizidine alkaloids in root cultures of *Senecio vulgaris* L. *Planta*, **211**, 239–245.

Griffin, W.J. and Lin, G.D. (2000) Chemotaxonomy and geographical distribution of tropane alkaloids. *Phytochemistry*, **53**, 623–637.

Griffiths, R.R., Evans, S.M., Heishman, S.J. et al. (1990) Low-dose caffeine discrimination in humans. *J. Pharmacol. Exp. Ther.*, **246**, 21–29.

Hartmann, T., Sander, H. and Adolf, R. (1988) Metabolic links between the biosynthesis of pyrrolizidine alkaloids and polyamines in root cultures of *Senecio vulgaris*. *Planta*, **175**, 82–90.

Hartmann, T. and Dierich, B. (1998) Chemical diversity and variation of pyrrolizidine alkaloids of the senecionine type: biological need or coincidence? *Planta*, **206**, 443.

Hashimoto, T., Hayashi, A., Amano, Y. et al. (1991) Hyoscyamine 6 beta-hydroxylase, an enzyme involved in tropane alkaloid biosynthesis, is localized at the pericycle of the root. *J. Biol. Chem.*, **266**, 4648–4653.

He, Z., Ma, W.-Y., Hashimoto, T. et al. (2003) Induction of apoptosis by caffeine is mediated by the p53, Bax, and caspase 3 pathways. *Cancer Res.*, **63**, 4396–4401.

Helferich, W. and Winter, C.K. (2001) *Food Toxicology*. CRC Press, Boca Raton, FL.

Hertog, M.G.L., Sweetman, P.M., Fehily, A.M., Elwood, P.C. and Kromhout, D. (1997) Antioxidant flavonols and ischaemic heart disease in a Welsh population of men: The Caerphilly Study. *Am. J. Clin. Nutr.*, **65**, 1489–1494.

Hesse, M. (2002) *Alkaloids: Nature's Curse or Blessing?* Wiley-VCH, New York.

Hibi, N., Higashiguchi, S. and Hashimoto, T. (1994) Gene expression in tobacco low-nicotine mutants. *Plant Cell*, **6**, 723–735.

Hobhouse, H. (1985) *Seeds of Change: Five Plants that Transformed Mankind*. Sedgwick and Jackson, London.

Huan, J.Y., Miranda, C.L., Buhler, D.R. et al. (1998) The roles of CYP3A and CYP2B isoforms in hepatic bioactivation and detoxification of the pyrrolizidine alkaloid senecionine in sheep and hamsters. *Toxicol. Appl. Pharmacol.*, **151**, 229–235.

IARC. (1991) *IARC Monographs on the Evaluation of Carcinogenic Risks to Humans, Vol. 51: Coffee, Tea, Maté, Methylxanthines and Glyoxal*. IARC, Lyons.

James, J.E. (1997) Is habitual caffeine use a preventable cardiovascular risk factor? *Lancet*, **349**, 279–281.

James, L.F., Panter, K.E., Gaffield, W. et al. (2004) Biomedical applications of poisonous plant research. *J. Agric. Food Chem.*, **52**, 3211–3230.

Jamieson, R.W. (2004) The essence of commodification: caffeine dependencies in the early modern world. *J. Soc. Hist.*, **35**, 269–294.

Kanegae, T., Kajiya, H., Amano, Y. et al. (1994) Species-dependent expression of the *hyoscyamine 6 β-hydroxylase* gene in the pericycle. *Plant Physiol.*, **105**, 483–490.

Kanner, J., Harel, S. and Granit, R. (2001) Betalains – a new class of dietary cationized anti-oxidants. *J. Agric. Food Chem.*, **49**, 5178–5185.

Kato, M., Mizuno, K., Crozier, A. et al. (2000) Caffeine synthase gene from tea leaves. *Nature*, **406**, 956–957.

Katoh, A. and Hashimoto, T. (2004) Molecular biology of pyridine nucleotide and nicotine biosynthesis. *Front Biosci.*, **9**, 1577–1586.

Keeler, R.F. (1986) Teratology of Steroidal Alkaloids. In S.W. Pelletier (ed.), *Alkaloids: Chemical and Biological Perspectives*, Vol. 4. Wiley, New York, pp. 389–425.

Korpan, Y.I., Nazarenko, E.A., Skryshevskaya, I.V. et al. (2004) Potato glycoalkaloids: true safety or false sense of security? *Trends Biotechnol.*, **22**, 147–151.

Koshiishi, C., Crozier, A. and Ashihara, H. (2001) Profiles of purine and pyrimidine nucleotides in fresh and manufactured tea leaves. *J. Agric. Food Chem.*, **49**, 4378–4382.

Koyama, Y., Tomoda, M., Kato, J. et al. (2003) Metabolism of purine bases, nucleosides and alkaloids in theobromine-forming *Theobroma cacao* leaves. *Plant Physiol. Biochem.*, **41**, 997–984.

Lobert, S., Vulevic, B. and Correia, J.J. (1996) Interaction of vinca alkaloids with tubulin: a comparison of vinblastine, vincristine, and vinorelbine. *Biochemistry*, **35**, 6806–6814.

Lopez, T.A., Cid, M.S. and Bianchini, M.L. (1999) Biochemistry of hemlock (*Conium maculatum* L.) alkaloids and their acute and chronic toxicity in livestock. A review. *Toxicon*, **37**, 841–865.

Macel, M., Vrieling, K. and Klinkhamer, P.G. (2004) Variation in pyrrolizidine alkaloid patterns of *Senecio jacobaea*. *Phytochemistry*, **65**, 865–873.

Mazzafera, P. (1991) Trigonelline in coffee. *Phytochemistry*, **30**, 2309–2310.

McCormack, J.J. (1990) *Pharmacology of Anti-tumor Bisindole Alkaloids from Catharanthus*, Vol. 37. Academic Press, Inc., Orlando, FL, pp. 205–226.

Meijer, A.H., Verpoorte, R. and Hoge, J.H.C. (1993) Regulation of enzymes and genes involved in terpenoid indole alkaloid biosynthesis in *Catharanthus roseus*. *J. Plant Res.*, **3**, 145–164.

Menke, F.L., Champion, A., Kijne, J.W. *et al.* (1999) A novel jasmonate- and elicitor-responsive element in the periwinkle secondary metabolite biosynthesis gene *Str* interacts with a jasmonate- and elicitor-inducible AP2-domain transcription factor, ORCA2. *EMBO J.*, **18**, 4455–4463.

Mizuno, K., Kato, M., Irino, F. *et al.* (2003) The first committed step reaction of caffeine biosynthesis: 7-methylxanthosine synthase is closely homologous to caffeine synthases in coffee (*Coffea arabica* L.). *FEBS Lett.*, **547**, 56–60.

Mizusaki, S., Tanabe, Y., Noguchi, H. *et al.* (1972) *N*-methylputrescine oxidase from tobacco roots. *Phytochemistry*, **11**, 2757–2762.

Moll, S., Anke, S., Kahmann, U. *et al.* (2002) Cell-specific expression of homospermidine synthase, the entry enzyme of the pyrrolizidine alkaloid pathway in *Senecio vernalis*, in comparison with its ancestor, deoxyhypusine synthase. *Plant Physiol.*, **130**, 47–57.

Morton, J.F. (1989) Tannins as a carcinogen in bush tea, tea, maté and khat. In R.W. Hemingway and J.J. Karchesy (eds), *Chemistry and Significance of Condensed Tannins*. Plenum Press, New York, pp. 403–415.

Mosli-Waldhauser, S.S. and Baumann, T.W. (1996) Compartmentation of caffeine and related purine alkaloids depends exclusively on the physical chemistry of their vacuolar complex formation with chlorogenic acids. *Phytochemistry*, **42**, 985–996.

Musk, A.W. and de Klerk, N.H. (2003) History of tobacco and health. *Respirology*, **8**, 286–290.

Nakajima, K. and Hashimoto, T. (1999) Two tropinone reductases, that catalyze opposite stereospecific reductions in tropane alkaloid biosynthesis, are localized in plant root with different cell-specific patterns. *Plant Cell Physiol.*, **40**, 1099–1107.

Neuwinger, H.D. (1998) Alkaloids in Arrow Poisons. In M.F. Roberts and M. Wink (eds), *Alkaloids: Biochemistry, Ecology, and Medicinal Applications*. Plenum Press, New York, pp. 45–83.

Ober, D., Harms, R., Witte, L. *et al.* (2003) Molecular evolution by change of function. Alkaloid-specific homospermidine synthase retained all properties of deoxyhypusine synthase except binding the eIF5A precursor protein. *J. Biol. Chem.*, **278**, 12805–12812.

Ober, D. and Hartmann, T. (1999) Homospermidine synthase, the first pathway-specific enzyme of pyrrolizidine alkaloid biosynthesis, evolved from deoxyhypusine synthase. *Proc. Natl. Acad. Sci. USA*, **96**, 14777–14782.

Ogita, S., Uefuji, H., Yamaguchi, Y. *et al.* (2003) Producing decaffeinated coffee plants. *Nature*, **423**, 823.

Ouwerkerk, P.B. and Memelink, J. (1999) A G-box element from the *Catharanthus roseus* strictosidine synthase (*Str*) gene promoter confers seed-specific expression in transgenic tobacco plants. *Mol. Gen. Genet.*, **261**, 635–643.

Ouwerkerk, P.B., Trimborn, T.O., Hilliou, F. *et al.* (1999) Nuclear factors GT-1 and 3AF1 interact with multiple sequences within the promoter of the *Tdc* gene from Madagascar periwinkle: Gt-1 is involved in UV light-induced expression. *Mol. Gen. Genet.*, **261**, 610–622.

Park, S.-U., Yu, M. and Facchini, P.J. (2003) Modulation of berberine bridge enzyme levels in transgenic root cultures of California poppy alters the accumulation of benzophenanthridine alkaloids. *Plant Mol. Biol.*, **51**, 153–164.

Phillips, B.J., Hughes, J.A., Phillips, J.C. *et al.* (1996) A study of the toxic hazard that might be associated with the consumption of green potato tops. *Food Chem. Toxicol.*, **34**, 439–448.

Piccillo, G.A., Mondati, E.G. and Moro, P.A. (2002) Six clinical cases of *Mandragora autumnalis* poisoning: diagnosis and treatment. *Eur. J. Emerg. Med.*, **9**, 342–347.

Rang, H.P., Dale, M.M. and Ritter, J.M. (1999) *Pharmacology*, 4th edn. Churchill-Livingstone, Edinburgh.

Robbins, M.C., Petterson, D.S. and Brantom, P.G. (1996) A 90-day feeding study of the alkaloids of *Lupinus angustifolius* in the rat. *Food Chem. Toxicol.*, **34**, 679–686.

Ruiz, M.A. and Sotelo, A. (2001) Chemical composition, nutritive value, and toxicology evaluation of Mexican wild lupins. *J. Agric. Food Chem.*, **49**, 5336–5339.

Saito, K., Suzuki, H., Takamatsu, S. et al. (1992) Acyltransferases for lupin alkaloids in *Lupinus hirsutus*. *Phytochemistry*, **32**, 87–91.

Samanani, N., Yeung, E.C. and Facchini, P.J. (2002) Cell type-specific protoberberine alkaloid accumulation in *Thalictrum flavum*. *J. Plant Physiol.*, **159**, 1189–1196.

Sato, F., Hashimoto, T., Hackiya, A. et al. (2001) Metabolic engineering of plant alkaloid biosynthesis. *Proc. Natl. Acad. Sci. USA*, **98**, 367–372.

Scatizzi, A., Dimaggio, A., Rizzi, D. et al. (1993) Acute-renal-failure due to tubular-necrosis caused by wildfowl-mediated hemlock poisoning. *Ren. Fail.*, **15**, 93–96.

Schimming, T., Tofern, B., Mann, P. et al. (1998) Distribution and toxonomic significance of calystegines in the Colvolvulaceae. *Phytochemistry*, **49**, 1989–1995.

Schmeller, T. and Wink, M. (1998) Utilization of Alkaloids in Modern Medicine. In M.F. Roberts, and M. Wink (eds), *Alkaloids: Biochemistry, Ecology, and Medicinal Applications.* Plenum Press, New York, pp. 435–459.

Scholl, Y., Hoke, D. and Drager, B. (2001) Calystegines in *Calystegia sepium* derived from the tropane alkaloid pathway. *Phytochemistry*, **58**, 883–889.

Sewram, V., De Stefani, E., Brennan, P. and Boffetta, P. (2003) Maté consumption and the risk of squamous cell esophageal cancer in Uruguay. *Cancer Epidemiol. Biomarkers Prev.*, **12**, 508–513.

Shoji, T., Yamada, Y. and Hashimoto, T. (2000) Jasmonate induction of putrescine N-methyltransferase genes in the root of *Nicotiana sylvestris*. *Plant Cell Physiol.*, **41**, 831–839.

Sinclair, S.J., Murphy, K.J., Birch, C.D. et al. (2000) Molecular characterization of quinolinate phosphoribosyltransferase (QPRtase) in *Nicotiana*. *Plant Mol. Biol.*, **44**, 603–617.

St-Pierre, B., Flota-Vazquez, F.A. and De Luca, V. (1999) Multicellular compartmentation of *Catharanthus roseus* alkaloid biosynthesis predicts intercellular translocation of a pathway intermediate. *Plant Cell*, **11**, 887–900.

Strack, D., Vogt, T. and Schliemann, W. (2003) Recent advances in betalain research. *Phytochemistry*, **62**, 247–269.

Suzuki, H., Murakoshi, I. and Saito, K. (1994) A novel O-tigloyltransferase for alkaloid biosynthesis in plants. Purification, characterization, and distribution in *Lupinus* plants. *J. Biol. Chem.*, **269**, 15853–15860.

Suzuki, H., Koike, Y., Murakoshi, I. et al. (1996) Subcellular localization of acyltransferases for quinolizidine alkaloid biosynthesis in *Lupinus*. *Phytochemistry*, **42**, 1557–1562.

Suzuki, K., Yamada, Y. and Hashimoto, T. (1999) Expression of *Atropa belladonna* putrescine N-methyltransferase gene in root pericycle. *Plant Cell Physiol.*, **40**, 289–297.

Uefuji, H., Ogita, S., Yamaguchi, Y. et al. (2003) Molecular cloning and functional chatacterization of three distinct N-methyltransferases involved in the caffeine biosynthetic pathway in coffee plants. *Plant Physiol.*, **132**, 372–380.

Upmeier, B., Gross, W., Koster, S. and Barz, W. (1988) Purification and properties of S-adenosyl-L-methionine:nicotinic acid-N-methyltransferase from cell suspension cultures of *Glycine max* L. *Arch. Biochem. Biophys.*, **262**, 445–454.

van der Fits, L. and Memelink, J. (2000) ORCA3, a jasmonate-responsive transcriptional regulator of plant primary and secondary metabolism. *Science*, **289**, 295–297.

van der Fits, L., Zhang, H., Menke, F.L.H. *et al.* (2000) A *Catharanthus roseaus* BPF-1 homologue interacts with an elicitor-responsive region of the secondary metabolite biosynthetic gene *Str* and is induced by elicitor via a JA-independent signal transduction pathway. Plant Mol. Biol., **40**, 675–685.

Ward, N., Whitney, C. and Avery, D. (1991) The analgesic effects of caffeine in headache. *Pain*, **44**, 151–155.

Watson, A.A., Fleet, G.W., Asano, N. *et al.* (2001) Polyhydroxylated alkaloids – natural occurrence and therapeutic applications. *Phytochemistry*, **56**, 265–295.

WHO (1982) Anthocyanins. In *Toxicological Evaluation of Certain Food Additives*, W.H.O. Geneva, W.H.O. Food Additives Series. J.E.C.F.A., pp. 42–49.

Wink, M. (1998) A Short History of Alkaloids. In M.F. Roberts and M. Wink (eds), *Alkaloids: Biochemistry, Ecology, and Medicinal Applications*. Plenum Press, New York, pp. 11–44.

Wink, M. and Hartmann, T. (1982) Localization of the enzymes of quinolizidine alkaloid biosynthesis in leaf chloroplasts of *Lupinus polyphyllus*. Plant Physiol., **70**, 74–77.

Yovo, K., Huguet, F., Pothier, J. *et al.* (1984) Comparative pharmacological study of sparteine and its ketonic derivative lupanine from seeds of *Lupinus albus*. Planta Med., **50**, 420–424.

Yun, D.J., Hashimoto, T. and Yamada, Y. (1992) Metabolic engineering of medicinal plants: transgenic *Atropa belladonna* with an improved alkaloid composition. Proc. Natl. Acad. Sci. USA, **89**, 11799–11803.

Zhang, L., Ding, R., Chai, Y. *et al.* (2004) Engineering tropane biosynthetic pathway in *Hyoscyamus niger* hairy root cultures. Proc. Natl. Acad. Sci. USA, **101**, 6786–6791.

Zheng, X.-Q., Koyama, Y., Nagai, C. and Ashihara, H. (2004) Biosynthesis, accumulation and degradation of theobromine in developing *Theobroma cacao* fruits. J. Plant Physiol., **161**, 363–369.

Chapter 5
Acetylenes and Psoralens

Lars P. Christensen and Kirsten Brandt

5.1 Introduction

Through epidemiological investigations it is well-known that a high consumption of vegetables and fruits protect against certain types of cancer and other important diseases (Greenvald *et al.* 2001; Kris-Etherton *et al.* 2002; Maynard *et al.* 2003; Gundgaard *et al.* 2003; Trichopoulou *et al.* 2003). In order to explain the health promoting effects of fruit and vegetables focus has primarily been on vitamins, minerals and antioxidants, but still we do not know which components are responsible for these effects of food plants. One of the possible explanations is the hypothesis that plants contain other bioactive compounds that provide benefits for health, even though they are not essential nutrients (Brandt *et al.* 2004).

Plants contain a great number of different secondary metabolites, many of which display biological activity and are used in plant defence against insects, fungi and other microorganisms. Many bioactive substances with known effects on human physiology and disease have been identified through studies of plants used in, for example, traditional medicine. Some of these compounds occur also in food plants, although many of these bioactive compounds are normally considered undesirable in human food due to their 'toxic' effects. However, a low daily intake of these 'toxins' may be an important factor in the search for an explanation of the beneficial effects of fruit and vegetables on human health (Brandt *et al.* 2004).

The acetylenes and linear furanocoumarins (psoralens) are examples of bioactive secondary metabolites that have been considered undesirable in plant foods due to their 'toxic' effects. Some acetylenes are known to be potent skin sensitizers and irritants, and neurotoxic in high concentrations, but have also been shown to have a pronounced selective cytotoxic activity against various cancer cells. Due to their role in plant defence many acetylenes and psoralens are considered natural pesticides or in some cases phytoalexins since their formation is often induced in plants as a response to external stimuli. Psoralens are photoactivated secondary metabolites that have been used since ancient times to treat human skin disorders. However, the use of these furanocoumarins in medicine has been associated with increased incidence of skin cancer, and a number of studies have also demonstrated that the furanocoumarins can be carcinogenic, mutagenic, photodermatitic and to have reproductive toxicity.

Although many acetylenes and psoralens are toxic when ingested in relatively high amounts, they may have beneficial effects in low concentrations and hence could explain some of the beneficial effects of the food plants where they appear.

This chapter highlights the present state of knowledge on naturally occurring acetylenes and psoralens in the edible parts of more or less common food plants, including their biochemistry, bioactivity and possible relevance for human health.

5.2 Acetylenes in common food plants

5.2.1 Distribution and biosynthesis

Acetylenes form a distinct group of relatively chemically reactive natural products, which have been found in about 24 families of the higher plants, although they seem to occur regularly in only seven families, namely Apiaceae (=Umbelliferae), Araliaceae, Asteraceae (=Compositae), Campanulaceae, Olacaceae, Pittosporaceae and Santalaceae (Bohlmann et al. 1973). The majority of the naturally occurring acetylenes have been isolated from the Asteraceae (Bohlmann et al. 1973; Christensen and Lam 1990; 1991a,b; Christensen 1992), and today more than 1400 different acetylenes and related compounds have been isolated from higher plants, including thiophenes, dithiacyclohexadienes (thiarubrines), thioethers, sulphoxides, sulphones, alkamides, chlorohydrins, lactones, spiroacetal enol ethers, furans, pyrans, tetrahydropyrans, isocoumarins, aromatic and aliphatic acetylenes. Despite the large structural variation among naturally occurring acetylenes, a comparison of their structures with those of oleic (**100**), linoleic (**101**), crepenynic (**102**) and dehydrocrepenynic (**103**) acids (see e.g. Figures 5.1 and 5.2) makes it reasonable to assume that most acetylenes are biosynthesized with the latter acids as precursors. Many feeding experiments with ^{13}C-, ^{14}C- and ^{3}H-labelled precursors have confirmed this assumption and further that they are built up from acetate and malonate units (Bu'Lock and Smalley 1962; Bu'Lock and Smith 1967; Bohlmann et al. 1973; Jones et al. 1975; Hansen and Boll 1986; Barley et al. 1988; Bohlmann 1988; Jente et al. 1988; Christensen and Lam 1990). Further evidence of C_{18}-acids as precursors in the biosynthesis of most acetylenes has recently been obtained through identification of a gene coding for a fatty acid acetylenase (triple bond forming enzyme), which occurs in the same plant species that produce acetylenes and can be induced by fungal infection (Cahoon et al. 2003).

Food plants so far known to produce acetylenes in their utilized plant parts are listed in Table 5.1, and include important vegetables such as carrot (*Duacus carota*), celery (*Apium graveolens*), lettuce (*Latuca sativa*), parsley (*Petroselinum crispum*), Jerusalem artichoke (*Helianthus tuberosus*), Jerusalem tomato (*Lycopersicon esculentum*) and aubergine (*Solanum melongena*), although the majority of food plants belong to plant families that do not normally produce acetylenes. Still, since many food plants from families that produce acetylenes have not yet been investigated in this respect, and these compounds do occur occasionally in other families, their occurrence in food is probably much more widespread than documented in the present review.

The most common acetylenes isolated from food plants are aliphatic acetylenes (Figures 5.1 and 5.3, and Table 5.1). Aliphatic acetylenes of the falcarinol-type (especially compounds 1, 2, 4 and 5) are widely distributed in the Apiaceae and Araliaceae plant families (Bohlmann et al. 1973; Hansen and Boll 1986), and consequently nearly all acetylenes found in the utilized/edible parts of food plants of the Apiaceae, such as carrot, caraway (*Carum carvi*), celery, celeriac (*Apium graveolens* var. *rapaceum*), fennel (*Feoniculum vulgare*), parsnip (*Pastinaca sativa*) and parsley are of the falcarinol-type

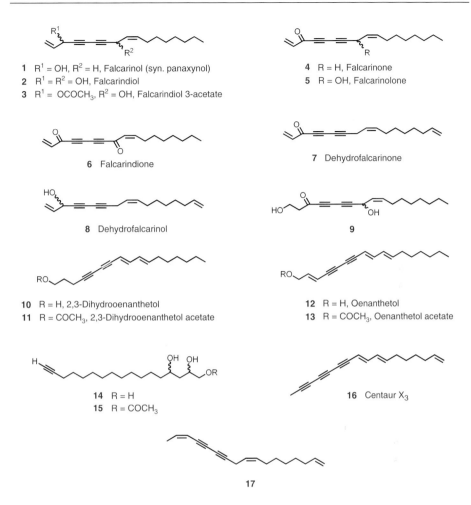

Figure 5.1 Aliphatic C_{17}-acetylenes isolated from the utilized parts of well-known major and/or minor food plants.

(Table 5.1). Some of these acetylenes have also been isolated from Solanaceous food plants such as tomatoes and aubergines, where they appear to be phytoalexins (see Section 5.2.2.1). The biosynthesis of polyacetylenes of the falcarinol-type follows the normal biosynthetic pathway for aliphatic C_{17}-acetylenes (Bohlmann et al. 1973; Hansen and Boll 1986), with dehydrogenation of oleic acid leading to the C_{18}-acetylenes crepenynic acid and dehydrocrepenynic acid, which is then transformed to C_{17}-acetylenes by β-oxidation (Figure 5.2). Further oxidation and dehydrogenation leads to acetylenes of the falcarinol-type as outlined in Figure 5.2.

In the Asteraceae, acetylenes are widely distributed and structurally very diverse, including aliphatic acetylenes, acetylenic thiophenes, aromatics, isocoumarins and spiroacetal enol ethers (Table 5.1 and Figures 5.3, 5.4 and 5.5).

Spiroacetal enol ethers are characteristic of the tribe Anthemideae of the Asteraceae, and it is therefore not surprising that the acetylenes isolated from the utilized parts of *Chrysanthemum coronarium* (garland chrysanthemum) and *Matricaria chamomilla*

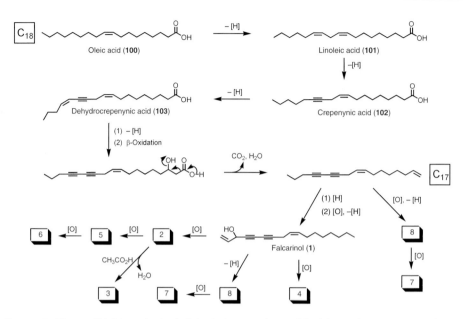

Figure 5.2 The possible biosynthesis of aliphatic C_{17}-acetylenes of the falcarinol-type. [O] = oxidation, [H] = reduction, −[H] = dehydrogenation (oxidation followed by the loss of water).

(camomile) are spiroacetal enol ethers (Table 5.1). The related species *Artemisia dracunculus* (tarragon) is, however, characterized by the presence of aromatic and isocoumarin acetylenes, especially in the underground parts, which are not used for food (Greger 1979; Engelmeier *et al.* 2004). From the aerial parts of tarragon that are used both as a vegetable and condiment an aromatic acetylene (**40**) and an isocoumarin acetylene (**41**) have been isolated together with several aliphatic acetylenes (**8, 18–20**) (Figures 5.1, 5.3 and 5.4, and Table 5.1).

The five-membered C_{12}-spiroacetal enol ethers (Figure 5.4) present in the above mentioned food plants are possibly biosynthesized from the C_{18}-triyne-ene acid **104** by an α-oxidation followed by two β-oxidations leading to the C_{13}-triyne-ene alcohol **105**, which is then transformed into the spiroacetal enol ethers **45** and **46** by further oxidation and ring closure as shown in Figure 5.5. Oxidation of compounds **45** and **46** leads directly to the spiroacetal enol ethers **42–44, 47** and **48**, whereas oxidation and decarboxylation followed by addition of CH_3SH or its biochemical equivalent leads to the spiroacetal enol ethers **49–57** (Figures 5.4 and 5.5). The majority of aromatic and isocoumarin acetylenes isolated from higher plants appear to follow almost the same biosynthetic route as the spiroacetal enol ethers starting from the C_{18}-triyne-ene acid **104**. The possible biosynthesis of aromatic and isocoumarin acetylenes has been described by Bohlmann *et al.* (1973) and Christensen (1992).

The roots of *Arctium lappa* (edible burdock, Gobo), which are used for food in Japan, are especially rich in both aliphatic acetylenes and acetylenic dithiophenes (Figure 5.6 and Table 5.1). Most of the aliphatic acetylenes isolated from *A. lappa* (**17, 23–28, 35–38**) are widely distributed in Asteraceae, whereas the dithiophenes (**58–69**) isolated from this species are not very common as only a few have been isolated from other plant species

Table 5.1 Acetylenes in major and minor food plants, their primary uses and plant part utilized

Family/species	Common name	Plant part used for foods[a]	Uses[b]	Acetylenes in used plant parts	References
Apiaceae (=Umbelliferae)					
Aegopodium podagraria L.	Bishop's weed, ground elder	L, St	V	1, 2, 9	Bohlmann et al. 1973; Kemp 1978; Degen et al. 1999
Anethum graveolens L.	Dill	L, S	C, V	1, 2, 13	Bohlmann et al. 1973; Degen et al. 1999
Anthriscus cerefolium (L.) Hoffm.	Chervil, salad chervil, French parsley	L, S	C, V	1, 2	Degen et al. 1999;[c]
Anthriscus sylvestris Hoffm.	Cow parsley	L	V	2	Nakano et al. 1998;[c]
Apium graveolens L. var. *dulce*	Celery	L, S	C, V	1, 2	Avalos et al. 1995;[c]
Apium graveolens L. var. *rapaceum*	Celeriac, knob celery, celery root	R	V	1, 2, 4, 5	Bohlmann 1967; Bohlmann et al. 1973;[c]
Bunium bulbocastanum L.	Great earthnut	T, L, F	C, V	1, 4, 5	Bohlmann et al. 1973
Carum carvi L.	Caraway	R, L, S	C	1, 2, 5, 6	Bohlmann et al. 1961, 1973; Degen et al. 1999
Centella asiatica L.	Asiatic or Indian pennywort	L	V	[d]	Nakano et al. 1998
Chaerophyllum bulbosum L.	Turnip-rooted chervil	R, L	V	1, 4	[c]
Coriandrum sativum L.	Coriander, cilantro	L, S	C, V	[d]	Nakano et al. 1998
Crithmum maritimum L.	Samphire, marine fennel	L	V	1, 2	Cunsolo et al. 1993
Cryptotaenia canadensis (L.) DC.	Hornwort, white or wild chervil	R, L, St, F	V	1, 2	Eckenbach et al. 1999
Cryptotaenia japonica Hassk.	Japanese hornwort, Mitsuba	R, L, St	V	[d]	Nakano et al. 1998
Daucus carota L.	Carrot	R, L	V	1–3, 5	Crosby and Aharonson 1967; Bentley et al. 1969; Garrod et al. 1978; Yates and England 1982; Lund and White 1990; Lund 1992; Degen et al. 1999; Czepa and Hofmann 2003, 2004; Hansen et al. 2003

(Continued)

Table 5.1 Continued

Family/species	Common name	Plant part used for foods[a]	Uses[b]	Acetylenes in used plant parts	References
Ferula assa-foetida L.	Asafoetida, giant fennel	R, S, Sh	C	5	Bohlmann et al. 1973
Ferula communis L.	Common giant fennel	L, S	C, V	2	Appendino et al. 1993
Foeniculum vulgare Mill.	Fennel	L, S	C, V	1, 2	Nakano et al. 1998; Degen et al. 1999
Heracleum sphondylium L.	Common cow parsnip, hogweed	L, Sh	V	1, 2	Degen et al. 1999;[c]
Levisticum officinale Koch.	Lovage, garden lovage	L, S	C, V	2, 5	—[c]
Myrrhis odorata (L.) Scop.	Sweet cicely, sweet chervil	R, L, S	C	—[d]	Bohlmann et al. 1973
Oenanthe javanica (Blume) DC.	Water dropwort, water celery	L, St, Sh	V	1, 2	Yates and Fenster 1983; Fujita et al. 1995
Pastinaca sativa L.	Parsnip	R, L	V	1, 2, 4, 5	Bohlmann et al. 1973; Degen et al. 1999
Petroselinum crispum (Mill.) Nyman ex A.W. Hill. (=*P. sativum* Hoffm.)	Parsley	L	C, V	1, 2, 4, 5	Bohlmann et al. 1973; Degen et al. 1999;[c]
Petroselinum crispum (Mill.) Nyman ex A.W. Hill. var. *tuberosum*	Hamburg parsley, turnip-rooted parsley	R, L	C, V	1, 2	Nitz et al. 1990;[c]
Pimpinella major (L.) Hud.	Greater burnet saxifrage	R, L, S	C	1, 2	Degen et al. 1999
Pimpinella saxifraga L.	Burnet saxifrage	R, L, S	C	30	Schulte et al. 1970
Sium sisarum L.	Skirret, chervin	R	V	5	Bohlmann et al. 1961, 1973
Trachyspermum ammi (L.) Spr.	Ajowan, ajwain	L, S	C	10–13	Bohlmann et al. 1973
Asteraceae (=Compositae)					
Arctium lappa L.	Edible burdock, Gobo	R	V	17, 23–28, 35–38, 58–69	Schulte et al. 1967; Bohlmann et al. 1973; Washino et al. 1986a,b; Washino et al. 1987; Takasugi et al. 1987; Christensen and Lam 1990
Artemisia dracunculus L.	Tarragon, esdragon	L	C, V	8, 18–20, 40, 41	Jakupovic et al. 1991

Species	Common name	Part	C, V	Numbers	References
Bellis perennis L.	Common daisy	L, F		21, 22	Avato and Tava 1995; Avato et al. 1997
Chrysanthemum coronarium L.	Garland chrysanthemum, Shungiku (Japanese), Kor tongho (Chinese)	L	V	42–57	Bohlmann and Fritz 1979; Tada and Chiba 1984; Sanz et al. 1990; Christensen 1992;[c]
Cynara scolymus L.	Globe artichoke	L, F	V	26, 27	Bohlmann et al. 1973;[c]
Cichorium endivia L.	Endive, escarole	L	V	29, [e]	—[c]
Cichorium intybus L. var. foliosum	Chicory, witloof chicory	R, L	V	29, [e]	Rücker and Noldenn 1991
Helianthus tuberosus L.	Jerusalem artichoke	T	V	7	Bohlmann et al. 1962, 1973
Lactuca sativa L.	Lettuce	L	V	16	Bentley et al. 1969[b]; Bohlmann et al. 1973
Matricaria chamomilla L. (=Chamomilla recutita (L.) Rausch.)	Chamomile, German chamomile	F	C	45, 46	Reichling and Becker 1977; Repčák et al. 1980, 1999; Redaelli et al. 1981; Holz and Miething 1994
Campanulaceae					
Platycodon grandiflorum (Jacq.) A. DC.	Balloon flower, Chinese bell flower	R, L	V	32–34	Tada et al. 1995; Ahn et al. 1996
Lauraceae					
Persea americana Mill.	Avocado	Fr	V	14[f], 15[f]	Adikaram et al. 1992; Oberlies et al. 1998
Leguminosae					
Lens culinaris Medik.	Lentil	P	V	71[f], 76[f]	Robeson 1978; Robeson and Harborne 1980
Vicia faba L.	Broad bean, fava bean	P, S	V	70–78[f]	Fawcett et al. 1968, 1971; Hargreaves et al. 1976a,b; Mansfield et al. 1980; Robeson and Harborne 1980; Buzi et al. 2003
Solanaceae					
Lycopersicon esculentum Mill.	Tomato	Fr	V	1[f], 2[f], 31[f]	De Wit and Kodde 1981; Elgersma et al. 1984
Solanum melongena L.	Eggplant, aubergine	Fr	V	2[f], 39[f]	Imoto and Ohta 1988

[a] R, roots; T, tubers; L, leaves; St, stems; Sh, shoots; F, flowers; Fr, fruits; P, pods; S, seeds. [b] V, vegetable; C, condiment or flavouring (Yamaguchi 1983; Pemberton and Lee 1996; Rubatzky et al. 1999). [c] Christensen, L.P., unpublished results. [d] Acetylenes detected but not identified. [e] Further acetylenes with a triyn-ene or diyn-ene chromophore have been detected but not identified. [f] Mainly isolated from infected plant tissue.

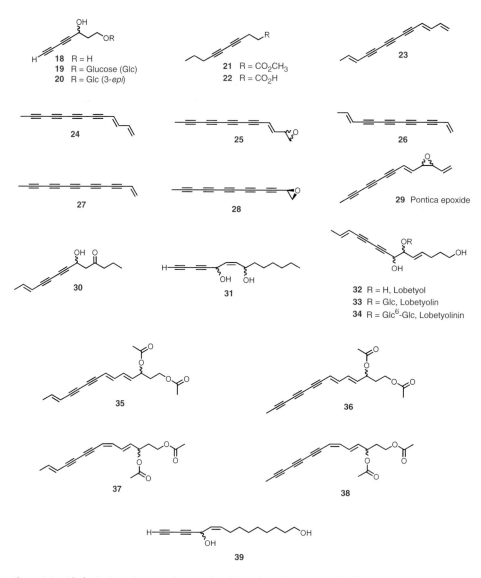

Figure 5.3 Aliphatic C_7 to C_{15} acetylenes isolated from the utilized parts of well-known major and/minor food plants.

(Bohlmann et al. 1973; Christensen and Lam 1990; 1991a). Dithiophenes are common in certain tribes of the Asteraceae and are biosynthesized from the C_{13}-polyacetylene pentayn-ene (**27**) followed by the addition of two times H_2S or its biochemical equivalent as shown in Figure 5.7. Of particular interest is the occurrence of acetylenic dithiophenes linked to a guaianolide sesquiterpene lactone (**68**, **69**) in edible burdock. The biosynthesis of these dithiophenes probably involve reaction with arctinal (**58**) and a sesquiterpene lactone (Figure 5.7) to afford oxetane-containing adducts that after cleavage afford diols by hydrolysis and subsequent oxidation then leads to the ketols **68** and **69** (Washino et al.

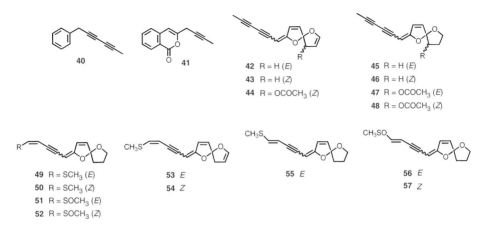

Figure 5.4 Aromatic, isocoumarin and spiroacetal enol ether acetylenes isolated from the utilized parts of food plants from the tribe Anthemideae (Asteraceae).

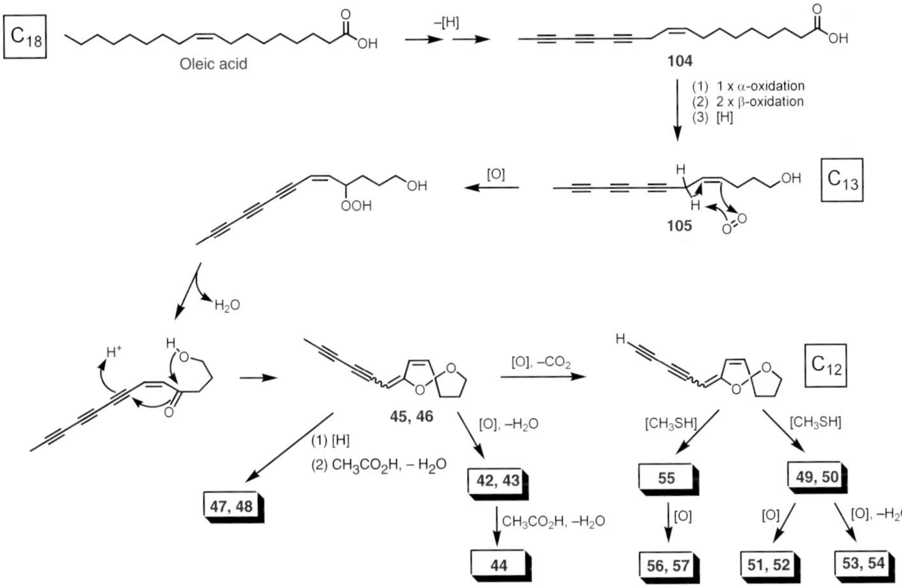

Figure 5.5 The possible biosynthesis of spiroacetal enol ethers isolated from the utilized parts of food plants of the tribe Anthemideae (Asteraceae). [O] = oxidation, [H] = reduction, −[H] = dehydrogenation (oxidation followed by the loss of water), [CH$_3$SH] = addition of CH$_3$SH or the biochemical equivalent to CH$_3$SH.

1987). Further details about the biosynthesis of acetylenic thiophenes can be found in Bohlmann *et al.* (1973).

The aliphatic acetylenes isolated from the utilized plant part of *A. lappa* and other food plants, including endive (**29**), chicory (**29**) and balloon flower (**32–34**) (Table 5.1) are biosynthesized from C$_{18}$-acetylenic acids by two β-oxidations leading to C$_{14}$-acetylenic

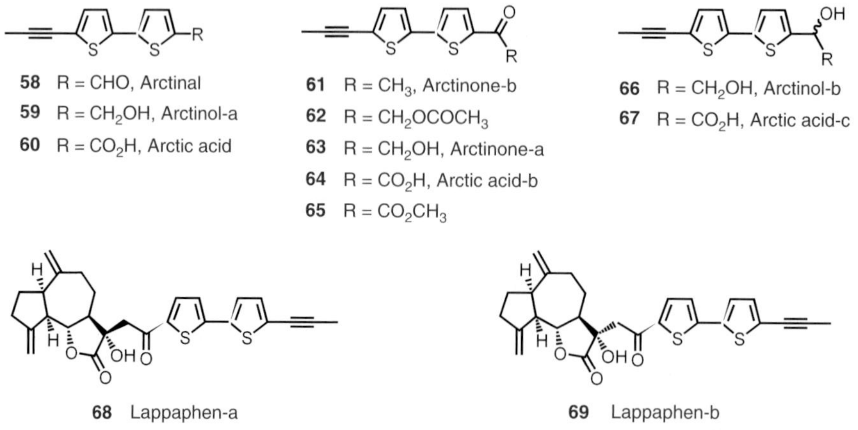

58 R = CHO, Arctinal
59 R = CH2OH, Arctinol-a
60 R = CO2H, Arctic acid
61 R = CH3, Arctinone-b
62 R = CH2OCOCH3
63 R = CH2OH, Arctinone-a
64 R = CO2H, Arctic acid-b
65 R = CO2CH3
66 R = CH2OH, Arctinol-b
67 R = CO2H, Arctic acid-c
68 Lappaphen-a
69 Lappaphen-b

Figure 5.6 Dithiophenes isolated from the roots of *Arctium lappa* (edible burdock).

C18
Oleic acid
–[H]
(1) 2 x β-oxidation
(2) Oxidative rearrangement
(3) –H2O
C14
(1) [O], –CO2
(2) –[H]
C13
(1) 2 × [H2S]
(2) – [H]
H2O
Arctinone-b (61)
[O]
Arctinone-a (63)
CH3CO2H
H2O
[H]
CO2
C12
Arctinal (58)
SQL
[CH3]
60
64
62
66
59
68, 69
65
67

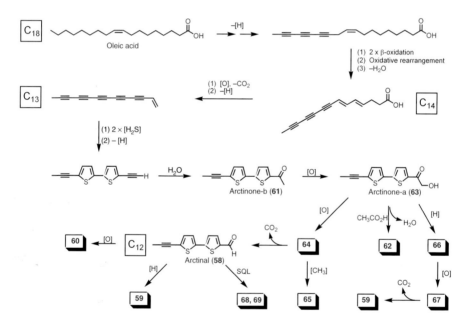

Figure 5.7 The possible biosynthesis of dithiophenes isolated from the roots of *Arctium lappa* (edible burdock). [O] = oxidation, [H] = reduction, –[H] = dehydrogenation (oxidation followed by the loss of water), [H_2S] = addition of H_2S or the biochemical equivalent to H_2S, [CH_3] = methylation due to the cofactor *S*-adenosyl methionine, SQL = sesquiterpene lactone (a guaianolide).

precursors (Figure 5.8). In Figure 5.8 the formation of some aliphatic C_{13}-polyacetylenes (**23–28**) and C_{14}-acetylenes (**33, 34, 35** and **37**) from a C_{14}-acetylenic precursor (**106**) with a diyne-diene chromophore is shown. Further dehydrogenation of the C_{14}-acetylene **106** would, for example, lead to the corresponding C_{14}-acetylene with a triyne-diene chromophore, which again is the precursor for other aliphatic acetylenes such as **29, 36** and **38.**

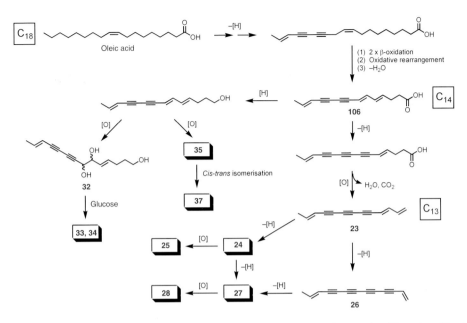

Figure 5.8 The possible biosynthesis of some aliphatic C_{13}- and C_{14} acetylenes isolated from the utilized parts of food plants. [O] = oxidation, [H] = reduction, −[H] = dehydrogenation (oxidation followed by the loss of water).

Further information about the biosynthesis of aliphatic acetylenes including those isolated from food plants can be found in Bohlmann *et al.* (1973), Christensen and Lam (1990; 1991a,b) and Christensen (1992).

A special type of furanoacetylenes (e.g. 70–78) has been isolated from leaves and edible parts of broad bean (*Vicia faba*) and/or lentil (*Lens culinaris*) (Table 5.1 and Figure 5.9) infected by fungi such as *Botrytis cinerea* and *B. fabae*. These furanoacetylenes are considered as phytoalexins as they do not seem to be present in healthy plant tissue (Mansfield *et al.* 1980; Robeson and Harborne 1980), although they may in fact be present in minute amounts. From incorporation studies with [^{13}C]-acetate precursors (Cain and Porter 1979; Al-Douri and Dewick 1986) it has been shown that the furanoacetylenes (e.g. 70–78) are most likely biosynthesized from the C_{18}-acetylene dehydrocrepenynic acid followed by two β-oxidation sequences at the carboxyl end leading to the C_{14}-acetylene 107. Further dehydrogenation, oxidation and isomerization steps then finally lead to the furanoacetylenes 70–78 as illustrated in Figure 5.10.

5.2.2 Bioactivity

5.2.2.1 Antifungal activity

Falcarinol (1) and falcarindiol (2) seem to have a defensive role in carrots against invading fungi. Falcarinol inhibits germination of *Botrytis cinerea* spores and its concentration is greatly increased when carrots are infected with this fungus (Harding and Heale 1981). *Botrytis cinerea* attacks carrots during storage but not when they are

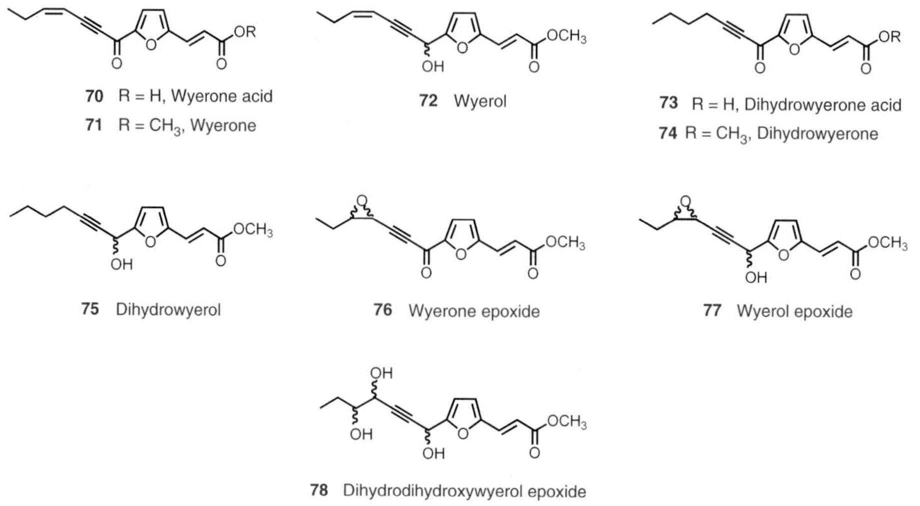

Figure 5.9 Examples of furanoacetylenes isolated from edible parts of broad bean (*Vicia faba*) and/or lentil (*Lens culinaris*) upon infection with *Botrytis* spp.

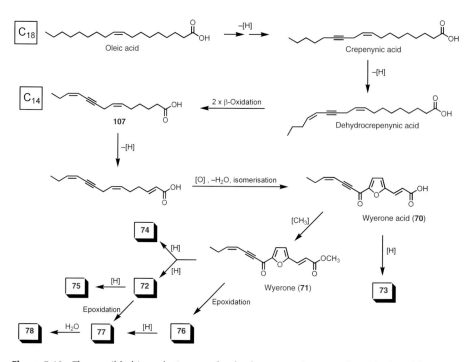

Figure 5.10 The possible biosynthetic route for the furanoacetylenes produced by broad bean (*Vicia faba*) and/or lentil (*Lens culinaris*) upon infection with *Botrytis* spp.

fresh (Harding and Heale 1980). Another fungus that attacks carrots during storage is *Mycocentrospora acerina*. It has been shown that falcarindiol is highly toxic towards this fungus (Garrod *et al.* 1978) and that carrot cultivars with high levels of this compound are less susceptible to this disease (Olsson and Svensson 1996). Falcarinol and falcarindiol have also been identified as antifungal compounds in many other Apiaceae plant species inhibiting spore germination of different fungi in concentrations ranging from 20 to 200 μg/mL (Christensen 1998; Hansen and Boll 1986). Polyacetylenes of the falcarinol-type seem to act as sort of pre-infectional compounds in the species producing them and hence may play an important role in protecting these plants from fungal attack.

The families Solanaceae and Lauraceae do not normally produce acetylenes (Bohlmann *et al.* 1973). However, when healthy tomato fruits and leaves (Solanaceae) are infected with leaf mould (*Cladosporium fulvum*), they accumulate the acetylenic phytoalexins falcarinol, falcarindiol and (6*Z*)-tetradeca-6-ene-1,3-diyne-5,8-diol (**31**) (De Wit and Kodde 1981) (Table 5.1). These compounds are also detected in tomato plants upon infection with *Verticillium alboatrum* (Elgersma *et al.* 1984). Whether healthy tomato fruits and leaves in fact contain small amounts of polyacetylenes, which undergo post-infectional increases in response to fungal attack, is not known. Also aubergines of the Solanaceae family have been shown to be capable of producing polyacetylenes (**2**, **39**) when exposed to phytoalexin elicitors (Imoto and Ohta 1988) (Table 5.1). Avocado anthracnose, caused by the fungus *Colletotrichum gloeosporioides*, is a major disease factor to post-harvest rotting in avocado fruit (*Persea americana*, Lauraceae). Unripe fruit show no evidence of incipient decay lesions but the decay process may develop rapidly during ripening, indicating the presence of latent infection. This characteristic behaviour has been shown to be attributed to the presence of a significant amount of antifungal acetylenes (**14**, **15**) and related alkadienes (Table 5.1), which suppress the vegetative growth of the fungus (Adikaram *et al.* 1992; Oberlies *et al.* 1998). During ripening this antifungal activity is gradually lost presumably due to degradation of the active compounds.

The production of acetylenes in plant species belonging to a family where acetylenes are not normally produced is also known from the food plants broad bean and lentil (*Lens culinaris*) of the Leguminosae plant family as mentioned in Section 5.2.1. Hence the production of antifungal acetylenes in food plants belonging to families that do not normally produce acetylenes could be a much more common phenomenon than first anticipated.

5.2.2.2 Neurotoxicity

The neurotoxic effects of some acetylenes have long been known. Acetylenes with this activity include the fish poisons ichthyothereol (**108**) and ichthyothereol acetate (**109**) (Cascon *et al.* 1965), which occur regularly in the tribes Heliantheae and Anthemideae of the Asteraceae (Bohlmann *et al.* 1973; Christensen and Lam 1991a; Christensen 1992). It has been suggested that the toxicity of the fish poisons **108** and **109** (Figure 5.11) is due to their ability to uncouple oxidative phosphorylation and inhibiting ATP-dependent contractions (Towers and Wat 1978).

The acetylenes oenanthotoxin (**110**) and cicutoxin (**111**) (Figure 5.11) isolated from water-hemlock (*Cicuta virosa* L.), spotted water-hemlock (*Cicuta maculata* L.), and from the hemlock water dropwort, *Oenanthe crocata* L. (Apiaceae) (Anet *et al.* 1953; Konoshima

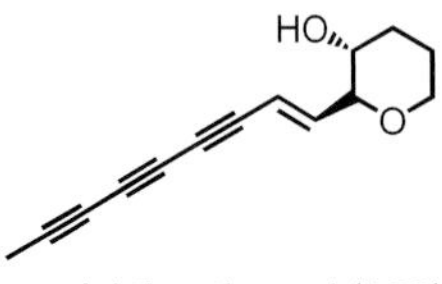

Ichthyothereol (**108**)

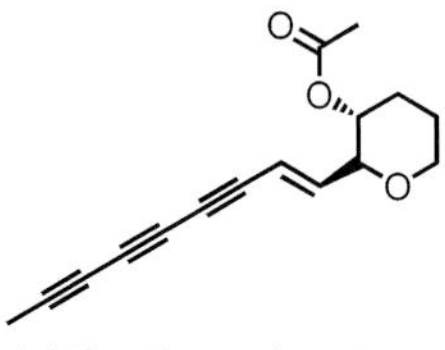

Ichthyothereol acetate (**109**)

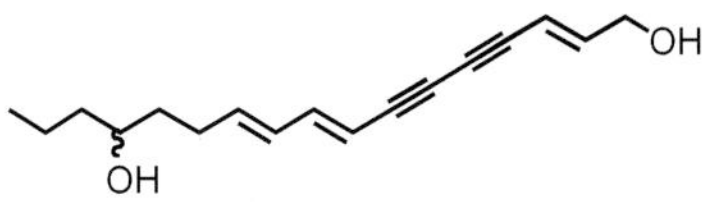

Oenanthotoxin (**110**)

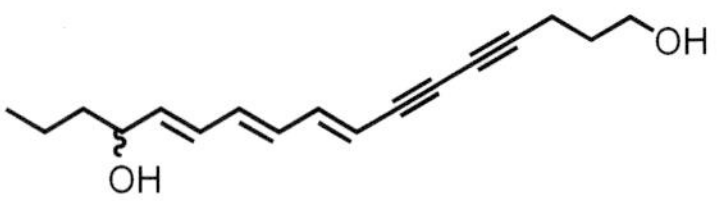

Cicutoxin (**111**)

Figure 5.11 Examples of highly neurotoxic acetylenes isolated from non-food plants.

and Lee 1986; Wittstock *et al.* 1995) are extremely poisonous causing violent convulsions and death, and they have been responsible for the death of numerous human beings and livestock (Anet *et al.* 1953). (However, it is not unusual to see feral Canada geese (Branta canadensis) gorge on Hemlock Water Dropwort without apparent ill effect, and the population is increasing – M.N. Clifford, personal obervation). These poisonous acetylenes are, however, closely related to the acetylenes **10–13** present in dill (*Anethum graveolens*) and/or ajowan (*Trachyspermum ammi*) (Table 5.1 and Figure 5.1). Whether the acetylenes **10–13** have similar toxic effects as oenanthotoxin and cicutoxin is not known.

Less well-known are the effects of falcarinol (**1**), which produces pronounced neurotoxic symptoms upon injection into mice with an LD_{50} of 100 mg/kg whereas the related falcarindiol (**2**) does not seem to have any acute effect (Crosby and Aharonson 1967). The type of neurotoxic symptoms produced by falcarinol is similar to those of oenanthotoxin and cicutoxin, although it is much less toxic.

5.2.2.3 Allergenicity

Many plants containing aliphatic C_{17}-acetylenes have been reported to cause allergic contact dermatitis and irritant skin reactions (Hausen 1988), primarily due to occupational exposure (e.g. nursery workers). The relation between the clinical effect and the content of polyacetylenes has been investigated in *Schefflera arboricola* (Hayata) Merrill (Araliaceae) (dwarf umbrella tree), and the results showed that falcarinol (**1**) is a potent contact allergen, whereas related polyacetylenes such as falcarindiol (**2**) and falcarinone (**4**) had no effect on the skin (Hansen *et al.* 1986). Falcarinol has been shown to be responsible for most of the allergic skin reactions caused by plants of the Apiaceae and Araliaceae (Hausen *et al.* 1987; Hausen 1988; Murdoch and Demster 2000; Machado *et al.* 2002). The allergenic properties of falcarinol indicate that it is very reactive towards mercapto and amino groups in proteins, forming a hapten-protein complex (antigen) (Figure 5.12). The reactivity of falcarinol towards proteins is probably due to its hydrophobicity and its ability to form an extremely stable carbocation with the loss of water, as shown in Figure 5.12, thereby acting as a very reactive alkylating agent towards various biomolecules. This mechanism may also explain its anti-inflammatory and antibacterial effect (see Section 5.2.2.4), its cytotoxicity (see Section 5.2.2.5) and perhaps its bioactivity in general.

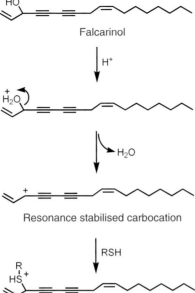

Figure 5.12 The possible reaction of falcarinol with biomolecules, which may explain its interaction with the immune system leading to allergenic reactions (type IV). The bioactivity of acetylenes of the falcarinol-type, in particular their cytotoxic activity, may be explained by a similar mechanism. RSH = thiol residue of a biomolecule, for example, a protein.

Allergic contact dermatitis from common vegetables of the Apiaceae, such as carrots, celery and parsley is known (Murdoch and Demster 2000; Machado *et al.* 2002) but rare, probably due to their relatively low concentrations of allergenic acetylenes compared with ornamental and wild plant species (Hausen 1988), or possibly a desensitizing effect of oral intake.

5.2.2.4 Anti-inflammatory, anti-platelet-aggregatory and antibacterial effects

Falcarinol (1) and falcarindiol (2) have shown anti-inflammatory and anti-platelet-aggregatory effects (Teng *et al.* 1989; Appendino *et al.* 1993; Alanko *et al.* 1994). For falcarinol it has been suggested that this pharmacological action is related to the ability of the compound to modulate prostaglandin catabolism by inhibiting the prostaglandin-catabolizing enzyme 15-hydroxy-prostaglandin dehydrogenase (Fujimoto *et al.* 1998).

Panaxydol (112)

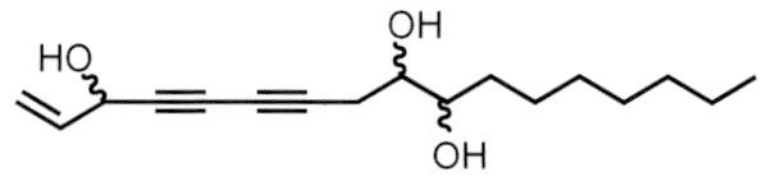

Panaxytriol (113)

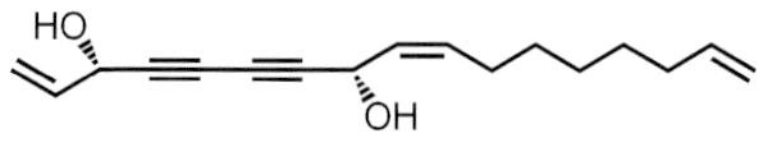

Dehydrofalcarindiol (114)

Figure 5.13 Examples of highly cytotoxic acetylenes of the falcarinol-type isolated from medicinal plants.

Falcarinol and related C_{17}-acetylenes have also shown antibacterial and antituberculosis activity (Kobaisy et al. 1997). These pharmacological activities indicate mechanisms by which falcarinol and related polyacetylenes could have positive effects on human health.

5.2.2.5 Cytotoxicity

Panax ginseng C. A. Meyer (Araliaceae) is one of the most famous and valuable herbal drugs in Asia. The active principles in *P. ginseng* have for many years been considered to be saponins (ginsenosides) and studies on the constituents of this plant have therefore mainly focused on these compounds. However, since the anticancer activity of hexane extracts of the roots of *P. ginseng* was discovered in the beginning of 1980s (Shim et al. 1983), the lipophilic portion of this plant has been intensively investigated. This had led to the isolation and identification of several cytotoxic polyacetylenes, including falcarinol (**1**), panaxydol (**112**) and panaxytriol (**113**) (Figures 5.1 and 5.13) of which only falcarinol (synonym panaxynol) appears to be present in food plants (Fujimoto and Satoh 1988; Ahn and Kim 1988; Matsunaga et al. 1989, 1990; Christensen 1998). The acetylenes **1**, **112** and **113** have been found to be highly cytotoxic against numerous cancer cell lines (Ahn and Kim 1988; Matsunaga et al. 1989, 1990) showing the strongest cytotoxic activity towards human gastric adenocarcinoma (MK-1) cancer cells with an ED_{50} of 0.108, 0.059 and 0.605 μM, respectively (Matsunaga et al. 1990). Furthermore, falcarinol, panaxydol and panaxytriol have been shown to inhibit the cell growth of normal cell cultures such as human fibroblasts (MRC-5), although the ED_{50} against normal cells was around 20 times higher than for cancer cells. In particular, panaxytriol did not inhibit the growth of MRC-5 cells by 50% even at concentrations higher than 2.5 μM (Matsunaga et al. 1990). The possible selective in vitro cytotoxicity of the acetylenes **1**, **112** and **113** against cancer cells compared with normal cells, indicate that they may be useful in the treatment of cancer.

From the aerial parts of the medicinal plant *Dendropanax arboreus* (angelica tree) several aliphatic polyacetylenes have been isolated of which falcarinol, falcarindiol (**2**), dehydrofalcarinol (**8**) and dehydrofalcarindiol (**114**) (Figures 5.1 and 5.13) were found to exhibit in vitro cytotoxicity against human tumour cell lines, with falcarinol showing the strongest activity (Bernart et al. 1996). Preliminary in vivo evaluation of the cytotoxic activity of falcarinol, dehydrofalcarinol and dehydrofalcarindiol using a LOX melanoma mouse xenograft model demonstrated some potential for in vivo antitumour activity of falcarinol and dehydrofalcarinol, with dehydrofalcarindiol showing the strongest therapeutic effect (Bernart et al. 1996), although the effect was not significant.

The mechanism for the inhibitory activity of falcarinol and related C_{17}-acetylenes of the falcarinol-type is still not known but may be related to their reactivity and hence their ability to interact with various biomolecules as mentioned earlier in Section 5.2.2.3. This is in accordance with a recent in vitro study, which showed that the suppressive effect of falcarinol on cell proliferation of various tumour cells (K562, Raji, Wish, HeLa, Calu-1 and Vero) probably was due to its ability to arrest the cell cycle progression of the tumour cells into various phases of their cell cycle (Kuo et al. 2002).

As falcarinol, falcarindiol and related C_{17}-acetylenes are common in the Araliaceae and Apiaceae one might expect that more species within these families exhibit cytotoxic activity, including food plants. The strong selective cytotoxic activity of falcarinol and related C_{17}-acetylenes towards different cancer cells indicates that they may be valuable in the treatment or prevention of different types of cancer and could contribute to the health promoting properties of food plants that contain these compounds. Also the cytotoxic activity of dehydrofalcarinol and dehydrofalcarindiol are interesting, since these compounds are widely distributed in several tribes of the Asteraceae (Bohlmann et al. 1973; Christensen 1992). So far only dehydrofalcarindiol has been isolated from tarragon (Table 5.1), but they may be present in other food plants of this family, and hence may contribute to the health promoting properties of some members of the Asteraceae.

5.2.2.6 Falcarinol and the health-promoting properties of carrots

Many studies have shown that a high content of natural β-carotene in blood is correlated with a low incidence of several types of cancer, although intervention studies have shown that supplementation with β-carotene does not protect against development of this disease (Greenberg et al. 1996; Omenn et al. 1996). In most European countries and North America more than 50% of the β-carotene intake is provided by carrots (O'Neil et al. 2001). However, in these regions carrot consumption appears to be better correlated with the intake of α-carotene (O'Neil et al. 2001). Several studies have found stronger negative correlations of lung cancer with intake of α-carotene rather than β-carotene (Ziegler et al. 1996; Michaud et al. 2000; Wright et al. 2003), confirming the central role of carrots. However, a beneficial effect of any compound primarily found in carrots, not only carotenoids, would give the same correlations. As shown in Table 5.1, carrots contain a group of bioactive polyacetylenes, of which falcarinol (1) clearly is the most bioactive of these, as described in the previous sections.

In the human diet carrots are the major dietary source of falcarinol, although it may also be supplied by many other plant food sources (Table 5.1). A recent in vitro study aiming to screen for potentially health promoting compounds from vegetables showed that falcarinol, but not β-carotene, could stimulate differentiation of primary mammalian cells in concentrations between 0.004 and 0.4 µM falcarinol. Toxic effects were found above >4 µM falcarinol (Figure 5.14), while β-carotene had no effect even at 400 µM (Hansen et al. 2003). This biphasic effect (hormesis) of falcarinol on cell proliferation is fully in accordance with the hypothesis that most toxic compounds have beneficial effects

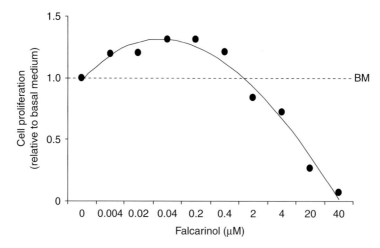

Figure 5.14 Effects of increasing concentrations of falcarinol on proliferation, measured by incorporation of [methyl-^{3}H]thymidine into mammary epithelial cells prepared from prepubertal Friesian heifers and grown in 3-dimensional collagen gels (redrawn from Hansen *et al.* 2003). Stimulatory as well as inhibitory effects on cell proliferation were significantly different from those obtained in basal medium (BM). No effect on proliferation was observed for β-carotene when tested in the same bioassay (Hansen *et al.* 2003).

at certain lower concentrations (Calabrese and Baldwin 1998). Therefore falcarinol appears to be one of the bioactive components in carrots and related vegetables that could explain their health promoting properties, rather than carotenoids or other types of primary and/or secondary metabolites. This hypothesis is further supported by recent studies on the bioavailability of falcarinol in humans (Hansen-Møller *et al.* 2002; Haraldsdóttir *et al.* 2002; Brandt *et al.* 2004). When for example falcarinol was administered orally via carrot juice (13.3 mg falcarinol/L) in volumes of 300, 600 and 900 mL, it was rapidly absorbed, reaching a maximum concentration in serum between 0.004 and 0.01 µM 2 h after dosing as shown in Figure 5.15 (Haraldsdóttir *et al.* 2002; Brandt *et al.* 2004). This is within the range where the in vitro data indicate that a potentially beneficial physiological effect would be expected (Figure 5.14) and also the inhibitive effect on the proliferation of cancer cells described in Section 5.2.2.5.

This effect has been studied in an established rat model for colon cancer by injections of the carcinogen azoxymethane in the inbred rat strain BDXI by feeding with carrot or purified falcarinol (with the same, physiologically relevant, intake of falcarinol) (Brandt *et al.* 2004; Kobaek-Larsen *et al.* 2005). Eighteen weeks after the first azoxymethane injection, the rats were killed and the colon examined for tumours and their microscopic precursors aberrant crypt foci (ACF) (McLellan and Bird 1988). The carrot and falcarinol treatments showed a significant tendency to reduced numbers of (pre)cancerous lesions with increasing size of lesion as shown in Figure 5.16. These results further suggest that the protective effect of carrot can be explained to a high degree by its content of falcarinol, and that the traditional view of acetylenes in food as generally undesirable toxicants may need to be revised, indicating a need to reinvestigate the significance of other acetylenes described in this review.

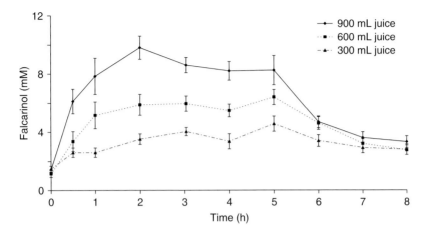

Figure 5.15 Concentration of falcarinol in plasma of 14 volunteers as a function of time after ingestion of a breakfast meal with 300, 600 and 900 mL carrot juice, respectively, containing 16, 33 and 49 μmol falcarinol, respectively. Means ± SEM.

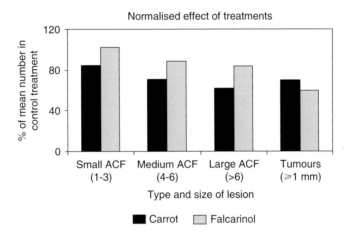

Figure 5.16 Effect of treatments with carrot or falcarinol on the average numbers per animal of four types of (pre)cancerous lesions in rat colons, each size class representing increasingly advanced steps on the progression towards cancer. The size of aberrant crypt foci (ACF) was measured as the number of crypts found on a corresponding area of normal colon tissue. The smallest tumours correspond to an ACF size of ∼20. The trend for reduced relative numbers with increasing size of lesion was significant at $P = 0.028$.

5.3 Psoralens in common food plants

5.3.1 Distribution and biosynthesis

Furanocoumarins (furocoumarins) are subdivided into linear furanocoumarins (psoralens) and angular furanocoumarins (angelicins). Linear furanocoumarins (LFCs) or psoralens were identified in the late 1940s as the cause of the photosensitization properties of the plants that contain them (Fahmy et al. 1947). LFCs occur widely in nature as constituents of many plant species, particular in the families Apiaceae, Leguminosae, Moraceae and Rutaceae. The structural variation within the LFCs is not large but still over 40 different LFCs have been isolated from food plants, although not necessarily from the edible parts (Wagstaff 1991).

Food plants that produce LFCs in their utilized plant parts are listed in Table 5.2 and include important vegetables such as carrots, celery, celeriac, parsnip and parsley of the Apiaceae and various citrus plants of the Rutaceae. The LFCs found in the utilized parts of Apiaceae food plants are primarily psoralen (**79**), xanthotoxin (**81**), bergapten (**91**) and isopimpinellin (**93**) of which celery, parsnip and parsley appear to contribute to the highest intake of psoralens, as these important vegetables are known to produce relatively high amounts of LFCs in their edible parts (Beier and Nigg 1992). The structural variation of the psoralens found in utilized parts of citrus species is much larger than those found in Apiaceae food plants. The most common LFCs found in citrus plants appear to be bergaptol (**80**), bergapten (**91**), bergamottin (**86**) and its oxygenated derivatives (**87, 88, 97–99**), although lemon (*Citrus limon*) and lime (*Citrus aurantifolia*) also are characterized by the presence of imperatorin (**89**) and isoimperatorin (**83**) and its oxygenated derivatives (**84, 85**) (Table 5.2, Figures 5.17 and 5.18).

The coumarin marmesin (**118**) (Figure 5.17) is one of the key precursors in the biosynthesis of the LFCs. This compound has been isolated from many plants, including members of the Apiaceae and Rutaceae, and appears to play a key role in the post-harvest resistance of celery and parsley to pathogens (Afek et al. 1995; 2002). Marmesin contain a coumarin subunit (umbelliferone) and a C_5 subunit. The biosynthetic pathway from umbelliferone (**116**) via demethylsuberosin (**117**) to marmesin has been delineated (Hamerski and Matern 1988a; Stanjek et al. 1999b) and is shown in Figure 5.17. The biosynthesis of umbelliferone follows the normal pathway for *ortho*-hydroxylated coumarins starting from *p*-coumaric acid (**115**) (Figure 5.17). Further metabolism of marmesin produces psoralen, in which three of the original carbon atoms have been lost. The conversion of marmesin to psoralen has been proven to proceed via oxidation in the 4′ position followed by *syn*-elimination releasing water and acetone as shown in Figure 5.17 (Stanjek et al. 1999a). The subsequent hydroxylation of psoralen in the 5- or 8-position then yields bergaptol (**80**) and xanthotoxol (8-hydroxypsoralen), respectively, which has been demonstrated in vitro for bergaptol (Hamerski and Matern 1988b). Bergaptol and xanthotoxol are then further processed by *O*-methylation to bergapten (**91**) and xanthotoxin (**81**), respectively, and the corresponding *O*-methyltransferases responsible for this methylation have been isolated and identified (Hehmann et al. 2004).

The hydroxylations to yield 5,8-dihydroxypsoralen are required for the formation of 8-hydroxy-5-methoxypsoralen (**92**) and isopimpinellin (**93**) and their prenylated derivatives (**94–96**) as shown in Figure 5.17. However, the order of hydroxylations, *O*-methylations and prenylations of these 5,8-oxygenated derivatives remains unresolved (Hehmann et al. 2004). Xanthotoxol, which is widely distributed in the Apiaceae and Rutaceae, but to date has not been isolated from the edible parts of food plants, appears to be the precursor for xanthotoxin and the prenylated and geranylated derivatives **82, 89** and **90** as shown in Figure 5.17. Bergaptol is the precursor for the prenylated derivative isoimperatorin (**83**) and its oxygenated derivatives (**84, 85**) as well as the geranylated bergamottin (**86**)

Table 5.2 Linear furanocoumarins (psoralens) in well-known major and minor food plants, their primary uses and plant part utilized

Family/species	Common name	Plant part used for foods[a]	Uses[b]	Psoralens in used plant parts	References
Apiaceae (=Umbelliferae)					
Aegopodium podagraria L.	Bishop's weed, ground elder	L, St	V	81, 91	Degen et al. 1999
Anethum graveolens L.	Dill	L, S	C, V	81, 91	Ceska et al. 1987; Degen et al. 1999
Anthriscus cerefolium (L.) Hoffm.	Chervil, salad chervil, French parsley	L, S	C, V	81, 91	Degen et al. 1999
Angelica archangelica L. (=A. officinalis Hoffm.)	Garden or European angelica	R, L, St, S	C, V	79, 81, 83, 89, 91–93	Patra et al. 1976; Zobel and Brown 1991a,b,c
Apium graveolens L. var. dulce	Celery	L, S	C, V	79, 81,91, 93	Innocenti et al. 1976; Beier et al. 1983a,b; Avalos et al. 1995; Diawara et al. 1995; Nigg et al. 1997; Lombaert et al. 2001
Apium graveolens L. var. rapaceum	Celeriac, knob celery	R	V	79, 81,91, 93	Järvenpää et al. 1997
Carum carvi L.	Caraway	R, L, S	C	81, 91	Ceska et al. 1987; Degen et al. 1999
Coriandrum sativum L.	Coriander, cilantro	L, S	C, V	81, 91	Ceska et al. 1987
Daucus carota L.	Carrot	R, L	V	81, 91,[c]	Ceska et al. 1986a; Degen et al. 1999
Foeniculum vulgare L.	Fennel	L, S	C, V	81, 91	Degen et al. 1999
Heracleum lanatum	Cow parsnip, masterwort	R, L, S	C, V	81, 91	Steck 1970; Camm et al. 1976; Zobel and Brown 1991c
Heracleum sphondylium L.	Common cow parsnip, hogweed	L, Sh	V	81, 91	Degen et al. 1999
Levisticum officinale Koch.	Lovage, garden lovage	L, S	C, V	79,81, 91	Zobel and Brown 1991a;[d]

(Continued)

Table 5.2 Continued

Family/species	Common name	Plant part used for foods[a]	Uses[b]	Psoralens in used plant parts	References
Pastinaca sativa L.	Parsnip	R, L	V	81, 91	Ceska et al. 1986b; Zobel and Brown 1991c; Degen et al. 1999; Lombaert et al. 2001
Petroselinum crispum (Mill.) Nyman ex A.W. Hill. (=*P. sativum* Hoffm.)	Parsley	L	C, V	79, 81, 83–85, 89, 91, 93	Innocenti et al. 1976; Beier et al. 1983a; Beier et al. 1994; Manderfeld et al. 1997; Degen et al. 1999; [d]
Pimpinella anisum L.	Anise	L, S	C	81, 91	Ceska et al. 1987
Pimpinella major (L.) Hud.	Greater burnet saxifrage	R, L, S	C	79, 81, 91	Zobel and Brown 1991a; Degen et al. 1999
Rutaceae					
Citrus aurantiifolia (Christm.) Swingle	Lime	Fr	B, C, V	82, 83–86, 89–91, 93, 94, 96	Stanley and Vannier 1967; Nigg et al. 1993; Dugo et al. 1999
Citrus aurantium L.	Bitter orange, sour orange	Fr	B, C, V	80, 86–88, 91, 97–99	Dugo et al. 1999, 2000; Guo et al. 2000
Citrus grandis (L.) Osbeck	Pummelo	Fr	B, C, V	86–88, 97–99	Guo et al. 2000
C. limon (L.) Burm. f.	Lemon	Fr	B, C, V	83–86, 89–91, 94–96	Shu et al. 1975; Verzera et al. 1999; Dugo et al. 1998, 1999; Andrea et al. 2003
Citrus paradisi Macfad.	Grapefruit	Fr	B, C, V	80, 86–88, 91, 97–99	Shu et al. 1975; He et al. 1998; Dugo et al. 1999, 2000; Guo et al. 2000; Ohnishi et al. 2000; Tassaneeyakul et al. 2000
Citrus sinensis (L.) Osbeck	Sweet orange	Fr	B, C, V	80	Fisher and Trama 1979; [d]

[a]R, roots; T, tubers; L, leaves; St, stems; Sh, shoots; Fr, fruits; S, seeds. [b]V, vegetable; C, condiment or flavouring; B, beverage (Yamaguchi 1983; Pemberton and Lee 1996; Rubatzky et al. 1999). [c]According to some investigations carrots lack the ability to produce psoralens (Beier et al. 1983a; Ivie et al. 1982). [d]Christensen, L.P., unpublished results.

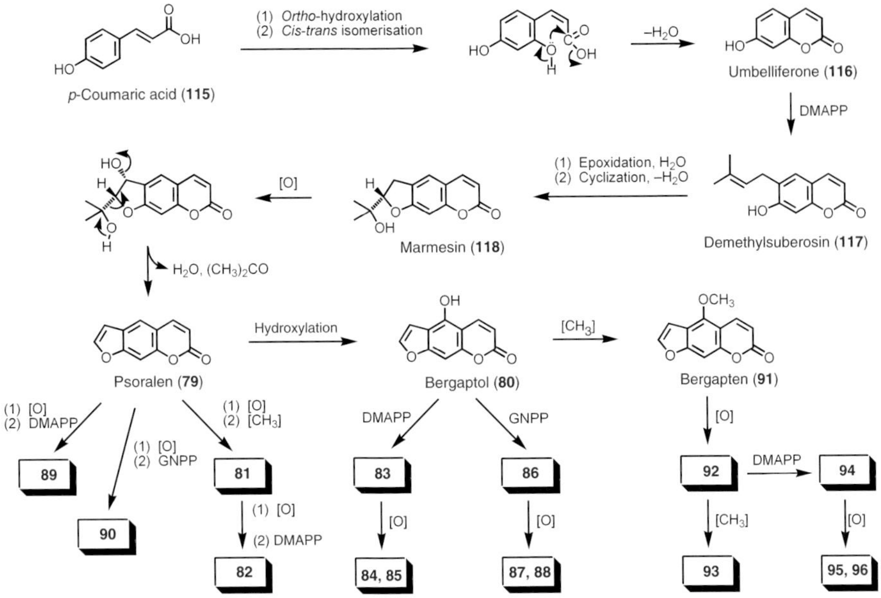

Figure 5.17 The possible biosynthetic route for selected linear furanocoumarins (psoralens) isolated from the utilized parts of major and/or minor food plants. [O] = oxidation, [CH$_3$] = methylation due to the cofactor S-adenosyl methionine, DMAPP = dimethylallyl pyrophosphate, GNPP = geranyl pyrophosphate.

and its oxygenated derivatives (**87**, **88**) (Figure 5.17). Looking at the structures of the LFCs paradisin A (**97**), paradisin B (**98**) and paradisin C (**99**) isolated from various citrus plants (Table 5.2 and Figure 5.18) it is obvious that these dimers are derived from epoxybergamottin hydrate (**88**).

5.3.2 Bioactivity

5.3.2.1 Phototoxic effects

The biological activities of LFCs are extremely diverse, due to their ability to intercalate into RNA (Talib and Banerjee 1982) and in particular DNA (Parsons 1980; Beier and Nigg 1992) where they form covalent bonds in the presence of UVA light (320–400 nm). The photoreaction of LFCs with DNA is the final result of a multistage process. The initial step is the formation of an intercalative complex between the LFCs and the nucleic acids of the DNA in a dark reaction. On exposure to UVA radiation, the intercalated LFCs can react by a [2 + 2]cycloaddition, at either the 3,4 double bond of the pyrone ring or the 4′,5′ double bond of the furan ring, with the 5,6 double bond of pyrimidine bases in DNA resulting in mono- and/or diadducts yielding interstrand cross-links (Parsons 1980; Christensen and Lam 1990), as shown in Figure 5.19.

Because LFCs are potent photoactive compounds, they have been used clinically to treat human skin disorders such as skin depigmentation (vitiligo), psoriasis, mycosis fungoides, polymorphous photodermatitis and eczema. About 40 mg of psoralen (**79**) administered

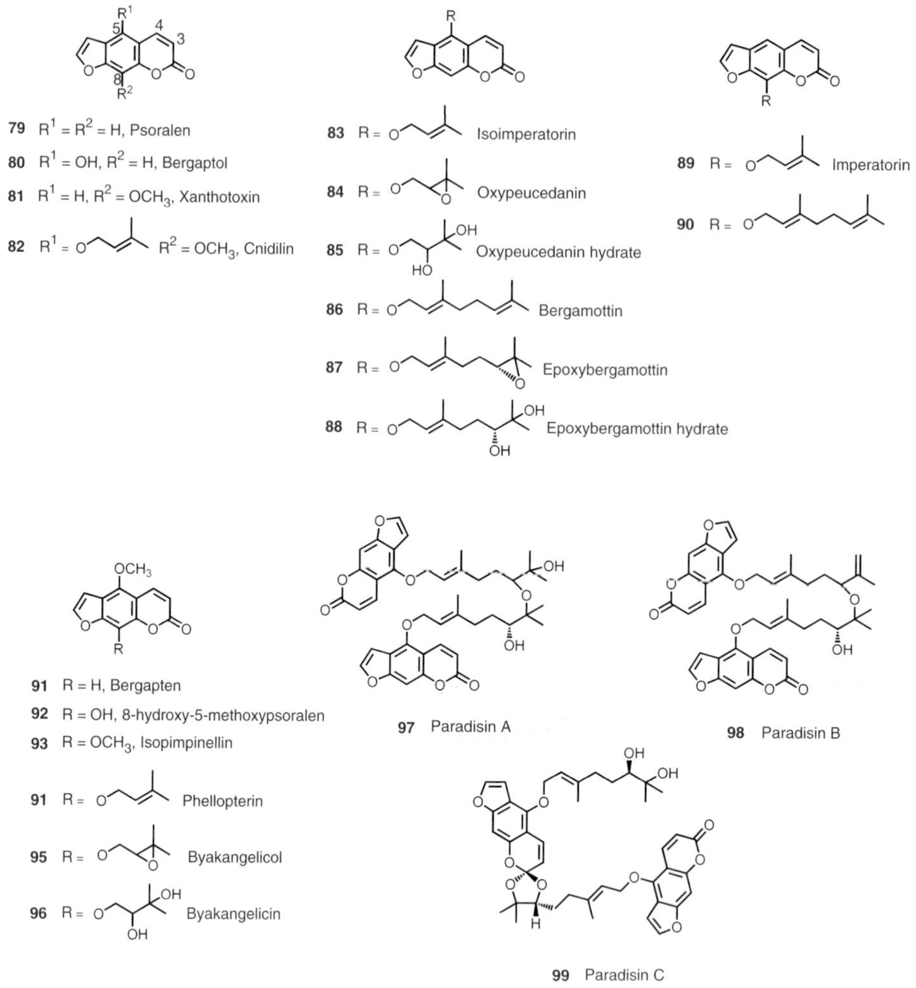

79 R^1 = R^2 = H, Psoralen
80 R^1 = OH, R^2 = H, Bergaptol
81 R^1 = H, R^2 = OCH$_3$, Xanthotoxin
82 R^1 = O~~, R^2 = OCH$_3$, Cnidilin

83 R = O~~ Isoimperatorin
84 R = O~~ Oxypeucedanin
85 R = O~~OH, HO Oxypeucedanin hydrate
86 R = O~~ Bergamottin
87 R = O~~ Epoxybergamottin
88 R = O~~OH, OH Epoxybergamottin hydrate

89 R = O~~ Imperatorin
90 R = O~~

91 R = H, Bergapten
92 R = OH, 8-hydroxy-5-methoxypsoralen
93 R = OCH$_3$, Isopimpinellin

97 Paradisin A

98 Paradisin B

94 R = O~~ Phellopterin
95 R = O~~ Byakangelicol
96 R = O~~OH, OH Byakangelicin

99 Paradisin C

Figure 5.18 Linear furanocoumarins (psoralens) isolated from the utilized parts of well-known major and/or minor food plants.

orally combined with UVA light is normally referred to as PUVA therapy, but other psoralens such as xanthotoxin (**81**) are also frequently used in PUVA treatment (Beier and Nigg 1992; Pathak and Fitzpatrick 1992). PUVA treatment controls excess cell division in the skin by virtue of its ability to damage DNA. The lesions in DNA caused by PUVA lead to inhibition of DNA synthesis, erythema production and increased skin pigmentation as well as mutation, chromosome aberrations, inactivation of viruses and inhibition of tumour transmitting capacity of certain tumour cells. Consequently, the use of furanocoumarins in medicine has been associated with higher incidence of skin cancer and other forms of cancer. A number of studies have demonstrated that concentrations of LFCs used for medical treatment are carcinogenic, mutagenic and photodermatitic. For example PUVA therapy has caused papillomas, keratoanthomas and squamous cell carcinomas in

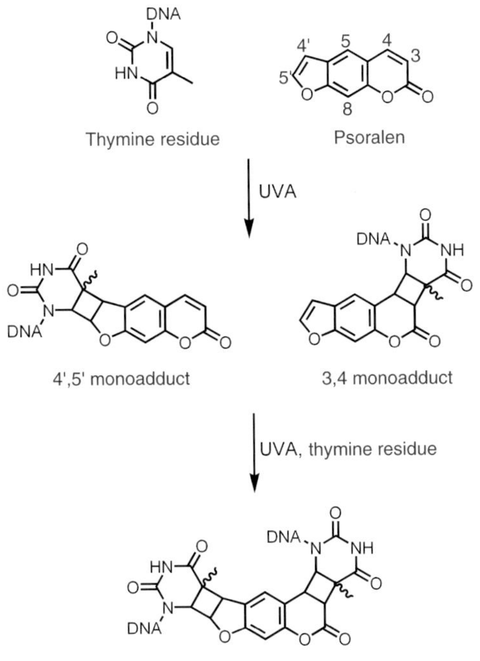

Figure 5.19 The possible mechanism for the reaction of linear furanocoumarins (psoralens) with nucleobases in DNA (e.g. thymine residues) under the influence of UVA radiation resulting in the formation of interstrand cross-links in DNA that may lead to photodermatitis and/or other health disorders such as cancer.

mice (Hannuksela et al. 1986), and there are strong indications that PUVA has resulted in photocarcinogenicity in humans (Stern et al. 1979; Young, 1990); and furthermore PUVA therapy has been linked to genital cancer in men (Beier and Nigg 1992).

The LFCs that appears to have most biochemical importance are psoralen, xanthotoxin, bergapten (**91**) and isopimpinellin (**93**). Usually, the sum of the first three psoralens is used as an estimate of the phototoxicity of the plant (Diawara et al. 1995). Isopimpinellin is much less bioactive, and is not considered to be phototoxic, nor are the angular furanocoumarins. Celery has been among the most extensively studied vegetables for LFCs, such as psoralen, xanthotoxin and bergapten, because of the potential for high concentrations of these compounds in particular in diseased plants. Healthy celery may contain up to 84 mg/kg (Finkelstein et al. 1994), which means that even for daily consumers of celery, the intake is an order of magnitude lower than when used for therapy, but microbial infection of celery may raise the content of LFCs up to 1 g/kg (Scheel et al. 1963). Skin exposure to plant material with such high concentrations of LFCs can induce photosensitization. Photodermatitis of the fingers, hands and forearms is therefore a known occupational risk for celery handlers and field workers (Beier et al. 1983a; Beier and Nigg 1992; Nigg et al. 1997). These skin disorders are referred to as celery dermatitis, celery itch, or celery blisters, and can be caused by LFCs of both healthy and diseased celery. Other food plants that are known to cause photodermatitis due to their relatively high content of LFCs are parsley, parsnip and various citrus species (Beier and Nigg 1992; Izumi and Dawson 2002).

5.3.2.2 Inhibition of human cytochrome P450

Grapefruit juice has been shown to interact with various clinically important drugs that differ in their chemical and pharmacological properties but are all extensively metabolized by a form of cytochrome P450 3A (CYP3 A4) in humans. The main interaction of grapefruit juice with these drugs has been shown to be an enhancement of their bioavailabilities through its inhibition of CYP3 A4 (Dresser and Bailey 2003). The causative components in grapefruit juice have been identified to be the LFCs **86–88** and **97–99** (Figure 5.18), and, in particular, the major LFC of grapefruit juice epoxybergamottin hydrate (**88**) and the LFC dimers **97–99** (He *et al.* 1998; Guo *et al.* 2000; Ohnishi *et al.* 2000; Tassaneeyakul *et al.* 2000; Guo and Yamazoe 2004). The inhibition of CYP3 A4 in the gastrointestinal mucosa in vivo thus increases the circulating levels of a various drugs including cyclosporine, felodipine, nisoldifine, midazolam, diazepam, terfenadine and quinidine (Clifford 2000; Ohnishi *et al.* 2000) to levels that cause undesirable or even dangerous side effects. The recognition of this phenomenon has led to the recommendation that patients receiving such drugs should avoid grapefruit and related fruits and associated products in accordance with the presence of the causative agents in other citrus fruits (Table 5.2). Other LFCs, including psoralen (**79**), xanthotoxin (**81**), and bergapten (**91**), have been shown to inhibit two isoforms of P450 in vitro (2A6 and 2B1), and it is possible that some LFCs other than epoxybergamottin hydrate and related psoralens also are active against CYP3 A4 (Clifford 2000).

5.3.2.3 Reproductive toxicity

It has recently been demonstrated that administration of high levels of LFCs such as bergapten (**91**) and xanthotoxin (**81**) in the diet of female rats reduced birth-rates, number of implantation sites, pups, corpora lutea, full and empty uterine weight, and circulating estrogen levels in a dose-dependent manner (Diawara and Kulkosky 2003), and a strong aversion to LCF-containing food, indicating nausea or other discomfort (Scalera 2002). Bergapten and xanthotoxin further induced mRNA of the liver enzyme CYP1 A1 and UGT1 A6, suggesting that enhanced metabolism of estrogens by the treatment of these psoralens may explain the reproductive toxicity and the observed reduction of ovarian follicular function and ovulation. The results from the investigation of Diawara and Kulkosky (2003) suggest, in addition to the other known health hazards associated with PUVA therapy, a potential risk to female and male fertility and capacity. This is in accordance with previous studies (Diawara *et al.* 1997a,b). Consequently, the potential risks to humans should be evaluated to ensure continued safe use of LFCs in PUVA therapy and in the diet, and the possibility of hormesis should be investigated, to determine if these compounds may benefit reproduction at the low intake levels that can be obtained from food.

5.3.2.4 Antifungal and antibacterial effects

The LFCs psoralen (**79**), xanthotoxin (**81**), bergapten (**91**) and isopimpinellin (**93**) (Figure 5.18) are antibacterial when combined with UV light (Beier *et al.* 1983a; Manderfeld *et al.* 1997; Ulate-Rodríguez *et al.* 1997), whereas compounds **79** and **81** also have some antibacterial activity without UV light (Beier *et al.* 1983a). The antifungal properties of LFCs are well-known, in particular those present in celery (**79, 81, 91** and **93**, Table 5.2), where they act as phytoalexins (Beier and Oertli 1983; Beier *et al.* 1983a; Heath-Pagliuso *et al.* 1992). Their phytoalexin behaviour have resulted in increased levels in celery in response to infection by fungi (Heath-Pagliuso *et al.* 1992) as well as general elicitors such as Hg^{2+} and Cu^{2+}-ions, UV light, fungicides, herbicides and polyamines (Beier *et al.* 1983a; Nigg *et al.* 1997), although some types of pesticides have little effect on raising the LFC levels in celery (Trumble *et al.* 1992). Further, atmospheric pollution also appears to affect the LFCs content up to 540% of normal levels (Dercks *et al.* 1990). The increased levels of LFCs in diseased celery or celery subjected to environmental pollution, therefore, constitute a health risk not only for workers and handlers but also to consumers, as described in Section 5.3.2.1, if the material is used in combined dishes where the unpleasant taste of high concentrations of LFCs is masked by other food ingredients.

5.4 Perspectives in relation to food safety

As is evident from the present review, food plants (mostly vegetables) contain a variety of highly bioactive acetylenes and linear furanocoumarins (LFCs), of which many occur in concentrations high enough to potentially affect our health, either negatively or positively. By implication this must also be the case for the food we eat, although hardly any data are available from actual foodstuffs, for example, cooked vegetables or other processed food. The reassuring epidemiological evidence that high vegetable intake is not a risk factor shows that there is no cause for immediate concern regarding unknown toxic effects of these compounds. However, it is both scientifically unsatisfying and potentially a threat to public health that at present we do not know if vegetables are good for health because of these bioactive compounds or despite them.

With regard to food safety, it is necessary to prevent the occurrence of harmful concentrations of toxicants in food. However, it is at least equally important to ensure that modern methods of food production do not inadvertently lead to reductions or loss of important health promoting components. In recent years several food constituents have changed status from toxicants to beneficial compounds, at least at moderate intake levels, for example glucosinolates, phytoestrogens and even ethanol (Kris-Etherton *et al.* 2002). For each of these compounds, significant efforts have at times been expended to try to minimize or even eliminate the dietary exposure, while the food safety efforts are now much more targeted towards control of (relatively infrequent) situations that may lead to harmful, excessive intake. A similar tendency to consider data in the context of hormesis is developing in other areas of toxicology (Calabrese and Baldwin 2003; Calabrese 2004).

Except for the example of falcarinol, no systematic attempts have been made to determine the net impact of these compounds on health. Until now, almost all experiments have been 'one-sided' designs where either toxic or beneficial effects could be assessed but not both. Another problem has been that the dose-response dependencies have often been assumed to be linear, so the range of concentrations used in the studies did not cover levels corresponding to the actual normal intake from food, since the possibility of hormesis has only recently received significant attention. A third issue, which is generally applicable for these natural bioactive compounds, is the ability of humans to detect and reject high concentrations of acetylenes or LFCs, due to their unpleasant effect on the taste of food: elevated levels of falcarinol and falcarindiol in carrots result in a distinctly bitter taste, causing rejection of such lots by food producers with effective quality control procedures (Czepa and Hofmann 2003; 2004). Rats also resist eating feed with high levels of LFCs (Scalera 2002; Diawara and Kulkosky 2003). Education of food professionals and the general public about the importance of diligence when selecting and preparing vegetables could thus be a very effective method to ensure food safety: simply checking the taste, before you eat, sell or serve the food, in particular if the product shows signs of rot, mould or other stress, could prevent most known or presumed cases of harm to humans from natural bioactive plant compounds.

Future research in relation to food safety must therefore be based on the improved understanding of:

1. The dose-response relationships between bioactive compounds (or their mixtures) and relevant biological responses.
2. The actual intake and bioavailability of the bioactive compounds or other measures of actual food related exposure (amount entering sensitive tissues).
3. The factors influencing the concentrations in food throughout the supply chain, from plant production through harvest, storage and processing until handling by the consumer.

In the meantime until this has been achieved, food safety is best served by active resistance against the tendencies to over-excitement about single findings, indicating either positive or negative effects on health. At the present (low) state of knowledge we must insist on adhering to the proven message about food and health: promote a diverse diet with many different vegetables, including attention to simple, logical quality criteria, such as 'don't eat it if the taste is not as good as normal'. Setting and enforcement of low safety limits, or commercial promotion of poorly documented health claims, will at best be very costly, with little chance of improving the benefits of food on health. Under some circumstances either trend may even reduce food safety, in particular if the science behind them is subsequently found to be inadequate.

Only when we thoroughly understand the consequences can we start to try to improve the role of acetylenes and LFCs in out diet, whether this implies promotion of reductions or increases in the intake, or a combination of specific targeted changes.

References

Adikaram, N.K.B., Ewing, D.F., Karunaratne, A.M. et al. (1992) Antifungal compounds from immature avocado fruit peel. *Phytochemistry*, 31, 93–96.

Afek, U., Aharoni, N. and Carmeli, S. (1995) The involvement of marmesin in celery resistance to pathogens during storage and the effect of temperature on its concentration. *Phytophatholology*, 85, 711–714.

Afek, U., Orenstein, J., Carmeli, S. et al. (2002) Marmesin, a new phytoalexin associated with resistance of parsley to pathogens after harvesting. *Postharv. Biol. Technol.*, 24, 89–92.

Ahn, J.C., Hwang, B., Tada, H. et al. (1996) Polyacetylenes in hairy roots of *Platycodon grandiflorum*. *Phytochemistry*, 42, 69–72.

Ahn, B.-Z. and Kim, S.-I. (1988) Beziehung zwischen Struktur und cytotoxischer Aktivität von Panaxydol-Analogen gegen L1210 zellen. *Arch. Pharm.*, **321**, 61–63.

Al-Douri, N.A. and Derwick, P.M. (1986) Biosynthesis of the furanoacetylene phytoalexin wyerone in *Vicia faba*. *Z. Naturforsch.*, **41c**, 34–38.

Alanko, J., Kurahashi, Y., Yoshimoto, T. *et al.* (1994) Panaxynol, a polyacetylene compound isolated from oriental medicines, inhibits mammalian lipoxygenases. *Biochem. Pharmacol.*, **48**, 1979–1981.

Andrea, V., Nadia, N., Theresa, R.M. *et al.* (2003) Analysis of some Italian lemon liquors (Limoncello). *J. Agric. Food Chem.*, **51**, 4978–4983.

Anet, E.F.L.J., Lythgoe, B., Silk, M.H. *et al.* (1953) Oenanthotoxin and cicutoxin. isolation and structure. *J. Chem. Soc.*, 309–322.

Appendino, G., Tagliapietra, S. and Nano, G.M. (1993) An anti-platelet acetylene from the leaves of *Ferula communis*. *Fitoterapia*, **64**, 179.

Avalos, J., Fontan, G.P. and Rodriguez, E. (1995) Simultaneous HPLC quantification of two dermatoxins, 5-methoxypsoralen and falcarinol, in healthy celery. *J. Liq. Chromatogr.*, **18**, 2069–2076.

Avato, P. and Tava, A. (1995) Acetylenes and terpenoids of *Bellis perennis*. *Phytochemistry*, **40**, 141–147.

Avato, P., Vitali, C., Mongelli, P. *et al.* (1997) Antimicrobial activity of polyacetylenes from *Bellis perennis* and their synthetic derivatives. *Planta Med.*, **63**, 503–507.

Barley, G.C., Jones, E.R.H. and Thaller, V. (1988) Crepenynate as a precursor of falcarinol in carrot tissue culture. In J. Lam, H. Breteler, T. Arnason *et al.* (eds), *Chemistry and Biology of Naturally-Occurring Acetylenes and Related Compounds (NOARC)*. Elsevier, Amsterdam, pp. 85–91.

Beier, R.C., Ivie, G.W. and Oertli, E.H. (1983a) Psoralens as phytoalexins in food plants of the family of the Umbelliferae – significance in relation to storage and processing. In J.W. Finley (ed.), *Xenobiotics in Foods and Feeds*; ACS Symposium Series, Vol. 234. American Chemical Society, Washington DC, pp. 295–310.

Beier, R.C., Ivie, G.W., Oertli, E.H. *et al.* (1983b) HPLC analysis of linear furanocoumarins (psoralens) in healthy celery (*Apium graveolens*). *Food Chem. Toxicol.*, **21**, 163–165.

Beier, R.C. Ivie, G.W. and Oertli, E.H. (1994) Linear furanocoumarins and graveolone from the common herb parsley. *Phytochemistry*, **36**, 869–872.

Beier, R.C. and Nigg, H.N. (1992) Natural toxicants in food. In H.N. Nigg and D. Siegler (eds), *Phytochemical Resources for Medicine and Agriculture*. Plenum Press, New York, pp. 247–367.

Beier, R.C. and Oertli, E.H. (1983) Psoralen and other linear furanocoumarins as phytoalexins in celery. *Phytochemistry*, **22**, 2595–2597.

Bentley, R.K., Bhattacharjee, D., Jones, E.R.H. *et al.* (1969a) Natural acetylenes. Part XXVIII. C_{17}-Polyacetylenic alcohols from the Umbellifer *Daucus carota* L. (carrot): Alkylation of benzene by acetylenyl(vinyl)carbinols in the presence of toluene-*p*-sulphonic acid. *J. Chem. Soc. (C)*, 685–688.

Bentley, R.K., Jones, E.R.H. and Thaller, V. (1969b) Natural acetylenes. Part XXX. Polyacetylenes from *Lactuca* (lettuce) species of the Liguliflorae sub family of the Compositae. *J. Chem. Soc. (C)*, 1096–1099.

Bernart, M.W., Cardellina II, J.H., Balaschak, M.S. *et al.* (1996) Cytotoxic falcarinol oxylipins from *Dendropanax arboreus*. *J. Nat. Prod.*, **59**, 748–753.

Bohlmann, F. (1967) Notiz über die Inhaltsstoffe von Petersilie- und Sellerie-Wurzeln. *Chem. Ber.*, **100**, 3454–3456.

Bohlmann, F. (1988) Naturally-occurring acetylenes. In J. Lam, H. Breteler, T. Arnason *et al.* (eds), *Chemistry and Biology of Naturally-Occurring Acetylenes and Related Compounds (NOARC)*. Elsevier, Amsterdam, pp. 1–19.

Bohlmann, F., Arndt, C., Bornowski, H. et al. (1962) Neue Polyine aus dem Tribus Anthemideae. Chem. Ber., 95, 1320–1327.

Bohlmann, F., Arndt, C., Bornowski, H. et al. (1961) Über Polyine aus der Familie der Umbelliferen. Chem. Ber., 94, 958–967.

Bohlmann, F., Burkhardt, T. and Zdero, C. (1973) Naturally Occurring Acetylenes. Academic Press, London.

Bohlmann, F. and Fritz, U. (1979) Neue Lyratolester aus Chrysanthemum coronarium. Phytochemistry, 18, 1888–1889.

Brandt, K., Christensen, L.P., Hansen-Møller, J. et al. (2004) Health promoting compounds in vegetables and fruits. A systematic approach for identifying plant components with impact on human health. Trends Food Sci. Technol., 15, 384–393.

Bu'Lock, J.D. and Smalley, H.M. (1962) The biosynthesis of polyacetylenes. Part V. The role of malonate derivatives, and the common origin of fatty acids, polyacetylenes, and acetate-derived phenols. J. Chem. Soc., 4662–4664.

Bu'Lock, J.D. and Smith, G.N. (1967) The origin of naturally-occurring acetylenes. J. Chem. Soc. (C), 332–336.

Buzi, A., Chilos, G., Timperio, A.M. et al. (2003) Polygalacturonase produced by Botrytis fabae as elicitor of two furanoacetylenic phytoalexins in Vicia faba pods. J. Plant Pathol., 85, 111–116.

Cahoon, E.B., Schnurr, J.A. and Huffman, E.A. (2003) Fungal responsive fatty acid acetylenases occur widely in evolutionarily distant plant families. Plant J., 34, 671–683.

Cain, R.O. and Porter, A.E.A. (1979) Biosynthesis of the phytoalexin wyerone in Vicia faba. Phytochemistry, 18, 322–323.

Calabrese, E.J. (2004) Hormesis: from marginalization to mainstream. A case for hormesis as the default dose-response model in risk assessment. Toxicol. Appl. Pharmacol., 197, 125–136.

Calabrese, E.J. and Baldwin, L.A. (1998) Hormesis as a biological hypothesis. Eviron. Health Perspect., 106 (Suppl. 1), 357–362.

Calabrese, E.J. and Baldwin, L.A. (2003) The hormetic dose-response model is more common than the threshold model in toxicology. Toxicol. Sci., 71, 246–250.

Camm, E.L., Wat, C.-K. and Towers, G.H.N. (1976) An assessment of the roles of furanocoumarins in Heracleum lanatum. Can. J. Bot., 54, 2562–2566.

Cascon, S.C., Mors, W.B., Tursch, B.M. et al. (1965) Ichthyothereol and its acetate, the active polyacetylene constituents of Ichthyothere terminalis (Spreng.) Malme, a fish poison from the lower Amazon. J. Am. Chem. Soc., 87, 5237–5241.

Ceska, O., Chaudhary, S.K. and Warrington, P.J. (1986a) Furocoumarins in the cultivated carrot, Daucus carota. Phytochemistry, 25, 81–84.

Ceska, O., Chaudhary, S.K., Warrington, P.J. et al. (1986b) Naturally-occurring crystals of photocarcinogenic furocoumarins on the surface of parsnip roots sold as food. Experientia, 42, 1302–1304.

Ceska, O., Chaudhary, S.K., Warrington, P.J. et al. (1987) Photoactive furocoumarins in fruits of some umbellifers. Phytochemistry, 26, 165–169.

Christensen, L.P. (1992) Acetylenes and related compounds in Anthemideae. Phytochemistry, 31, 7–49.

Christensen, L.P. (1998) Biological activities of naturally occurring acetylenes and related compounds from higher plants. Recent Res. Devel. Phytochem., 2, 227–257.

Christensen, L.P. and Lam, J. (1990) Acetylenes and related compounds in Cynareae. Phytochemistry, 29, 2753–2785.

Christensen, L.P. and Lam, J. (1991a) Acetylenes and related compounds in Heliantheae. Phytochemistry, 30, 11–49.

Christensen, L.P. and Lam, J. (1991b) Acetylenes and related compounds in Astereae. Phytochemistry, 30, 2453–2476.

Clifford, M.N. (2000) Miscellaneous phenols in foods and beverages – nature, occurrence and dietary burden. *J. Sci. Food Agric.*, **80**, 1126–1137.

Crosby, D.G. and Aharonson, N. (1967) The structure of carotatoxin, a natural toxicant from carrot. *Tetrahedron*, **23**, 465–472.

Cunsolo, F., Ruberto, G., Amico, V. *et al.* (1993) Bioactive metabolites from sicilian marine fennel, *Crithium maritimum*. *J. Nat. Prod.*, **56**, 1598–1600.

Czepa, A. and Hofmann, T. (2003) Structural and sensory characterization of compounds contributing to the bitter off-taste of carrots (*Daucus carota* L.) and carrot pure. *J. Agric. Food Chem.*, **51**, 3865–3873.

Czepa, A. and Hofmann, T. (2004) Quantitative studies and sensory analyses on the influence of cultivar, spatial tissue distribution, and industrial processing on the bitter off-taste of carrots (*Daucus carota* L.) and carrot products. *J. Agric. Food Chem.*, **52**, 4508–4514.

Degen, T., Buser, H.-R. and Städler, E. (1999) Patterns of oviposition stimulants for carrot fly in leaves of various host plant. *J. Chem. Ecol.*, **25**, 67–87.

Dercks, W., Trumble, J.T. and Winther, C. (1990) Impact of atmospheric pollution on linear furanocoumarin content in celery. *J. Chem. Ecol.*, **16**, 443-454.

De Wit, P.J.G.M. and Kodde, E. (1981) Induction of polyacetylenic phytoalexins in *Lycopersicon esculentum* after inoculation with *Cladosporium fulvum* (syn. *Fulvia fulva*). *Physiol. Plant Pathol.*, **18**, 143–148.

Diawara, M.M., Allison, T., Kulkosky, P. *et al.* (1997a) Mammalian toxicity of bergapten and xanthotoxin, two compounds used in skin photochemotherapy. *J. Nat. Toxins*, **6**, 183–192.

Diawara, M.M., Allison, T., Kulkosky, P. *et al.* (1997b) Psoralen-induced growth inhibition in Wistar rats. *Cancer Lett.*, **114**, 159–160.

Diawara, M.M. and Kulkosky, P.J. (2003) Reproductive toxicity of the psoralens. *Pediatr. Pathol. Mol.*, **22**, 247–258.

Diawara, M.M., Trumble, J.T., Quiros, C.F. *et al.* (1995) Implications of distribution of linear furanocoumarins within celery. *J. Agric. Food Chem.*, **43**, 723–727.

Dresser, G.K. and Bailey, D.G. (2003) The effects of fruit juices on drug disposition: a new model for drug interactions. *Eur. J. Clinic. Invest.*, **33** (Suppl. 2), 10–16.

Dugo, P., Mondello, L., Dugo, L. *et al.* (1998) On the genuineness of citrus essential oils. Part LIII. Determination of the composition of the oxygen heterocyclic fraction of lemon essential oils (*Citrus limon* (L.) Burm. F) by normal-phase high performance liquid chromatography. *Flavour Fragr. J.*, **13**, 329–334.

Dugo, P., Mondello, L., Dugo, L. *et al.* (2000) LC-MS for the identification of oxygen heterocyclic compounds in citrus essential oils. *J. Pharm. Biomed. Anal.*, **24**, 147–154.

Dugo, P., Mondello, L., Sebastiani, E. *et al.* (1999) Identification of minor oxygen heterocyclic compounds of citrus essential oils by liquid chromatography-atmospheric pressure chemical ionisation mass spectrometry. *J. Liq. Chromatogr. Relat. Technol.*, **22**, 2991–3005.

Eckenbach, U., Lampman, R.L., Seigler, D.S. *et al.* (1999) Mosquitocidal activity of acetylenic compounds from *Cryptotaenia canadensis*. *J. Chem. Ecol.*, **25**, 1885–1893.

Elgersma, D.M., Weijman, A.C.M., Roeymans, H.J. *et al.*(1984) Occurrence of falcarinol and falcarindiol in tomato plants after infection with *Verticillium albo-atrum* and charcterization of four phytoalexins by capillary gas chromatography-mass spectrometry. *Phytopathol. Z.*, **109**, 237–240.

Engelmeier, D., Hadacek, F., Hofer, O. *et al.* (2004) Antifungal 3-butylisocoumarins from Asteraceae-Anthemideae. *J. Nat. Prod.*, **67**, 19–25.

Fahmy, I.R., Abushady, H., Schonberg, A. *et al.* (1947) A crystalline principle from *Ammi majus* L. *Nature*, **160**, 468–469.

Fawcett, C.H., Firn, R.D. and Spencer, D.M. (1971) Wyerone increase in leaves of broad bean (*Vicia faba* L.) after infection by *Botrytis fabae*. *Physiol. Plant Pathol.*, **1**, 163–166.

Fawcett, C.H., Spencer, D.M., Wain, R.L. *et al.* (1968) Natural acetylenes. Part XXVII. An antifungal acetylenic furanoid keto-ester (wyerone) from shoots of the broad bean (*Vicia faba* L.; fam. Papilionaceae), *J. Chem. Soc.* (C), 2455–2462.

Finkelstein, E., Afek, U., Gross, E. *et al.* (1994) An outbreak of phytophotodermatitis due to celery. *Int. J. Dermatol.*, 33, 116–118.

Fisher, J.F. and Trama, L.A. (1979) High-performance liquid chromatographic determination of some coumarins and psoralens found in citrus peel oil. *J. Agric. Food Chem.*, 27, 1334–1337.

Fujimoto, Y., Sakuma, S., Komatsu, S. *et al.* (1998) Inhibition of 15-hydroxyprostaglandin dehydrogenase activity in rabbit gastric antral mucosa by panaxynol isolated from oriental medicines. *J. Pharm. Pharmacol.*, 50, 1075–1078.

Fujimoto, Y. and Satoh, M. (1988) A new cytotoxic chlorine-containing polyacetylene from the callus of *Panax ginseng*. *Chem. Pharm. Bull.*, 36, 4206–4208.

Fujita, T., Kadoya, Y., Aota, H. *et al.* (1995) A new phenylpropanoid glucoside and other constituents of *Oenanthe javanica*. *Biosci. Biotechnol. Biochem.*, 59, 526–528.

Garrod, B., Lewis, B.G. and Coxon, D.T. (1978) *Cis*-heptadeca-1,9-diene-4,6-diyne-3,8-diol, an antifungal polyacetylene from carrot root tissue. *Physiol. Plant Pathol.* 13, 241–246.

Greenberg, E.R., Baron, J.A., Karagas, M.R. *et al.* (1996) Mortality associated with low plasma concentration of β-carotene and the effect of oral supplementation. *JAMA*, 275, 699–703.

Greenvald, P., Clifford C.K. and Milner, J.A. (2001) Diet and cancer prevention. *Eur. J. Cancer*, 37, 948–965.

Greger, H. (1979) Aromatic acetylenes and dehydrofalcarinone derivatives within the *Artemisia dracunculus* group. *Phytochemistry*, 18, 1319–1322.

Gundgaard, J., Nielsen, J.N., Olsen, J. *et al.* (2003) Increased intake of fruit and vegetables: estimation of impact in terms of life expectancy and healthcare costs. *Public Health Nutr.*, 6, 25–30.

Guo, L.-Q., Fukuda, K., Ohta, T. *et al.* (2000) Role of furanocoumarin derivatives on grapefruit juice-mediated inhibition of human CYP3A activity. *Drug Metabol. Disposit.*, 28, 766–771.

Guo, L.-Q. and Yamazoe, Y. (2004) Inhibition of cytochrome P450 by furanocoumarins in grapefruit juice and herbal medicines. *Acta Pharmacol. Sin.*, 25, 129–136.

Hamerski, D. and Matern, U. (1988a) Elicitor-induced biosynthesis of psoralens in *Ammi majus* L. suspension cultures. Microsomal conversion of demethylsuberosin into (+)-marmesin and psoralen. *Eur. J. Biochem.*, 171, 369–375.

Hamerski, D. and Matern, U. (1988b) Biosynthesis of psoralens. Psoralen 5-monooxygenase activity from elicitor-treated *Ammi majus* cells. *FEBS Lett.*, 239, 263–265.

Hannuksela, M., Stenback, F. and Lahti, A. (1986) The carcinogenic properties of topical PUVA. *Arch. Dermatol. Res.*, 278, 347–351.

Hansen, L. and Boll, P. M. (1986) Polyacetylenes in Araliaceae: their chemistry, biosynthesis and biological significance. *Phytochemistry*, 25, 285–293.

Hansen, L., Hammershøy, O. and Boll, P.M. (1986) Allergic contact dermatitis from falcarinol isolated from *Schefflera arboricola*. *Contact Dermatitis*, 14, 91–93.

Hansen-Møller, J., Hansen, S.L., Christensen, L.P. *et al.* (2002) Quantification of polyacetylenes by LC–MS in human plasma after intake of fresh carrot juice (*Daucus carota* L.). In *International Symposium on Dietary Phytochemicals and Human Health*, Salamanca, Spain, 18–20 April 2002, pp. 203–204.

Hansen, S.L., Purup, S. and Christensen, L.P. (2003) Bioactivity of falcarinol and the influence of processing and storage on its content in carrots (*Daucus carota* L.). *J. Sci. Food Agric.*, 83, 1010–1017.

Haraldsdóttir, J., Jespersen, L., Hansen-Møller, J. *et al.* (2002) Recent developments in bioavailability of falcarinol. In K. Brandt and B. Åkesson (eds), *Health Promoting Compounds in Vegetables and*

Fruit; Proceedings of workshop in Karrebæksminde, Denmark, 6–8 November; DIAS report – Horticulture, **29**, 24–28.

Harding, V.K. and Heale, J.B. (1980) Isolation and identification of the antifungal compounds accumulation in the induced resistance response of carrot slices to *Botrytis cinerea. Physiol. Plant Pathol.*, **17**, 277–289.

Harding, V.K. and Heale, J.B. (1981) The accumulation of inhibitory compounds in the induced resistance response of carrot root slices to *Botrytis cinerea. Physiol. Plant Pathol.*, **18**, 7–15.

Hargreaves, J.A., Mansfield, J.W. and Coxon, D.T. (1976a) Conversion of wyerone to wyerol by *Botrytis cinerea* and *B. fabae* in vitro. *Phytochemistry*, **15**, 651–653.

Hargreaves, J.A., Mansfield, J.W., Coxon, D.T. *et al.* (1976b) Wyerone epoxide as a phytoalexin in *Vicia faba* and its metabolism by *Botrytis cinerea* and *B. fabae* in vitro. *Phytochemistry*, **15**, 1119–1121.

Hausen, B.M. (1988) Allergiepflanzen – Pflanzenallergene: Handbuch und Atlas der allergieinduzierenden Wild- und Kulturpflanzen. Ecomed Verlagsgesellschaft mbH, Landsberg/München.

Hausen, B.M., Bröhan, J., König, W.A. *et al.* (1987) Allergic and irritant contact dermatitis from falcarinol and didehydrofalcarinol in common ivy (*Hedera helix* L.). *Contact Dermatitis*, **17**, 1–9.

He, K., Iyer, K.R., Hayes, R.N. *et al.* (1998) Inactivation of cytochrome P450 3A4 by bergamottin, a component of grapefruit juice. *Chem. Res. Toxicol.*, **11**, 252–259.

Heath-Pagliuso, S., Matlin, S.A., Fang, N. *et al.* (1992) Stimulation of furanocoumarin accumulation in celery and celeriac tissues by *Fusarium oxysporum* f. sp. *Apii*. *Phytochemistry*, **31**, 2683–2688.

Hehmann, M., Lukačin, R., Ekiert, H. *et al.* (2004) Furanocoumarin biosynthesis in *Ammi majus* L. Cloning of bergaptol *O*-methyltransferase. *Eur. J. Biochem.*, **271**, 932–940.

Holz, W. and Miething, H. (1994) Wirkstoffe in wässrigen Kamillenaufgüssen. 2, Mitteilung: Freisetzung von Ätherisch-Öl-Komponenten. *Pharmazie*, **49**, 53–55.

Imoto, S. and Ohta, Y. (1988) Elicitation of diacetylenic compounds in suspension cultured cells of eggplant. *Plant Physiol.*, **86**, 176–181.

Innocenti, G., Dall'Acqua, F. and Caporale, G. (1976) Investigations of the content of furocoumarins in *Apium graveolens* and in *Petroselinum sativum*. *Planta Med.*, **29**, 165–170.

Ivie, G.W., Beier, R.C. and Holt, D.L. (1982) Analysis of the garden carrot (*Daucus carota* L.) for linear furocoumarins (psoralens) at the sub parts per million level. *J. Agric. Food Chem.*, **30**, 413–416.

Izumi, A.K. and Dawson, K.L. (2002) Zabon phytophotodermatitis: First case reports due to *Citrus maxima. J. Am. Acad. Dermatol.*, **46** (5 Suppl.), 146–147.

Jakupovic, J., Tan, R.X., Bohlmann, F. *et al.* (1991) Acetylenes and other constituents from *Artemisia dracunculus*. *Planta Med.*, **57**, 450–453.

Järvenpää, E.P., Jestoi, M.N. and Huopalahti, R. (1997) Quantitative determination of phototoxic furocoumarins in celeriac (*Apium graveolens* L. var. *rapeceum*) using supercritical fluid extraction and high performance liquid chromatography. *Phytochemical Anal.*, **8**, 250–256.

Jente, R., Richter, E., Bosold, F. *et al.* (1988) Experiments on biosynthesis and metabolism of acetylenes and thiophenes. In J. Lam, H. Breteler, T. Arnason *et al.* (eds), *Chemistry and Biology of Naturally-Occurring Acetylenes and Related Compounds (NOARC)*. Elsevier, Amsterdam, pp. 187–199.

Jones, E.R.H., Thaller, V. and Turner, J.L. (1975) Natural acetylenes. Part XLVII. Biosynthetic experiments with the fungus *Lepista-diemii* (Singer) Biogenesis of C_8 acetylenic cyano acid diatretyne 2. *J. Chem. Soc. Perkin Trans.*, I, 424–428.

Kemp, M.S. (1978) Falcarindiol: an antifungal polyacetylene from *Aegopodium podagraria*. *Phytochemistry*, **17**, 1002.

Kobaisy, M., Abramowski, Z., Lermer, L. *et al.* (1997) Antimycobacterial polyynes of Devil's Club (*Oplopanax horridus*), a North American native medicinal plant. *J. Nat. Prod.*, **60**, 1210–1213.

Kobaek-Larsen, M., Christensen, L.P., Vach, W. *et al.* (2005) Inhibitory effect of feeding with carrots or (−)-falcarinol on development of azoxymethane-induced preneoplastic lesions in the rat colon. *J. Agric. Food Chem.*, **53**, 1823–1827.

Konoshima, T. and Lee, K.-H. (1986) Antitumor agents. Cicutoxin, an antileukemic principle from *Cicuta maculata*, and the cytotoxicity of the related derivatives. *J. Nat. Prod.*, **49**, 1117–1121.

Kris-Etherton, P.M., Etherton, T.D., Carlson, J. *et al.* (2002) Recent discoveries in inclusive food-based approaches and dietary patterns for reduction in risk for cardiovascular disease. *Curr. Opin. Lipid*, **13**, 397–407.

Kuo, Y.-C., Lin, Y.L., Huang, C.-P. *et al.* (2002) A tumor cell growth inhibitor from *Saposhnikovae divaricata*. *Cancer Invest.*, **20**, 955–964.

Lombaert, G.A., Siemens, K.H., Pellaers, P. *et al.* (2001) Furanocoumarins in celery and parsnips: method and multiyear Canadien survey. *J. AOAC Int.*, **84**, 1135–1143.

Lund, E.D. (1992) Polyacetylenic carbonyl compounds in carrots. *Phytochemistry*, **31**, 3621–3623.

Lund, E.D. and White, J.M. (1990) Polyacetylenes in normal and water-stressed 'Orlando Gold' carrots (*Daucus carota*). *J. Sci. Food Agric.*, **51**, 507–516.

Machado, S., Silva, E. and Massa, A. (2002) Occupational allergic contact dermatitis from falcarinol. *Contact Dermatitis*, **47**, 113–114.

Manderfeld, M.M., Schafer, H.W., Davidson, P.M. *et al.* (1997) Isolation and identification of antimicrobial furocoumarins from parsley. *J. Food Protect.*, **60**, 72–77.

Mansfield, J.W., Porter, A.E.A. and Smallman, R.V. (1980) Dihydrowyerone derivatives as components of the furanoscetylenic phytoalexin response of tissues of *Vicia faba*. *Phytochemistry*, **19**, 1057–1061.

Matsunaga, H., Katano, M., Yamamoto, H. *et al.* (1990) Cytotoxic activity of polyacetylene compounds in *Panax ginseng* C. A. Meyer. *Chem. Pharm. Bull.*, **38**, 3480–3482.

Matsunaga, H., Katano, M., Yamamoto, H. *et al.* (1989) Studies on the panaxytriol of *Panax ginseng* C. A. Meyer. Isolation, determination and antitumor activity. *Chem. Pharm. Bull.*, **37**, 1279–1281.

Maynard, M., Gunnell, D., Emmett, P. *et al.* (2003) Fruit, vegetables, and antioxidants in childhood and risk of adult cancer: the Boyd Orr cohort. *J. Epidemiol. Community Health*, **57**, 218–225.

McLellan, E.A. and Bird, R.P. (1988) Aberrant crypts – potential preneoplastic lesions in the murine colon. *Cancer Res.*, **48**, 6187–6192.

Michaud, D.S., Feskanich, D., Rimm, E.B. *et al.* (2000) Intake of specific carotenoids and risk of lung cancer in 2 prospective US cohorts. *Am. J. Clin. Nutr.*, **72**, 990–997.

Murdoch, S.R. and Dempster, J. (2000) Allergic contact dermatitis from carrot. *Contact Dermatitis*, **42**, 236.

Nakano, Y., Matsunaga, H., Saita, T. *et al.* (1998) Antiproliferative constituents in Umbelliferae plants II. Screening for polyacetylenes in some Umbelliferae plants, and isolation of panaxynol and falcarindiol from the root of *Heracleum moellendorffii*. *Biol. Pharm. Bull.*, **21**, 257–261.

Nigg, H.N., Nordby, H.E., Beier, R.C. *et al.* (1993) Phototoxic coumarins in limes. *Food Chem. Toxicol.*, **31**, 331–335.

Nigg, H.N., Strandberg, J.O., Beier, R.C. *et al.* (1997) Furanocoumarins in florida celery varieties increased by fungicide treatment. *J. Agric. Food Chem.*, **45**, 1430–1436.

Nitz, S., Spraul, M.H. and Drawert, F. (1990) C_{17} Polyacetylenic alcohols as the major constituents in roots of *Petroselinum crispum* Mill. ssp. *tuberosum*. *J. Agric. Food Chem.*, **38**, 1445–1447.

Oberlies, N.H., Rogers, L.L., Martin, J.M. *et al.* (1998) Cytotoxic and insecticidal constituents of the unripe fruit of *Persea americana*. *J. Nat. Prod.*, **61**, 781–785.

Ohnishi, A., Matsuo, H., Yamada, S. *et al.* (2000) Effect of furanocoumarin derivatives in grapefruit juice on the uptake of vinblastine by Caco-2 cells and on the activity of cytochrome P450 3A4. *Br. J. Pharmacol.*, **130**, 1369–1377.

Olsson, K. and Svensson, R. (1996) The influence of polyacetylenes on the susceptibility of carrots to storage diseases. *J. Phytopathol.*, **144**, 441–447.

Omenn, G.S., Goodmann, G.E., Thornquist, M.D. *et al.* (1996) Effects of a combination of β-carotene and vitamin A on lung cancer and cardiovascular disease. *N. Eng. J. Med.*, **334**, 1150–1155.

O'Neil, M.E., Carroll, Y., Corridan, B. *et al.* (2001) A European carotenoid database to assess carotenoid intakes and its use in a five-country comparative study. *Br. J. Nutr.*, **85**, 499–507.

Parsons, B.J. (1980) Psoralen photochemistry. *Photochem. Photobiol.*, **32**, 813–821.

Pathak, M.A. and Fitzpatrick, T.B. (1992) The evolution of photochemotherapy with psoralens and UVA (PUVA) – 2000 BC to 1992 AD. *J. Photochem. Photobiol.* B, **14**, 3–22.

Patra, A., Ghosh, A. and Mitra, A.K. (1976) Triterpenoids and furanocoumarins of the fruits of *Angelica archangelica*. *Indian J. Chem.*, **14B**, 816–817.

Pemberton, R.W. and Lee, N.S. (1996) Wild food plants in South Korea; market presence, new crops, and exports to the United States. *Econ. Bot.*, **50**, 57–70.

Reichling, J. and Becker, H. (1977) Ein Beitrag zur Analytik des ätherischen Öls aus Flores Chamomillae. *Dtsch Apoth. Ztg.*, **117**, 275–277.

Redaelli, C., Formentini, L. and Santaniello, E. (1981) High performance liquid chromatography of *cis* and *trans* ene-yne-dicycloethers (spiroethers) in *Matricaria chamomilla* L. flowers and in camomile extracts. *J. Chromatogr.*, **209**, 110–112.

Repčák, M., Halásová, J., Hončariv, R. *et al.* (1980) The content and composition of the essential oil in the course of anthodium development in wild camomile (*Matricaria chamomilla* L.). *Biol. Plant.*, **22**, 183–189.

Repčák, M., Imrich, J. and Garčár, J. (1999) Quantitative evaluation of the main sesquiterpenes and polyacetylenes of *Chamomilla recutita* essential oil by high-performance liquid chromatography. *Phytochem. Anal.*, **10**, 335–338.

Robeson, D.J. (1978) Furanoacetylene and isoflavonoid phytoalexins in *Lens culinaris*. *Phytochemistry*, **17**, 807–808.

Robeson, D.J. and Harborne, J.B. (1980) A chemical dichotomy in phytoalexin induction within the tribe *Vicieae* of the Leguminosae. *Phytochemistry*, **19**, 2359–2365.

Rubatzky, V.E., Quiros, C.F. and Simon, P.W. (1999) *Carrots and Related Vegetable Umbelliferae.* CABI Publishing, New York.

Rücker G. and Noldenn, U. (1991) Polyacetylenes from the underground parts of *Cichorium intybus*. *Planta Med.*, **57**, 97–98.

Sanz, J.F., Falcó, E. and Marco, J.A. (1990) New acetylenes from *Chrysanthemum coronarium* L. *Liebigs Ann. Chem.*, 303–305.

Scalera, G. (2002) Effects of conditioned food aversions on nutritional behavior in humans. *Nutr. Neurosci.*, **5**, 159–188.

Scheel, L., Perone, V., Larkin, R. *et al.* (1963) The isolation and characterisation of two phototoxic furanocoumarins (psoralens) from diseased celery. *Biochemistry*, **2**, 1127–1132.

Schulte, K.E., Rücker, G. and Backe, W. (1970) Polyacetylenes from *Pimpinella*-species. *Arch. Pharm.*, **303**, 912–919.

Schulte, K.E., Rücker, G. and Boehme, R. (1967) Polyacetylene als Inhaltsstoffe der Klettenwurzeln. *Arzn.- Forsch.*, **17**, 829–833.

Shim, S.C., Koh, H.Y. and Han, B.H. (1983) Polyacetylene compounds from *Panax ginseng* C. A. Meyer. *Bull. Korean Chem. Soc.*, **4**, 183–188.

Shu, C.K., Waldradt, J.P. and Taylor, W.L. (1975) Improved method for bergapten determination by high-performance liquid chromatography. *J. Chromatogr.*, **106**, 271–282.

Stanjek, V., Miksch, M., Lüer, P. *et al.* (1999a) Biosynthesis of psoralen: mechanism of a cytochrome P450 catalyzed oxidative bond cleavage. *Angew. Chem. Int. Ed.*, **38**, 400–402.

Stanjek, V., Piel, J. and Boland, W. (1999b) Biosynthesis of furanocoumarins: mevalonate-independent prenylation of umbelliferone in *Apium graveolens* (Apiaceae). *Phytochemistry*, **50**, 1141–1145.

Stanley, W.L. and Vannier, S.H. (1967) Psoralens and substituted coumarins from expressed oil lime. *Phytochemistry*, **6**, 585–596.

Steck, W. (1970) Leaf furanocoumarins of *Heracleum lanatum*. *Phytochemistry*, **9**, 1145–1146.

Stern, R.S., Thibodeau, L.A., Kleinerman, R.A. et al. (1979) Risk of cutaneous carcinoma in patients treated with oral methoxalen photochemotherapy for psoriasis. *N. Eng. J. Med.*, **300**, 809–813.

Tada, M. and Chiba, K. (1984) Novel plant growth inhibitors and an insect antifeedant from *Chrysanthemum coronarium* (Japanese name: Shungiku). *Agric. Biol. Chem.*, **48**, 1367–1369.

Tada, H., Shimomura, K. and Ishimaru, K. (1995) Polyacetylenes in *Platycodon grandiflorum* hairy root and Campanulaceous plants. *J. Plant Physiol.*, **145**, 7–10.

Takasugi, M., Kawashima, S., Katsui, N. et al. (1987) Two polyacetylenic phytoalexins from *Arctium lappa*. *Phytochemistry*, **26**, 2957–2958.

Talib, S. and Banerjee, A.K. (1982) Covalent attachment of psoralen to a single site on vesicular stomatitis virus genome RNA blocks expression of viral genes. *Virology*, **118**, 430–438.

Tassaneeyakul, W., Guo, L.-Q., Fukuda, K. et al. (2000) Inhibition selectivity of grapefruit juice components on human cytochromes P450. *Arch. Biochem. Biophys.*, **378**, 356–363.

Teng, C.-M., Kuo, S.-C., Ko, F.-N. et al. (1989) Antiplatelet actions of panaxynol and ginsenosides isolated from ginseng. *Biochim. Biophys. Acta*, **990**, 315–320.

Towers, G.H.N. and Wat, C.-K. (1978) Biological activity of polyacetylenes. *Rev. Latinoamer. Quim.*, **9**, 162–170.

Trichopoulou, A., Naska, A., Antoniou, A. et al. (2003) Vegetable and fruit: the evidence in their favour and the public health perspective. *Int. J. Vitamin Nutr. Res.*, **73**, 63–69.

Trumble, J.T., Millar, J.G., Ott, D.E. et al. (1992) Seasonal patterns and pesticidal effects on the phototoxic linear furanocoumarins in celery, *Apium graveolens* L. *J. Agric. Food Chem.*, **40**, 1501–1506.

Ulate-Rodríguez, J., Schafer, H.W., Zottola, E.A. et al. (1997) Inhibition of *Listeria monocytogenes*, *Escherichia coli* O157:H7, and *Micrococcus luteus* by linear furanocoumarins in a model food system. *J. Food Protect.*, **60**, 1050–1054.

Verzera, A., Dugo, P., Mondello, L. et al. (1999) Extraction technology and lemon oil composition. *Ital. J. Food Sci.*, **11**, 361–370.

Wagstaff, D.J. (1991) Dietary exposure to furocoumarins. *Reg. Toxicol. Pharmacol.*, **14**, 261–271.

Washino, T., Kobayashi, H. and Ikawa, Y. (1987) Structures of lappaphen-a and lappaphen-b, new guianolides linked with a sulfur-containing acetylenic compound, from *Arctium lappa* L. *Agric. Biol. Chem.*, **51**, 1475–1480.

Washino, T., Yoshikura, M. and Obata, S. (1986a) New sulfur-containing acetylenic compounds from *Arctium lappa*. *Agric. Biol. Chem.*, **50**, 263–269.

Washino, T., Yoshikura, M. and Obata, S. (1986b) Polyacetylenic compounds of *Arctium lappa* L. *Nippon Nogeik Kaishi*, **60**, 377–383.

Wittstock, U., Hadacek, F., Wurz, G. et al. (1995) Polyacetylenes from water hemlock, *Cicuta virosa*. *Planta Med.*, **61**, 439–445.

Wright, M.E., Mayne, S.T., Swanson, C.A. et al. (2003) Dietary carotenoids, vegetables, and lung cancer risk in women: the Missouri Women's Health Study (United States). *Cancer Causes Control*, **14**, 85–96.

Yamaguchi, M. (1983) *World Vegetables: Principles, Production and Nutritive Values*. AVI Publishing Company Inc.; Westport; CT.

Yates, S.G. and England, R.E. (1982) Isolation and analysis of carrot constituents: myristicin, falcarinol, and falcarindiol. *J. Agric. Food Chem.*, **30**, 317–320.

Yates, S.G. and Fenster, J.C. (1983) Myristicin, falcarinol, and falcarindiol as chemical constituents of water celery (*Oenanthe javanica* DC). *Abstr. Pap. Am. Chem. Soc.*, **186**, 149.

Young, A.R. (1990) Photocarcinogenicity of psoralens used in PUVA treatment – present status in mouse and man. *J. Photochem. Photobiol.* B, **6**, 237–247.

Ziegler, R.G., Colavito, E.A., Hartge, P. *et al.* (1996) Importance of α-carotene, β-carotene, and other phytochemicals in the etiology of lung cancer. *J. Nat. Cancer Inst.*, **88**, 612–615.

Zobel, A.M. and Brown, S.A. (1991a) Dermatitis-inducing psoralens on the surfaces of seven medicinal plant species. *J. Toxicol. Cut. Ocular Toxicol.*, **10**, 223–231.

Zobel, A.M. and Brown, S.A. (1991b) Furanocoumarin concentrations in fruits and seeds of *Angelica archangelica*. *Environ. Exp. Bot.*, **31**, 447–452.

Zobel, A.M. and Brown, S. (1991c) Psoralens on the surface of seeds of Rutaceae and fruits of Umbelliferae and Leguminosae. *Can. J. Bot.*, **69**, 485–488.

Chapter 6
Functions of the Human Intestinal Flora: The Use of Probiotics and Prebiotics

Kieran M. Tuohy and Glenn R. Gibson

6.1 Introduction

Recent molecular studies have confirmed earlier suggestions that the human gut microflora comprises several hundred bacterial species (Moore and Holdeman 1974; Wilson and Blitchington 1996). Indeed, many bacteria resident in the gut flora are new to science and have so far escaped laboratory cultivation (Suau *et al.* 1999). As such, little is known about the role played by particular commensal species in important gut functions such as colonization resistance, stimulation of the immune system, production of short chain fatty acids and regulation of mucosal growth and differentiation (Figure 6.1). To date, studies on host–microbe interactions in the gut have been dominated by work on specific human pathogens. Modern molecular techniques, however, are now opening up the possibility of characterizing the complete gut microflora, quantitatively monitoring population fluxes of phylogenetically related groups of bacteria in situ, and determining gene expression in vivo (Amann *et al.* 1995; Wilson and Blitchington 1996; Hooper *et al.* 2001).

The concerted activities of the microbial flora make the hindgut the most metabolically active organ in the body. Carbohydrates and proteins, provided by the diet or through indigenous sources, are fermented anaerobically to produce organic acids and gases. Through the formation of such end products gut bacteria are able to impact markedly upon host health and welfare. The gut flora also contains components which may excrete toxins or other deleterious compounds. There is interest therefore in attempting to alleviate gut disorder by influencing the composition and activities of the resident microflora. Both probiotics and prebiotics do so by increasing populations seen as 'beneficial'.

6.2 Composition of the gut microflora

Colonization of the gastrointestinal tract by the human gut microflora is largely determined by host physiological factors. Gastric pH, redox potential gradients, digestive secretions (e.g. bile acids, lysozyme, pepsin, trypsin) and peristalsis all play a role in limiting the numbers and species diversity of the microflora colonizing the upper regions of the gastrointestinal tract. Although facultative anaerobic species of the genera *Lactobacillus*,

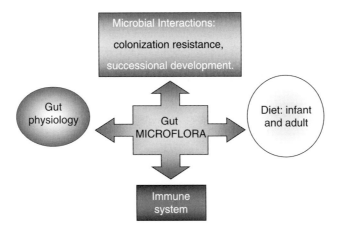

Figure 6.1 Schematic representation of the role played by the gut microflora in maintaining host health through interactions with cells of the mucosa, immunological interactions, interactions between microorganisms and interactions with dietary constituents.

Streptococcus, Enterococcus and yeasts such as *Candida albicans*, have been isolated from the stomach and upper regions of the small intestine, it is difficult to determine whether such organisms actually colonize these regions of the gut (Gorbach 1993). A number of important human pathogens have evolved ways of getting around the harsh environments of the upper gut. *Helicobacter pylori*, for instance, burrows into the mucous layer covering the walls of the stomach and releases ammonia and calcium carbonate to neutralize local pH, thus evading the unfavourable acidic conditions in the stomach lumen. The close association between *H. pylori* and the mucosal cells of the stomach is complex. *H. pylori* infection involves epithelial cell attachment, vacuolating toxin, lipopolysaccharide immune stimulation, natural DNA transformation and hijacking of host cell signal transduction pathways (Ge and Taylor 1999; Moran 1999; Evans and Evans 2001). This organism is thought to be the causative agent of gastric ulcers and may play a role in the onset of stomach cancer.

Microbial numbers and species diversity increases in the distal small intestine, with facultative anaerobes as well as more strict anaerobic species such as bacteroides, clostridia, Gram positive cocci and bifidobacteria reaching population levels of up to 10^{7-8} colony forming units (CFU)/mL contents. The colon is the main site of microbial colonization in the gut, and the microflora is dominated by the strict anaerobes. This microflora is made up of *Bacteroides* spp., *Eubacterium* spp., *Clostridium* spp., *Fusobacterium* spp, *Peptostreptococcus* spp., and *Bifidobacterium* spp., with lower population levels of anaerobic streptococci, lactobacilli, methanogens and sulphate-reducing bacteria (Figure 6.2). Climax microbial populations occur (up to 10^{12} cells/g) and estimates of diversity range from 400 to 500 different bacterial species. The facultative anaerobes such as lactobacilli, streptococci/enterococci and the *Enterobacteriaceae* occur in population levels about 100–1000-fold lower than strict anaerobes (Moore and Holdeman 1975; Conway 1995).

Recent molecular studies on the composition of the human gut microflora have largely confirmed the picture of the gut microflora as generated through traditional

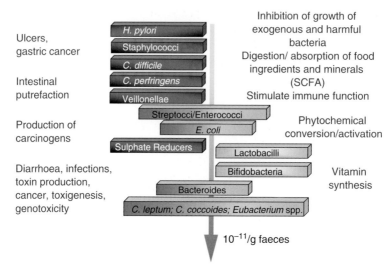

Figure 6.2 Schematic representation of the the gut microflora, indicating relative population levels and some positive health interactions associated with the gut microflora (in dark grey) as well as some harmful activities (in light grey).

microbiological culture techniques. Moreover, analysis of 16S rDNA profiles obtained directly from faeces have greatly expanded estimates of species diversity within the gut microflora. Suau et al. (1999) showed that 95% of clones obtained from total 16S rDNA sequences isolated from a single adult faecal sample belonged to one of three phylogenetic groupings, namely, the *Bacteroides* group, the *Clostridium coccoides* group and the *Clostridium leptum* group. The *C. leptum* and *C. coccoides* groups contain many genera previously identified as constituting dominant members of the gut microflora, for example *Eubacterium* spp., *Ruminococcus* spp., *Butyrovibrio* spp., *Faecalibacterium prausnitzii*, as well as the *Clostridium* species. Only 24% of clones corresponded to previously identified bacterial species, showing that the vast majority of bacteria present in the gut microflora (about 70%) corresponded to novel bacterial lineages, the majority of which fell within the three dominant groupings. Our on-going work characterizing 16S rDNA diversity within the microflora of infants, adult volunteers and in samples taken from the elderly is confirming this high degree of diversity within individual faecal samples, with samples taken from elderly people showing a significant increase in species diversity compared with microflora profiles in younger persons (unpublished data). Similarly, using community 16S rDNA fingerprinting techniques such as temperature gradient gel electrophoresis (TGGE), it has been shown that all individuals carry their own unique gut microflora profile and that this profile is largely stable over extended periods of time (Zoetendal et al. 1998). Through TGGE and universal bacterial primers targeting 16S rRNA and rDNA, Zoetendal et al. (1998) showed that a unique microflora profile exists in faecal samples taken from 16 healthy individuals. TGGE separates PCR amplicons according to their sequence variation. Here, using universal bacterial primers the authors were able to derive a 16S rDNA/rRNA profile for the dominant members of the faecal microflora (with numbers greater or equal to 10^9 cells/g faeces) present in each individual. In two of the volunteers this microflora profile showed a high degree of stability over a 6–7 month period. Although each individual had a unique microflora profile, some TGGE bands were found to be commonly shared between individuals. Upon band excision, nested PCR and DNA sequencing, these common bands were found to be related to species previously described as dominant members of the gut microflora, that is showed highest sequence homology to *Ruminococcus obeum*, *Eubacterium hallii* and *Faecalibacterium prausnitzii*.

Such diversity at the species level has also been illustrated in studies using traditional microbiological culture. Focusing on *Lactobacillus* spp. and *Bifidobacterium* spp., McCartney *et al.* (1996) showed that different individuals carry a unique collection of strains and that species and strain diversity differed between individuals. Similarly, Reuter (2001) reviewed the existing evidence on the composition of the *Lactobacillus* and *Bifidobacterium* moieties of the gut microflora. Agreeing with other workers, Reuter (2001) described distinct populations of lactobacilli and bifidobacteria within individuals and identified species truly autochthonous to the gut, for example *Lactobacillus gasseri*, *L. reuteri* and a non-motile variant of *L. ruminis* (formally *Catenabacterium catenaforme*), while different combinations of bifidobacterial strains predominated in different age groups, notably, infants, compared with adult volunteers.

6.3 Successional development and the gut microflora in old age

It is becoming apparent that the acquisition of the gut microflora impacts greatly on health and disease not only in the neonate but throughout life. The human gut is sterile at birth and microflora acquisition begins during delivery, from the mothers' vaginal and faecal microflora, and from the environment, for example through direct human contact and the hospital surroundings. Early colonizers of the gut include facultative anaerobic species such as streptococci, staphylococci, lactobacilli and the *Enterobacteriaceae*. Such primary colonizers in the days following birth are thought to create a low redox potential in the gut allowing colonization by more strictly anaerobic species such as the bifidobacteria, certain *Clostridium* spp. and to a lesser extent, *Bacteroides* spp. In the neonate, diet has a great impact on microflora composition and plays an important role in the successional development of the early gut microflora. Traditionally, breast fed infants are characterized by carrying a microflora dominated by bifidobacteria. Conversely, infants fed on milk formula develop a much more complex microflora during suckling comprising mainly *Bacteroides* spp., *Clostridium* spp., *Bifidobacterium* spp. and the *Enterobacteriaceae*. Human breast milk contains a number of microbially active components. As well as being a 'complete infant food', human milk contains a range of peptides, lipids, immunoglobulins (IgA), anti-microbial compounds like lysozyme and lactoferrin and a range of oligosaccharides, some of which may act as prebiotics, stimulating the growth of bifidobacteria in the infant gut (Mountzouris *et al.* 2002). The composition of breast milk changes in the profile of such microbially active components during the life of the infant. Adequate nutrition in infancy and especially breast feeding has been linked to a number of important health outcomes including reduced incidence of gastrointestinal infection in childhood, reduced risk of heart disease and even intelligence (Das 2002; Smith *et al.* 2003). The gut microflora in the neonate also plays an important role in the development of mucosal physiology and maturation as well as in the education of the naïve immune system. In both groups of infants, diversity of the microflora greatly increases with the supplementation of solid food, and by the time weaning is complete, a microflora similar to that of the adult in species numbers and diversity is achieved. During this time, there may be increased risk of gastrointestinal infections since the microflora is in flux and may be less capable of impeding gastrointestinal colonization by exogenous microorganisms (Edwards and Parrett 2002). It is now becoming clear that once established, the adult gut microflora is relatively stable over extended periods of time. External factors capable of upsetting microflora balance include medication, especially antibiotics, radiotherapy, gastrointestinal infections, stress and major dietary change.

There are a limited number of studies examining the composition of the gut microflora in the aged and how age affects the gut microflora. However, a general picture is emerging indicating that numbers of bifidobacteria decrease in the elderly population with a concomitant increase in numbers of enterobacteria, lactobacilli and certain clostridium species (Mitsuoka 1992). Recent and on-going studies are also revealing that the gut microflora of the elderly may have a higher degree of species diversity than that of younger people (Tuohy et al. 2004). By constructing a library of 16S rRNA gene fragments cloned from the faecal samples of ten elderly volunteers (mean age 78 years) a comparative phylogenetic profile of the elderly gut microflora was compiled (unpublished data). This showed that the elderly shared the same dominant groups of bacteria with younger adults, that is *C. leptum*, *C. coccoides*, *Bacteroides*, *Enterobacteriaceae*, *Lactobacillus* and *Bifidobacterium* but also the *Sporomusa*, *Acholeplasma-Anaeroplasma* and *Atopobium* groups were among the dominant microflora in the majority of volunteers (Saunier and Doré 2002). The immune system is also affected by old age, with a reduction in efficacy related to dysfunction in specific immunological parameters. Similarly, both basal and peak gastric acid output are reduced in the elderly (Baron 1963) and may account for the relatively common incidence of small bowel bacterial overgrowth observed in this population group (Husebye et al. 1992; Wallis et al. 1993; Lovat, 1996).

6.4 Modulation of the gut microflora through dietary means

Over recent years three scientific rationales have been put forward as a means of modulating the human gut microflora towards improved host health. Probiotics have been defined as live microbial food supplements, which have a beneficial effect on the intestinal balance of the host animal (Fuller 1989). Prebiotics on the other hand, are non-viable food components, which evade digestion in the upper gut, reach the colon intact and then are selectively fermented by bacteria seen as beneficial to gastrointestinal health, namely the bifidobacteria and/or lactobacilli (Gibson and Fuller 2000). The synbiotic approach combines both probiotics and prebiotics and aims to stimulate the growth and/or activity of indigenous bifidobacteria and lactobacilli, while also presenting proven probiotic strains to the host (Gibson and Fuller 2000). All three have received much scientific and commercial interest in recent years and a range of such microbially active food products are now available.

6.4.1 Probiotics

A number of desirable features have been identified for the selection of efficacious probiotic strains (Reid 1999; Dunne *et al.* 2001; Saarela *et al.* 2002).

(1) The probiotic should be of host animal origin. Although most probiotics for human consumption are of human origin, a number of probiotic strains for animal use, particularly companion animals and pigs are derived from human isolates. This raises questions over the suitability of such strains, since it is likely that the most effective and best adopted probiotic strains will be those originally isolated from the gastrointestinal tract of the target animal.

(2) The probiotic strain must be safe. The probiotic strains should not cause disease, be associated with disease states or even related to recognized bacterial pathogens. Generally, the lactobacilli and bifidobacteria, have a long history of safe use in foods and are not closely related to recognized human pathogenic bacteria. The use of some other bacteria, such as enterococci and *Clostridium butyricum*, as probiotics raises some health concerns. Enterococci, in particular, have been associated with nosocomial infections in hospitals. One important aspect of this is their possible possession of virulence genes and ability to transfer such traits to other bacteria within the human gut.

(3) Probiotic strains must be amenable to industrial processes. Many early production problems, especially in delivering specific numbers of viable cells in the final product have been overcome. However, in the case of the more strictly anaerobic strains of *Bifidobacterium* spp., problems remain in guaranteeing specific numbers of viable cells in probiotic products at the point of sale.

(4) To persist to the colon, probiotic strains must show a degree of resistance to gastric acid, mammalian enzymes and bile secretions. It may be of less importance if the mode of probiotic activity is stimulation of the immune system, where the requirement for cell viability and metabolic activity in the gastrointestinal tract is not as relevant.

(5) To maximize the colonization potential of a probiotic strain, the ability to adhere to human mucosal cells may be an advantage. The ability of certain probiotic strains to hinder adhesion of pathogens, or their toxins, to human cells has been proposed as one possible mode of probiotic action. However, despite positive results in anti-pathogen adhesion assays in vitro, such activity has not been demonstrated in the human gastrointestinal tract. Indeed, we know little about which species of commensal bacteria adhere to the gut wall and the health outcomes of such interactions.

(6) The ability of the probiotic strain to persist in the gut has been identified as one important prerequisite of probiotic efficacy. Indeed, a number of probiotic products claim that their strains colonize the human gastrointestinal tract. However, it is more likely that after cessation of probiotic feeding, the vast majority of probiotic strains fall below detection. This is not unexpected since the human gut microflora provides a robust barrier to the establishment of exogenous micro-organisms.

(7) Some probiotics produce anti-microbial agents targeting important gastrointestinal pathogens which is a desirable characteristic. Many lactic acid bacteria and bifidobacteria have been shown to produce bacteriocin-like molecules with different spectrums of activity.

(8) Certain probiotic strains have been shown to stimulate the immune system in a beneficial, non-inflammatory manner. Such strains have also been shown to relieve the symptoms of allergic conditions such as atopic eczema and bovine milk protein intolerance in human feeding studies. The mechanisms underlying such beneficial modulation of immune response have not yet been fully elucidated.

(9) Efficacious probiotic strains should also impact on metabolic activities such as cholesterol assimilation, lactase production and vitamin production.

Probiotic intervention has been investigated for its effectiveness against a range of gastrointestinal diseases and disorders. Examples are given below.

6.4.1.1 Probiotics in relief of lactose maldigestion

It has been estimated that about two thirds of the world's adult population suffers from lactose maldigestion, with the prevalence particularly high in Africa and Asia. In Europe, lactose maldigestion varies from about 2% in Scandinavia to about 70% in Sicily (Vesa *et al.* 2000). Individuals with lactose maldigestion may tolerate lactose present in yoghurt to a much greater degree than the same dose in raw milk (Marteau *et al.* 2002). This is thought to be due to enzymic lactase activity expressed by the bacteria.

6.4.1.2 Use of probiotics to combat diarrhoea

Probiotic strains have been evaluated for their anti-diarrhoeal capabilities, with some proving more successful than others (Table 6.1). *L. rhamnosnus* GG has repeatedly been shown to reduce the duration of diarrhoea by about 50% in patients with acute infantile diarrhoea caused by rotavirus (Isolauri *et al.* 1999). The mechanisms of action have not been fully elucidated but may involve fortification of the mucosal integrity and/or stimulation of specific anti-rotavirus IgA. Other probiotic strains have been shown to reduce the incidence of rotaviral diarrhoea, that is, *B. bifidum* given in conjunction with *S. thermophilus* in standard milk formula (Saavedra *et al.* 1994).

Medical intervention with antibiotics leads to a perturbation of the gut microflora which in many cases leads to diarrhoea. About 20% of patients who receive antibiotics experience diarrhoea (Marteau *et al.* 2002). Compromising the gut microflora through the use of antibiotics reduces its ability to prevent the growth of pathogenic micro-organisms within the gut, which can lead to antibiotic-associated diarrhoea. This is particularly true of *Clostridium difficile* diarrhoea. *C. difficile* is a common gut inhabitant usually present in low numbers until antibiotic therapy compromises the gut microflora allowing it to outcompete other bacteria, leading to overgrowth, toxin production and diarrhoea. In a recent meta-analysis, both lactobacilli (*L. acidophilus, L. bulgaricus, L. rhamnosus* GG) and the yeast *Saccharomyces boulardii* were proven to be effective in preventing antibiotic associated diarrhoea (D'Souza *et al.* 2002). Traveller's diarrhoea includes different gastrointestinal infections sharing common symptoms such as diarrhoea, intestinal pain and bloating and sometimes vomiting. Table 6.1 summarizes the results of trials conducted to investigate the protective activities of probiotic dietary supplements against traveller's diarrhoea. In the majority of cases the cause of illness, whether bacterial or viral, has not been identified.

Table 6.1 Examples of human feeding studies showing a positive effect of probiotic consumption on the symptoms of gastrointestinal infections, for example diarrhoea

Disorder	Probiotic	Effect	Reference
Infantile diarrhoea	*Lactobacillus* GG	Reduced duration of rotaviral diarrhoea	Isolauri *et al.* (1991)
			Isolauri *et al.* (1994)
			Kaila *et al.* (1992)
			Majamaa *et al.* (1995)
	L. reuteri	Reduced duration of rotaviral diarrhoea	Shornikova *et al.* (1997)
	B. bifidum and *S. thermophilus*	Prevented rotaviral diarrhoea	Saavedra *et al.* (1994)
	L. casei and *L. acidophilus* or *S. boulardii*	Both lactobacilli and the yeast reduced incidence and duration of diarrhoea	Gaon *et al.* (2003)
Antibiotic-associated-diarrhoea/ *Clostridium difficile*-associated-diarrhoea	*B. longum*	Decreased course of erythromycin-induced diarrhoea	Colombel *et al.* (1987)
	Lactobacillus GG	Decreased course of erythromycin-induced diarrhoea, and other side effects of erythromycin	Siitonen *et al.* (1990)
	Streptococcus faecuim	Decreased diarrhoea-associated anti-tubercular drugs administered for pulmonary TB	Borgia *et al.* (1982)
	S. boulardii	Reduced incidence of diarrhoea	Surawicz *et al.* (1989)
			McFarland *et al.* (1995)
	Lactobacillus GG	Improves/terminates colitis	Gorbach *et al.* (1987)
	Lactobacillus GG	Eradicated associated diarrhoea	Biller *et al.* (1995)
Travellers' diarrhoea	*L. acidophilus* and *B. bifidum*	Decreased frequency, not duration of diarrhoea	Black *et al.* (1989)
	Lactobacillus GG	Decreased incidence of diarrhoea	Oksanen *et al.* (1990)
			Hilton *et al.* (1996)
	S. bourardii	Reduced diarrhoea in acute amoebiosis and increased cyst passage	Mansour-Ghanaei *et al.* (2003)

The mode of action of these probiotic strains is likely to be a fortification of the compromised gut microflora, thus restoring colonization resistance and/or direct inhibitory effects against the diarrhoea-causing pathogen. Probiotic strains have been shown to inhibit pathogenic bacteria both in vitro and in vivo in animal models through a number of different mechanisms (Paubert-Braquet *et al.* 1995; Asahara *et al.* 2001; Likotrafiti *et al.*

2004). These include production of directly inhibitory compounds such as bacteriocins and short chain fatty acids (SCFA). SCFA may also be inhibitory through reduction of luminal pH, and by inducing competition for nutrients and adhesion sites on the gut wall and modulation of the immune response (Tuohy *et al.* 2003). Critical to our future understanding of how probiotics work is an appreciation of the cross-talk between probiotics and cells of the mucosa and lymphatic systems. Recent studies employing the power of DNA microarrays through a transcriptomics approach are beginning to elucidate this microbial-host cellular communication (Hooper *et al.* 2001; Williams *et al.* 2003; Zoetendal *et al.* 2004). Molecular biology is also providing the tools necessary to study the in vivo activity of probiotic strains within the human gut microflora, for example, molecular marker systems such as *lux* and *gfp* (Oozeer *et al.* 2002). This ability to show in vivo activity of a probiotic strain, for example the ability to synthesize a bacteriocin under the physiological conditions of the gut microflora, goes to the root of identifying how probiotics work.

6.4.1.3 Probiotics for the treatment of inflammatory bowel disease

Inflammatory bowel disease (IBD) refers to a group of disorders (ulcerative colitis, Crohn's disease and pouchitis), all of unknown aetiology but characterized by chronic or recurrent inflammation of the alimentary mucosa. This inflammation is thought to arise from three underlying pathogenic factors: a genetic predisposition, immune dysregulation and environmental triggers (Shanahan 2004). Diet and the gut microflora may constitute such environmental triggers and a number of bacterial groups, especially the sulphate-reducing bacteria, which convert dietary and endogenous sulphate into toxic derivatives (e.g. H_2S) have been postulated to play a role in the onset or maintenance of IBD (Roediger *et al.* 1997; Pitcher *et al.* 2000; Fite *et al.* 2004). This environmental input into IBD is the target for intervention with probiotic biotherapeutics. The probiotic approach in some cases is proving as successful as existing therapies. Indeed, it is fair to say that existing therapies for IBD are limited in that they treat the symptoms of disease, that is, mucosal inflammation, and fall well short of a cure. Surgery and intestinal resection are often the end result in cases where relapse frequently occurs. Probiotics, especially the lactobacilli and bifidobacteria have been shown in various animal models of ulcerative colitis to reduce mucosal inflammation and inflammatory markers (Shanahan 2004). In human feeding studies carried out with diverse probiotic preparations, symptom relief and reduction in the incidence of relapse have been achieved. The non-pathogenic *E. coli* strain Nissle 1917 has proven as effective in maintaining remission from symptoms of ulcerative colitis as standard treatments (e.g. mesalazine) and has proven more effective in preventing relapse in Crohn's disease patients than placebo treatments (Rembacken *et al.* 1999). *S. boulardii* has also shown some success in relieving IBD symptoms, reducing stool frequency and disease activity in active Crohn's disease and in reducing the risk of Crohn's relapse (Guslandi *et al.* 2000). VSL#3 is a mixture of four lactobacilli (*L. acidophilus, L. bulgaricus, L. casei* and *L. plantarum*), three bifidobacteria (*B. breve, B. infantis* and *B. longum*) and *S. thermophilus* which has shown probably the most convincing results. This probiotic mixture has proven effective in reducing the recurrence of chronic relapsing pouchitis. VSL#3 at 6 g/day significantly reduced relapse recurrence (15%) compared with a placebo group (100%) over a nine-month period (Gionchetti *et al.* 2000). VSL#3 has also proven effective in preventing pouchitis in patients having received ileo-pouch anal anastomosis for ulcerative colitis compared with a placebo (Gionchetti *et al.* 2003). The mechanisms of action of this probiotic mix have recently been investigated using in vitro assays of mucosal integrity and gene expression. VSL#3, unlike the gram negative biotherapeutic *E. coli* Nissle 1917 and pathogen *Salmonella dublin*, did not induce the cytokine IL-8 production in intestinal epithelial cells. Induction of IL-8 by the gram negative strains was also much reduced when co-cultures of the VSL#3 and the gram negative strains were presented to the intestinal epithelial cells. VSL#3 also increased transepithelial resistance (a marker of mucosal integrity) and stabilized TER in co-culture with *S. dublin* (Otte and Podolsky 2004).

6.4.1.4 Impact of probiotics on colon cancer

Colo-rectal cancer (CRC) is among the biggest killers of all cancers in the UK, responsible for more than 12 000 deaths annually. The best treatments available, use surgery and anti-cancer drugs but only save two out of every five patients. About three-quarters of all CRC cases are sporadic with no familial or other disease association. The interaction between diet and the gut microflora is central to both cancer risk and protection from disease. CRC occurs after initial environmental insult to the genetic material of the mucosa, that is, DNA damage from the intestinal contents. Members of the gut microflora are capable of producing a range of toxic and carcinogenic compounds from dietary components. For example, microbial activities are responsible for conversion of cooked food mutagens to direct carcinogenic derivatives such as 2-amino-3,6-dihydro-3-methyl-$7H$-imidazo[4,5-f]quinoline-7-one. Other microbial metabolic activities such as nitrate and sulphate reduction, bile acid deconjugation and amino acid fermentation lead to the production of toxic sulphur and nitrogenous compounds such as hydrogen sulphide, N-nitroso compounds, ammonia, phenols and cresols, and secondary bile acids (Rowland 1995). Indeed, it is now accepted that the human gut microflora has an intimate association with the onset and development of CRC. Although the species responsible for these activities have not always been identified, it is recognised that bifidobacteria and lactobacilli do not produce toxic or carcinogenic metabolites. Indeed, probiotic bacteria have been investigated for their ability to modulate microbial biomarkers of CRC in both animal and human feeding studies (Burns and Rowland 2000). Probiotic strains, both lactobacilli and bifidobacteria have been shown in animal or human feeding studies to reduce production of some of these toxic metabolites (Rafter 2002a). Pool-Zobel *et al.* (1996) investigated the ability of different probiotic strains to protect against DNA damage in rats dosed with the colonic carcinogens N-methyl-N-nitro-N-nitrosoguanidine and 1,2-dimethylhydrazine. Most of the probiotics strongly inhibited DNA damage, with the lactobacilli and bifidobacteria providing increased protection compared with the dairying strain *Streptococcus thermophilus*. In other animal studies probiotic supplementation has been shown to reduce the frequency and size of aberrant crypt foci (pre-neoplastic lesions) and induce mucosal apoptosis (Arimochi *et al.* 1997; Fukui *et al.* 2001). Promising, although sometimes equivocal, results have been observed in feeding studies in both healthy individuals and patients with colon cancer. Probiotic supplementation has been shown to impact on biomarkers of colon cancer (e.g. faecal water genotoxicity, urinary mutagenicity and proliferation of rectal mucosal crypts) in these studies (Burns and Rowland 2000; Rafter 2002b; Oberreuther-Moschner *et al.* 2004).

6.4.1.5 Impact of probiotics on allergic diseases

One of the major physiological benefits of probiotics is the enhancement of immune function. Therefore, the concept of feeding probiotics to individuals with suboptimal immune function, such as the young, elderly, immuno-compromised individuals and those with depleted microflora post-antibiotic treatment, is rational. It has been demonstrated that in atopic infants, with proven cow milk allergies, the clinical course of atopic dermatitis could be greatly improved following a probiotic-supplemented elimination diet (Kalliomaki et al. 2001). Rosenfeldt et al. (2003) investigated the therapeutic nature of a combination of *L. rhamnosus* 19070-2 and *L. reuteri* DSM 122460 in the management of atopic dermatitis in children. In all, 56% of children treated with the probiotics showed reduced clinical severity of eczema compared with 15% in the placebo-fed group. Another *L. rhamnosus* strain (*L. rhamnosus* GG) has also been shown to down-regulate the immunoinflammatory response in individuals with milk-hypersensitivity, while acting as an immunostimulator in healthy subjects (Pelto et al. 1998).

6.4.1.6 Use of probiotics in other gut disorders

Irritable bowel syndrome (IBS) is a major problem both medically and economically, with an estimated 8–22% affected. The causes differ between patients and include the use of drugs (antibiotics), ovarian hormones, operations, fibre deficiency, food intolerances, stress, microbial infections (e.g. with *Candida albicans*) and a depletion of beneficial gut bacteria. Therefore, ingestion of probiotics is likely to restore numbers of beneficial bacteria and reduce IBS symptoms. Human feeding studies in IBS patients to date have yielded mixed results. O'Sullivan and O'Morain (2000) found that *L. rhamnosus* GG had little effect on gastrointestinal pain, urgency or bloating in IBS patients in a double-blind, placebo-controlled cross-over study. *L. plantarum* 299V was found to have a beneficial effect on patients with IBS (Niedzielin et al. 2001). All patients who received the probiotic reported a reduction in abdominal pain compared with 11 out of 20 individuals in the placebo group. There was significant improvement in all symptoms in 95% of probiotic-treated patients compared with 15% of patients in the placebo group. More recently, Kim et al. (2003) investigated the usefulness of the probiotic mix VSL#3 in treating the symptoms of diarrhoea-predominant IBS. Twenty five patients were asked to consume VSL#3 or a matching placebo (starch) twice daily for 8 weeks. Although the probiotic had no significant effect on intestinal transit time, abdominal pain, flatulence or defecation urgency, dietary supplementation with VSL#3 did bring about significant relief from abdominal bloating.

Gastroenteritis is a common symptom in autistic spectrum disorders and there is some evidence that an altered gut microflora may even play a role in autistic pathology. In many patients diagnosed with autism, excessive antibiotic therapy in early life is a common feature. Antibiotic therapy can bring about significant changes within the gut microflora, and prolonged modification of the gut microflora may occur after prolonged antibiotic intake. Following initial observations that the antibiotic vancomycin (Sandler et al. 2000), which is active against intestinal clostridia as well as other bacteria, can bring about short lived improvement in autistic symptoms in some individuals, Finegold et al. (2002) examined the composition of the gut microflora of autistic children compared with healthy controls. Using traditional microbiological culture techniques, these authors found that autistic children carried higher numbers of clostridia, and the autistic children also harboured a different, more diverse collection of clostridial species. Work in our laboratory has recently confirmed that autistic children have higher counts of clostridia (of the *Clostridium perfringens/histolyticum* subgroup) than healthy children (Parracho et al. 2004). Probiotic therapies may hold promise not only in the relief of gastrointestinal symptoms associated with autism, often due to long-term, multiple antibiotic usage, but in normalizing the autistic gut microflora in terms of composition and in respect to the profile of metabolites produced by the microflora (e.g. metabolites derived from tryptophan metabolism, which are thought to play a psychoactive role in autism). Indeed, there is anecdotal evidence from clinical practice and primary care givers of autistic subjects, that probiotic intake does provide some relief from autistic symptoms (Bingham 2003). There is now a need for well-controlled blinded placebo intervention studies using probiotics to establish any impact on the disease. Similarly, further characterization of the gut microflora of clinically diagnosed groups within the autistic spectrum of disorders is needed, both at the level of species diversity (using direct molecular methodologies) and microbial metabolic activity (e.g. identification of autism-specific microbial metabolites and gastrointestinal biomarkers of disease).

6.4.1.7 Future probiotic studies

Improved studies on probiotic efficacy should result from genotypic approaches that allow fed strains to be discriminated from those indigenous to the gut (Tannock 1999). Most commercial probiotics have been tested in vitro for their resistance to gastric acidity, bile salts, etc. but there are few data on in situ survivability and metabolic activity (e.g. whether probiotic strains produce particular bacteriocins within the human gut microflora). Maintenance of product viability and integrity during processing and after feeding is a major issue for probiotic approaches. It is largely agreed that probiotics may have effects in the small and large intestine, but it is unclear how robust the strains are therein.

Survival of probiotics under various physicochemical conditions that are imposed during both processing and after intake (in the gastrointestinal tract) varies between strains. This needs further explanation and inter-species differences more fully determined and explained. This will lead to focusing upon the most reliable strains. For probiotics, nomenclature (taxonomical changes), poor stability and inaccurate labelling make sound conclusions difficult. Similarly there is no consensus on what constitutes a transient, persistent or colonizing effect within the gut flora.

Traditionally, resistance to low pH and bile acids, production of anti-microbial compounds and amenability to industrial food processing have been seen as important criteria for the selection of probiotic strains (Gibson and Fuller 2000; Dunne et al. 2001). Similarly, probiotic strains for human use must be safe and of human origin (Salminen et al. 1998). However, there is a need for comparative studies on existing and emerging probiotic strains to identify characteristic probiotic capabilities of different products. There is also a lack of information about the genetic determinants responsible for many important probiotic traits. Comparative studies on probiotic strains (especially those for which the genome sequence is available) and identification of underlying genetic determinants encoding probiotic aspects are essential (Klaenhammer et al. 2002). Techniques such as subtractive hybridization of differentially expressed genes or of whole genomes of closely related probiotic strains may facilitate the search for the genes encoding important probiotic traits (Reckseidler *et al.* 2001; Akopyants *et al.* 1998; Soares, 1997). This could also be employed to develop high throughput screening programmes to identify novel probiotic strains using DNA microarrays technology (Kuipers *et al.* 1999). Identification of the genetic determinants responsible for important probiotic activities would also facilitate any move towards the construction of efficient genetically modified probiotic strains designed to treat specific disease states.

6.4.2 Prebiotics

Carbohydrates said to be prebiotics have been variably tested for modulating the gut flora activities (Cummings *et al.* 2001). For example, fructo-oligosaccharides, galacto-oligosaccharides and lactulose are recognized for their bifidogenic effects in laboratory, animal and human trials carried out in multiple centres (Gibson *et al.* 2000). Some data are conflicting but these materials appear to be the current market leaders, particularly in Europe. In Japan a much wider list of prebiotics exists which includes soya-oligosaccharides, xylo-oligosaccharides, isomalto-oligosaccharides, gentio-oligosaccharides, lactosucrose and gluco-oligosaccharides. These are currently being tested in Europe and elsewhere. Resistant starches and some sugar alcohols have also been proposed as prebiotics (Crittenden *et al.* 2001). New prebiotics with multiple functionality are also under development (Rastall and Maitin 2002). With new advances in molecular based diagnostic procedures for characterizing the gut flora responses to dietary change, a more reliable database of effects should ensue (Tuohy *et al.* 2003).

To help define how prebiotics operate, there is a need for more structure to function studies. A selective fermentation is one requirement for an efficient prebiotic, with certain oligosaccharides seemingly preferring the bifidobacteria. However, it is not clear why this is the case or why certain linkages induce selective changes in a mixed microbial ecosystem. As more information on the biochemical, physiological and ecological capabilities of the target organisms is generated, such information will become more apparent.

For prebiotics, there are problems if the gut flora does not contain their target microorganisms and there are a more limited range of available products than there are probiotics. Effects of prebiotics would be easier to define if their influence on the gut microflora could be standardized, that is, are any changes that ensue consistent across populations?

6.4.2.1 Modulation of the gut microflora using prebiotics

As mentioned above, the fructans (inulin and fructo-oligosaccharides), galacto-oligosaccharides and lactulose are the leading prebiotics available on the European market and they are the oligosaccharides for which the strongest scientific evidence supporting a prebiotic activity exists. All three have been repeatedly shown to be selectively fermented by the bifidobacteria and lactobacilli in vitro (using models of the human gut microflora), in animal feeding studies, and, in vivo, in healthy volunteers (Table 6.2). This modulation of the gut microflora, whereby prebiotic ingestion leads to an increase in relative numbers of faecal *Bifidobacterium* spp. in particular, and sometimes a reduction in bacteria seen as possibly detrimental to human health, that is, certain species of clostridia, have been shown

Table 6.2 Prebiotic modification of the human gut microflora in vivo as measured by changes in numbers of faecal bacteria (FOS = fructo-oligosaccharides; GOS = galacto-oligosaccharides; IMO = isomalto-oligosaccharides; SOS = soy-oligosaccharides)

Prebiotic	Daily dose & duration	No. of subjects	Study design	Microflora modulation	Reference
Inulin (long-chain)	8 g/day, 14 days	9	Placebo-controlled cross-over	Increase in bifidobacteria, small increase in clostridia	Tuohy et al. (2001)
Inulin	Up to 34 g/day 64 days	8	Feeding study	Increase in bifidobacteria	Kruse et al. (1999)
Inulin	15 g/day 15 days	4	Cross-over placebo controlled	Increase in bifidobacteria	Gibson et al. (1995)
Inulin (and lactose)	20–40 g/day 19 days	25 elderly	Parallel placebo-controlled	Increase in bifidobacteria, decrease in enterococci and enterobacteria	Kleessen et al. (1997)
FOS	15 g/day 15 days	4	Placebo-controlled cross-over	Increase in bifidobacteria, decrease in *Bacteroides*, clostridia and fusobacteria	Gibson et al. (1995)
FOS and PHGG biscuits	6.6 g/day FOS 3.4 g/day PHGG for 21 days	31	Placebo-controlled cross-over	Increase in bifidobacteria	Tuohy et al. (2001)
FOS	0–20 g/day 7 days	40	Placebo-controlled parallel	Increase in bifidobacteria (optimal dose 10 g/day)	Bouhnik et al. (1999)
FOS	4 g/day 42 days	12	Feeding study	Increase in bifidobacteria	Buddington et al. (1996)
FOS + GOS	0.04 g/L 0.08 g/L	3 × 30 infants 30 infants	Placebo-controlled parallel	Increase in bifidobacteria Increase in lactobacilli	Moro et al. (2002)

(Continued)

Table 6.2 Continued

Prebiotic	Daily dose & duration	No. of subjects	Study design	Microflora modulation	Reference
FOS + GOS	10 g/L 28 days	15 infants (preterm)	Placebo-controlled parallel	Increase bifidobacteria	Boehm et al. (2002)
IMO	13.5 g/day 14 days	6 healthy 18 senile	Feeding study	Increase in bifidobacteria	Kohmoto et al. (1988)
IMO	5–20 g/day 12 days	14	Variable dose feeding study	Increase in bifidobacteria with Increased IMO DP	Kaneko et al. (1994)
Lactulose	10 g/day 26 days	2 × 10	Placebo-controlled parallel	Increase in bifidobacteria	Tuohy et al. (2002)
Lactulose	3 g/day 14 days	8	Feeding study	Increase in bifidobacteria, decrease in lactobacilli	Terada et al. (1992)
Lactulose	2 × 10 g/day For 4 weeks	36	Placebo-controlled	Increase in bifidobacteria and lactobacilli	Ballongue et al. (1997)
Lactulose	5 g/L and 10 g/L For 3 weeks	6 infants	Case control	Increase in bifidobacteria decrease in coliforms	Nagendra et al. (1995)
GOS	0–10 g/days 8 weeks	12	Placebo-controlled Parallel	Increase in bifidobacteria and lactobacilli	Ito et al. (1990)
GOS	2.5 g/days 3 weeks	12	Feeding study	Increase in bifidobacteria, decrease in *Bacteroides* and clostridia	Ito et al. (1993)
SOS	10 g/day 3 weeks	6	Placebo-controlled cross-over	Increase in bifidobacteria	Hayakawa et al. (1990)

using both classical cultural microbiological techniques and more modern phylogenetically rigorous molecular techniques (Gibson et al. 1995; Tuohy et al. 2001, 2002). This distinction is important considering the huge species diversity of the gut microflora, up to 500 different species, and the shortcomings of selective microbiological culture techniques which have for long limited our ability to study the microbial ecology of the gut in any real sense.

These three oligosaccharide preparations have also been shown to impact favourably on human health, providing relief from gastrointestinal disease symptomology, improving global metabolic parameters such as mineral absorption and lipid metabolism and in bringing about changes within the gut seen as protective against colon cancer. Currently however, there is a gap in our understanding about how prebiotics work. Although we can readily observe a change within the microbial ecology of the gut upon prebiotic ingestion, that is, increased faecal bifidobacteria, and we can show improved human health, for example mineral absorption or reduced risk of colon cancer, we do not have enough information to describe how the two observations may be linked. Studies of a very fundamental nature examining how species of gut bacteria or groups of gut bacteria interact at the molecular level with cells of the mucosa and gastrointestinal lymphatic tissue are needed. Similarly, we need to be able to directly measure the activities of specific bacteria within the gut microflora. The techniques for such studies are now becoming available with advances in the '-omics' technologies.

6.4.2.2 Health effects of prebiotics

Improving global metabolic parameters. The small intestine has traditionally been seen as the main site of mineral absorption in humans. However, the importance of the colon as a site of nutrient absorption is becoming increasingly recognized, and a number of prebiotic oligosaccharides have been examined for their ability to improve this function (Frank 1998). In animal feeding studies, fructo-oligosaccharides have been shown to increase calcium and magnesium absorption (Scholtz-Ahrens et al. 2000; Coudray et al. 2003). Other prebiotics, such as resistant starches and lactulose, have also been shown to increase calcium absorption in rats (Brommage et al. 1993; Greger 1999). One section of the population for which enhanced mineral absorption or retention is particularly important is post-menopausal women. Osteoporosis affects about a third of women post-menopause in Europe. Maximizing peak bone mass during adolescence is central to preventing the adverse effects of bone mineral leaching in later life. This has led to a range of calcium-fortified foods becoming available in the marketplace. However, only about 50% of dietary calcium is absorbed. Prebiotic oligosaccharides are proving efficacious functional foods suitable for enhancing calcium absorption both during adolescence and in post-menopausal women. Delzenne et al. (1995) showed that a 10% dietary supplementation with either inulin, or inulin-derived fructo-oligosaccharides resulted in a significant increase in apparent retention of calcium, magnesium and iron in ovariectomized rats. This animal model simulates bone demineralization, which occurs due to hormonal changes after menopause. Tahiri et al. (2003) showed that dietary supplementation with fructo-oligosaccharides can improve uptake of calcium in late menopausal women. Inulin and inulin-derived fructo-oligosaccharides have also been observed to improve calcium absorption in adolescent girls (Griffin et al. 2002). However, it is worth noting that different prebiotics, as evidenced from recent findings with fructans of different chain length, may differ in their ability to enhance calcium absorption (Kruger *et al.* 2003). The mechanism or mechanisms by which prebiotics enhance mineral absorption within the gut are not fully understood. Ohta *et al.* (1998) showed that feeding with fructo-oligosaccharides increased the ratio of a calcium binding protein (calbindin-9kD) in the colon compared with the small intestine of rats. However, it is not known how fructo-oligosaccharides should regulate colonocyte production of calbindin-9kD. Prebiotics do, however, result in increased short chain fatty acid production upon colonic fermentation and these may impact on mucosal gene expression. In particular, butyrate has been shown to regulate mucosal cell proliferation, differentiation and gene expression.

Prebiotics have also been suggested to modify serum triglyceride levels and cholesterol in animal models and in humans. However, due to the complexity of human lipid metabolism, comprehensive investigations are difficult to undertake and few studies have been conducted. Human studies have often given conflicting results (Delzenne and Williams 2002). Data for the consumption of inulin and fructo-oligosaccharides tend to show either no effect or a slight decrease in circulating triacylglycerols and plasma cholesterol concentrations (Davidson and Maki 1999), suggesting that these prebiotics have no detrimental influence in subjects with minor hypercholesterolaemia or hypertriglyceridaemia. Pereira and Gibson (2002a) reviewed the possible routes through which prebiotics may impact on lipid metabolism. In vitro work by these authors has led to the identification of *Lactobacillus* strains with enhanced cholesterol assimilatory activities (Pereira and Gibson 2002b).

Gastrointestinal disease symptomology
Treatment of IBD mainly relies upon the attenuation of the local inflammation in the digestive tract by means of steroids or anti-tumour necrosis factor. Probiotics, as mentioned above have shown promise in the treatment of IBD symptoms. Butyrate has been shown to maintain periods of remission by promoting mucosal cell proliferation and accelerating the healing process (Breuer *et al.* 1997; Bamba *et al.* 2002). It has been postulated that prebiotics may thus impact on the symptoms of IBD via their impact on probiotic bifidobacteria and lactobacilli indigenous to the gut or through their production of SCFAs upon colonic fermentation. However, few studies have been conducted on the efficacy of prebiotic supplementation in ulcerative colitis, and most of these have been carried out in animal models of colitis. Inulin has been shown to reduce disease severity in the distal colon rats where colitis was induced by dextran sodium sulphate (Videla *et al.* 1998). Using the same model system, these authors showed that rats fed inulin showed increased colonic lactobacilli, a normalization of luminal pH and an extension of microbial saccharolytic activity towards the distal colon while both disease severity and duration were reduced (Videla *et al.* 1998; 1999; 2001). In another animal model where colitic symptoms were induced using intracolonic trinitrobenzene sulphonic acid (TNBS), the efficacy of prebiotic therapy and possible mode of prebiotic action were investigated. TNBS-induced colitic rats were infused either intragastrically or intracolonically with fructo-oligosaccharides (1 g/day), a mixture of 10^{11} CFU/day of probiotic bacteria (*L. acidophilus, L. casei* subsp. rhamnosus and *B. animalis*), SCFAs lactate and/or butyrate, SCFA plus probiotic mix (at $10^{9.5}$ CFU/day) or a saline placebo (Cherbut *et al.* 2003). The prebiotic fructo-oligosaccharides was shown to increase numbers of lactic acid bacteria within the rat caecum, reduce luminal pH through increased lactate and butyrate production and significantly reduce the gross score of inflammation, myeloperoxidase activity

(a specific enzyme marker of polymorphononuclear neutrophil primary granules) and ulcerative colitis associated anorexia. Intragastric infusion of the probiotic bacteria gave similar effects. The short chain fatty acids infused intracolonically resulted in a significant reduction in inflammatory indices. However, with lower doses of SCFA, similar to those observed naturally within the large intestine, addition of the probiotic strains was necessary to reproduce the significant improvements in disease severity observed through intervention with fructo-oligosaccharides. It thus appears that, at least in this model of ulcerative colitis, both the microflora modulatory effects of prebiotics (increasing numbers of probiotic bacteria within the gut) and the metabolic products of prebiotic fermentation within the colon (such as lactate and butyrate) may play a role in alleviation of ulcerative colitis symptoms. The authors concluded that in these experiments stimulation of the lactic acid bacteria growth within the colon of the colitic rats was an essential step for the colitis-reducing effect of fructo-oligosaccharides (Cherbut et al. 2003).

Initial studies in our laboratory indicate that the gut microflora of patients with ulcerative colitis responds in a similar manner to that of healthy individuals upon prebiotic ingestion, that is, increased faecal bifidobacteria and/or lactobacilli (Kolida unpublished data). Germinated barley foodstuff (GBF), which is a mixture of glutamine-rich protein and hemicellulose-rich fibres, has been shown to alleviate colitic symptoms in animal models of ulcetative colitis and in colitic patients (Bamba et al. 2002; Araki et al. 2000). This foodstuff, although rich in readily fermented fibre, has not been assessed for its prebiotic capabilities. That is to say, that although a beneficial health effect has been observed upon ingestion of GBF, the impact of GBF fermentation upon the gut microflora, and in particular on relative numbers of bifidobacteria and lactobacilli, has not been determined.

Patients with ulcerative colitis are at greater risk of developing colon cancer, another chronic disease of the gut for which modulation of the colonic environment through probiotic and prebiotic intervention is showing promise (Shanahan 2003). The mechanistic link between ulcerative colitis and colon cancer remains to be determined, but it is possible that probiotic or prebiotic therapies of proven efficacy in reducing ulcerative colitis symptomology may also provide heightened protection from colon cancer in these individuals. Confirmation of such hypothesis through dietary intervention studies are a long way off considering the length of disease development in colon cancer, which can span twenty years from initial environmental insult and DNA damage through to tumour detection.

Reducing the risk of colon cancer
As mentioned above, the human gut microflora plays an important role in colon cancer with components serving as cancer risk factors and others thought to protect against cancer development. Prebiotic oligosaccharides too have shown potential as cancer protectants. Data mainly from animal models of colon cancer, have shown that prebiotic oligosaccharides through their impact on the colonic environment can protect against DNA damage, reduce the formation of toxic and carcinogenic compounds, reduce the size and incidence of aberrant crypt foci (pre-cancerous lesions of the colonic mucosa) and impact favourably in tumour development.

Hughes and Rowland (2001) showed that long chain inulin and inulin-derived fructo-oligosaccharides significantly increased apoptosis in the colonic crypts of rats challenged with the colonic carcinogen 1,2-dimethylhydrazine. Within the colonic crypt, apoptosis is the means by which older, differentiated epithelial cells are sloughed off and replaced by newly generated cells. It may serve as a means of eliminating mutated and damaged cells and thus protect against colon cancer development in the earliest of stages. Long chain inulin has been shown to reduce the number of aberrant crypt foci (ACF) in mature rats challenged with the carcinogen azoxymethane. This reduction in ACF, which are generally accepted as reliable markers for colon carcinogenesis, occurred in a dose-dependant manner, with a 65% reduction in the occurrence of ACF in rats fed 10 g inulin per 100 g diet (Verghese et al. 2002a). These authors then went on to show that long chain inulin at 10% (w/w) diet suppressed tumour development in this rat model particularly at the promotion stage (Verghese et al. 2002b). The mechanisms by which prebiotics offer protection from colon cancer may be multi-factorial but are likely to be derived from their fermentative metabolism by the gut microflora. Butyrate in particular, which is produced during carbohydrate fermentation within the colon, is a potent regulator of colonocyte proliferation and differentiation (Williams et al. 2003). Butyrate is also directly involved in the prevention of cancer through its role in hyperacetylation of histone proteins (Tran et al. 1998) and through its role in regulating mucosal apoptosis (Ruemmele et al. 2003). The cancer preventative activities of prebiotics may also be attributed to their impact on bacterial numbers within the colon. Species of bifidobacteria and lactobacilli have been shown to reduce DNA damage in carcinogen treated animal models of colon cancer (Pool-Zobel et al. 1996). This protection from DNA damage was shown to be dose-dependant and only mediated by viable bacterial cells. Prebiotics are a proven means of stimulating numbers and activity of viable bifidobacteria and lactobacilli in situ in the colon. Other mechanisms of protection from carcinogenesis may include stimulation of mucosal enzyme activities such as glutathione transferase or regulation of the inflammatory immune response. Both the metabolic end products of prebiotic metabolism and direct microbial interactions (e.g. a down-regulation of the inflammatory response by bifidobacteria or lactobacilli) may be responsible for such activities (Challa et al. 1997; Perdigon et al. 1998; Burns and Rowland 2000). Recent studies with gnotobiotic animals have shown that dietary intervention with chicory-derived inulin and fructo-oligosaccharide greatly impacts on mucosal architecture and microbial colonization. Compared with a standard rodent diet, the prebiotic-supplemented diets resulted in higher villi and deeper crypts within the mucosa, a thicker mucus layer and increased number of mucin-secreting goblet cells within the colonic mucosa. The predominant mucin type was also affected by prebiotic ingestion, with sulphomucins predominating in the fructan-fed rats compared with sialomucins predominating in the rats fed standard chow. Importantly, there was also a significant difference in the number of bifidobacteria associated with the colonic wall of the fructan-fed rats compared with the controls (Kleesen et al. 2003). The mechanisms by which bifidobacteria adhere to the colonic mucosa have not as yet been identified. However, the demonstration that prebiotics, in this case the chicory-derived prebiotics, can modulate mucosal architecture and mucosal adhesion of bifidobacteria may have important implications when considering the mechanisms by which prebiotics appear to protect against colon cancer.

6.4.3 Synbiotics

A synbiotic is a probiotic combined with a prebiotic. This may be a rational way in which to progress dietary intervention studies. Use of an appropriate (selectively fermented) carbohydrate should help to fortify the live addition in the gut, whilst the dual advantages of both approaches may also be realized. To help deliver probiotics to the lower bowel, encapsulation is possible. No products exist that use prebiotics as the encapsulation material, but an appropriate choice of molecular weight may help persistence throughout the colon.

Several studies have been carried out in humans on the effectiveness of synbiotics. Bouhnik *et al.* (1996) examined the ability of a synbiotic mix containing inulin and *Bifidobacterium* spp. to modulate the gut microflora of healthy volunteers. Although an overall increase in numbers of faecal bifidobacteria upon synbiotic ingestion, the authors concluded that no additional increase was observed solely due to the prebiotic component. The synbiotic approach has proven successful, however, in other studies. A fermented milk product containing yoghurt starter strains and *Lactobacillus acidophilus* plus 2.5% fructo-oligosaccharides was shown to decrease total serum cholesterol levels, and reducing the ratio of low density lipoprotein-cholesterol to high density lipoprotein, an alteration in lipid metabolism seen as protective against coronary heart disease (Schaafsma *et al.* 1998). Further evidence that synbiotic formulations may prove more effective than their constituent probiotic and prebiotic moieties come from studies with animal models of colon cancer. Synbiotic products containing *B. longum* and lactulose or inulin have been shown to reduce the number and size of ACF in azoxymethane-challenged rats, with the synbiotic outperforming the prebiotic or *Bifidobacterium* alone (Rowland *et al.* 1998; Gallagher and Khil, 1999). Femia *et al.* (2002) showed that rats fed Synergy 1 (a mixture of fructo-oligosaccharides and inulin) or Synergy 1 and *L. rhamnosus* GG and *B. lactis* BB12, developed fewer colonic tumours upon azoxymethane challenge than did rats fed the probiotic strains alone.

6.5 In vitro and in vivo measurement of microbial activities

An array of model systems of the gastrointestinal microbial environment of varying degrees of complexity have been developed and validated in recent years (Rumney and Rowland 1992; Molly *et al.* 1994). Such models, based around continuous flow culture, enable us to look at the human gut microflora under laboratory conditions and investigate such microflora-associated activities as fermentation of dietary constituents (e.g. dietary fibre, proteins and prebiotics). These models of the colonic microflora are invaluable in the development of efficacious or novel prebiotics. They may also be useful in conducting initial studies on the effect of antibiotics on the complex gut microflora and in DNA transfer studies between members of the gut microflora and genetically modified foods (Tuohy *et al.* 2002; Payne *et al.* 2003).

There is a need for more realistic models incorporating mammalian cells with the human gut microflora. One approach would be increased use of human flora-associated (HFA) animal models (Rumney and Rowland 1992). However, these systems are expensive and their relevance to the human situation may be limited to more general studies, for example, fermentation patterns and DNA transfer in the gut.

To further understand the mechanisms underlying the interaction between human cells and both beneficial and deleterious members of the human gut microflora, human cells may be incorporated into model systems. Traditional human cell culture and ex vivo tissue

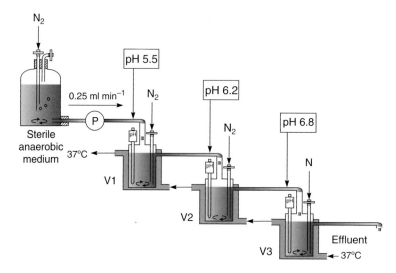

Figure 6.3 A three-stage compound continuous flow model of the human colonic microflora (after Macfarlane et al. 1998).

samples from hospital patients provide a starting point about which in vitro models allowing cultivation of human cells and members of the gut microflora (especially anaerobic species) may be developed.

A variety of model systems exist for determining the effects of probiotics and prebiotics on the gut flora. These help to better inform and plan well-conducted human/animal trials. There are various limitations and advantages. For example, multiple stage chemostat systems allow a prediction on the site of interaction in the gastrointestinal tract and are useful for 'challenge' tests not possible in humans, for example, with pathogens or genetically engineered strains (Figure 6.3). Laboratory animals can be used to determine immunological effects. In vitro cell lines are useful for attachment studies and cytokine expression work. Biopsy collections give information on microbiology at the mucosal interface. Useful biomarkers of functionality (organic acids, bioactive molecules) should be used in concert with reliable indices of microflora change. Various complementary systems have been developed and should be applied with the research hypothesis in mind.

6.6 Molecular methodologies for assessing microflora changes

There has been a move towards more molecular based assessment of gut microflora changes in response to probiotics and prebiotics. This applies to laboratory, model and human studies. The research has been driven by the subjective approach of conventional culture-based microbiology, as well as the extremely complex community structure of the gut. Examples are given below.

6.6.1 Fluorescent in situ hybridization

Data generated through sequencing of 16S rRNA genes of bacterial isolates or whole community 16S rDNA analysis from environmental samples, have enabled the generation of oligonucleotide hybridization probes targeting important groups of bacteria in the gut. Coupling such probes with fluorescent microscopy or flow cytometry allows the direct quantification of phylogenetically related groups of bacteria in gastrointestinal samples. An array of probes targeting important groups of bacteria present in the gut microflora are frequently being applied to monitor changes in bacterial numbers in response to diet, age, disease states and antibiotic therapy (Amann *et al.* 1990; Harmsen *et al.* 2002). This enables both the culturable and non-culturable moieties of the human gut microflora to be quantified. As our understanding of gut microflora composition is expanding through isolation of novel bacterial species and whole community analysis of 16S rDNA sequences, so too does the range of 16S rRNA probes available to enumerate important groups of gut bacteria.

The recent application of flow cytometry to the enumeration of bacterial populations labelled with fluorescent probes has met with some success (Wallner *et al.* 1995; Zoetendal *et al.* 2002). The rapid sample handling of flow cytometry allows us to employ a greater number of oligonucleotide probes to monitor an increased range of phylogenetically related groups of gut bacteria in a relatively short period of time compared with fluorescent microscopy. None the less, as with dot blot hybridization, the application of flow cytometry at best allows the determination of relative percentage changes in 16S rRNA species compared with total 16S rRNA pools and not exact cell numbers. Sophisticated image analysis software and automated fluorescent microscopy may in the future combine the quantitative power of fluorescent in situ hybridization (FISH) analysed microscopically with high-throughput, labour-saving systems (Jansen *et al.* 1999).

6.6.2 DNA microarrays – microbial diversity and gene expression studies

The ability to construct arrays consisting of thousands of different oligonucleotide probes on a single glass slide or microchip combined with the wealth of information generated through the human and prokaryotic genome sequencing projects has opened up a range of exciting possibilities in the field of gut microbiology (Kuipers *et al.* 1999). DNA microarrays may be used to study host microbe cross-talk in the gut, identifying human genes expressed in response to probiotic strains and members of the gut microflora (Cummings and Relman 2000; Hooper *et al.* 2001). Microarrays may also be constructed with oligonucleotide probes targeting genes encoding important probiotic traits thus allowing rapid screening of putative probiotic isolates (Klaenhammer *et al.* 2002). Another possible application of DNA microarrays is in the rapid characterization of gut microflora composition, with arrays consisting of 16S rDNA gene sequences (Guschin *et al.* 1997; Rudi *et al.* 2002). This is an active area of research with membrane arrays consisting of up to 60 different oligonucleotide probes being described (Wang *et al.* 2002).

6.6.3 Monitoring gene expression – subtractive hybridization and in situ PCR/FISH

Subtractive hybridization allows the isolation of differentially expressed sequences through the elimination of common sequences present in two pools of cDNA (e.g. from mRNA pools present in colonocytes before and after incubation with a probiotic strain) (Hubank and Schatz 1999). Although lacking the scope of DNA microarrays, such techniques enable us to look at specific differences in gene expression or gene content between bacteria or cultured human cells (Soares 1997; Akopyants *et al.* 1998). Such studies constitute a starting point from which further development of more high-cost, high-throughput techniques such as DNA microarrays may be justified.

6.6.4 Proteomics

Modern proteomics allow us for the first time to look at the totality of proteins produced by cells. Techniques such as matrix-associated laser-desorption ionization-time of flight mass spectrometry, (MALDI-TOF-MS) enable changes in protein composition on the surface of bacterial and mammalian cells under different environmental conditions to be determined. By combining proteomic techniques with the wealth of data generated through genome sequencing projects and DNA microarrays, a fuller understanding of the biological functioning of cells may be obtained in the near future. In the context of gut microbiology, proteomics promises to bridge the gap between the genetic information encoded by probiotic strains and their expressed phenotypes. Similarly, techniques such as MALDI-TOF-MS may enable the identification of important probiotic traits such as mucosal attachment sites and immunologically active proteins that are expressed on the surface of probiotic strains.

In recent years, advances in molecular technologies based on rRNA have shed new light on the diversity of the gut microflora (Vaughan *et al.* 2000). In particular, rRNA gene sequencing studies have revealed the presence of a myriad of previously undiscovered species (Suau *et al.* 1999). It is now clear that a major gap in our knowledge exists concerning the diversity of organisms resident within the human gut. Some of the new diversity studies are likely to uncover hitherto unrecognized probiotics, and these may have advantages over existing strains that are in use. rRNA sequencing provides an excellent means of characterizing organisms in terms of resolving power but is time consuming. The ultimate aim is to characterize the 'microflora at a glance.' Technologies available include genetic probing strategies by microscopy, image analysis or flow cytometry; microarray developments; genetic fingerprinting studies; direct community analysis; RT-PCR, etc. These genotypic methods should be used in conjunction with conventional cultural techniques to improve our knowledge of the gut flora and its interactions. Some techniques are fully qualitative and give an overall picture of the diversity present, others are quantitative but require prior knowledge of the target organisms. Several complementary approaches and recognition of limitations is desirable.

Such genomic approaches have allowed improved probe design (in some cases at the species or strain levels) and are being increasingly applied to both probiotic and prebiotic research. One fundamental observation is that there are age-related changes in the gut microflora composition. Moreover, there may be geographical variation and little commonality between individuals. Nevertheless, the database on flora diversity studies has expanded markedly in recent years and provided much needed information on groups that are more likely to benefit from pro/prebiotic intake.

A further advance would be the use of MALDI-TOF-MS, which provides high resolution proteomic-based comparisons of whole bacterial cells. It is planned to use MALDI-TOF-MS to assemble a database of species-characteristic profiles to facilitate the rapid identification of gut anaerobes. Intact cells and single colonies can be analysed, and the automated acquisition of mass spectra from 96 well target plates (single run) will be ideally suited to the high-throughput required for examining population dynamics of large numbers of samples. Profiles of all species known to reside in the human gut would be generated. Use of a proteomic approach in conjunction with ongoing genomic data (extensive 16S rRNA gene sequencing) will greatly facilitate the recognition of gut microflora composition. Strains giving rise to unidentified proteomic profiles may be subjected to gene sequence analysis to facilitate their detailed phylogenetic characterization. This will permit a parallel updating of proteomic and genomic databases which will provide an invaluable resource for future gut ecological studies.

In terms of functionality, transcriptomics could look at activity through mRNA expression studies, whilst metabolomics, in concert with NMR spectroscopy could be used to assay, in an unambiguous manner, diagnostic sets of biomarkers including microbial metabolites. Transcriptomics and metabolomics are still in their infancy from the microbial perspective but progression is rapid and promises to allow activity measurements also to be encompassed in new studies. It is proposed that current and emerging molecular-based information be collated into a functional approach and directed towards disorders for which treatment is ill defined or even lacking but has the potential to be managed by pro, pre and synbiotics. The trials should be done in multiple countries and would be a good progression for current work which is developing the technology, generating new test materials, exploring mechanisms, determining safety, identifying best products, etc. The application of post-genomic principles in gut microbial studies will help to fully explore human gut microflora diversity, develop reliable model systems, test a new generation of purpose designed pro/prebiotic molecules with enhanced functionality and determine the effectiveness of dietary intervention in the clinical situation.

6.7 Assessing the impact of dietary modulation of the gut microflora – does it improve health, what are the likelihoods for success, and what are the biomarkers of efficacy?

The scientific rationale behind the opinion that bifidobacteria and lactobacilli are beneficial to human health and the mechanisms underlying the proven success of certain strains is still unclear. Identification of genes expressed by bacteria, both probiotic strains and pathogens, in response to cultured human cells may also provide an important supportive rationale for the beneficial modulation of the human gut microflora through dietary means (Graham and Clark-Curtiss 1999; Li et al. 2001). Of critical importance to the future development of probiotics and prebiotics is the establishment of mechanisms underlying the beneficial interaction of bifidobacteria and lactobacilli with host cells and other members of the gut microflora. There is a need to identify clear biomarkers of probiotic effect or improvement in health status, not only in patients suffering from gastrointestinal complaints but also in healthy volunteers. Existing human feeding studies have shown the ability of the probiotic/prebiotic and synbiotic approaches to effectively increase numbers of bifidobacteria and lactobacilli in the gut. The challenge now is to correlate such changes with real improvements in the gastrointestinal health status of patients. In particular, we need to establish the relationship between increased numbers of probiotic bacteria in the gut (bifidobacteria and lactobacilli) and health parameters such as immune stimulation, effect on biomarkers of colon cancer (faecal water genotoxicity, butyrate production, microbial enzyme activities), alleviation of the symptoms of IBS and IBD and degree of colonization resistance to gastrointestinal pathogens (e.g. through human flora-associated animal feeding studies). Such studies will also establish effective doses of probiotics, prebiotics and synbiotics required to bring about specific measurable improvements in health biomarkers. Many studies on probiotic recovery and prebiotic functionality have been carried out in healthy persons (Tuohy et al. 2003) and there is now a requirement to assess the clinical impact of this. Similarly, their feeding to companion animals and farm livestock may improve nutritional status but the health values are much less known. One major question for probiotics and prebiotics is therefore: 'what are the consequences of gut microflora modulation and how do they occur?' This accepts that the best products will modify the gut microflora composition but addresses the applied consequences of this. Moreover, given the lack of mechanistic data on their use it is imperative to generate hypothesis-driven research that determines functionality. A harnessing of multiple disciplines that exploit the best technologies available should now address these issues (Gibson and Fuller 2000). The long-term physiological effects of dietary intervention also need clarification.

6.8   Justification for the use of probiotics and prebiotics to modulate the gut flora composition

Diseases of the gastrointestinal tract are of major economic and medical concern. For example, reported infections from agents of food-borne disease such as *Campylobacter* spp., *E. coli* and *Salmonella* spp. continue to increase. This is further exacerbated by the continuous emergence of novel variants of established pathogens. Such acute infections are widespread and are said to affect almost everyone at some point in their lives. On a chronic basis, inflammatory bowel disease, colon cancer and irritable bowel syndrome have all been linked to intestinal microorganisms and their activities (Chadwick and Anderson 1995; Burns and Rowland 2000). The gut flora may also be linked with certain systemic states. The site of action, namely the human gut, is a relatively under explored ecosystem and yet affords the best opportunity for reducing the impact of food-related disease(s). This is amplified by the fact that few effective working therapies exist for most gut disorders, while often the approach is to attempt to manage conditions through non-specific approaches involving anti-inflammatory drugs or antibiotics. In the severest of cases, surgery may be required whilst some states like colorectal cancer can be fatal (Yancik et al. 1998; Becker 1999). As such, clinicians, patients and medical authorities are becoming increasingly interested in defining alternative approaches that may be either prophylactic or curative. Probiotics and prebiotics have a track record of being safe and a long history of use in humans (Adams and Marteau 1995) and are popular 'dietary intervention' tools for modulating the gut microflora composition and activities. It is suspected that pathogenic bacteria are the aetiological agents of many acute and chronic gut disorders and probiotics/prebiotics may exert suppressant effects on such components of the flora. Comparative studies in multiple centres have clear advantages, as long as the technology transfer is reliable. Hypothesis based research can help product development. Such developments ought to produce more targeted pro-, pre- and synbiotics to help specific disease states. The targets should be planned around situations where a defined aetiology is suspected or confirmed. The main intention is to address the health consequences of flora modulation through exploiting current technological developments. Ultimately, both the effects and the mechanisms behind them should be unravelled, that is provide consumers with the definitive health aspects and also give accurate information on why they occur.

References

Adams, M.R. and Marteau, P. (1995) Letter to the editor. On the safety of lactic acid bacteria from food. *Int. J. Food Microbiol.*, **27**, 263–264.

Akopyants, N.S., Fradkov, A., Diatchenko, L. et al. (1998) PCR-based subtractive hybridisation and differences in gene content among strains of *Helicobacter pylori*. *Proc. Natl. Acad. Sci. USA*, **95**, 13108–13113.

Amann, R.I., Binder, B.J., Olson, R.J. et al. (1990) Combination of 16S rRNA-targeted oligonucleotide probes with flow cytometry for analysing mixed microbial populations. *Appl. Environ. Microbiol.*, **56**, 1919–1925.

Amann, R.J., Ludwig, W. and Schleifer, K.H. (1995) Phylogenetic identification and *in situ* detection of individual microbial cells without cultivation. *Microbiol. Rev.*, **59**, 143–169.

Araki, Y., Andoh, A., Fujiyama, Y. et al. (2000) Effects of germinated barley foodstuff on microflora and short chain fatty acid production in dextran sulphate sodium-induced colitis in rats. *Biosci. Biochem.*, **64**, 1794–1800.

Arimochi, H., Kinouchi, T., Kataoka, K. et al. (1997) Effect of intestinal bacteria on formation of azoxymethane-induced aberrant crypt foci in the rat colon. *Biochem. Biophys. Res. Commun.*, **238**, 753–757.

Asahara, T., Nomoto, K., Shimizu, K. et al. (2001) Increased resistance of mice to *Salmonella enterica* serovar Typhimurium infection by synbiotic administration of bifidobacteria and transgalactosylated oligosaccharides. *J. Appl. Microbiol.*, **91**, 985–996.

Ballongue, J., Schumann, C. and Quignon, P. (1997) Effects of lactulose and lacticol on colonic microflora and enzymatic activity. *Scand. J. Gastroenterol.*, **32**, 41–44.

Bamba, T., Kanauchi, O., Andoh, A. et al. (2002) A new prebiotic from germinated barley for nutraceutical treatment of ulcerative colitis. *J. Gastroenterol. Hepatol.*, **17**, 818–824.

Baron, J.H. (1963) Studies of basal and peak acid output with an augmented histamine meal. *Gut*, **4**, 136–144.

Becker, J.M. (1999) Surgical therapy for ulcerative colitis and Crohn's disease. *Gastroenterol. Clin. North Am.*, **28**, 371–390.

Bingham, M.O. (2003) Dietary intervention in autistic spectrum disorders. *Food Sci. Technol. Bull.*, **1**(4), 1–11.

Boehm, G., Lidestri, M., Casetta, P. et al. (2002) Supplementation of a bovine milk formula with an oligosaccharide mixture increases counts of faecal bifidobacteria in preterm infants. *Arch. Dis. Child Fetal Neonatal Ed.*, **86**(3), F178–F181.

Bouhnik, Y., Flourie, B., Andrieux, C. et al. (1996) Effect of *Bifidobacterium* sp. fermented milk ingested with and without inulin on colonic bifidobacteria and enzymatic activities in healthy humans. *Eur. J. Clin. Nutr.*, 50, 269–273.

Bouhnik, Y., Vahedi, K., Achour, L. et al. (1999) Short-chain fructo-oligosaccharide administration dose-dependantly increases fecal bifidobacteria in healthy humans. *J. Nutr.*, 129, 113–116.

Breuer, R.I., Soergel, K.H., Lashner, B.A. et al. (1997) Short chain fatty acid rectal irrigation for left-sided ulcerative colitis: a randomised, placebo-controlled trial. *Gut*, 40, 485–491.

Brommage, R., Binacua, C., Antille, S. et al. (1993) Intestinal calcium absorption in rats is stimulated by dietary lactulose and other resistant sugars. *J. Nutr.*, 123, 2186–2194.

Buddington, R.K., Williams, C.H., Chen, S.C. et al. (1996) Dietary supplement of neosugar alters the faecal flora and decreases activities of some reductive enzymes in human subjects. *Am. J. Clin. Nutr.*, 63, 709–716.

Burns, A.J. and Rowland, I.R. (2000) Anti-carcinogenicity of probiotics and prebiotics. *Curr. Issues Intest. Microbiol.*, 1, 13–24.

Campieri, M., Rizzello, F., Venturi, A. et al. (2000) Combination of antibiotic and prebiotic treatment is efficacious in prophylaxis of post-operative recurrence of Crohn's disease: a randomised controlled study vs mesalamine. *Gastroenterol.*, 118, G4179.

Chadwick, V.S. and Anderson, R.P. (1995) The role of intestinal bacteria in etiology and maintenance of inflammatory bowel disease. In G.R. Gibson and G.T. Macfarlane (eds), *Human Colonic Bacteria: Role in Nutrition, Physiology and Pathology*. CRC Press, Boca Raton, FL, pp. 227–256

Challa, A., Rao, D.R., Chawan, C.B. et al. (1997) *Bifidobacterium longum* and lactulose suppress azoxymethane induced aberrant crypt foci in rats. *Carcinogenesis*, 18, 517–521.

Cherbut, C., Michel, C. and Lecannu, G. (2003) The prebiotic characteristics of fructo-oligosaccharides are necessary for reduction of TNBS-induced colitis in rats. *J. Nutr.*, 133, 21–27.

Conway, P.L. (1995) Microbial ecology of the human large intestine. In G.R. Gibson and G.T. Macfarlane (eds), *Human Colonic Bacteria: Role in Nutrition, Physiology and Pathology*. CRC Press, Boca Raton, FL, pp. 1–24.

Coudray, C., Demigne, C. and Rayssiguier, Y. (2003) Effects of dietary fibres on magnesium absorption in animals and humans. *J. Nutr.*, 133, 1–4.

Crittenden, R., Laitila, A., Forssell, P. et al. (2001) Adhesion of bifidobacteria to granular starch and its implications in probiotic technologies. *Appl. Environ. Microbiol.*, 76, 3469–3475.

Cummings, C.A. and Relman, D.A. (2000) Using DNA microarrays to study host:microbe interactions. *Emerg. Infect. Dis.*, 6, 513–525.

Cummings, J.H., Macfarlane, G.T. and Englyst, H.N. (2001) Prebiotic digestion and fermentation. *Am. J. Clin. Nutr.*, 73, 415S–420S.

Das, U.N. (2002) The lipids that matter from infant nutrition to insulin resistance. *Prostaglandins Leukot. Essent. Fatty Acids*, 67, 1–12.

Davidson, M.H. and Maki, K.C. (1999) Effects of dietary inulin on serum lipids. *J Nutr.*, 129, 1474S–1477S.

Delzenne, N., Aertssens, J., Verplaetse, H. et al. (1995) Effect of fermentable fructo-oligosaccharides on mineral, nitrogen and energy digestive balance in the rat. *Life Sci.*, 57, 1579–1587.

Delzenne, N.M. and Williams, C.M. (2002) Prebiotics and lipid metabolism. *Curr. Opin. Lipidol.*, 13, 61–67.

D'Souza, A.L., Rajkumar, C., Cooke, J. et al. (2002) Probiotics in prevention of antibiotic associated diarrhoea: meta-analysis. *Br. Med. J.*, 324, 1361.

Dunne, C., O'Mahony, L., Murphy, L. et al. (2001) *In vitro* selection criteria for probiotic bacteria of human origin: correlation with in vivo findings. *Am. J. Clin. Nutr.*, 73, 386S–392S.

Edwards C.A. and Parrett, A.M. (2002) Intestinal flora during the first months of life: new perspectives. *Br. J. Nutr.*, 88 (Suppl. 1), S11–S18.

Evans, D.J., Jr. and Evans, D.G. (2001) Helicobacter pylori CagA: analysis of sequence diversity in relation to phosphorylation motifs and implications for the role of CagA as a virulence factor. *Helicobacter*, **6**, 187–198.

Femia, A.P., Luceri, C., Dolara, P. *et al.* (2002) Antomutagenic activity of the prebiotic inulin enriched with oligofructose in combination with probiotics *Lactobacillus rhamnosus* and *Bifidobacterium lactis* on azoxymethane induced colon carcinogenesis in rats. *Carcinogenesis*, **23**, 1953–1960.

Finegold, S.M., Molitoris, D., Song, Y. *et al.* (2002) Gastrointestinal microbiota studies in late-onset autism. *Clin. Infect. Dis.*, **35**, S6–S16.

Fite, A., Macfarlane, G.T. and Cummings, J.H. (2004) Identification and quantitation of mucosal and faecal desulfovibrios using real time polymerase chain reaction. *Gut*, **53**, 523–529.

Frank, A. (1998) Prebiotics stimulate calcium absorption: a review. *Milchwissenschraft*, **53**, 427–429.

Fukui, M., Fujino, T., Tsutsui, K. *et al.* (2001) The tumor-preventing effect of a mixture of several lactic acid bacteria on 1,2-dimethylhydrazine-induced colon carcinogenesis in mice. *Oncol. Rep.*, **8**, 1073–1078.

Fuller, R. (1989) Probiotics in man and animals. *J. Appl. Bacteriol.*, **66**, 365–378.

Gallaher, D.D. and Khil, J. (1999) The effect of synbiotics on colon carcinogenesis in rats. *J. Nutr.*, **129**, 1483S–1487S.

Gaon, D., Garcia, H., Winter, L. *et al.* (2003) Effect of *Lactobacillus* strains and *Saccharomyces boulardii* on persistent diarrhea in children. *Medicina*, **63**, 293–298.

Ge, Z. and Taylor, D.E. (1999) Contributions of genome sequencing to understanding the biology of *Helicobacter pylori*. *Annu. Rev. Microbiol.*, **53**, 353–387.

Gibson, G.R. and Fuller, R. (2000) Aspects of *in vitro* and *in vivo* research approaches directed toward identifying probiotics and prebiotics for human use. *J. Nutr.*, **130**, 391S–395S.

Gibson, G.R., Beatty, E.R., Wang, X. *et al.* (1995) Selective stimulation of bifidobacteria in the human colon by oligofructose and inulin. *Gastroenterol.*, **108**, 975–982.

Gibson, G.R., Ottaway, P.B. and Rastall, R.A. (2000) *Prebiotics: New Developments in Functional Foods*. Chandos Publishing Ltd. Oxford.

Gionchetti, P., Rizzello, F., Venturi, A. *et al.* (2000) Oral bacteriotherapy as maintenance treatment in patients with chronic pouchitis: a double-blind, placebo-controlled trial. *Gastroenterology*, **119**, 305–309.

Gionchetti, P., Rizzello, F., Helwig, U. *et al.* (2003) Prophylaxis of pouchitis onset with probiotic therapy: a double-blind, placebo-controlled trial. *Gastroenterology* **124**, 1202–1209.

Gorbach, S.L. (1993) Perturbation of intestinal microflora. *Vet. Human Toxicol.*, **35** (Suppl. 1), 15–23.

Graham, J.E. and Clark-Curtiss, J.E. (1999) Identification of *Mycobacterium tuberculosis* RNAs synthesized in response to phagocytosis by human macrophages by selective capture of transcribed sequences (SCOTS). *Proc. Natl. Acad. Sci. USA*, **96**, 11554–11559.

Greger, J.L. (1999) Nondigestible carbohydrates and mineral bioavailability. *J. Nutr.*, **129**, 1434S–1435S.

Griffin, I.J., Davila, P.M. and Abrams, S.A. (2002) Non-digestible oligosaccharides and calcium absorption in girls with adequate calcium intakes. *Br. J. Nutr.*, **87**, S187–S191.

Guschin, D.Y., Mobarry, B.K., Proudnikov, D. *et al.* (1997) Oligonucleotide microchips as genosensors for determinative and environmental studies in microbiology. *Appl. Environ. Microbiol.*, **63**, 2397–2402.

Guslandi, M., Mezzi, G., Sorghi, M. *et al.* (2000) *Saccharomyces boulardii* in maintenance treatment of Crohn's disease. *Dig. Dis. Sci.*, **45**, 1462–1464.

Harmsen, H.J., Raangs, G.C., He, T. *et al.* (2002) Extensive set of 16S rRNA-based probes for detection of bacteria in human feces. *Appl. Environ. Microbiol.*, **68**, 2982–2990.

Hayakawa, K., Mizutani, J., Wada, K. et al. (1990) Effects of soybean oligosaccharides on human faecal flora. *Microb. Ecol. Health Dis.*, **3**, 293–303.

Hooper, L.V., Wong, M.H., Thelin, A. et al. (2001) Molecular analysis of commensal host-microbial relationships in the intestine. *Science,* **291**, 881–884.

Hopkins, M.J., Sharp, R. and Macfarlane, G.T. (2001) Age and disease related changes in intestinal bacterial populations assessed by cell culture, 16S rRNA abundance, and community cellular fatty acid profiles. *Gut,* **48**, 198–205.

Hubank, M. and Schatz, D.G. (1999) cDNA representational difference analysis: a sensitive and flexible method for identification of differentially expressed genes. *Methods Enzymol.*, **303**, 325–349.

Hughes, R. and Rowland, I.R. (2001) Stimulation of apoptosis by two prebiotic chicory fructans in the rat colon. *Carcinogenesis,* **22**, 43–47.

Hunter, J.O., Lee, A.J., King, T.S. et al. (1996) *Enterococcus faecium* strain PR88 – an effective probiotic. *Gut,* **38** (Suppl. 1), A62.

Husebye, E., Skar, V., Hoverstad, T. et al. (1992) Fasting hypochlorhydria with gram positive gastric flora is highly prevalent in healthy old people. *Gut,* **33**, 1331–1337.

Isolauri, E., Kaila, M., Mykkanen, H. et al. (1994) Oral bacteriotherapy for viral gastroenteritis. *Dig. Dis. Sci.,* **39**, 2595–2600.

Isolauri, E., Arvilommi, H. and Salminen, S. (1999) Gastrointestinal infections. In G.R. Gibson and M.B. Roberfroid (eds), *Colonic Microflora, Nutrition and Health.* Kluwer Academic Publishers, Netherlands, pp. 267–279.

Ito, M., Deguchi, Y., Miyamori A. et al. (1990) Effects of administration of galacto-oligosaccharides on the human fecal microflora, stool weight and abdominal sensation. *Microb. Ecol. Health Dis.,* **3**, 285–292.

Ito, M., Deguchi, Y., Matsumoto, K. et al. (1993) Influence of galacto-oligosaccharides on the human fecal microflora. *J. Nutr. Sci. Vitaminol.,* **39**, 635–640.

Jansen, G.J., Wildeboer-Veloo, A.C.M., Tonk, R.H.J. et al. (1999) Development and validation of an automated, microscopy-based method for enumeration of groups of intestinal bacteria. *J. Microbiol. Methods,* **37**, 215–221.

Kaila, M.,, Isolauri, E., Soppi, E. et al. (1992) Enhancement of the circulating antibody secreting cell response in human diarrhea by a human *Lactobacillus* strain. *Pediatr. Res.,* **32**, 141–144.

Kalliomaki, M., Salminen, S., Arvilommi, H. et al. (2001) Probiotics in primary prevention of atopic disease: a randomised placebo-controlled trial. *Lancet,* **357**, 1076–1079.

Kaneko, T., Kohmoto, T., Kikuchi H. et al. (1994) Effects of isomalto-oligosaccharides with different degrees of polymerisation on human fecal bifidobacteria. *Biosci. Biotechnol. Biochem.,* **58**, 2288–2290.

Kim, H.J., Camilleri, M., McKinzie, S. et al. (2003) A randomized controlled trial of a probiotic, VSL#3, on gut transit and symptoms in diarrhoea-predominant irritable bowel syndrome. *Aliment. Pharmacol. Ther.,* **17**, 895–904.

Klaenhammer, T., Altermann, E., Arigoni, F. et al. (2002) Discovering lactic acid bacteria by genomics. *Antonie van Leeuwenhoek,* **82**, 29–58.

Kleessen, B., Sykura, B., Zunft, H.J. et al. (1997) Effects of inulin and lactose on faecal microflora, microbial activity and bowel habit in elderly constipated persons. *Am. J. Clin. Nutr.,* **65**, 1397–1402.

Kleesen, B., Hartmann, L. and Blaut, M. (2003) Fructans in the diet cause alterations of intestinal mucosal architecture, released mucins and mucosa-associated bifidobacteria in gnotobiotic rats. *Br. J. Nutr.,* **89**, 597–606.

Kohomoto, T., Fukui, F., Takaku, H. et al. (1988) Effect of isomalto-oligosaccharides on human fecal flora. *Bifidobacteria Microflora,* **7**, 61–69.

Kruger, M.C., Brown, K.E., Collett, G., Layton, L. and Schollum, L.M. (2003) The effects of fructo-oligosaccharides with variuos degrees of polymerization on calcium bioavailability in the growing rat. *Exp. Biol. Med.*, **228**, 683–688.

Kruse, H.P., Kleesen, B. and Blaut, M. (1999) Effect of inulin on faecal bifidobacteria in human subjects. *Br. J. Nutr.*, **82**, 375–782.

Kuipers, O.P., de Jong, A. Holsappel, S. *et al.* (1999) DNA-microarrays and food-biotechnology. *Antonie van Leeuwenhoek*, **76**, 353–355.

Li, M.-S., Monahan, I.M., Waddell, S.J. *et al.* (2001) cDNA-RNA subtractive hybridisation reveals increased expression of mycocerosic acid synthase in intracellular *Mycobacterium bovis* BCG. *Microbiology*, **147**, 2293–2305.

Likotrafiti, E., Manderson, K.S., Fava, F. *et al.* (2004) Molecular identification and antipathogenic activities of putative probiotic bacteria isolated from faeces of healthy elderly individuals. *Microb. Ecol. Health Dis.*, **16**, 105–112.

Lovat, L.B. (1996) Age related changes in gut physiology and nutrition status, *Gut*, **38**, 306–309.

Macfarlane, G.T., Macfarlane, S. and Gibson, G.R. (1998) Validation of a three-stage compound continuous culture system for investigating the effect of retention time on the ecology and metabolism of bacteria in the human colon. *Microb. Ecol.*, **35**, 180–187.

Majamaa, H., Isolauri, E., Saxelin, M. *et al.* (1995) Lactic acid bacteria in the treatment of acute rotavirus gastroenteritis. *J. Pediatr. Gastroenterol. Nutr.*, **20**, 333–338.

Mansour-Ghanaei, F., Dehbashi, N., Yazdanparast, K. *et al.* (2003) Efficacy of *Saccharomyces boulardii* with antibiotics in acute amoebiasis. *World J. Gastroenterol.*, **9**, 1832–1833.

Marteau, P., Seksik, P. and Jian, R. (2002) Probiotics and intestinal health: a clinical perspective. *Br. J. Nutr.*, **88**, S51–S57.

McCartney, A.L., Wenzhi, W. and Tannock G.W. (1996) Molecular analysis of the composition of the bifidobacterial and lactobacillus microflora of humans. *Appl. Environ. Microbiol.*, **62**, 4608–4613.

Mitsuoka, T. (1992) Intestinal flora and aging. *Nutr. Rev.*, **50**, 438–446.

Molly, K., Van De Woestyne, M., De Smet, I. *et al.* (1994) Validation of the simulator of the human intestinal microbial ecosystem (SHIME) reactor using microorganism-associated activities. *Microb. Ecol. Health Dis.*, **7**, 191–200.

Moore, W.E.C. and Holdeman, L.V. (1974) Human fecal flora: The normal flora of 20 Japanese-Hawaiians. *Appl. Microbiol.*, **27**, 961–979.

Moran, A.P. (1999) *Helicobacter pylori* lipopolysaccharide-mediated gastric and extragastric pathology. *J. Physiol. Pharmacol.*, **50**, 787–805.

Moro, G., Minoli, I., Mosca, M. *et al.* (2002) Dosage-related bifidogenic effects of galacto- and fructo-oligosaccharides in formula-fed term infants. *JPGN*, **34**, 291–295.

Mountzouris, K.C., McCartney, A.L. and Gibson, G.R. (2002) Intestinal microflora of human infants and current trends for its nutritional modulation. *Br. J. Nutr.*, **87**, 405–420.

Nagendra, R., Viswanatha, S. and Arun Kumar, S. (1995) Effect of feeding milk formula containing lactulose to infants on faecal bifidobacterial flora. *Nutr. Res.*, **15**, 15–24.

Niedzielin, K., Kordecki, H. and Birkenfeld, B. (2001) A controlled, double-blind, randomized study on the efficacy of *Lactobacillus plantarum* 299V in patients with irritable bowel syndrome. *Eur. J. Gastroenterol. Hepatol.*, **13**, 1143–1147.

Oberreuther-Moschner, D.L., Jahreis, G., Rechkemmer, G. *et al.* (2004) Dietary intervention with the probiotics *Lactobacillus acidophilus* 145 and *Bifidobacterium longum* 913 modulates the potential of human faecal water to induce damage in HT29clone19A cells. *Br. J. Nutr.*, **91**, 925–932.

Ohta, A., Motohashi, Y., Sakai K. *et al.* (1998) Dietary fructo-oligosaccharides increase calcium absorption and levels of mucosal calbindin-D9k in the large intestine of gastrectomized rats. *Scand. J. Gastroenterol.*, **33**, 1062–1068.

Oozeer, R., Goupil-Feuillerat, N., Alpert, C.A. *et al.* (2002) *Lactobacillus casei* is able to survive and initiate protein synthesis during its transit in the digestive tract of human flora-associated mice. *Appl. Environ. Microbiol.*, **68**, 3570–3574.

O'Sullivan, M.A. and O'Morain, C.A. (2000) Bacterial supplementation in the irritable bowel syndrome. A randomised double-blind placebo-controlled crossover study. *Dig. Liver Dis.*, **32**, 294–301.

Ott, J.-M. and Podolsky, D.K. (2004) Functional modulation of entrocytes by gram-positive and pram-negative microorganisms. *Am. J. Physiol. Gastrointest. Liver Physiol.*, **286**, G613–G626.

Parracho, H., Bingham, M., McCartney, A.L. *et al.* (2004) Differences between the gut microflora of children with autistic spectrum disorders and that of healthy children. *J. Med. Microbiol.*, **54**, 987–991.

Paubert-Braquet, M., Xiao-Hu Gan, Gaudichon C. *et al.* (1995) Enhancement of host resistance against Salmonella typhimurium in mice fed a diet supplemented with yogurt or milks fermented with various *Lactobacillus casei* strains. *Int. J. Immunother.*, **XI**(4), 153–161.

Payne, S., Gibson, G., Wynne, A. *et al.* (2003) *In vitro* studies on colonization resistance of the human gut microbiota to *Candida albicans* and the effects of tetracycline and *Lactobacillus plantarum* LPK. *Curr. Issues Intest. Microbiol.*, **4**, 1–8.

Pelto, L., Isolauri, E., Lilius, E.M. *et al.* (1998) Probiotic bacteria down-regulate the milk-induced inflammatory response in milk-hypersensitive subjects but have an immunostimulatory effect in healthy subjects. *Clin. Exp. Allergy*, **28**, 1474–1479.

Perdigon, G., Valdez, J.C. and Rachid, M. (1998) Antitumour activity of yoghurt: study of possible immune mechanisms. *J. Dairy Res.*, **65**, 129–138.

Pereira, D.I. and Gibson, G.R. (2002a) Effect of consumption of probiotics and prebiotics on serum lipid levels in humans. *Crit. Rev. Biochem. Mol. Biol.*, **37**, 259–281.

Pereira, D.I. and Gibson, G.R. (2002b) Cholesterol assimilation by lactic acid bacteria and bifidobacteria isolated from human gut. *Appl. Environ. Microbiol.*, **68**, 4689–4693.

Pool-Zobel, B.L., Neudecker, C., Domizlaff, I. *et al.* (1996) *Lactobacillus* and *Bifidobacterium* mediated antigenotoxicity in the colon of rats. *Nutr. Res.*, **26**, 365–380.

Pitcher, M.C., Beatty, E.R. and Cummings, J.H. (2000) The contribution of sulphate reducing bacteria and 5-aminosalicylic acid to faecal sulphide in patients with ulcerative colitis. *Gut*, **46**, 64–72.

Rafter, J. (2002a) Lactic acid bacteria and cancer: mechanistic perspective. *Br. J. Nutr.*, **88**, S89–S94.

Rafter, J.J. (2002b) Scientific basis of biomarkers and benefits of functional foods for reduction of disease risk: cancer. *Br. J. Nutr.*, **88** (Suppl. 2), S219–S224.

Rastall, R.A. and Maitin, V. (2002) Prebiotics and synbiotics: towards the next generation. *Curr. Opp. Biotechnol.*, **13**, 490–496.

Reckseidler, S.L., DeShazer, D., Sokol, P.A. *et al.* (2001) Detection of bacterial virulence genes by subtractive hybridisation: identification of capsular polysaccharide of *Burkholderia pseudomallei* as a major virulence determinant. *Infect. Immun.*, **69**, 34–44.

Reid, G. (1999) The scientific basis for probiotic strains of *Lactobacillus*. *Appl. Environ. Microbiol.*, **65**, 3763–3766.

Rembacken, B.J., Snelling, AM., Hawkey, P.M. *et al.* (1999) Non-pathogenic *Escherichia coli* versus mesalizine for the treatment of ulcerative colitis: a randomised trial. *Lancet*, **354**, 635–639.

Ruemmele, F.M., Schwartz, S., Seidman, E.G. *et al.* (2003) Butyrate induced Caco-2 cell apoptosis is mediated via mitochondrial pathway. *Gut*, **52**, 94–100.

Reuter, G. (2001) The *Lactobacillus* and *Bifidobacterium* microflora of the human intestine: composition and succession. *Curr. Issues Intest. Microbiol.*, **2**, 43–53.

Roediger, W.E., Moore, J. and Babidge, W. (1997) Colonic sulfide in pathogenesis and treatment of ulcerative colitis. *Dig. Dis. Sci.*, **42**, 1571–1579.

Rosenfeldt, V., Benfeldt, E., Nielsen, S.D. *et al.* (2003) Effect of *Lactobacillus* strains in children with atopic dermatitis. *J. Allergy Clin. Immunol.*, **111**, 389–395.

Rowland, I.R. (1995) Toxicology of the colon: role of the intestinal microflora. In G.R. Gibson and G.T. Macfarlane (eds), *Human Colonic Bacteria: Role in Nutrition, Physiology and Pathology*. CRC Press, Boca Raton, FL, pp. 155–174.

Rowland, I.R., Rumney, C.J., Coutts, J.T. *et al.* (1998) Effect of *Bifidobacterium longum* and inulin on gut bacterial metabolism and carcinogen-induced aberrant crypt foci in rats. *Carcinogenesis*, **19**, 281–285.

Rudi, K., Flateland, S.L., Hanssen, J.F. *et al.* (2002) Development and evaluation of a 16S ribosomal DNA array-based approach for describing complex microbial communities in ready-to-eat vegetable salads packed in a modified atmosphere. *Appl. Environ. Microbiol.*, **68**, 1146–1156.

Rumney, C.J. and Rowland, I.R. (1992) *In vivo* and *in vitro* models of the human colonic flora. *Crit. Rev. Food Sci. Dis.*, **31**, 299–331.

Saarela, M., Lähteenmäki, L., Ctirrenden, R. *et al.* (2002) Gut bacteria and health foods – the European perspective. *Int. J. Food Microbiol.*, **78**, 99–117.

Saavedra, J.M., Bauman, N.A., Oung, I. *et al.* (1994) Feeding of *Bifidobacterium bifidum* and *Streptococcus thermophilus* to infants in hospital for prevention of diarrhoea and shedding of rotavirus. *Lancet*, **344**, 1046–1049.

Salminen, S., von Wright, A., Morelli, L. *et al.* (1998) Demonstration of safety of probiotics – a review. *Int. J. Food Microbiol.*, **44**, 93–106.

Sandler, R.H., Finegold, S.M., Bolte, E. R. *et al.* (2000) Short-term benefit from oral vancomycin treatment of regressive-onset autism. *J. Child Neurol.*, **15**, 429–435.

Saunier, K. and Doré, J. (2002) Gastrointestinal tract and the elderly: functional foods, gut microflora and healthy ageing. *Dig. Liver Dis.*, **34** (Suppl. 2), S19–S24.

Schaafsma, G., Meuling, W.J., van Dokkum, W. *et al.* (1998) Effects of a milk product, fermented by *Lactobacillus acidophilus* and with fructo-oligosaccharides added, on blood lipids in male volunteers. *Eur. J. Clin. Nutr.*, **52**, 436, 440.

Scholz-Ahrens, K. and Schrezenmeir, J. (2000) Inulin, oligofructose and mineral metabolism – experimental data and mechanism. *Br. J. Nutr.*, **87**, S186–S219.

Shanahan, F. (2003) Review article: colitis-associated cancer – time for new strategies. *Aliment. Pharmacol. Ther.*, **18** (Suppl. 1), 6–9.

Shanahan, F. (2004) Host-flora interactions in inflammatory bowel disease. *Inflamm. Bowel Dis.*, **10** (Suppl. 1), S16–S24.

Soares, M.B. (1997) Identification and cloning of differentially expressed genes. *Curr. Opin. Biotechnol.*, **8**, 542–546.

Smith, M.M., Durkin, M., Hinton, V.J. *et al.* (2003) Influence of breastfeeding on cognitive outcomes at age 6–8 years: follow-up of very low birth weight infants. *Am. J. Epidemiol.*, **158**, 1075–1082.

Suau, A., Bonnet, R., Sutren, M., Godon, J.-J., Gibson, G.R., Collins, M.D. and Doré, J. (1999) Direct analysis of genes encoding 16S rRNA from complex communities reveals many novel molecular species within the human gut. *Appl. Environ. Microbiol.*, **65**, 4799–4807.

Surawicz, C.M., Elmer, G.W., Speelman, P. *et al.* (1989) Prevention of antibiotic-associated diarrhea by *Saccharomyces boulardii*: a prospective study. *Gastroenterology*, **96**, 981–988.

Tahiri, M., Tressol, J.C., Arnaud, J. *et al.* (2003) Effect of short chain fructo-oligosaccharideson intestinal calcium absorption and calcium status in postmenauposal women: a stable-isotope study. *Am. J. Clin. Nutr.*, **77**, 449–457.

Tannock, G.W. (1999) Analysis of the intestinal microflora: a renaissance. *Antonie Van Leeuwenhoek*, **76**, 265–278.

Terada, A., Hara, H., Kataoka, M. *et al.* (1992) Effect of lactulose on the composition and metabolic activity of the human faecal microflora. *Microb. Ecol. Health Dis.*, **5**, 43–50.

Thompson, A., Lucchini, S. and Hinton, J.C. (2001). It's easy to build your own microarray! *Trends Microbiol.*, **9**, 154–156.

Tran, C.P., Familari, M., Parker, L.M. *et al.* (1998) Short-chain fatty acids inhibit intestinal trefoil factor gene expression in colon cancer cells. *Am. J. Physiol. Gastrointest. Liver Physiol.*, **275**, G85–G94.

Tuohy, K.M., Finlay, R.K., Wynne, A.G. *et al.* (2001a) A human volunteer study on the prebiotic effects of HP-Inulin – faecal bacteria enumerated using fluorescent *in situ* hybridisation (FISH). *Anaerobe,* **7**, 113–118.

Tuohy, K.M., Kolida, S., Lustenberger, A.M. *et al.* (2001b) The prebiotic effects of biscuits containing partially hydrolysed guar gum and fructo-oligosaccharides – a human volunteer study. *Br. J. Nutr.*, **86**, 341–348.

Tuohy, K., Rowland, I.R. and Rumsby, P.C. (2002a) Biosafety of marker genes – the possibility of DNA transfer from novel foods to gut microorganisms. In K.T. Atherton (ed.), *Genetically Modified Crops Assessing Safety.* Taylor & Francis, London, pp. 110–137.

Tuohy, K.M., Ziemer, C.J., Klinder, A. *et al.* (2002b) A human volunteer study to determine the prebiotic effects of lactulose powder on human colonic microbiota. *Microb. Ecol. Health Dis.*, **14**, 165–173.

Tuohy, K.M., Probert, H.M., Smejkal, C.W. *et al.* (2003) Using probiotics and prebiotics to improve gut health. *Drug Dis. Today,* **8**, 692–700.

Tuohy, K.M., Likotrafiti, E., Manderson, K. *et al.* (2004) Improving gut health in the elderly. In C. Remacle and B. Reusens (eds), *Functional Foods, Ageing and Degenerative Disease.* Woodhead, Cambridge, UK, pp. 394–415.

Vaughan, E.E., Schut, F., Heilig, H.G.H.J. *et al.* (2000) A molecular view of the intestinal ecosystem. *Curr. Issues Intest. Microbiol.*, **1**, 1–12.

Verghese, M., Rao, D.R., Chawan, C.B. *et al.* (2002a) Dietary inulin suppresses azoxymethane-induced preneoplastic aberrant crypt foci in mature fisher 344 rats. *J. Nutr.*, **132**, 2804–2808.

Verghese, M., Rao, D.R., Chawan, C.B. *et al.* (2002b) Dietary inulin suppresses azoxymethane-induced aberrant crypt foci and colon tumors at the promotion stage in young fisher 344 rats. *J. Nutr.*, **132**, 2809–2813.

Vesa, T.H., Marteau, P. and Korpela, R. (2000) Lactose intolerance. *J. Am. Coll. Nutr.*, **19**, 165S-175S.

Videla, S., Vilaseca, J., Garcia-Lafuente, A. *et al.* (1998) Dietary inulin prevents distal colitis induced by dextran sulphate sodium (DSS). *Gastroenterol.*, **114**, G4514.

Videla, S., Salas, A., Vilaseca, J. *et al.* (1999) Deranged luminal pH homeostasis in experimental colitis can be restored by a prebiotic. *Gastroenterol.*, **116**, G4098.

Videla, S., Vilaseca, J., Antolin, M. *et al.* (2001) Dietary inulin improves distal colitis induced by dextran sodium sulfate in the rat. *Am. J. Gastroenterol.*, **96**,1486–1493.

Wallis, M.R., Lipski, P.S., Mathers, J.C. *et al.* (1993) Duodenal brush-border mucosal glucose transport and enzyme activities in aging man and effect of bacterial contamination of the small intestine. *Dig. Dis. Sci.*, **38**, 403–409

Wallner, G., Erhart, R. and Amann, R. (1995) Flow cytometric analysis of activated sludge with rRNA-targeted probes. *Appl. Environ. Microbiol.*, **61**, 1859–1866.

Wang, R.F., Kim, S.J., Robertson, L.H. *et al.* (2002) Development of a membrane-array method for the detection of human intestinal bacteria in fecal samples. *Mol. Cell Probes,* **16**, 341–350.

Williams, E.A., Coxhead, J.M. and Mathers, J.C. (2003) Anti-cancer effects of butyrate: use of microarray technology to investigate mechanisms. *Proc. Nutr. Soc.,* **62**, 107–115.

Wilson, K.H. and Blitchington, R.B. (1996) Human colonic bacteria studied by ribosomal DNA sequence analysis. *Appl. Environ. Microbiol.*, **62**, 2273–2278.

Yancik, R., Wesley, M.N., Ries, L.A. *et al.* (1998) Comorbidity and age as predictors of risk for early mortality of male and female colon carcinoma patients: a population-based study. *Cancer,* **82**, 2123–2134.

Zoentendal, E.G., Akkermans, A.D.L. and De Vos, W.M. (1998) Temperature gradient gel electrophoresis analysis of 16S rRNA from human faecal samples reveals stable and host-specific communities of active bacteria. *Appl. Environ. Microbiol.*, **64**, 3854–3859.

Zoetendal, E.G., Ben-Amor, K., Harmsen, H.J. *et al.* (2002) Quantification of uncultured *Ruminococcus obeum*-like bacteria in human fecal samples by fluorescent in situ hybridization and flow cytometry using 16S rRNA-targeted probes. *Appl. Environ. Microbiol.*, **68**, 4225–4232.

Zoetendal, E.G., Collier, C.T., Koike, S. *et al.* (2004) Molecular ecological analysis of the gastrointestinal microbiota: a review. *J. Nutr.*, **134**, 465–472.

Chapter 7
Secondary Metabolites in Fruits, Vegetables, Beverages and Other Plant-Based Dietary Components

Alan Crozier, Takao Yokota, Indu B. Jaganath, Serena C. Marks, Michael Saltmarsh and Michael N. Clifford

7.1 Introduction

The current advice is that for optimum health people should consume on a daily basis five portions of fruit and vegetables each comprising at least 80 g (Williams 1995). The epidemiological evidence for the benefit of consuming diets that are high in fruit and vegetables is quite compelling. The evidence for specific vegetables, and indeed specific phytochemicals, is less convincing although epidemiological studies of cancer suggest that it is mainly the highly coloured green or yellow vegetables that are associated with reduced incidence and mortality rates.

The use of the terms 'fruit' and 'vegetable' is a culinary rather than a botanical distinction. In a botanical context 'vegetables' such as tomatoes, cucumbers, courgettes, peppers (capsicums) and avocado pears are classified as fruits. To avoid confusion, in this chapter foods will be classed according to their common usage. Some vegetables, such as potatoes, cassava and yams, serve as staple foods because they tend to have a high starch content. Such staples are not generally classed as 'vegetables' for the purposes of dietary guidelines. Outside the provision of a dietary staple, vegetables are used for dietary variety, colour and taste and can be purchased in a range of processed states including fresh, frozen, canned and dried. In contemporary Western society, fresh vegetables may also be processed to a high degree including washing and slicing or shredding. The most widely consumed vegetables in the world that are not used as staples are onions and tomatoes. Fruits and vegetables are generally low in fat, may contain significant amounts of dietary fibre (non-starch polysaccharides) and, apart from traditional vitamins and minerals, contain a wide range of compounds called phytochemicals that may have biological activity in humans.

In this chapter we identify the phytochemicals, including those contributing to colour, known to occur in fruits and vegetables which, together with the vitamins and minerals also present, may be the basis for their beneficial effects. Authoritative figures for contents of macronutrients, vitamins and minerals are given in food tables (McCance and Widdowson 1991) and these will not be repeated in this chapter. Similarly, vegetables in general are a source of fibre and the levels will not be identified for each individual species. These figures are also available in food tables.

In addition to the references cited throughout this Chapter, other information about the historical context of fruit and vegetables can be found in a number of books including (Simpson and Conner-Ogorzaly 1986; Tannahill 1988; Readers Digest Association Ltd 1996; Pavoro 1999; Whiteman 1999).

7.2 Dietary phytochemicals

There are basically four classes of phytochemicals: terpenoids, phenolics and polyphenolics, nitrogen containing alkaloids and sulphur compounds (see Chapters 1–4) whose members may exert positive effects on human health. It is a truism that the effect produced depends on the amount consumed. For example, within the class of glucosinolates, progoitrin can be converted to goitrin (Figure 7.1) which is a potent goitrogen. Glucobrassicin (Figure 7.1) is mutagenic at high doses. However, in the levels found in the normal diet both compounds may have protective rather than detrimental effects on health (Rhodes and Price 1998).

In the growing plant, phytochemicals have roles in metabolism and in the interaction of the plant with the environment. The occurrence of the phytochemicals of interest varies throughout the plant kingdom from the widespread carotenoids to the glucosinolates that are found only in the Cruciferae and a few members of some other families of dicotyledonous angiosperms (Fenwick *et al.* 1983). Carotenoids are ubiquitous in leaves and stems because they are an essential part of the photosynthetic process, involved in light harvesting and protecting against photo-oxidative damage (van de Berg *et al.* 2000). Sterols generally control membrane fluidity and permeability, while some have further specific roles (Piironen *et al.* 2000). Phenolic compounds, on the other hand, appear to be mainly associated with the defence of the plant against a range of attacks, from browsing animals through insects, bacteria and fungi to inhibiting the growth of competing plants. Flavonols in epidermal cells of leaves and the skins of fruit provide protection against the damaging effects of UVB irradiation. They also are involved in fertilization by promoting the growth of pollen tubes in the style of flowers. Polyphenols and carotenoids provide colour to stems,

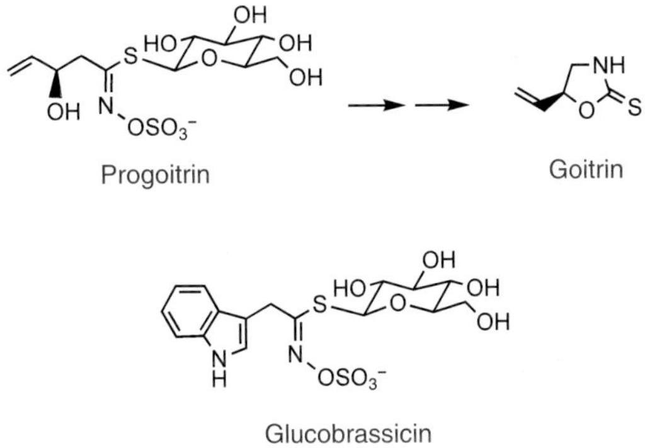

Figure 7.1 The glucosinolates goitrin, derived from progoitrin, and glucobrassicin can have either adverse or positive effects on health depending upon the amount consumed.

leaves, flowers and fruits. The carotenoids provide yellows with some orange and red while the polyphenolics, most notably the anthocyanins, are more numerous and provide a greater range of colours from orange to blue.

Glucosinolates have been shown to affect insect predation; some act as feeding deterrents to certain species of insects but other insects such as the large white butterfly larvae are attracted by specific glucosinolates. There is also evidence that glucosinolates have antifungal and antibacterial activities and their presence in the plant may contribute to resistance to infection by mildew and other fungi.

Members of the terpenoid family are diverse in structure and have an extremely wide range of actions. For example, the complex limonoid, azadirachtin (Figure 7.2) from the neem (*Azadirachta indica*) tree is a powerful insect antifeedant, whereas the monoterpene linalool (Figure 7.2) is an attractant bringing insects to fertilize flowers. Other terpenoids are the active ingredients in essential oils while diterpene resins block insect attack on conifer trees (Croteau *et al.* 2000). Limonoids are oxygenated triterpenoids that are found in only two plant families. The citrus limonoids are only found in the *Citrus* genus where in the leaves and young fruit they act as antifeedants (Hasegawa *et al.* 2000).

Salicylic acid (Figure 7.3) is produced rapidly in some plants as a signal molecule that initiates defence responses following attack by insects, fungi, bacteria and virus. Salicylic acid and acetylsalicylic acid (aspirin) levels have been monitored in a range of fruits, vegetables, herbs, spices and beverages (Venema *et al.* 1996). Acetylsalicylic acid was not detected in any of the samples and with the exception of some herbs and spices, most notably cinnamon (*Cinnamonum zeylandicum*), oregano (*Origanum vulgare*) and rosemary (*Rosmarinus officinalis*), salicylic acid levels were less than 1 mg/kg fresh weight. It is thought that the levels of salicylates present in most diets are much too low to have an impact on health (Janssen *et al.* 1996). However, the possibility of salicylic acid formation, along with many other phenolic acids, from dietary (poly)phenols by the gut micro flora

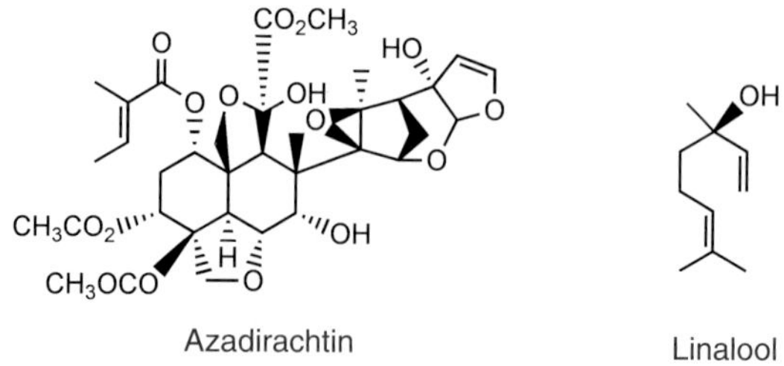

Figure 7.2 An example of the diverse structures of terpenoids.

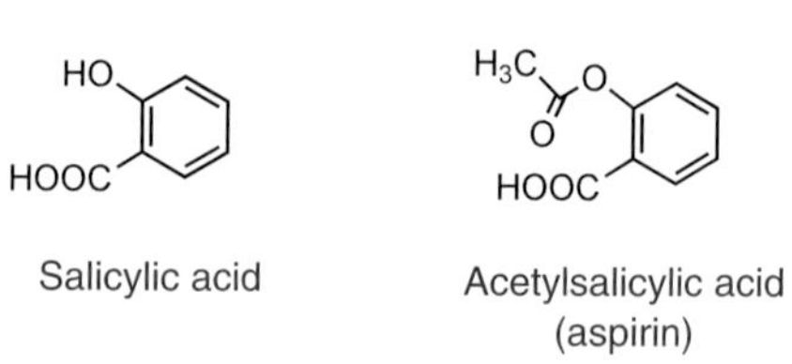

Figure 7.3 Salicylic acid and its acetylated derivative, aspirin.

leading to its appearance in the colon and human faecal water, from where a portion can be absorbed, has been noted (Jenner *et al.* 2005; Karlsson *et al.* 2005). The use of crops containing elevated levels of salicylic acid is something that could be achieved through genetic engineering. Although they would not necessarily be acceptable in the present political climate in Europe such crops would offer two advantages. First the plants themselves would be more resistant to pathogens and, second, consumers may be less prone to heart attacks and cerebral thrombosis as salicylic acid, like aspirin (Figure 7.3), would retard the production of prostaglandins which promote blood clotting.

It is important to be aware of the sheer numbers of phytochemicals. Attention tends to be focused on a few representatives of each class but 25 000 members of the terpenes have been identified, as have many tens of thousands of polyphenols and there are even 250 different sterols although these do not all occur in foods or even traditional medicines.

Where studies on plant material have been compared, it is not unusual for reported levels to vary by a factor of 20, with the difference in some cases being as much as 100-fold (van der Berg *et al.* 2000). The levels of phytochemicals recorded have to be considered in context and no single figure can be regarded as representative of a plant species. The phytochemical content of individual fruits and vegetables is affected by many factors including variety, soil, climatic conditions, agricultural methods, physiological stress under which plants are grown and degree of ripeness, storage conditions and length of storage before consumption.

Not all reports in the literature have recognized the resultant variability, and results from a single unnamed variety purchased in a supermarket have often been taken as representative of the species (van der Berg *et al.* 2000). Major cultivated crops in the developed world have been bred so that they are true to type; naming the variety is sufficient to define the plant. This is not necessarily the case where local varieties are still grown. Genotype influences not only the overall content of a class of phytochemicals, but also the proportions of individual chemicals. The form in which vegetables are available is also changing. In the past spinach was typically a mature leaf purchased loose from the greengrocer. The product available in a modern supermarket is pre-packed young leaves cut to a prescribed length.

The levels of phytochemical vary within the plant; within fruits many are concentrated in the skin, and within vegetables in the outer leaves. For example, the outer leaves of Savoy cabbage contain more than 150 times the level of lutein (Figure 7.4) and 200 times the level of β-carotene (Figure 7.4) present in the inner leaves (van der Berg *et al.* 2000). Glucosinolates are also heterogeneously distributed within the plant, and a study has shown a varying composition even in different parts of a swede root tuber (Adams *et al.* 1989).

7.3 Vegetables

In the following sections each vegetable will be discussed briefly. Particular phytochemicals will be highlighted, but this should not be taken as an indication that these are the only phytochemicals associated with the foodstuff as it is evident that the phytochemical content of some dietary fruits and vegetables have been investigated in detail while others have received very little, if any, attention.

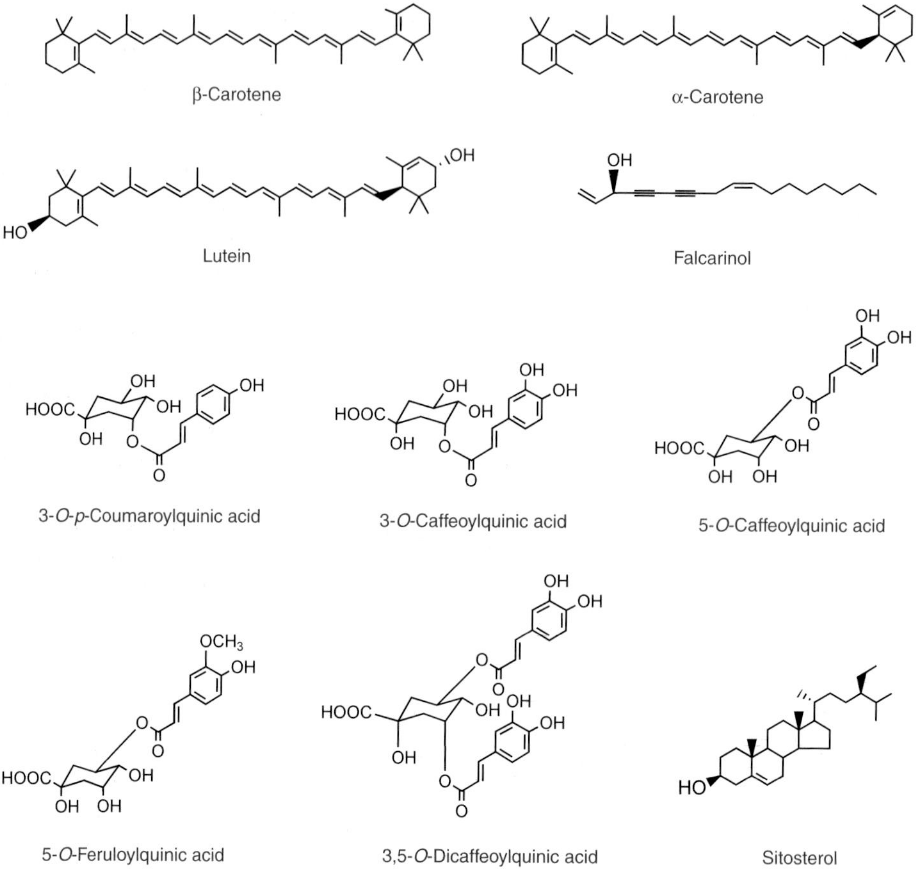

Figure 7.4 Carrots contain carotenoids, chlorogenic acids, phytosterols and the polyacetylene, falcarinol.

7.3.1 Root crops

Root crops include carrot (*Daucus carota*), turnip (*Brassica campestris*), swede (*Brassica napus*) (also known as rutabaga), parsnip (*Pastinaca sativa*) and Jerusalem artichoke (*Helianthus tuberosus*). These were important dietary components during the nineteenth and early twentieth century; turnip and swede were more common at the beginning of the twentieth century, but carrot is now the most popular root vegetable after the potato. Changes in dietary habits have also seen a decrease in the consumption of the leaves of root crops as vegetables. Originally, leaves were the only part of the beetroot that was consumed, and carrot leaves were eaten until the middle of the nineteenth century in the United Kingdom (Anon 1897). Young turnip leaves are still considered an early season delicacy in some countries, and beetroot leaves are still eaten in the United Kingdom. Interestingly, in comparison, the leaves of leaf beet are eaten as spinach and the root discarded.

Carrot was introduced into Europe from Arabia in the fourteenth century. The original carrots were purple and yellow, and the orange carrots we use today originate from selective breeding in Holland in the seventeenth century. Nowadays, different varieties of carrot are grown both for immediate consumption and for storage. The principal phytochemicals of interest in carrots are α-carotene and β-carotene (Figure 7.4) and carrots are a rich source, containing up to 650 mg/kg (van der Berg 2000). Carrots also contain sterols mainly as sitosterol (Piironen *et al.* 2000) and a range of chlorogenic acids including 3-*O*- and 5-*O*-caffeoylquinic acids, 3-*O*-*p*-coumaroylquinic acid, 5-*O*-feruloylquinic acid and 3,5-*O*-dicaffeoylquinic acids (Figure 7.4). The chlorogenic acids are found in orange, purple, yellow and white carrots and the level of 5-*O*-caffeoylquinic acid in purple carrots, at 540 mg/kg, being almost 10-fold higher than the amounts present in the other varieties (Alasalvar *et al.* 2001). Chlorogenic acids are considered a carrot root fly-attractant and root fly-resistant varieties have been bred to have low caffeoylquinic acid contents (Cole 1985). Roots of Apiaceae, including carrots, contain polyacetylenes such as falcarinol (see Chapter 5), with some evidence of absorption and anti-cancer properties (Koebaek-Larsen *et al.* 2005; Zidorn *et al.* 2005).

Beetroot (*Beta vulgaris*) contains β-carotene in the leaves while red pigmentation in the roots is due to betanin and isobetanin (Figure 7.5) which are betalains. Betalains were long thought to be related to anthocyanins, even though they contain nitrogen and are structurally quite distinct (Figure 7.5). Betalains are restricted to ten plant families, all of which are members of the order Caryophyllales which lack anthocyanins. Beetroot extract is used as a food colouring: E162.

Swedes and turnips are both brassicas in which the principal phytochemicals of interest are glucosinolates found throughout the plant but particularly in the root. Parsnip is a member of the Umbelliferae family and like many members of this family, including celery (*Apium graveolens*), contains psoralen (Figure 7.6) which in sensitive people can cause blistering on exposure to light (see Section 7.3.5 and Chapter 5).

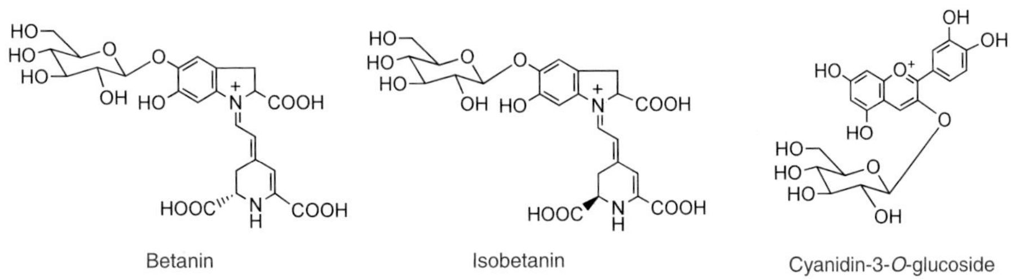

Betanin

Isobetanin

Cyanidin-3-*O*-glucoside

Figure 7.5 Betanin and isobetanin are the red pigments in beetroot and should not be confused with anthocyanins such as cyanidin-3-glucoside.

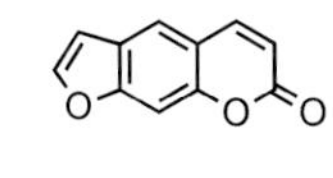

Psoralen

Figure 7.6 Some people are very sensitive to psoralen, which occurs in Umbelliferous plants and causes blistering of the skin on exposure to sunlight.

7.3.2 Onions and garlic

Members of the family *Alliaceae* have been an important part of the human diet for thousands of years; even in ancient times they were used extensively throughout the northern hemisphere. Onions (*Allium cepa*), leeks (*Allium porrum*) and garlic (*Allium sativum*) were important in the diet of the Romans who probably introduced them to the United Kingdom. Now a considerable array of varieties of red, white, yellow and silver skin onions, spring onions (scallions), shallots, leeks, chives (*Allium schoenoprasum*) and garlic are available throughout Europe and other varieties are eaten on other continents. It is believed that all members of the genus *Allium* are edible. World onion production is estimated to be in excess of 30 billion kg per annum. Garlic is reputed to have benefits for protection against cardiovascular diseases, cancer, microbial infections, asthma, diabetes and vampires! Although there is some evidence for the first two effects, the rest of the list is largely speculation or folklore. Garlic has been commercially exploited and is available as essential oil of garlic, and garlic pearls and other extracts in the form of tablets or capsules.

S-Alkyl cysteine sulphoxides are found in all members of the *Allium* genus (Figure 7.7). All species contain *S*-methyl cysteine sulphoxide but *S*-propyl cysteine sulphoxide is the major component in chives, while the 1-propenyl (vinyl-methyl) derivative predominates in onions and the allyl (methyl-vinyl or 2-propenyl) derivative, alliin, in garlic. When cutting onions conversion of *S*-1-propenyl cysteine sulphoxide to propanethial *S*-oxide (Figure 7.7) results in the well-known phenomenon of onion-induced kitchen tears as propanethial *S*-oxide (cycloalliin) is a lachrymatory factor. The concentration of the precursor and relative enzyme activity is greater near the roots, thus making the lower part of the bulb more lachymatory and justifying advice not to cut too close to the roots when slicing onions.

Fresh garlic has little smell but tissue damage by cutting, crushing or biting results in alliin being cleaved by the enzyme alliinase resulting in the formation of diallyl

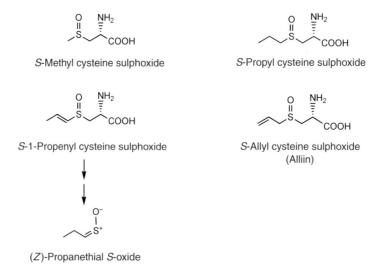

Figure 7.7 Some of the *S*-alkyl cysteine sulphoxides found in *Allium* species.

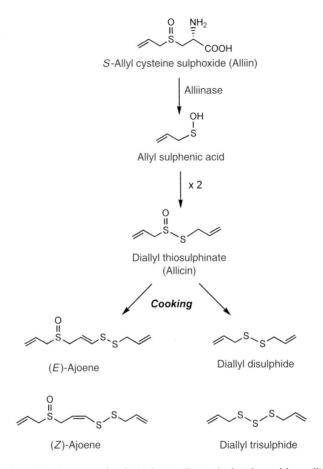

Figure 7.8 The characteristic aroma of garlic is due to allicin which is formed from alliin, in a reaction catalysed by alliinase which is released when the cloves are cut or crushed. When garlic is cooked, allicin breaks down to a number of products including sulphides and disulphides which are responsible for bad breath after eating garlic flavoured foods.

thiosulphinate (allicin) (Figure 7.8). Allicin gives crushed garlic its characteristic aroma. Alliin and alliinase are both stable when dry, so dried garlic retains the potential to release allicin when moistened and crushed. Nonetheless, the composition of dried garlic and assorted garlic powders and oils is very variable with reported values for the alliin content differing 10-fold.

Allicin is very unstable to heat, so cooking garlic results in its degradation to a number of compounds including diallyl sulphides and ajoenes (Figure 7.8). Bad breath, which follows the ingestion of garlic products, is due to a range of sulphide compounds including diallyl disulphide and diallyl trisulphide (Figure 7.8). Most studies on the potentially protective effects of garlic have used extracts or preparations rather than cooked garlic. There are very few investigations using raw garlic, arguably because the flavours and smells are so strong that double-blinded, placebo-controlled trials are not possible. There is epidemiological evidence associating reduced risk of colon cancer (Steinmetz et al. 1994) and coronary

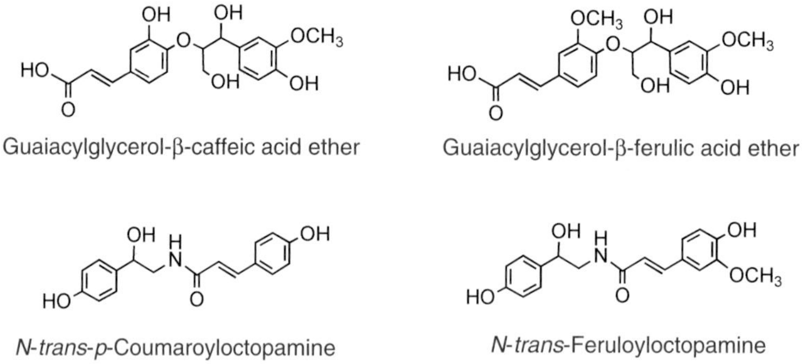

Guaiacylglycerol-β-caffeic acid ether

Guaiacylglycerol-β-ferulic acid ether

N-trans-p-Coumaroyloctopamine

N-trans-Feruloyloctopamine

Figure 7.9 Dry garlic skins contain hydroxycinnamate derivatives.

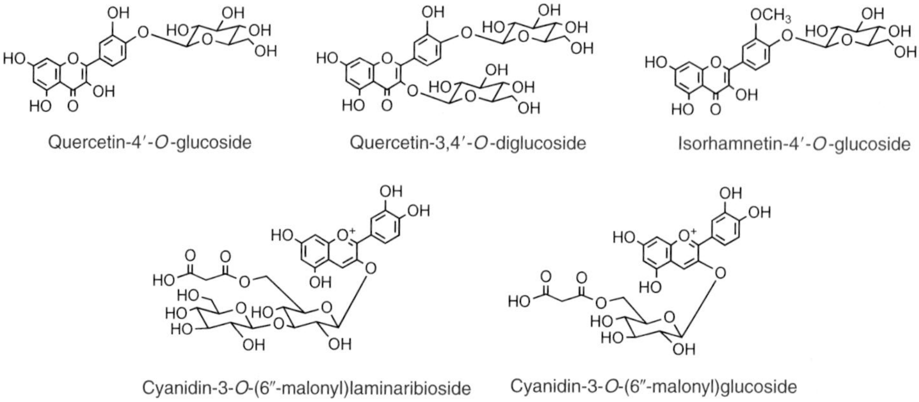

Quercetin-4'-O-glucoside

Quercetin-3,4'-O-diglucoside

Isorhamnetin-4'-O-glucoside

Cyanidin-3-O-(6''-malonyl)laminaribioside

Cyanidin-3-O-(6''-malonyl)glucoside

Figure 7.10 The main flavonol conjugates and anthocyanins in red onions.

heart disease (Keys 1980) with regular consumption of garlic. Five of six intervention studies using fresh garlic or freshly prepared extracts demonstrated a lowering of serum cholesterol, increased fibrinolytic activity and inhibition of platelet aggregation (Kleijnen et al. 1989). Another study also showed significant reductions in systolic blood pressure (Steiner et al. 1996). However, these effects were generally achieved at very high levels of intake, the equivalent of between 7 and 28 cloves per day, and this produces side effects including body odour, bad breath and flatulence that are unacceptable to many people. There are claims that allicin and ajoene are the protective agents in garlic but there is limited evidence to substantiate this view.

The dry skin of garlic, which is usually removed before cooking, exhibits antioxidant activity that has been attributed to the presence of the hydroxycinnamates, *N-trans-p*-coumaroyloctopamine, *N-trans*-feruloyloctopamine, guaiacylglycerol-β-caffeic acid ether and guaiacylglycerol-β-ferulic acid ether (Figure 7.9) (Ichikawa et al. 2003).

The main flavonols in onions are glycosylated derivatives, principally quercetin-4'-O-glucoside and quercetin-3,4'-O-diglucoside with smaller amounts of isorhamnetin-4'-O-glucoside (Figure 7.10) and other quercetin conjugates (Mullen et al. 2004). Yellow onions form one of the main sources of flavonols in the Northern European diet, the edible flesh containing between 280 and 490 mg/kg (Crozier *et al.* 1997). Even higher concentrations are found in the dry outer scales (Chu *et al.* 2000). By contrast leeks have been found to have only 10–60 mg/kg kaempferol and no quercetin. White onions are all but devoid of flavonols. Red onions like their yellow counterparts are rich in flavonols and also contain up to 250 mg/kg anthocyanins (Clifford 2000); among the major components are cyanidin-3-*O*-(6″-malonyl)glucoside and cyanidin-3-*O*-(6″-malonyl)laminaribioside (Donner *et al.* 1997) (Figure 7.10).

7.3.3 Cabbage family and greens

Members of the genus *Brassica* (in the family *Cruciferae*) have been cultivated for thousands of years although the main use in ancient times was probably for medicine. Carbonized seeds of brown mustard (*Brassica juncea*) have been found at a site in China dating to around 4000 BC (Fenwick *et al.* 1983). The Romans cultivated a number of members of the genus, including mustard, cabbage, kale and possibly broccoli and kohlrabi, and introduced the crop to the United Kingdom. Cauliflower is mentioned in the twelfth century and the most recent member of the family to be discovered was Brussels sprouts in around 1750. Varieties of only one species, *Brassica oleracea*, are the most commonly consumed vegetables in the United Kingdom and include broccoli, Brussels sprouts, cabbage, calabrese, cauliflower, kale and kohlrabi.

Within the brassica, all parts of the plant are consumed, roots (turnip, swede, kohlrabi), leaves (cabbage, kale), apical buds (Brussels sprouts), flower heads (broccoli and cauliflower) and seeds (mustard). As well as being consumed fresh, worldwide considerable tonnages of these crops are processed, mainly into sauerkraut, coleslaw and pickles. Fermented brassica crops are important constituents of Asia-pacific diets. All members of the genus contain glucosinolates. These break down on chewing as the enzyme myrosinase is released (see below) and yields compounds that are responsible for the spicy/hot flavour of mustard and a number of other cruciferous plants which are not brassicas, including radish (*Raphanus sativus*), horseradish (*Armoracia rusticana*), watercress (*Nasturtium officinale*) and rocket (*Eruca sativa*).

The glucosinolate sinalbin accumulates in white mustard (*Sinapis alba* syn. *Brassica hirta*) seed and when moistened and crushed the glucose moiety is cleaved by myrosinase and a sulphonated intermediate that is formed undergoes re-arrangement forming acrinylisothiocyanate (Figure 7.11) which is responsible for the hot pungent taste of the condiment. Black mustard (*Brassica nigra*) seeds contain sinigrin which is similarly hydrolysed to allylthiocyanate (Figure 7.11), considerably more volatile than acrinylisothiocyanate which gives black mustard powder a pungent aroma as well as a hot spicy taste.

While glucosinolates are desired for their intense flavour, as in mustard, their presence in leaf crops such as Brussels sprouts can make these foods less attractive to consumers, particularly the young. Glucosinolates are not biologically active per se, but once the glucose moiety has been removed by myrosinase, the resulting aglycone is unstable and rearranges forming active compounds including isothiocyanates, thiocyanates, nitriles and sometimes indole derivatives. In the plant, myrosinase is a membrane-bound enzyme, but when tissues are chewed or processed, cellular compartmentation breaks down and

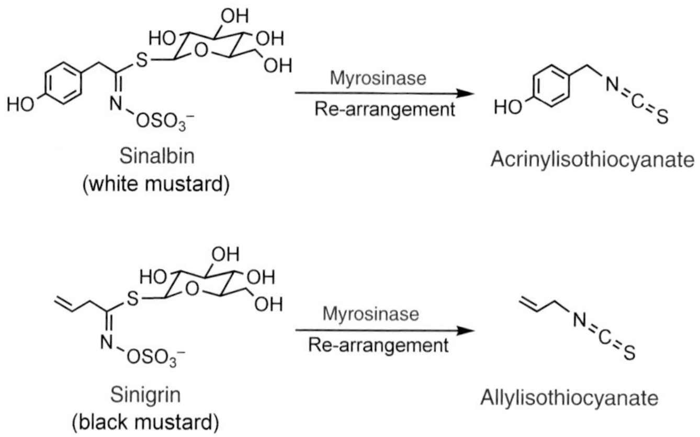

Figure 7.11 Glucosinolates in white and black mustard. When the seeds are crushed and moistened sinalbin is converted to acrinylisothiocyanate and sinigrin to allylisothiocyanate.

myrosinase comes into contact with glucosinolates that accumulate in the cell vacuole. Cooking will make inactive much of the myrosinase, which means that in most foodstuffs both intact glucosinolates and breakdown products will be ingested.

There have been reports of cabbage having a goitrogenic effect in animals fed on a diet low in iodine, and thyroid, liver and kidney enlargement in rats fed a diet rich in rapeseed. There is no evidence that direct consumption of brassicas causes goitre, but it has been suggested that endemic goitre in certain regions of Europe could be caused by the consumption of milk containing the goitrin precursor progoitrin (see Figure 7.1) from the ingestion of cruciferous forage or weeds, together with marginal or deficient iodine status. This question has not yet been resolved unequivocally.

Some glucosinolate derivatives, including sulphoraphane, are potent, selective inducers of phase II enzymes. The sprouting seedlings of some cultivars of broccoli and cauliflower contain 10–100 times the level of glucoraphanin, the glucosinolate precursor of sulforaphane (Figure 7.12), than mature plants (Fahey et al. 1997). Furthermore, these immature plants do not contain significant levels of the indole glucosinolates, such as glucobrassicin, and related indoles including indole-3-methanol (Figure 7.12) that can enhance tumorigenesis, although protective effects have also been reported. Broccoli sprouts, very similar to mustard and cress, are available commercially as a 'health food', and their consumption may provide scope for significantly increasing glucosinolate intake without large increases in the consumption of brassica vegetables. These products may be more palatable to those who dislike the taste of the mature form.

Broccoli florets contain glucosinolates as well as quercetin-3-O-sophoroside and kaempferol-3-O-sophoroside (Plumb et al. 1997). Several hydroxycinnamoyl derivatives are also present, the main ones being 1-O-sinapoyl-2-O-feruloylgentiobiose, 1,2-O-diferuloylgentiobiose, 1,2,2'-O-trisinapoylgentiobiose and 3-O-caffeoylquinic acid. Broccoli florets also contain a number of hydroxycinnamate esters of novel complex kaempferol and quercetin glycosides such as kaempferol-3-O-sophorotrioside-7-O-sophoroside (Figure 7.13) (Vallejo and Tomás-Barberán 2004). Compared with freshly harvested florets, broccoli that was film-wrapped and stored for seven days at 1°C and then kept at 15°C for

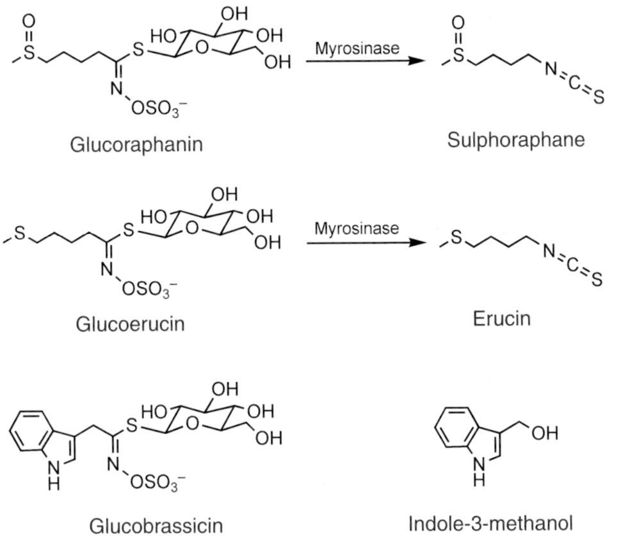

Figure 7.12 Glucosinolates and indole-3-methanol are found in brassica species. Note indole-3-methanol is also known as indole-3-carbinol.

three days, to simulate commercial transport and distribution and retail shelf life, showed a ~25% decline in glucosinolates and caffeoylquinic acids and a ~50% loss in sinapic acid derivatives (Vallejo *et al.* 2003).

Other members of this genus are consumed as roots (turnip and swede) and as salad leaves such as Chinese cabbage (*Brassica pekinensis*), rocket and watercress. The main glucosinolate in sprouts and leaves of rocket is glucoerucin which myrosinase converts to erucin (Figure 7.12) (Barillari *et al.* 2005). Other leaves are consumed cooked as greens. These include spinach (*Spinaceae oleraceae*) and the closely related spinach beet and Swiss chard (*Beta vulgaris*). Swiss chard has a high flavonoid content estimated at 2700 mg/kg, compared with spinach with 1000 mg/kg and red onion with 900 mg/kg (Gil *et al.* 1999). Spinach contains conjugates of *p*-coumaric acid and high levels of carotenoids; lutein contents have been determined ranging from 20 to 203 mg/kg and β-carotene from 8 to 240 mg/kg (van der Berg *et al.* 2000). Spinach is devoid of the common flavonol conjugates of quercetin and kaempferol but contains axillarin-4'-O-glucoside and spinacetin-3-O-gentobioside (Figure 7.14) (Kidmose *et al.* 2001) and other novel methoxyflavonol derivatives (Zane and Wender 1961) some of which have antimutagenic properties (Edenharder *et al.* 2001). Leaves of cauliflower contain an unusual spectrum of flavonols, in the form of kaempferol-3,7-O-diglucoside (Figure 7.13) and sinapoyl and feruloyl derivatives of kaempferol di-, tri- and tetra-glucosides (Llorach *et al.* 2003).

7.3.4 Legumes

Although widely consumed in fresh and processed forms there is relatively little information of the phytochemicals present in most legumes of dietary significance. The notable

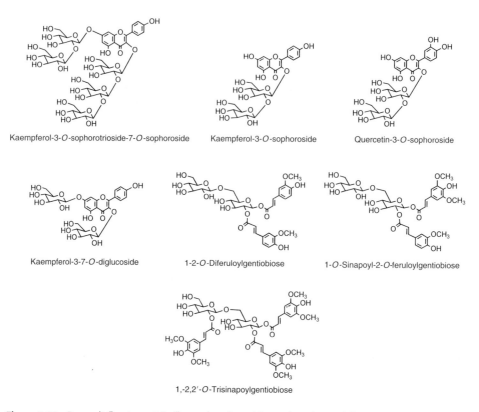

Figure 7.13 Broccoli florets contain flavonol sophorosides and conjugated derivatives of sinapic acid while kaempferol-3,7-O-diglucoside is among the flavonols present in cabbage leaves.

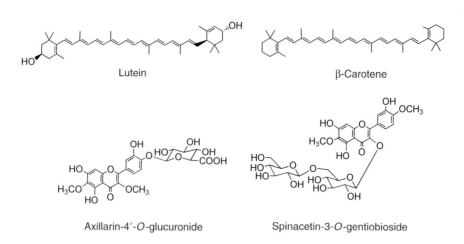

Figure 7.14 The carotenoids lutein and β-carotene and novel methoxylated flavonols occur in spinach.

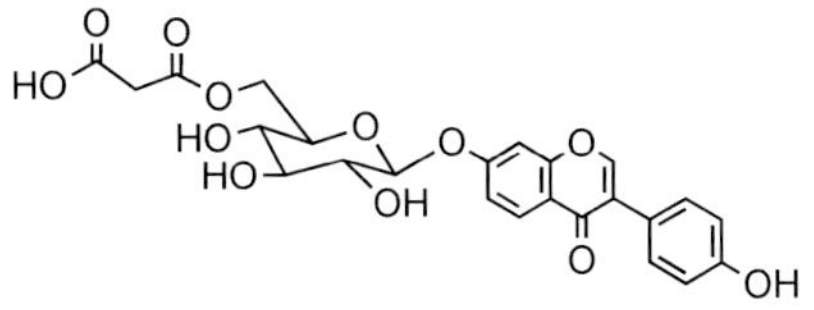

Daidzein-7-*O*-(6″-*O*-malonyl)glucoside

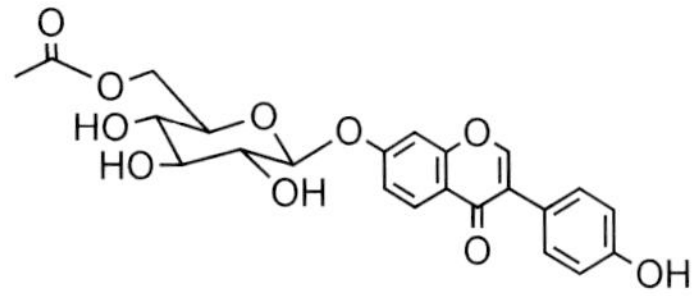

Daidzein-7-*O*-(6″-*O*-acetyl)glucoside

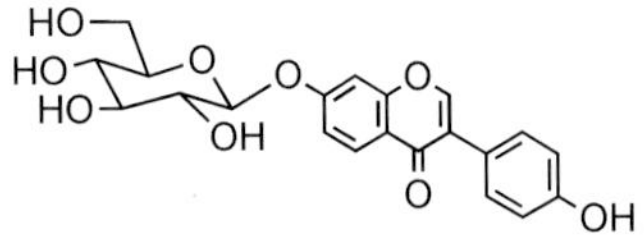

Daidzein-7-*O*-glucoside

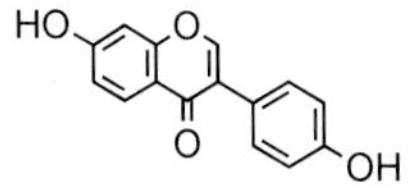

Daidzein

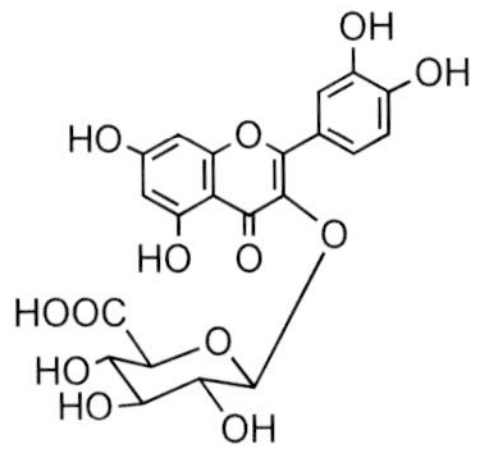

Quercetin-3-*O*-glucuronide

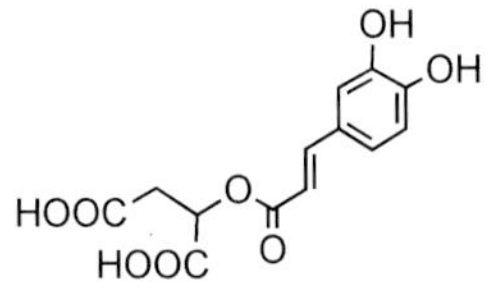

2-*O*-Caffeoyl-L-malic acid

Figure 7.15 The isoflavone daidzein is found in soyabeans along with its 7-*O*-glucoside and 7-*O*-(6″-malonyl)glucoside. Genistein and its 7-*O*-glucoside and 7-*O*-(6″-malonyl)glucoside are also present. French beans contain quercetin-3-*O*-glucuronide in high amounts while caffeoyl-l-malic acid is present in pods of *Vicia faba*.

exception is soyabean (*Glycine max*) which contains the isoflavones daidzein-7-*O*-(6″-*O*-malonyl)glucoside and genistein-7-*O*-(6″-*O*-malonyl)glucoside with lower quantities of the corresponding glucosides (6″-acetyl)glucosides and the aglycones (Figure 7.15) (Barnes *et al.* 1994). The levels of isoflavones in soybeans have been reported to range from 560 to 3810 mg/kg which is two orders of magnitude higher than the amounts detected in other legumes. Fermented soya products can be comparatively rich in the aglycones as hydrolysis of the glycosides can occur (Coward 1993). Products whose manufacture involves heating at 100°C, such as soya milk and tofu, contain reduced quantities of isoflavones, the principal components being daidzein and genistein glucosides which form as a result of degradation of the malonyl- and acetylglucosides (Liggins *et al.* 2001; Fletcher 2003).

As far as other legumes are concerned, peanuts (*Arachis hypogaea*) contain 5,7-dimethoxyisoflavone (see Section 7.7) broad beans (*Vicia faba*) are a relatively rich source of flavan-3-ols containing more than 150 mg/kg (de Pascual-Teresa *et al.* 2000) while French beans (*Phaseolus vulgaris*) can contain substantial quantities of quercetin-3-glucuronide (Hempel and Böhm 1996). Pinto beans and red kidney beans (*Phaseolus vulgaris*) contain in excess of 5 g/kg of proanthocyanidins principally as prodelphinidins and propelargonidins, most with a degree of polymerization >4 (Gu *et al.* 2004). In addition, pods of

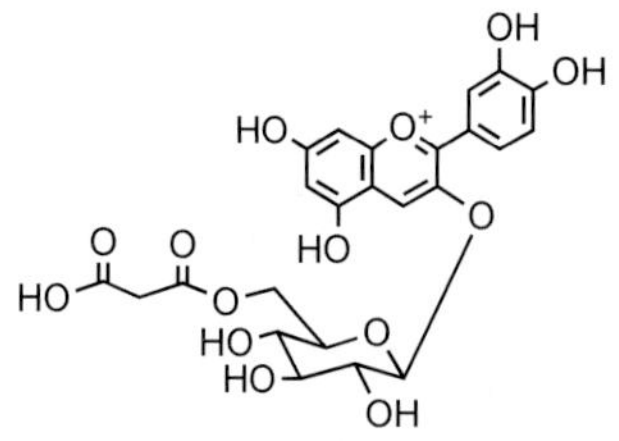

Cyanidin-3-*O*-(6″-*O*-malonyl)glucoside

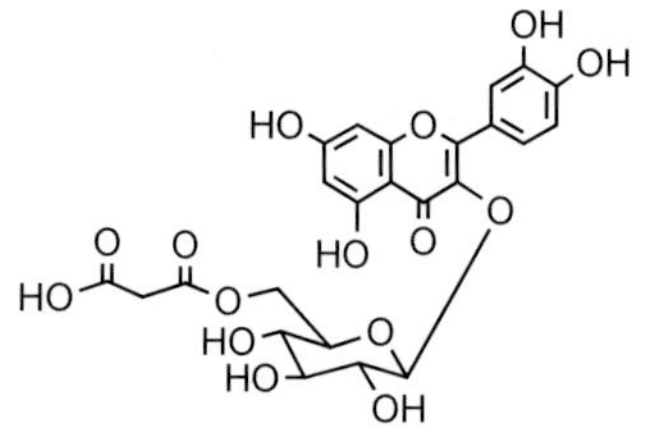

Quercetin-3-*O*-(6″-*O*-malonyl)glucoside

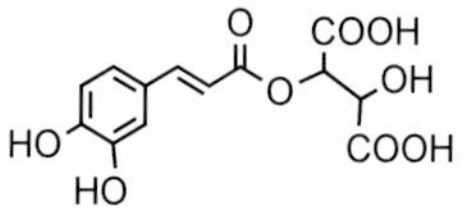

Caffeoyltartaric acid

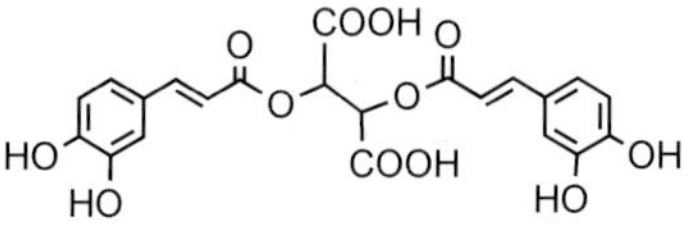

Dicaffeoyltartaric acid

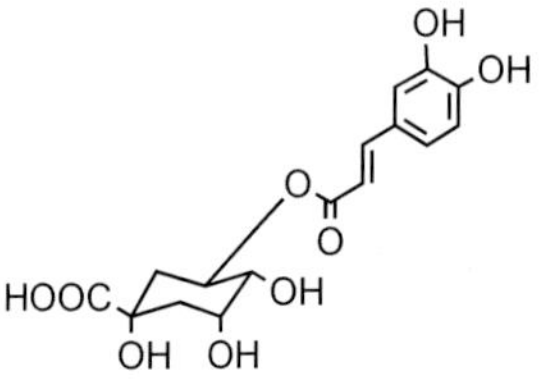

5-*O*-Caffeoylquinic acid

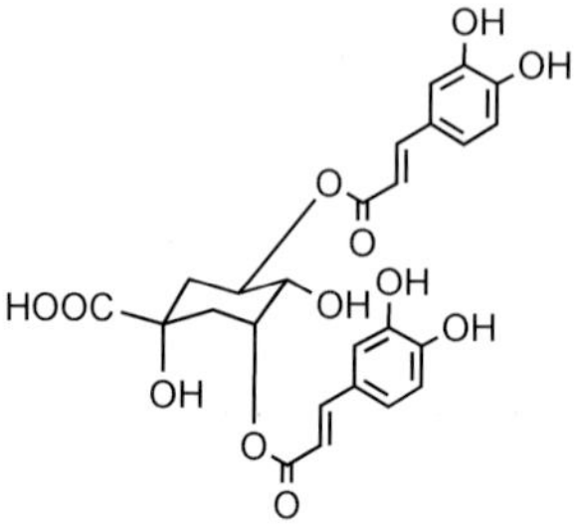

3,5-*O*-Dicaffeoylquinic acid

Figure 7.16 The major anthocyanin, flavonol conjugate and chlorogenic acids in Lollo Rosso lettuce.

Vicia faba contain caffeoyl-L-malic acid (phaseolic acid) (Figure 7.15) at up to 100 mg/kg (Winter and Herrmann 1986).

7.3.5 Lettuce

Lettuce (*Latuca sativa*) is a source of carotenoids containing both lutein and β-carotene, although the concentrations determined have varied by a factor of 60, the highest being 45 mg/kg for lutein (van den Berg *et al.* 2000). The red-leaved lettuce Lollo Rosso contains the anthocyanin cyanidin-3-*O*-(6″-malonyl)glucoside and several flavonols, including the major component quercetin-3-*O*-(6″-malonyl)glucoside, and the hydroxycinnamate derivatives caffeoyltartaric acid, dicaffeoyltartaric acid, 5-*O*-caffeoylquinic acid and 3,5-*O*-dicaffeoylquinic acid (Figure 7.16) (Ferreres *et al.* 1997). The levels of flavonols, measured as quercetin released by acid hydrolysis, were 911 ± 27 mg/kg fresh weight in the outer leaves and around half this amount in the inner leaves of Lollo Rosso. Other varieties have much lower flavonol levels with Round lettuce containing only 11 mg/kg and Iceberg, which is used widely in commercial salads and sandwiches, a mere 2 mg/kg (Crozier *et al.* 1997, 2000).

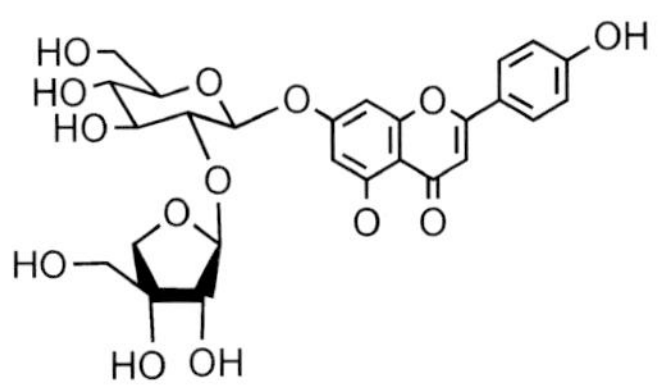

Apigenin-7-O-(2″-O-apiosyl)glucoside

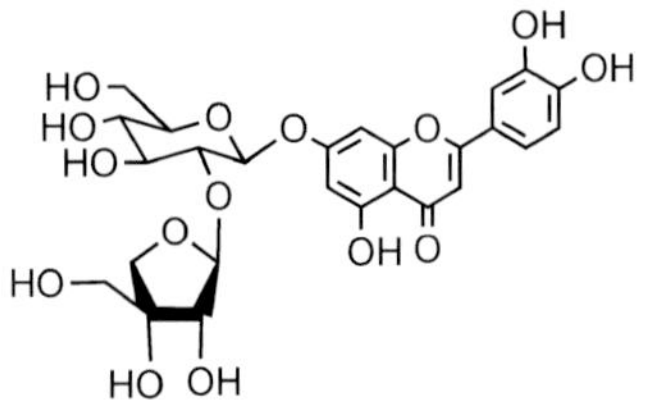

Luteolin-7-O-(2″-O-apiosyl)glucoside

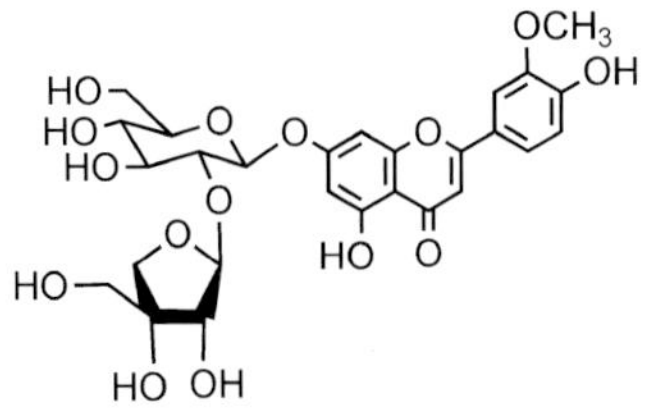

Chrysoeriol-7-O-(2″-O-apiosyl)glucoside

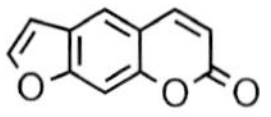

Psoralen 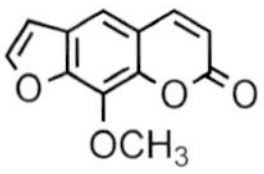   Xanthotoxin 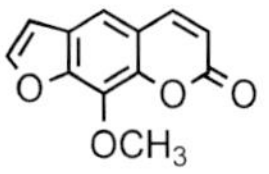   Bergapten

Figure 7.17 Celery contains conjugates of the flavones apigenin, luteolin and chrysoeriol. Following fungal infection furocoumarins, including psoralen, xanthotoxin and bergapten, accumulate.

7.3.6 Celery

Celery was cultivated as a medicine by the ancients and was not used as a food until 1623. Celery contains several flavone conjugates, including the 7-O-(2″-O-apiosyl)glucosides of apigenin, luteolin and chrysoeriol (Figure 7.17) (Herrmann 1976, 1988), although the amounts can be variable (Crozier *et al.* 1997). Fungal infection of celery results in the accumulation of psoralen and other furocoumarins such as the methoxylated derivatives xanthotoxin and bergapten (Figure 7.17) (see Chapter 5). People harvesting infected plants by hand can become very sensitive to the UVA component of ultraviolet light and develop a sunburn-type rash (phytophotodermatitis). The level of psoralen is considerably reduced by cooking, especially boiling. Psoralen is now used in the treatment of skin disorders such as psoriasis. Celeriac (*Apium graveolens* var. *rapaceum*) has a similar flavour to celery, but the root, rather than the stem, is eaten, either peeled and parboiled in salads or as a cooked vegetable.

7.3.7 Asparagus

Asparagus (*Asparagus officinalis*) is native to the Mediterranean. It was cultivated by the Romans both as food and medicine and has been cultivated in Northern Europe since the beginning of the first millennium. Asparagus contains β-carotene, the level varying with the colour of the spear. Makris and Rossiter (2001) also detected 280 mg/kg of quercetin-3-O-rutinoside in fresh asparagus spears and that boiling in water for one hour resulted in a 40% loss. The spears also contain the steroidal saponin, protodioscin (Figure 7.18) (Wang *et al.* 2003b).

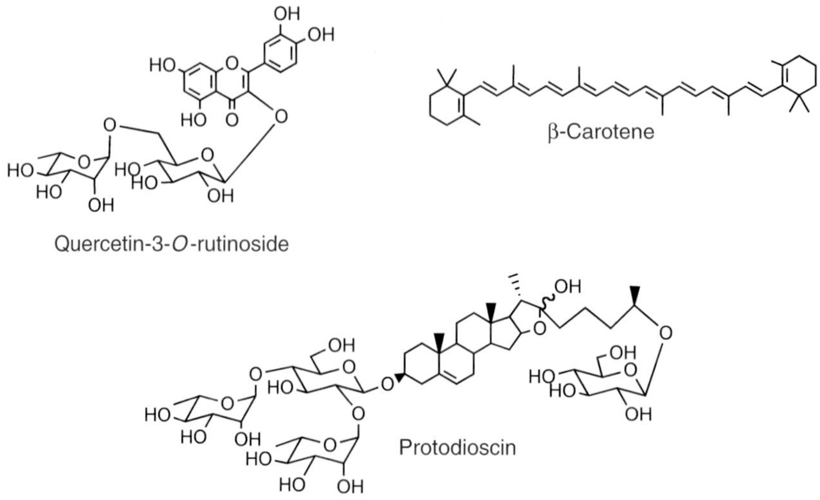

Figure 7.18 Quercetin-3-O-rutinoside, β-carotene and protodioscin occur in asparagus.

7.3.8 Avocados

Avocados (*Persea americana*) are found in archaeological deposits in Mexico, which date to 7000 BC. There are three cultivated varieties with differing oil content in the pulp. The West Indian variety, which can weigh up to a kilogram, has only 8–10% oil whereas the Mexican variety, which is smaller, contains about 30% oil. It thus has the highest energy content of any fruit pulp (with the possible exception of olives). The principal components of the lipid fraction are polyunsaturated fatty acids, such as docosahexaenoic acid. Avocado is becoming increasingly popular, being used in salads, sandwich fillings, dips and spreads. The flesh is very nutritious containing vitamins B, C and E, chlorophyll and the carotenoids lutein, zeaxanthin, α-carotene and β-carotene (Figure 7.19) (Lu *et al*. 2005). The avocado fruit also contains persenone A and B which inhibit superoxide and nitric oxide generation in mouse macrophage cells and possess anti-tumour properties (Kim *et al*. 2000). In addition the phytosterol, sitosterol, is present in avocado in substantial amounts (760 mg/kg) together with smaller quantities of campesterol (51 mg/kg) (Figure 7.19) (Duester 2001).

7.3.9 Artichoke

Artichoke (*Cynara scolymus*) is an ancient herbaceous perennial plant originating from Mediterranean North Africa. The artichoke head, an immature flower constitutes the edible part of this vegetable which is grown widely around the world with Italy and Spain being the leading producers. Artichoke heads contain antioxidants and the main phenolic compounds are 5-O-caffeoylquinic acid with smaller amounts of 1-O-caffeoylquinic acid, 1,4-O-dicaffeoylquinic acid, luteolin-7-O-glucoside, luteolin-7-O-rutinoside

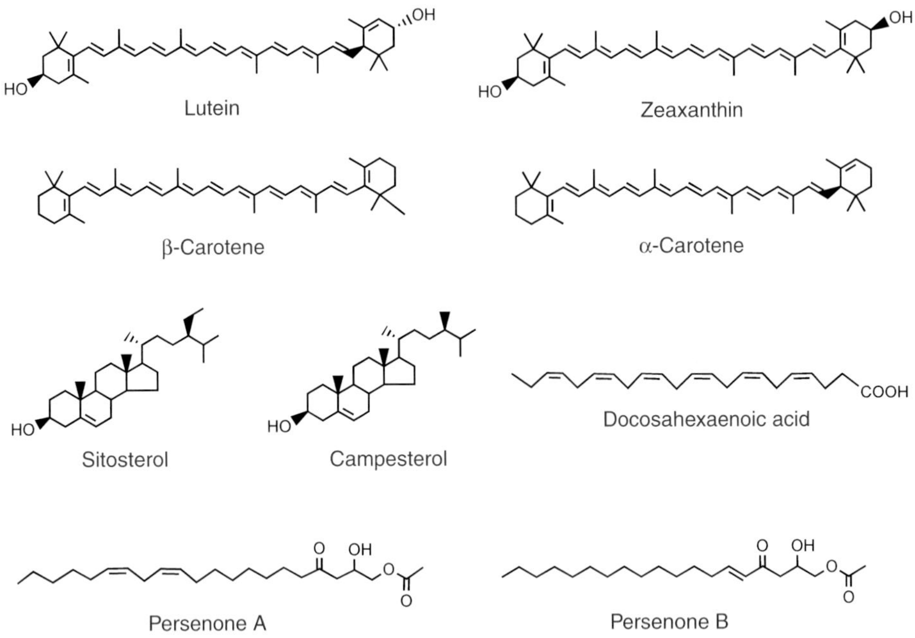

Figure 7.19 Avocados contain carotenoids, poly-unsaturated fats, phytosterols and persenone A and B.

apigenin-7-O-rutinoside and naringenin-7-O-rutinoside (Figure 7.20) (Wang et al. 2003). Wang et al. (2003) also reported the presence of 1,3-O-dicaffeoylquinic acid, but this was probably formed from 1,5-O-dicaffeoylquinic acid during extraction of the artichoke tissues with aqueous methanol (Clifford 2003).

7.3.10   Tomato and related plants

The tomato (*Lycopersicon esculentum*) (family Solanaceae) was introduced into Europe in the sixteenth century from South America but took nearly three centuries to become widely accepted as a foodstuff. The original tomato was yellow which is reflected in the Italian – pomodoro – pomo d'oro, golden fruit.

7.3.10.1   Tomatoes

Many different types of tomatoes are now available, ranging from the very small, cherry through plum tomatoes to the giant beefsteak type that can weigh more than one kilogram. Colours range from yellow through green to purple. Together with onions, tomatoes are the most widely consumed non-staple food. In the United Kingdom, tomatoes are eaten as an important component of salads, soups and sauces while tomatoes play a central role in what is seen as the traditional diet of Mediterranean countries.

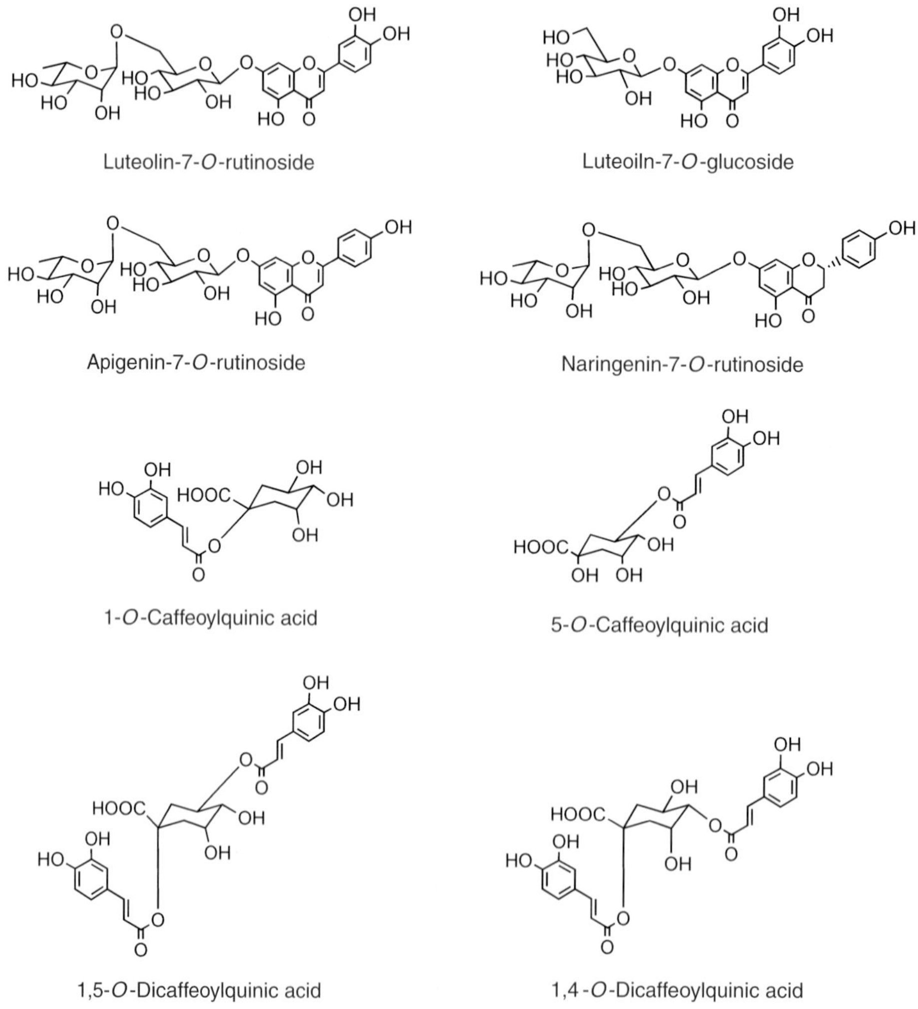

[Structures of Luteolin-7-O-rutinoside, Luteolin-7-O-glucoside, Apigenin-7-O-rutinoside, Naringenin-7-O-rutinoside, 1-O-Caffeoylquinic acid, 5-O-Caffeoylquinic acid, 1,5-O-Dicaffeoylquinic acid, 1,4-O-Dicaffeoylquinic acid]

Figure 7.20 Flavone and flavanone glycoside conjugates and chlorogenic acids occur in artichokes.

In the United States the consumption of tomato and tomato products is second only to potatoes.

Green tomatoes contain the steroidal alkaloid tomatine, which disappears as the fruit ripens. Tomatoes contain the carotenoids lycopene, β-carotene and lutein (Figure 7.21), which are produced in the flesh as the fruit ripens. Lycopene is quantitatively the most important carotenoid and extensive research has identified the cultural conditions to optimize levels. The content is affected by nitrogen, calcium and potassium in fertilizers, by light, temperature and irrigation, being reduced by excessive light and by temperatures over 32°C. Lycopene levels as high as 600 mg/kg are reported in the literature (van den Berg et al. 2000), but other references consider 200 mg/kg to be exceptionally high (Grolier et al. 2001). Tomatoes also contain flavonols, mainly as quercetin-3-O-rutinoside, which

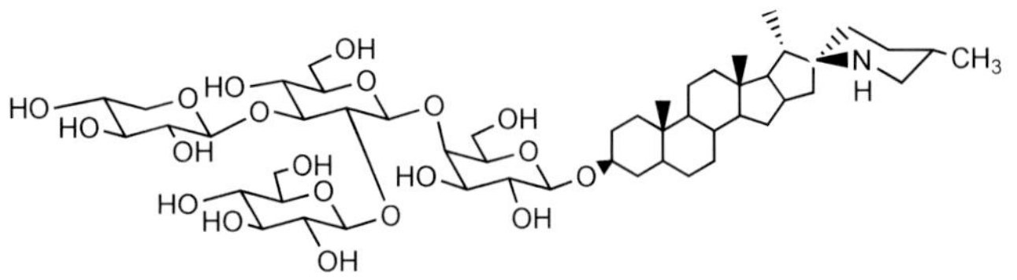

Tomatine

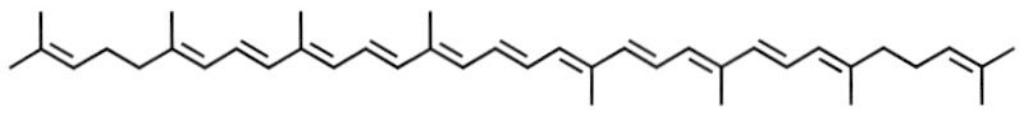

Lycopene

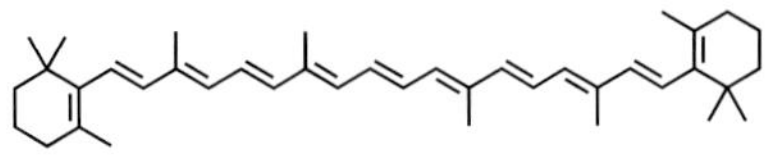

β-Carotene

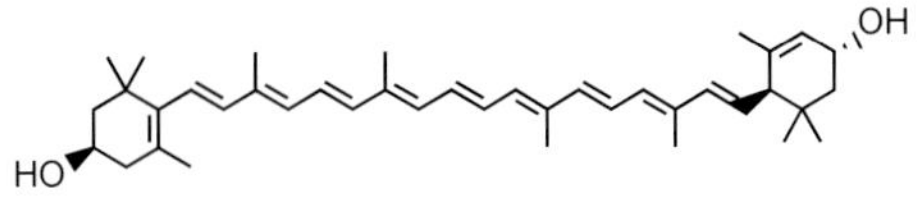

Lutein

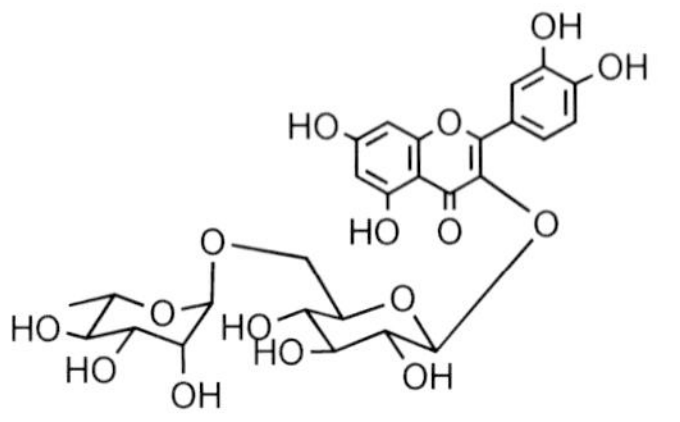

Quercetin-3-*O*-rutinoside

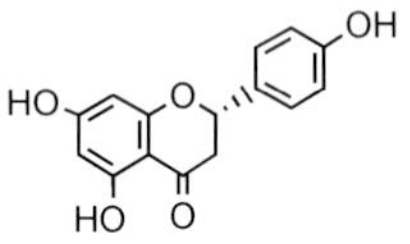

Naringenin

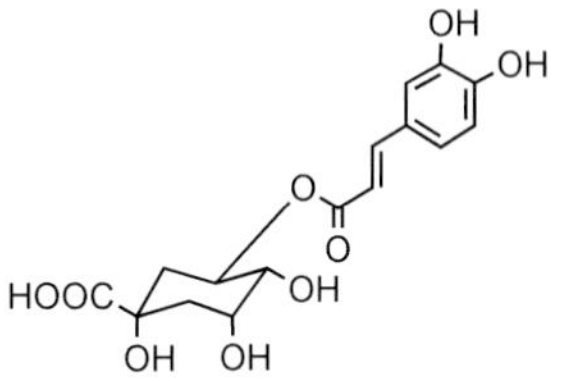

5-*O*-Caffeoylquinic acid

Figure 7.21 Green tomatoes contain the steroidal alkaloid tomatine. Levels decline in ripe fruit which contain the carotenoids lycopene, β-carotene and lutein. Quercetin-3-*O*-rutinoside, naringenin and 5-*O*-caffeoylquinic acid are also present.

accumulates in the skin, and because of their high skin:volume ratio, cherry tomatoes are an especially rich source (Stewart *et al.* 2000). In addition, the flavanone naringenin and 5-*O*-caffeoylquinic acid (Figure 7.21) have also been detected in tomato (Paganga *et al.* 1999).

7.3.10.2 Peppers and aubergines

Peppers (*Capsicum annuum*) and aubergines (*Solanum melongena*) are also fruits of members of the Solanaceae. Peppers are native to Mexico and were introduced to Europe from the West Indies by Columbus. There are two main types – bell peppers which tend to be large and sweet and are available in a range of colours from green through yellow to orange, red and purple and chilli peppers, of which there are many varieties, smaller and much hotter (see Section 7.5). Bell peppers are often eaten raw in salads and are used in many Mediterranean dishes. They contain a number of carotenoids, the main components being lutein and β-carotene. However, the overall level of carotenoids in bell peppers is typically only one tenth of the total carotenoid content of tomatoes. Special varieties have

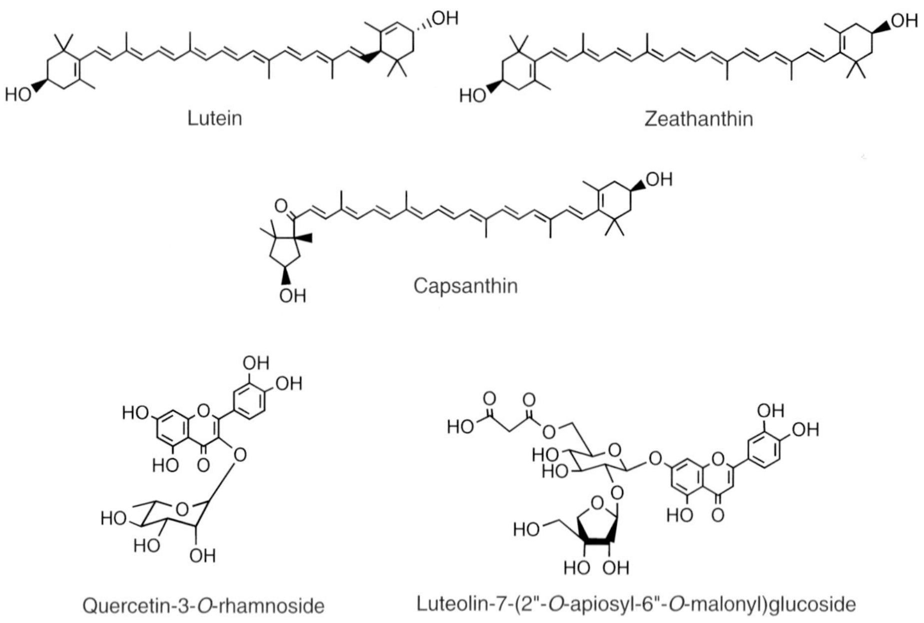

[Structures: Lutein, Zeathanthin, Capsanthin, Quercetin-3-O-rhamnoside, Luteolin-7-(2"-O-apiosyl-6"-O-malonyl)glucoside]

Figure 7.22 The colour of red, yellow and orange bell peppers is due to the respective accumulation of lutein, zeathanthin and capsanthin. Flavonol and flavone conjugates also occur.

been bred with vastly increased levels of different carotenoids, which result in different colours. Yellow bell peppers accumulate lutein and zeaxanthin is the major component in orange-coloured peppers, while capsanthin predominates in red varieties. Bell peppers also contain several hydroxycinnamate glucosides, flavonols and numerous flavones, including C-glycosides, with quercetin-3-O-rhamnoside and luteolin-7-O-(2"-O-apiosyl-6"-O-malonyl)glucoside being present in highest quantities (Figure 7.22) (Marín et al. 2004).

Aubergines are native to South East Asia and have been used as a vegetable in China for over 2000 years but only comparatively recently in Europe. They are low in energy but can absorb a great deal of fat during cooking. They contain anthocyanins in the form of delphinidin glycosides in the skin and other phenolics including 5-O-caffeoylquinic acid in the flesh.

7.3.11 Squashes

Marrow, pumpkin, squash, courgette (zucchini) are all *Cucurbita* species and members of the Cucurbitaceae, as are melons (*Cucumis melo*). Squashes were very important to early inhabitants of Southern and Central America, as important as corn and beans. Fossilized remains of squashes in Peru have been dated to 4000 BC. Originally the flowers, seeds and flesh were eaten. The seeds provided a source of sulphur-containing amino acids. Wild members of the family are thin-skinned and bitter. There are few data available on the phytochemical content of the flesh of squashes. Butternut squash has been found

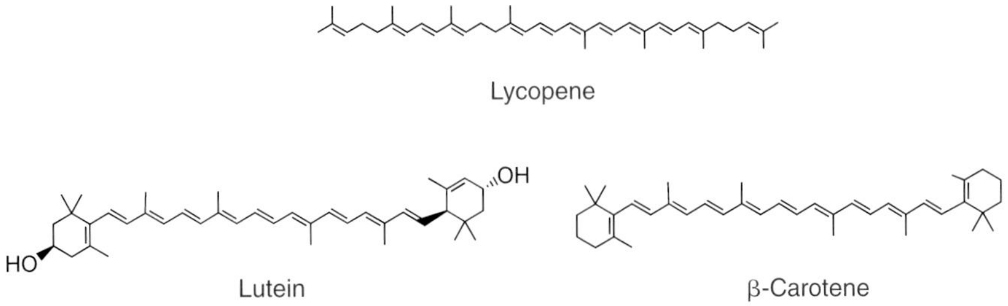

Figure 7.23 The carotenoids lycopene, lutein and β-carotene occur in pumpkins.

to contain 350 mg/kg total phenolics (Lister and Podivinsky 1998). Pumpkins, including Asian pumpkin (*Cucurbita moschata*) are reported to contain β-carotene and smaller amounts of lycopene and lutein (Figure 7.23) (Seo *et al.* 2005).

7.4 Fruits

As in the sections on vegetables, in the following sections each fruit will be discussed briefly. Particular phytochemicals will be highlighted, but this should not be taken as an indication that these are the only phytochemicals associated with the foodstuff.

7.4.1 Apples and pears

Small, bitter, crab-apples are very widely distributed throughout the world and have been eaten since prehistoric times. However, the first apples resembling modern apples (*Malus* × *domestica*) probably grew on the slopes of the Tien Shan between China and Kazakhstan. The Romans first cultivated the fruit, grew at least a dozen varieties and are believed to have introduced it to Northern Europe including Britain. The Pilgrim Fathers took pips to America. Cox's Orange Pippin was first grown in England in 1826 and later that century Granny Smith was grown in Australia. More than 7000 named varieties are now known worldwide. Apples and pears are the most commonly consumed fruits in the United Kingdom, after bananas (Henderson *et al.* 2002)

Apples are a good source of flavonoids and phenolic compounds containing 2310–4880 mg/kg (Podsedek *et al.* 1998). The principal ingredients include 5-O-caffeoylquinic acid, 4-O-p-coumaroylquinic acid, caffeic acid, phloretin-2′-O-glucoside (phloridzin), phloretin-2′-O-(2″-O-xylosyl)glucoside, quercetin-3-O-glucoside, quercetin-3-O-galactoside, quercetin-3-O-rhamnoside, (−)-epicatechin and its procyanidin dimers, B_1 and B_2 (Figure 7.24) and oligomers (Clifford *et al.* 2003; Kahle *et al.* 2005). The procyanidins have been shown to have an average degree of polymerization of between 3.1 and 8.5 (Sanoner *et al.* 1998). Cider apples generally contain higher concentrations of procyanidins than dessert apples (Guyot *et al.* 2003) with an average degree of polymerization between 4.2 and 50.3 (Sanoner *et al.* 1999). These compounds have a major influence on taste – too much, as in the wild apple, and the fruit is inedible – too little and it is insipid (Haslam, 1998). The main contributors to the antioxidant capacity of apples are 5-O-caffeoylquinic

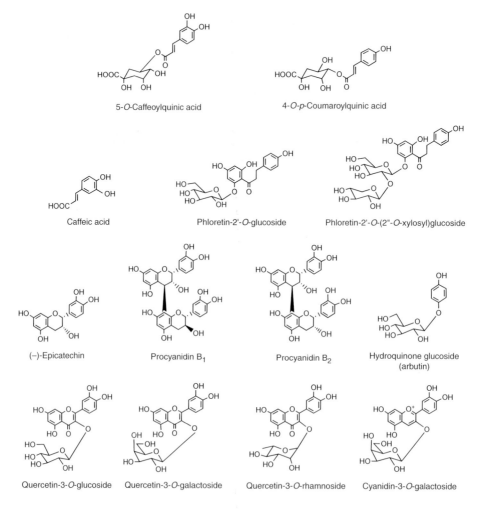

Figure 7.24 Hydroxycinnamate derivatives and flavonoids found in apples. Pears have a similar phenolic profile but do not contain phloretin conjugates while arbutin does not occur in apples.

acid, caffeic acid, and (−)-epicatechin (Bandoniene et al. 2000). The red colour of some cultivars of apples is due to the presence of the anthocyanin, cyanidin-3-O-galactoside (Figure 7.24) (Wu and Prior 2005).

Pears (*Pyrus communis*) were cultivated by the Phoenicians and later by the Romans. There are now in excess of 500 named varieties worldwide. The total phenolic content of some cultivars of pears has been shown to be between 1235 and 2500 mg/kg in the peel and 28–81 mg/kg in the flesh (Galvis-Sánchez et al. 2003). The phenolic composition of pears is very similar to that of apples containing 5-O-caffeoylquinic acid, 4-O-p-coumaroylquinic acid, procyanidins and quercetin glycosides. The main difference in the phenolic content of apples and pears is the presence of hydroquinone glucoside (arbutin) (Figure 7.24) in pears and the hydroxychalcones in apples (Spanos and Wrolstad 1992). The average degree of polymerization of procyanidins in some varieties of pears has been shown to be as high as 44 (Ferreira *et al.* 2002). Apples and pears are among the main sources of proanthocyanidins in the diet (Santos-Buelga and Scalbert 2000).

7.4.2 Apricots, nectarines and peaches

The apricot (*Prunus armeniaca*) was introduced into Europe from China by silk merchants and arrived in England during the reign of Henry VIII. Peaches (*Prunus persica*) originate from the mountainous regions of Tibet and Western China. The fruit was cultivated by the Chinese as early as 2000 BC and reached Greece around 300 BC. They were taken along the silk route to Persia and from there were introduced to Greece and Rome. They were grown by the Romans in the first century and introduced to Mexico by the Spaniards in the 1500s. It is now cultivated commercially in many countries with the fruit being consumed fresh, canned, frozen, dried and processed into jelly jam and juices. There is increasing usage of nectarine (*P. persica* var. *nectarina*), a smooth-skinned variety of peach. Peaches and nectarines contain cyanidin-3-*O*-glucoside, cyanidin-3-*O*-rutinoside, quercetin-3-*O*-glucoside and quercetin-3-*O*-rutinoside. 3-*O*-Caffeoylquinic acid also occurs in stone fruits and in larger amounts than 5-*O*-caffeoylquinic acid (Clifford 2003). They also contain (+)-catechin, (−)-epicatechin and proanthocyanidins including procyanidin B_1 (Figure 7.25), the levels of which decline with thermal processing and storage in cans (Hong *et al.* 2004). Apricots and peaches both contain carotenoids principally in the form of β-carotene. Enzymic browning during processing lowers the carotenoid content.

7.4.3 Cherries

Sweet cherries were known to the Egyptians and Chinese. Sour cherries were cultivated by the Greeks. Modern varieties are either pure-bred sweet (*Prunus avium*) or sour (*Prunus cerasus*) or hybrids of the two. Both contain anthocyanins, mainly cyanidin-3-*O*-rutinoside, with lower levels of other anthocyanins, including cyanidin-3-*O*-glucoside and peonidin-3-rutinoside (Wu and Prior 2005). Like peaches, they also contain hydroxycinnamates including 3-*O*-caffeoylquinic acid and 3-*O*-*p*-coumaroylquinic acid (Figure 7.26) (Mozetic *et al.* 2002).

7.4.4 Plums

Plums were first cultivated by the Assyrians and were extensively hybridized by the Romans. They were introduced to Northern Europe by the crusaders. Prunes are plums that have been dried without being allowed to ferment. Numerous varieties of plums, mainly *Prunus domestica*, are cultivated world-wide and they are a rich source of anthocyanins in the form of cyanidin-3-*O*-glucoside and cyanidin-3-*O*-rutinoside which are also found in peaches. They also contain significant quantities of 3-*O*- and 5-*O*-caffeoylquinic acid and procyanidins with degrees of polymerization up to and greater than ten (Tomás-Barberán *et al.* 2001; Gu *et al.* 2004). Dried plums lack anthocyanins but 3-*O*- and 5-*O*-caffeoylquinic acid

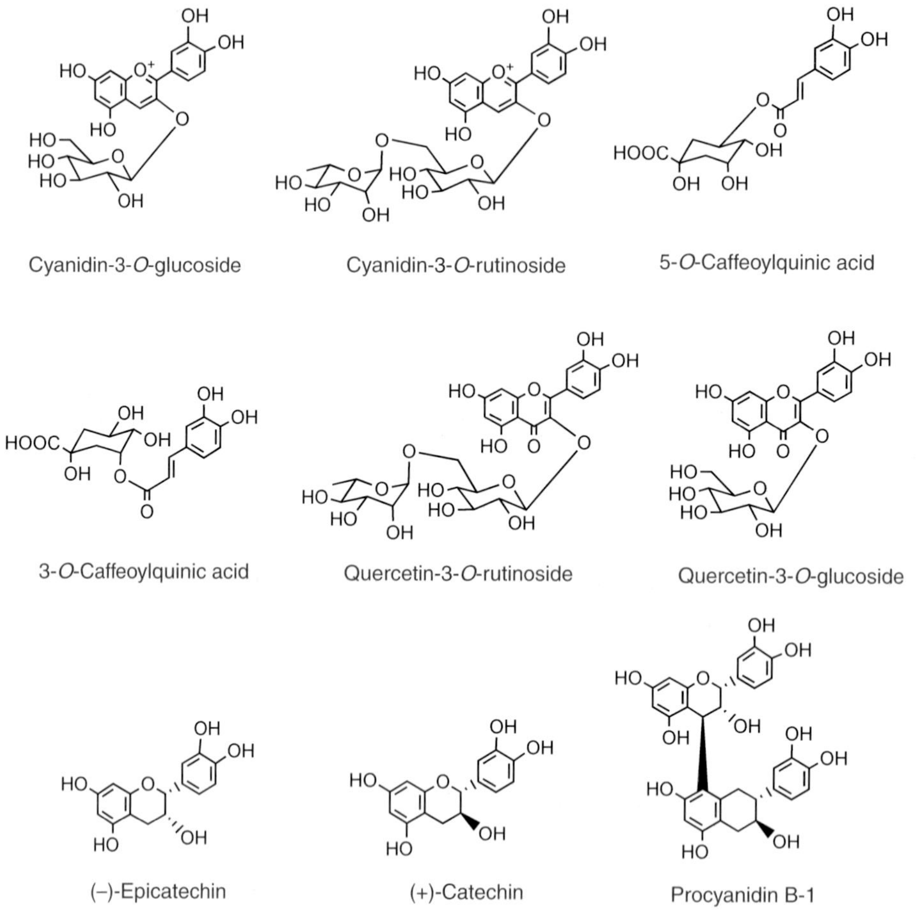

[Chemical structures shown:]

Cyanidin-3-O-glucoside   Cyanidin-3-O-rutinoside   5-O-Caffeoylquinic acid

3-O-Caffeoylquinic acid   Quercetin-3-O-rutinoside   Quercetin-3-O-glucoside

(−)-Epicatechin   (+)-Catechin   Procyanidin B-1

Figure 7.25 Chlorogenic acids and flavonoids detected in peaches and nectarines.

are present together with the non-phenolic compounds 5-(hydroxymethyl)-2-furaldehyde and sorbic acid (Figure 7.27) (Fang et al. 2002).

7.4.5 Citrus fruits

With the exception of grapefruit (*Citrus paradisi*), citrus fruits originate from Asia. Orange (*Citrus sinensis*) and tangerine (*Citrus reticulata*) originated in China and were brought to Rome by Arab traders. The Romans and Greeks knew only of the bitter orange (*Citrus aurantium*). Sweet oranges were brought to Europe from India in the seventeenth century. Lemons (*Citrus limon*) originated in Malaysia or India. They were introduced into Assyria where they were discovered by the soldiers of Alexander the Great who took them back to Greece. The crusaders introduced them into Europe. Limes (*Citrus aurantifolia*) originated in India and were introduced as a crop into the West Indies. The original grapefruit was

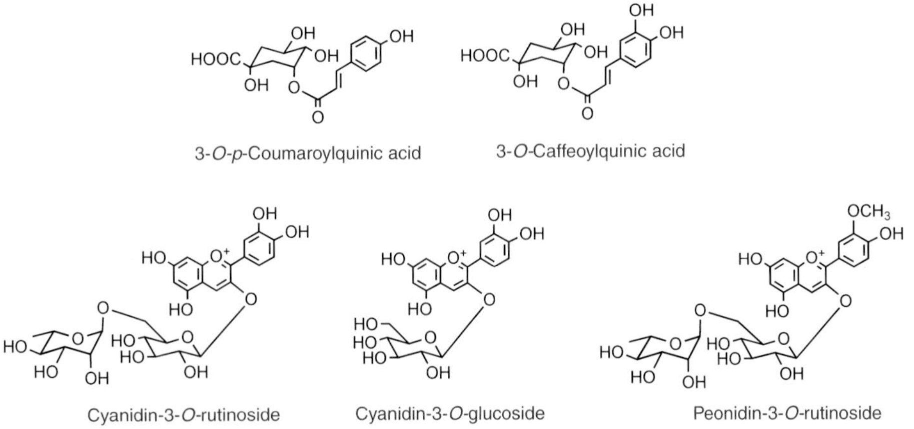

Figure 7.26 Hydroxycinnamates and the main anthocyanins in cherries.

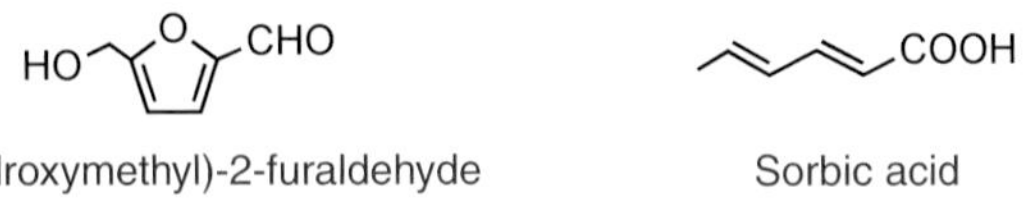

Figure 7.27 In addition to a range of phenolic compounds, plums contain sizable amounts of 5-(hydroxymethyl)-2-furaldehyde and sorbic acid.

discovered in Polynesia and introduced to the West Indies where it was developed and brought to Europe in the seventeenth century.

Citrus fruits are significant sources of flavonoids, principally flavanones, which are present in both the juice and the tissues that are ingested when fruit segments are consumed. The tissues are a particularly rich source but are only consumed as an accidental adjunct to the consumption of the pulp. It is difficult to estimate dietary intake in such cases because it is so heavily dependent on the amount of tissue surrounding the segments after peeling. p-Coumaroyl and feruloyl conjugates of glucaric acid occur in citrus peel (Risch and Herrmann 1988). Citrus peel, and to a lesser extent the segments, also contain the conjugated flavanone naringenin-7-O-rutinoside (narirutin) as well as hesperetin-7-O-rutinoside (hesperidin) (Figure 7.28), which is included in dietary supplements and is reputed to prevent capillary bleeding. Naringenin-7-O-neohesperidoside (naringin) from grapefruit peel and hesperetin-7-O-neohesperidoside (neohesperidin) from bitter orange are intensely bitter flavanone glycosides. Orange juice contains polymethoxylated flavones such as nobiletin, scutellarein, sinensetin and tangeretin which are found exclusively in citrus species. The relative levels of these compounds can be used to detect the illegal adulteration of orange juice with juice of tangelo fruit (*Citrus reticulata*). A further distinguishing feature is that β-cryptoxanthin and its fatty acid esters are present in higher amounts relative to β-carotene (Figure 7.28) in tangelo juice than orange juice (Pan *et al.* 2002).

Citrus fruits also contain significant amounts of terpenoids, with the volatile monoterpenes (+)- and (−)-limonene being responsible for the fragrance of oranges and lemons,

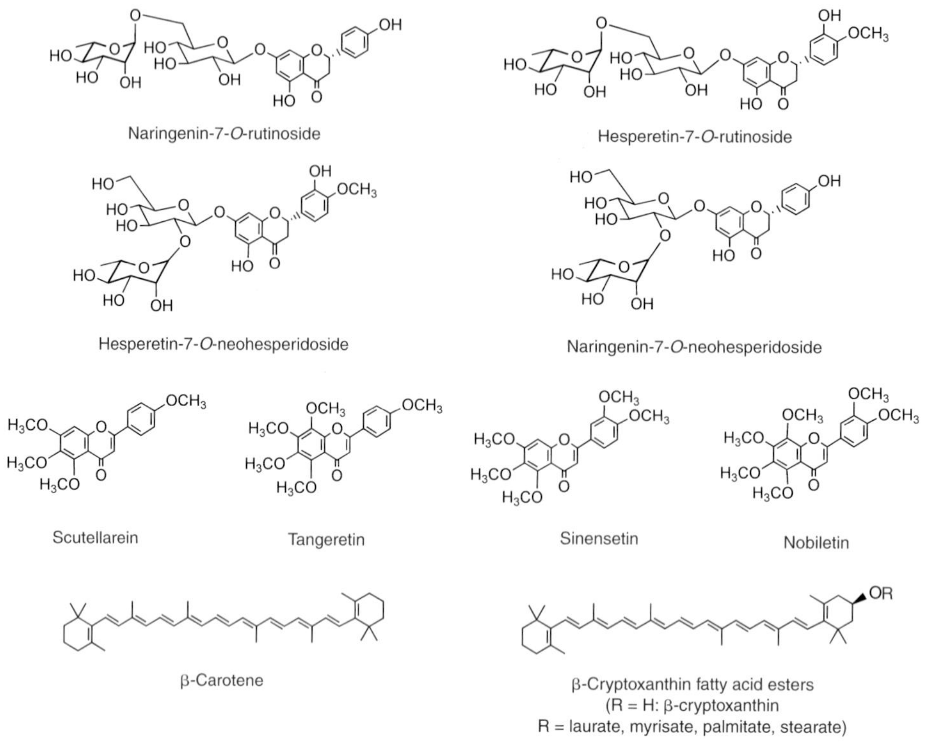

Figure 7.28 Flavanone conjugates, polymethoxylated flavones, β-carotene, β-cryptoxanthin and its fatty acid esters are found in citrus fruit.

respectively (Figure 7.29). Citrus fruits also contain the more complex limonoids which are modified triterpenoids. The bitterness due to limonoids is an important economic problem in commercial citrus juice production. Among the more than 30 limonoids that have been isolated from citrus species, limonin (Figure 7.29) is the major cause of limonoid bitterness in citrus juices. Nomilin (Figure 7.29) is also a bitter limonoid that is present in grapefruit juice and other citrus juices, but its concentration is generally very low so its contribution to limonoid bitterness is minor. As the fruit ripens, the concentration of limonoid aglycones, such as limonoate A-ring lactone (Figure 7.29) declines and bitterness decreases. This natural debittering process was known for over a century, but the mechanism was not understood until the discovery of limonoid glucosides in citrus fruit in 1989, when it was shown that limonoid aglycones are converted to their respective glucosides in fruit tissues and seeds during the later stages of fruit growth and ripening. In contrast to their aglycones, limonoid glucosides, such as limonin-17-O-glucoside (Figure 7.29), are practically tasteless (Hasegawa et al. 2000). In planta, limonoids appear to act as insect anti-feedants; however, they also have a variety of medicinal effects in animals and humans including some anticarcinogenic effects on in vitro human cancer cell lines and animal tests. Other limonoid properties include antifungal, bactericidal and antiviral effects.

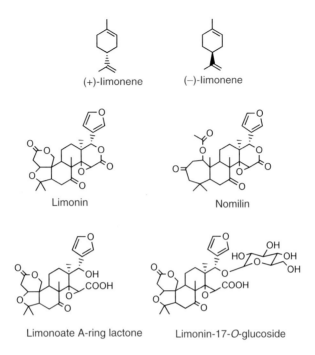

Figure 7.29 Citrus fruit are a rich source of terpenoids including the C_{10} diterpenes (+)- and (−)-limonene and a number of more complex limonoids which are distinctive in that the glucosides are tasteless while the aglycones have a bitter taste.

7.4.6 Pineapple

Pineapple (*Ananas comosus*) is a member of the family Bromeliacea, cultivated in Southern America and was first brought to Europe by Columbus. It is now grown in a number of tropical and subtropical countries and is consumed fresh, canned and processed to give juice. The fruit is notable for the presence of a proteolytic enzyme, bromelain, which is used to prevent a proteinaceous haze in chill-proof beer when refrigerated. Other than the identification of S-sinapyl-L-cysteine, N-L-γ-glutamyl-S-sinapyl-L-cysteine and S-sinapylgluthathione (Wen *et al.* 1999) there are few reports on the occurrence of phenolics in pineapple or pineapple juice although stems contain glyceryl esters of caffeic and p-coumaric acids (Figure 7.30) (Takata and Scheuer 1976).

7.4.7 Dates

Dates (*Phoenix dactylifera*) are probably the oldest cultivated fruit, having been cultivated for over 5000 years. It is a major crop in the Middle East and there are over 2000 cultivars. As with other palms, the sap is tapped for fermentation into 'toddy' which is also distilled. A well-managed tree will produce 400–600 kg dates per year from the age of 5 years for up to 60 years. Al Farsi *et al.* (2005) report that dates contain unspecified carotenoids and anthocyanins together with protocatechuic acid, vanillic acid, syringic acid and ferulic acid

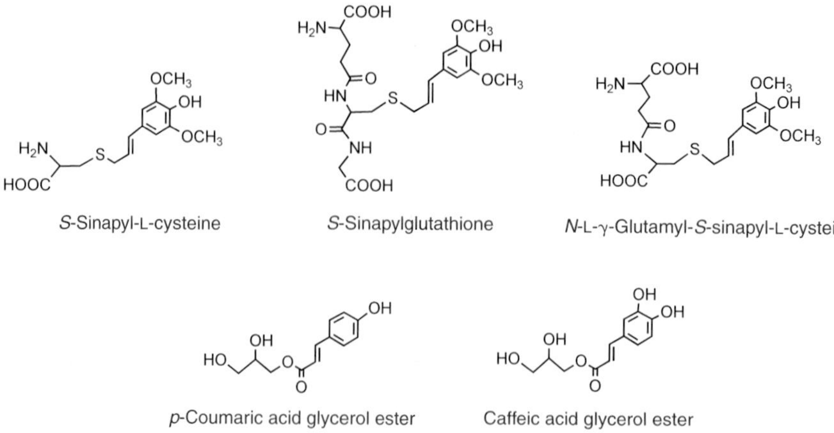

Figure 7.30 Pineapple juice contains sinapic acid conjugates and glyceryl esters of p-coumaric acid and caffeic acid.

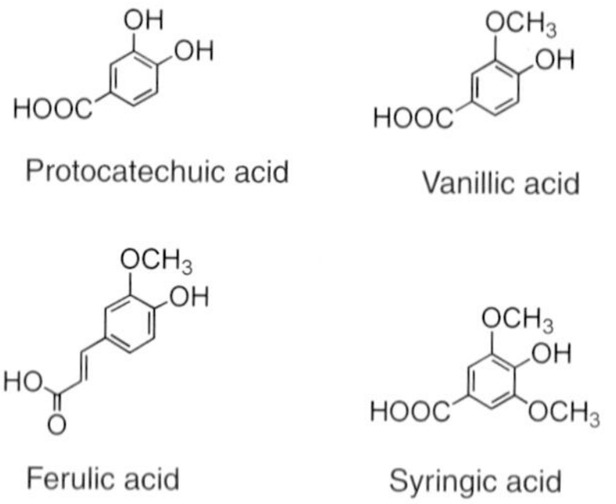

Figure 7.31 Among the compounds found in dates are phenolic acids and uncharacterized anthocyanins and carotenoids. Dates also contain procyanidin oligomers and polymers as well as luteolin, quercetin and apigenin glycosides.

(Figure 7.31). Green dates contain three isomeric caffeoylshikimic acids which appear to form glucosides (Harborne et al. 1974).

A recent study using HPLC-MS^2 has detected a number of flavonoid glycosides and procyanidins in dates with a reddish brown colour and firm texture at the khalal stage of maturity. Procyanidin oligomers through to decamers were identified along with higher molecular weight polymers, undecamers through to hepatadecamers. A total of 19 glycosides of luteolin, quercentin and apigenin were also detected. These included methylated and sulphated forms of luteolin and quercetin present as mono-, di- and triglycosylated conjugate, principally as O-linked conjugates and a single apigenin-di-C-hexoside (Hong et al. 2006).

7.4.8 Mango

Mango (*Mangifera indica*) has been eaten for over 6000 years in India and Malaysia and was introduced to South America and the West Indies in the eighteenth century. It is

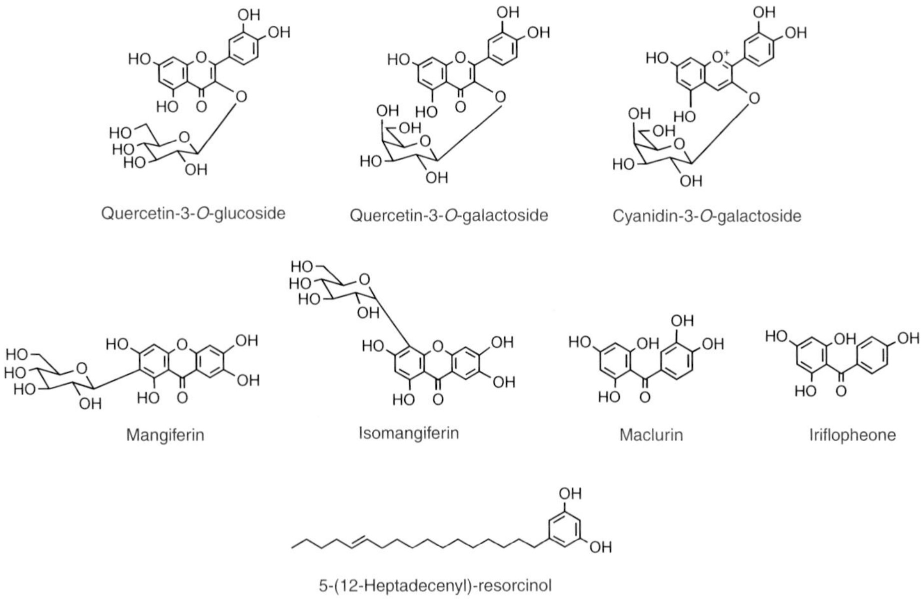

Figure 7.32 Compounds detected in mango fruit.

a good source of β-carotene and vitamin C and currently is one of the more important tropical fruits in the European and American markets where it is sold fresh and as a range of mango products including purée, chutneys, pickles and canned slices. The peel of the fruit contains higher levels of total phenols than the pulp and both peeled and unpeeled fruits are used to prepare purées. The red colour of ripe mango peel is due to cyanidin-3-O-galactoside. The peels also contains several quercetin and kaempferol glycosides, the principal flavonols being quercetin-3-O-glucoside and quercetin-3-O-galactoside, the xanthone C-glucoside, mangiferin and smaller amounts of its isomer isomangiferin and an array of gallotannins, and C-glucosides and galloyl derivatives of the benzophenones, maclurin and iriflopheone (Figure 7.32) (Schieber et al. 2003; Berardini et al. 2004). Mango latex also contains the contact allergen, 5-(12-heptadecenyl)-resorcinol (Figure 7.32) (Cjocaru et al. 1986) which may contaminate the peel but not normally the fruit itself. Mango extracts are used widely in traditional medicines for treating a number of conditions including diarrhoea, diabetes and skin infections (Núñez-Sellés et al. 2002), and mangiferin is reported to inhibit bowel carcinogenesis in rats (Yoshimi et al. 2001).

7.4.9 Papaya

Papaya (*Carica papaya*) is native to Central America but is now grown widely in the tropics. It fruits all year round. Compared with other fruits it is high in carotenes especially β-crytoxanthin. The unripe fruit is a source of the enzyme papain which is used as a meat tenderizer and a beer clarifier. The ripe fruit contains a number of phytoalexins including danieleone (Figure 7.33) (Echeverri et al. 1997).

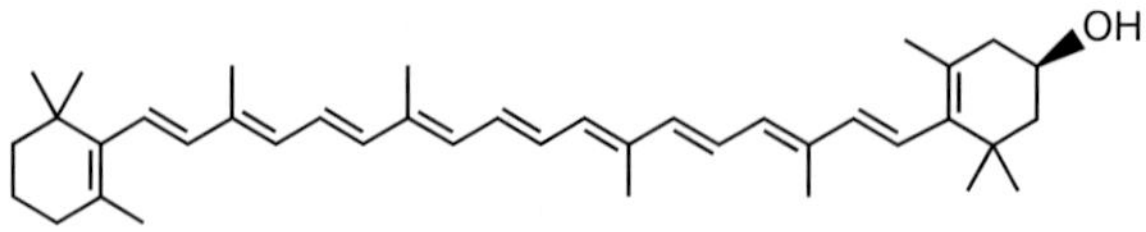

β-Cryptoxanthin

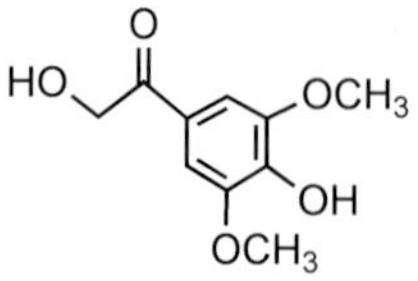

Danielone

Figure 7.33 β-Crytoxanthin is the main carotenoid in papaya fruit which also contains danieleone, a phytoalexin.

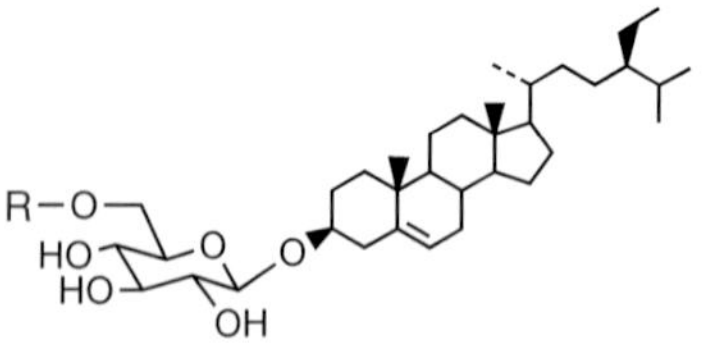

(6'-O-Palmitoylglucosyl)sitosterol (R = palmitate)
(6'-O-Linoleoylglucosyl)sitosterol (R = linoleate)

Figure 7.34 Sitosterol derivatives with potential anti-tumour activity are found in latex released when figs are picked.

7.4.10 Fig

The fig is among the oldest known fruit crops, its seeds having been found in early Neolithic sites dating to 7000 BC. It was probably cultivated from about 2700 BC in Egypt and Mesopotamia. The genus *Ficus* contains over 1000 species, the most important of which as a commercial fruit crop is *Ficus carica* which is widely used as a food and as a medicine in the Middle East. The latex released on picking fruits has anti-tumour activity, and there is evidence that the bioactive components are (6'-O-palmitoylglucosyl)sitosterol and (6'-O-linoleoylglucosyl)sitosterol (Figure 7.34) (Rubnov et al. 2001).

7.4.11 Olive

The olive tree (*Olea europa*) has been cultivated for thousands of years. The oil is the most important constituent, but olives also contain phenolics, including vanillic acid, ferulic acid, the flavones luteolin and apigenin together with substantial amounts of the glucoside, oleuropein (Figure 7.35), which is bitter and is commonly neutralized by treatment with caustic soda before the olives can be eaten. Olives contain up to 40% oil of which, typically, three quarters is a monounsaturated fatty acid oleic acid (C18 : 1), 14% saturated fatty acid (mainly palmitic acid, C16 : 0) and 9% polyunsaturated fatty acids. Olive oil is

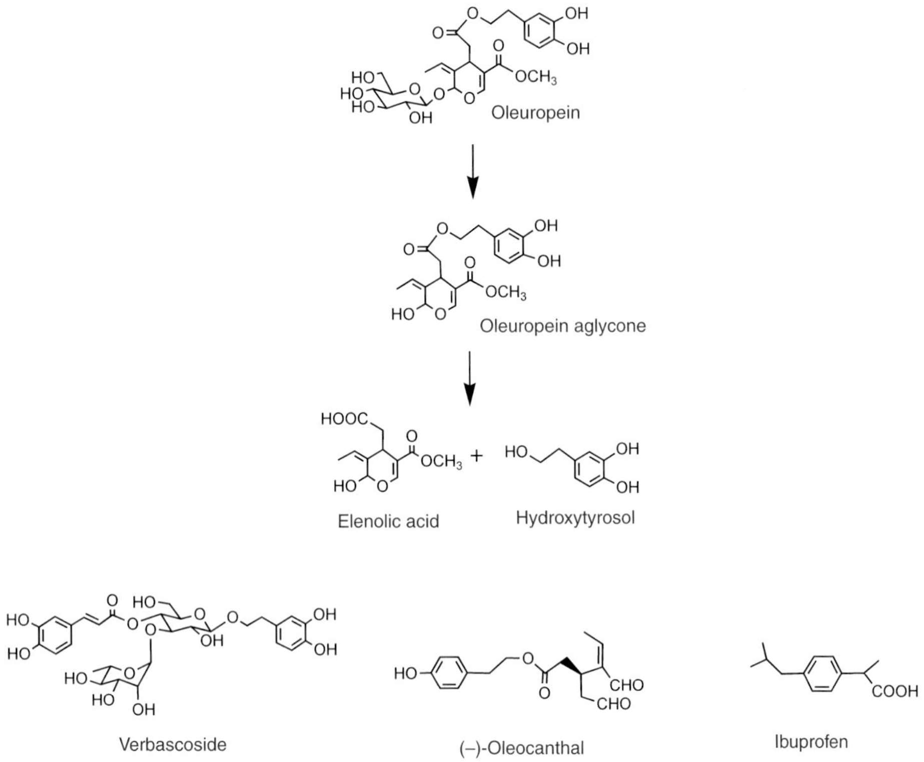

Figure 7.35 The bitter taste in unripe olives is due to oleuropein. The levels fall as the fruits mature and oleuropein aglycone accumulates. Verbascoside is the main hydroxycinnamate derivative in olives. Olive oil contains oleuropein aglycone and hydroxytyrosol both of which are strong antioxidants. Virgin olive oil contains (−)-oleocanthal which has ibuprofen-like anti-inflammatory properties.

also rich in oleic acid and the main phenol is oleuropein aglycone which is produced by enzymatic degradation. The aglycone contains a hydroxytyrosol group which is the antioxidant moiety. The oil also contains hydroxytyrosol itself, derived from oleuropein but in smaller amounts than the aglycone (Figure 7.35). Hydroxytyrosol and oleuropein are found in a number of species within the Oleaceae family, but only olives and olive oil are significant dietary components (Soler-Rivas *et al.* 2000). Besides oleuropein, *Olea europa* contains other phenolic glucosides including verbascoside (Figure 7.35), a heterosidic ester of caffeic acid and hydroxytyrosol, which is almost ubiquitous in the Oleaceae. Small-fruit cultivars of olive are characterized by high oleuropein and low verbascoside contents while large-fruit cultivars have low oleuropein and high verbascoside contents (Amiot *et al.* 1986).

Newly pressed, extra-virgin olive oil contains (−)-oleocanthal, a compound with anti-inflammatory action similar to that of ibuprofen, the non-steroidal anti-inflammatory drug (Beauchamp *et al.* 2005). Although structurally dissimilar (Figure 7.35), both compounds inhibit the same cyclooxygenase enzymes in the prostoglandin biosynthesis pathway. Daily ingestion of 50 mL of extra-virgin olive oil corresponds to about 10% of the recommended intake of ibuprofen for adult pain relief. Ibuprofen is associated with a reduced risk of developing some cancers and of platelet aggregation in the blood as well as with secretion of amyloid-b42 peptide in a mouse model of Alzheimer's disease. A Mediterranean diet rich in olive oil, is believed to confer various health benefits some of which appear to overlap with those attributed to non-steroidal anti-inflammatory drugs (Beauchamp et al. 2005).

7.4.12 Soft fruits

This section includes those fruits that in strict botanical terms are berries but which are commonly known as currants as well as agglomerates which perversely are widely referred to as berries. A wide range of berries are consumed. Most are cultivated but some are picked from the wild. The range includes strawberry (*Fragaria* × *ananassa*), raspberry (*Rubus idaeus*), blackberry (*Rubus* spp.), blueberry (*Vaccinium corymbosum*), elderberry (*Sambucus nigra*), cranberry (*Vaccinium oxycoccus*), gooseberry (*Ribes grossularia*) and the black (*Ribes nigrum*), red (*Ribes rubrum*) and white currants. Soft fruits make up only a tiny part of the diet in the United Kingdom but are more important in some Nordic countries. They tend to be susceptible to decay and have to be processed to extend the shelf life. Until the introduction of canning in the mid nineteenth century preservation was almost impossible, but now a range of methods is available including processing into jam.

The modern strawberry is the descendant of the tiny woodland strawberry that was grown by the Romans. Modern cultivated strawberries derive from a cross between an American and a Chilean variety that occurred around 1750. Raspberries are native to Europe and have been cultivated since the Middle Ages. Cloudberries (*Rubus chamaemorus*) are relatives of the raspberry, grown either side of the Arctic Circle. Blackberries have been eaten since Neolithic times and the Greeks prized them for the medicinal value of the leaves.

A number of crosses have been made between raspberries and blackberries including the loganberry (*Rubus loganbaccus*) and the Tayberry. The blueberry is native to North America and is cultivated both there and in Europe. Cranberries grow wild in both Northern Europe and Northern United States. Native Americans prized them for both their nutritional and medicinal properties and are said to have introduced the first Europeans to cranberries to help them prevent scurvy. Cranberry juice is currently used for preventing urinary infections (Schenker 2001). The cranberry is *Vaccinium oxycoccus* while *Vaccinium macrocarpon* is the large or American cranberry which is grown commercially in America as well as Europe. Gooseberries were popular in Mediaeval England but were not cultivated until the sixteenth century. They are little consumed outside the United Kingdom.

The currants grow wild throughout northern Europe but were not cultivated until the sixteenth century. Anthocyanin-deficient whitecurrants are rarely grown now and redcurrants are generally only grown for jelly. The main end products of blackcurrant cultivation are juice drinks and jam. Consumption is thus relatively low.

Anthocyanins provide the distinctive and vibrant palate of colours found in berries. The structures of the main anthocyanins in berries are summarized in Figure 7.36 and listed

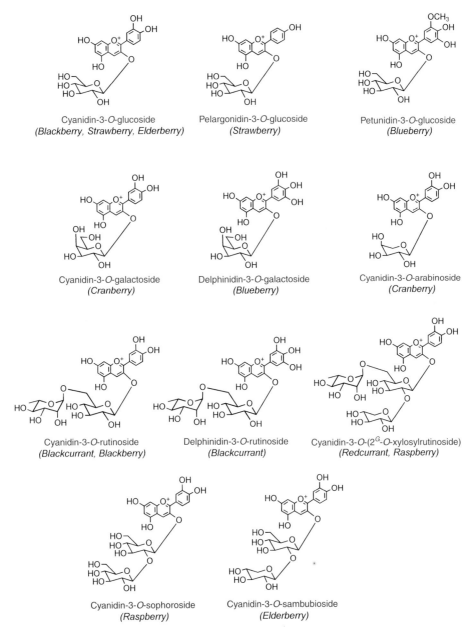

Figure 7.36 The major anthocyanins in berries.

in more detail along with ellagitannins and flavonols in Table 7.1. There is much variety and while some fruits, such as cranberry, blackberry and elderberry, contain derivatives of only one type of anthocyanin (i.e. cyanidin), a wide array of anthocyanins is found in blueberry and blackcurrant. In general the anthocyanin profile of a tissue is characteristic,

Table 7.1 Summary of anthocyanins, ellagitannins and flavonols in berries. Major components indicated in bold font but there are varietal differences in the relative levels of individual anthocyanins

Common name	Genus and species	Family	Phenolics	Reference
Blackcurrant	*Ribes nigrum*	Grossulariaceae	**Del-3-Rut; Del-3-Glc; Cy-3-Rut; Cy-3-Glc** Peo-3-Rut, Malv-3-Rut, Malv-3-Rut, Malv-3-Glc Myr-3-Rut, Myr-3-Glc, Q-3-Rut, Q-3-Glc	Määttä et al. (2003) Degénéve (2004) Wu et al. (2004)
Redcurrant	*Ribes rubrum*	Grossulariaceae	**Cy-3-Rut, Cy-3-Xyl-Rut, Cy-3-Glc-Rut, Cy-3-Sop** Cy-3-Glc, Cy-3-Samb Q-3-Glc, Q-3-Rut	Määttä et al. (2003) Degénéve (2004) Wu et al. (2004)
Strawberry	*Fragaria × ananassa*	Rosaceae	**Pel-3-Glc, Pel-3-GlcMal** sanguiin H-6 Q-3-GlcAc, K-3-Glc	Degénéve (2004) Cerdá et al. (2005)
Blackberry	*Rubus* spp.	Rosaceae	**Cy-3-Glc, Cy-3-Rut** Cy-3-Xyl, Cy-3-GlcMal **lambertianin C** Q-3-Gal, Q-3-Glc, Q-3-Xyl, Q-3-Rut, Q-3-XylGlcAC	Cho et al. (2004) Degénéve (2004)

Red raspberry	*Rubus idaeus*	Rosaceae	**Cy-3-Sop, Cy-3-Glc-Rut, Cy-3-Glc** Cy-3,5-DiGlc, Cy-3-Samb, Cy-3-Rut, Pel-3-Glc, Pel-3-Sop, Pel-3-Glc-Rut, Pel-3-Rut **sanguiin H-6,** lambertianin C Q-3-Rut, Q-3-Glc, Q-3-GlcAC	Mullen *et al.* (2002 a,b)
Blueberry	*Vaccinium corymbosum*	Ericaceae	**Del-3-Gal, Del-3-Ara, Cy-3-Gal, Pet-3-Gal,** **Pet-3-Ara, Pet-3-GlcAc, Peo-3-Gal, Malv-3-Gal,** **Malv-3-Arab** Del-3-Glc, Del-3-GlcAc, Cy-3-Glc, Cy-3-Arab, Pet-3-Glc, Peo-3-Arab, Malv-3-Glc, Malv-3-GlcAc Q-3-Gal, Q-3-Glc, Q-3-Xyl, Q-3-Rut, Myr-3-Glc, Myr-3-Gal	Prior *et al.* (2001) McGhie *et al.* (2003) Cho *et al.* (2004) Degénéve (2004)
Cranberry	*Vaccinium macrocarpum*	Ericaceae	**Cy-3-Gal, Cy-3-Ara, Peo-3-Glc, Peo-3-arab** Cy-3-Glc, Malv-3-Glc, Malv-3-Arab Myr-3-Gal, Q-3-Gal, Q-3-Rham	Prior *et al.* (2001) Degénéve (2004)
Elderberry	*Sambucus nigra*	Caprifoliaceae	**Cy-3-Samb, Cy-3-Glc** Cy-3,5-DiGlc, Cy-3-Samb-5-Glc	Wu *et al.* (2004)

Abbreviations: Cyanidin (Cy); Pelargonidin (Pel); Peonidin (Peo), Petunidin (Pet), Malvidin (Malv), Quercetin (Q), Myricetin (Myr), Kaempferol (K); Glucoside (Glc); Acetylglucoside (GlcAc); Malonylglucoside (GlcMal); Diglucoside (DiGlc); Sophoroside (Sop); Xyloside (Xyl); Acetylxyloside (XylAc); Arabinoside (Ara); Acetylarabinoside (AraAc); Glucuronide (GlcAC); Xylosylglucuronide (XylGlcAC); Galactoside (Gal); Rhamnoside (Rham), Rutinoside (Rut); Sambubioside (Samb).

and it has been used in taxonomy, and for the detection of adulteration of juices and wines. Blackcurrants are characterized by the presence of the rutinosides and glucosides, with the rutinosides being the most abundant. Other anthocyanins and flavonol conjugates have been noted, but at much lower concentration. Whilst redcurrants are very closely related to blackcurrants, they contain mainly cyanidin diglycosides with cyanidin monoglucosides present only as minor component. Strawberries, blackberries and red raspberries are all from the Rosaceae family but they have a diverse anthocyanin content. The major anthocyanins in raspberries and blackberries are derivatives of cyanidin, while in strawberries pelargonidin glycosides predominate. The major components in blueberries are delphinidin-3-O-galactoside and petunidin-3-O-glucoside; however, many minor anthocyanins are also present. Cranberries belong to the Ericaceae, the same family as blueberries, but have cyanidin-based compounds as their major anthocyanins. As with cranberries, blackberries and raspberries, the major anthocyanins in elderberries are cyanidin-based, with cyanidin-3-O-sambubioside and cyanidin-3-O-glucoside predominating (Table 7.1, Figure 7.36).

Flavonols and other flavonoids are commonly quantified as the aglycone after acid or enzyme hydrolysis to remove sugar residue (Hertog et al. 1992). Using this approach the myricetin, quercetin and kaempferol content of edible berries had been estimated (Hakkinen et al. 1999). Quercetin was found to be highest in bog whortleberry (*Vaccinium uliginosum*) (158 mg/kg), bilberry (*Vaccinium myrtillus*) (17–30 mg/kg) and in elderberries. In blackcurrant cultivars, myricetin was the most abundant flavonol (89–203 mg/kg), followed by quercetin (70–122 mg/kg) and kaempferol (9–23 mg/kg). In comparison, the total anthocyanin content of red raspberries is ~600 mg/kg (Mullen et al. 2002). Specific flavonol glycosides that have been identified include quercetin-3-O-glucoside, quercetin-3-O-rutinoside quercetin-3-O-galactoside and quercetin-3-O-xylosylglucuronide, myricetin-3-O-glucoside, myricetin-3-O-galactoside and myricetin-3-O-rutinoside (Table 7.1, Figure 7.37).

Berries can contain substantial amounts of the flavan-3-ol monomers (+)-catechin and (−)-epicatechin as well as dimers, trimers and polymeric proanthocyanidins. The concentration of the polymers is usually greater than the monomers, dimers and trimers, and overall cranberries are a particularly rich source of these compounds (Table 7.2).

The hydroxybenzoate, ellagic acid (Figure 7.38) has been reported to be present in berries, particularly raspberries (5.8 mg/kg), strawberries (18 mg/kg) and blackberries (88 mg/kg) (Amakura et al. 2000). Indeed ellagic acid has been described as being responsible for >50% of total phenolics quantified in strawberries and raspberries (Häkkinen et al. 1999). In reality, however, free ellagic acid levels are generally low, although substantial quantities are detected along with gallic acid after acid treatment of extracts as products of ellagitannin breakdown. For instance, red raspberries, the health benefits of which are often promoted on the basis of a high ellagic acid content, contain ~1 mg/kg of ellagic acid compared with ~300 mg/kg of ellagitannins, mainly in the form of sanguiin H-6 and lambertianin C (Figure 7.38) (Mullen et al. 2002b). Berries also contain a variety of hydroxycinnamates including caffeoyl/feruloyl esters, usually in low concentrations although blueberries have been reported to contain 0.5–2.0 g/kg of 5-O-caffeoylquinic acid (Schuster and Herrmann 1985).

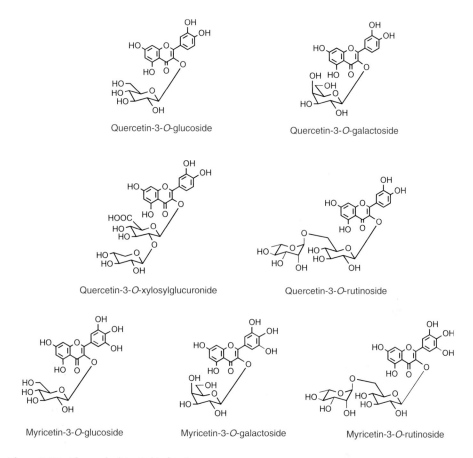

Figure 7.37 Flavonols detected in berries.

7.4.13 Melons

Melons (*Cucumis melo*) are relatives of cucumbers. The first melons were bitter, but they were bred to produce sweeter fruit and introduced into Europe from Africa by the Moors. They reached France in the fifteenth century and were taken to the New World by Columbus. Melons and cantaloupes contain high levels of carotenes. Watermelons (*Citrullus lanatus*) are distant relatives of melons, widely spread throughout Africa. They were known to the Egyptians and wild watermelons grow in the Kalahari Desert. Watermelons were introduced to Europe in the fifteenth century. Watermelon can contain high levels of carotenes, particularly lycopene; 23–72 mg/kg have been reported (van den Berg *et al.* 2000).

7.4.14 Grapes

Grapes were among the earliest cultivated crops. The Egyptians, Greeks and Romans all made wine from them and the Romans bred many new varieties. Concord grapes

Table 7.2 Concentration of flavan-3-ol monomers, dimers and trimers and total proanthocyanidins in berries

Berry	Monomers	Dimers	Trimers	Total PAs	Type	Reference
Cranberry	73 ± 15	259 ± 61	189 ± 13	4188 ± 750	A, PC	Gu et al. (2004)
Blueberry	40 ± 15	72 ± 18	54 ± 12	1798 ± 508	PC	Gu et al. (2004)
Blackcurrant	9 ± 2	29 ± 4	30 ± 3	1478 ± 280	PC, PD	Gu et al. (2004)
Strawberry	42 ± 7	65 ± 13	65 ± 12	1450 ± 250	PP, PC	Gu et al. (2004)
Redcurrant	13	20	15	608	—	Wu et al. (2004)
Red raspberry	44 ± 34	115 ± 100	57 ± 5.7	302 ± 230	PP, PC	Gu et al. (2004)
Blackberry	37 ± 22	67 ± 29	36 ± 19	270 ± 170	PC	Gu et al. (2004)

Data expressed as mg/kg fresh weight ± standard deviation. PA – proanthocyanidins; PC – procyanidins; PD – prodelphidins; PP – propelargonidins; A – indicates existence of A-type proanthocyanidins.

(*Vitis labrusca*) are characterized by a red-coloured flesh as well as skin. They are grown in America and are a different species from the European grape *Vitis vinifera*. Fresh red *V. vinifera* grapes contain in the region of four grams of phenolic material per kilo. There is substantial variation in the levels of phenolics in red grapes that reflects a number of factors including the variety of grape, with small thick-skinned grapes such as Cabernet Sauvignon, which are characterized by a high skin:volume ratio, having a higher phenolic content than 'thinner-skinned' varieties such as Grenache with a low skin:volume ratio. There is a trend towards higher phenolic levels in wines made from grapes grown in sunnier climates, such as Chile, Argentina and Australia, rather than cooler regions, such as northern Italy and northern France. In planta, flavonols, at least, are located principally in epidermal cells where they serve as UV protectants with their levels increasing in response to exposure to sunlight. In keeping with this role, there is a report that Pinot Noir grapes from sun-exposed clusters contain seven times more quercetin glycosides than shaded berries (Price *et al.* 1995). The flavonols in red grapes are conjugates of myricetin, quercetin, kaempferol and isorhamnetin. The anthocyanin content is quite complex with the main components in Cabernet Sauvignon grapes being malvidin-3-*O*-glucoside, malvidin-3-*O*-(6″-*O*-*p*-coumaroyl)glucoside, malvidin-3-*O*-(6″-*O*-acetyl)glucoside and delphinidin-3-*O*-glucoside while the presence of significant amounts malvidin-3,5-*O*-diglucoside is an indication of a hybrid grape (Burns *et al.* 2001, 2002a). The seeds of red grapes contain substantial quantities of (+)-catechin, (−)-epicatechin, procyanidin oligomers and polymers mainly with a degree of polymerization >10 (Gu *et al.* 2004). The grapes also contain gallic acid, several *p*-coumaroyl derivatives and caftaric acid. The phytoalexin *trans*-resveratrol-3-*O*-glucoside (*trans*-piceid) also occurs but in low and variable amounts that are probably dictated by cultivar and disease pressure (Burns *et al.* 2001, 2002b). Other stilbenes include *trans*-astringin, and the resveratrol oligomers ε-viniferin and pallidol (Landrault *et al.* 2002). The structures of some of the diverse phenolics found in red grapes are presented in Figure 7.39.

Table grapes are picked earlier and do not ripen to the same extent as grapes used to make wines. They are therefore likely to contain much lower levels of flavonoids and

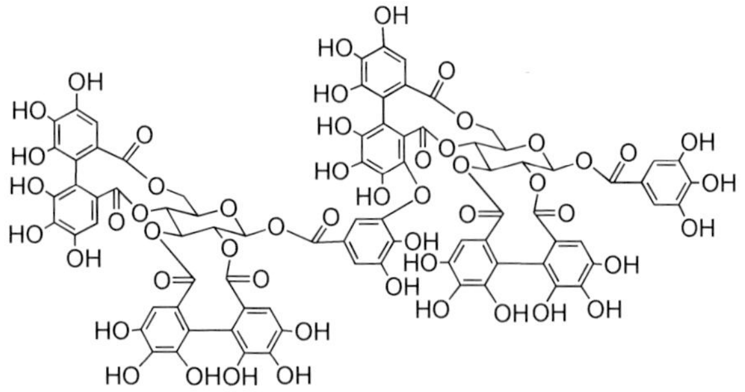

Sanguiin H-6

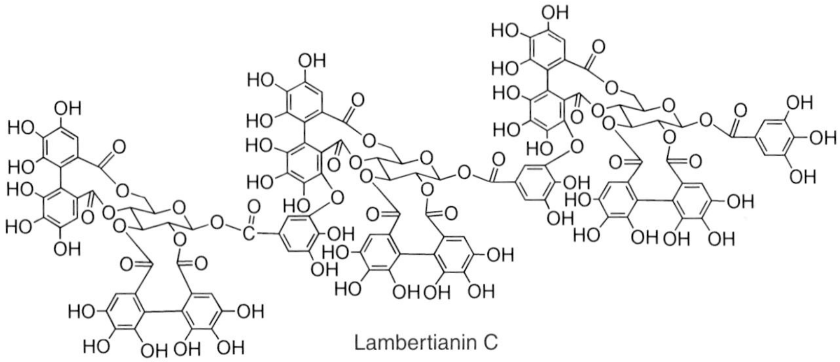

Lambertianin C

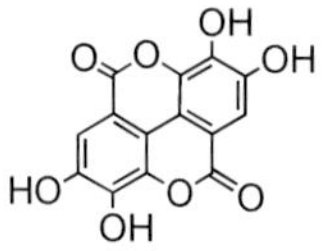

Ellagic acid

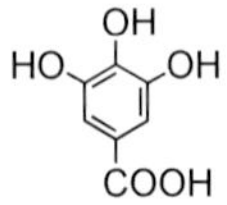

Gallic acid

Figure 7.38 Raspberries contain high concentrations of two ellagitannins, sanguiin H-6 and lambertianin C. When extracts are treated with acid the ellagitannins are breakdown releasing substantial quantities of ellagic acid and gallic acid.

phenolic compounds. Nowadays red grapes for table use are usually seedless varieties and so will contain much lower levels of flavan-3-ols and their procyanidin oligomers and have a much lower antioxidant capacity than grapes used to make red wine (Table 7.3). White grapes contain much lower levels of phenolics than red grapes. Although similar caftaric acid levels have been reported, white grapes lack anthocyanins and contain only trace levels of flavonols and, if seedless, the flavan-3-ol content will also be seriously diminished.

Raisins are grapes that have been dried in full sun, whereas sultanas are dried in partial shade and treated with sulphur compounds to prevent darkening. This results in significant degradation of caftaric acid and coutaric acid as well as flavan-3-ols and procyanidins. In contrast flavonols are not affected to the same degree (Karadeniz *et al.* 2000). Both raisins and sultanas are generally made from the Thompson seedless grape. Currants are

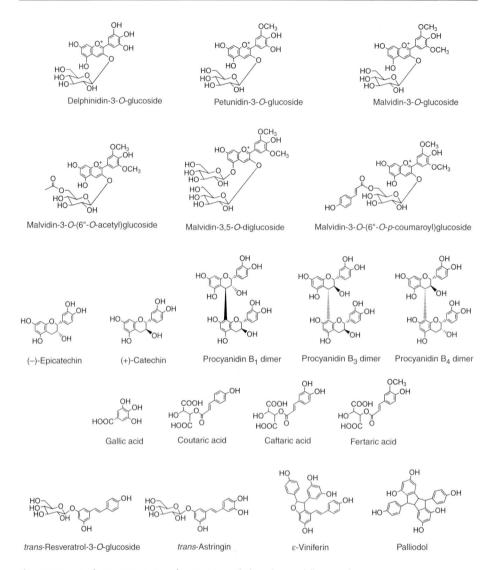

Figure 7.39 Red grapes contain a diverse array of phenolics and flavonoids.

dried small black seedless grapes originally grown in the region of Corinth, from which they derive their name; they were originally known in the United Kingdom as raisins of Corinth. The brown colour of raisins is due to a combination of pigments produced by polyphenol oxidase activity and non-enzymic reactions.

7.4.15 Rhubarb

Botanically, rhubarb (*Rheum rhaponticum*) is a vegetable not a fruit. It was originally cultivated some 2000 years ago in Northern Asia as a medicinal and ornamental plant. It is

Table 7.3 Total antioxidant capacity and phenolic content of red wine grapes and red table grapes (Borges and Crozier, unpublished)

Samples	Total antioxidant capacity[a]	Total phenolic content[b]
Red Wine Grapes		
Zinfandel (*Chile*)	21.8 ± 0.5	16.2 ± 0.3
Syrah (*Chile*)	29.0 ± 0.5	22.4 ± 0.5
Merlot (*Chile*)	36.0 ± 0.8	26.3 ± 0.2
Pinot Noir (*Chile*)	38.3 ± 0.5	28.3 ± 0.4
Cabernet Sauvignon (*Chile*)	52.3 ± 0.6	36.2 ± 0.7
Mean	**35.5 ± 5.1**[a]	**25.9 ± 3.3**[a]
Cabernet Sauvignon without seeds	23.3 ± 0.0	15.1 ± 0.1
Red Table Grapes		
Flame (*Egypt*)	12.9 ± 0.2	12.8 ± 0.3
Flame (*USA*)	9.4 ± 0.1	6.9 ± 0.3
Flame (*Egypt*)	13.6 ± 0.1	10.7 ± 0.3
Flame (*Mexico*)	18.9 ± 1.3	12.3 ± 0.1
Crimson (*South Africa*)	2.9 ± 0.1	2.6 ± 0.1
Crimson (*Spain*)	7.1 ± 0.2	13.8 ± 0.4
Ruby (*Chile*)	4.7 ± 0.2	4.1 ± 0.2
Red Globe (*Chile*)	9.8 ± 0.6	6.5 ± 0.2
Mean	**9.9 ± 1.8**[b]	**8.7 ± 1.5**[b]

[a]Data expressed as mean concentration of Fe^{2+} produced (mmol/kg fresh weight) ± SE. [b]Data expressed as mean mmol gallic acid equivalents/kg ±SE. In each column mean values with different subscripts are significantly different at $p > 0.05$.

rich in salicylates and contains up to 2000 mg/kg anthocyanins. The leaves are poisonous due to high levels of oxalic acid which are not present in the edible stem. Several stilbenes have been isolated from rhubarb including *trans*-resveratrol, which is a strong anti-cancer agent, and piceatannol and rhapontigenin (Figure 7.40). Rhubarb also contains anthraquinones such as chrysophanol, emodin, aloe-emodin and rhein (Figure 7.40) which may contribute to the toxicity of the leaf (Clifford 2000). Rhizomes of *Rheum rhizoma* also contain a range of gallotannins and condensed tannins (Kashiwada *et al.* 1986) although they have not been found in the edible part of the plant.

7.4.16 Kiwi fruit

Kiwi fruit (*Actinidia deliciosa*) was first grown in China, imported to the United Kingdom in the nineteenth century and seeds from Kew were sent to New Zealand in 1906. Kiwi

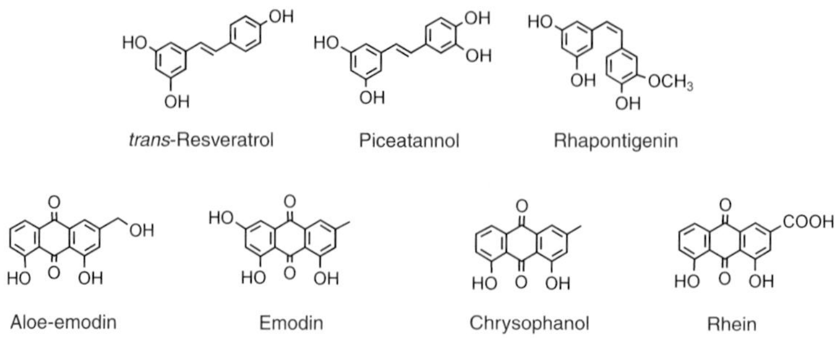

Figure 7.40 Stilbenes and anthraquinones that occur in rhubarb.

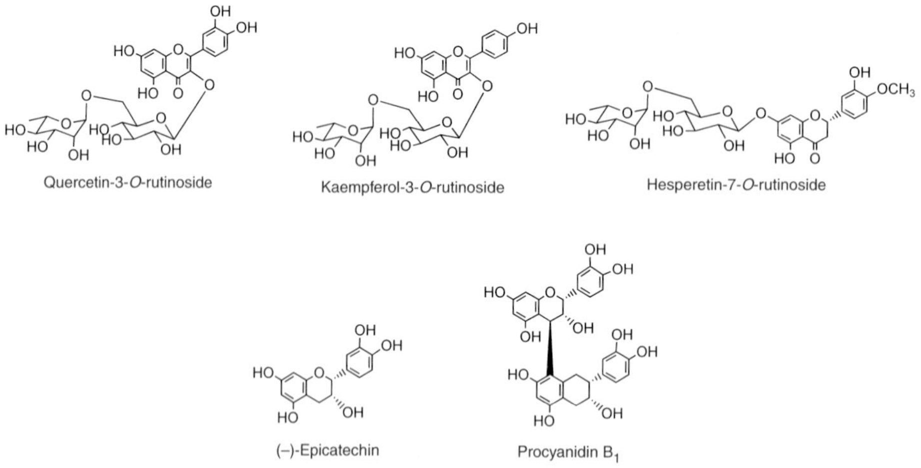

Figure 7.41 Kiwi fruit contain flavonol and flavanone conjugates and flavan-3-ols.

fruit contain vitamin C and like avocados they are rich in chlorophyll which is unusual for fruits. Lymphocytes collected from volunteers after the consumption of a kiwifruit juice supplement are less susceptible to oxidative DNA damage as determined by the Comet assay and this potentially protective effect is not entirely attributable to vitamin C (Collins et al. 2001). Kiwi fruit contain flavonols, including kaempferol-3-O-rutinoside and quercetin-3-O-rutinoside, the flavanone hesperetin-7-O-rutinoside together with (−)-epicatechin and the procyanidin B_2 dimer (Figure 7.41) (Dégenéve 2004).

7.4.17 Bananas and plantains

The original banana grew in South East Asia but contained many bitter black seeds so that they would have been almost inedible. They are recorded in the reports of Alexander the Great in India where they were introduced by 600 BC. Cultivation of bananas

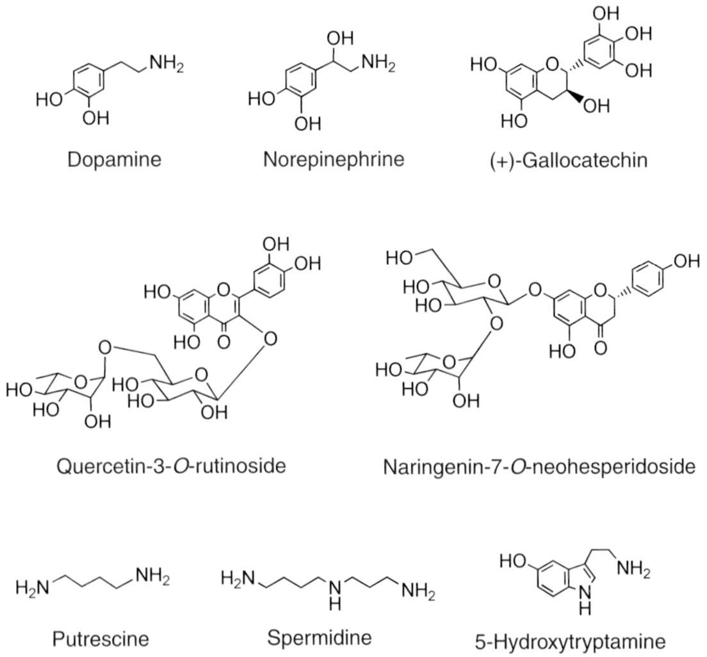

Figure 7.42 Amines and flavonoids detected in bananas.

(*Musa cavendishii*) commenced in the West Indies in the seventeenth century. There are both sweet and cooking bananas, the latter sometimes called plantains. Cooking bananas contain more starch and less sugar than the dessert varieties and are a staple food in East Africa. Bananas are reported to contain lutein, α-carotene and β-carotene (van den Berg *et al.* 2000) and high concentrations of the catecholamine dopamine which is a strong antioxidant, together with norepinephrine (noradrenaline), (+)-gallocatechin, naringenin-7-*O*-neohesperidoside and quercetin-3-*O*-rutinoside (Figure 7.42). Typically, much higher levels of these components are found in the peel than the pulp (Kanazawa and Sakakibara 2000). Green bananas contain 5-hydroxytryptamine (serotonin), spermidine and putrescine (Figure 7.42), but the levels are reduced by the time the fruit becomes edible (Adão and Glória 2005).

7.4.18 Pomegranate

The pomegranate (*Punica granatum* L.) is native from Iran to the Himalayas in Northern India and was cultivated and naturalized over the whole Mediterranean region since ancient times. It is widely cultivated throughout India and the drier parts of Southeast Asia, Malaya, the East Indies and tropical Africa. Spanish settlers introduced the tree into California in 1769 and it is now grown for its fruits mainly in the drier parts of California and Arizona. The fruit has a tough, leathery rind which is typically yellow overlaid with light or deep pink or rich red. The interior is separated by membranous walls and white, spongy, bitter

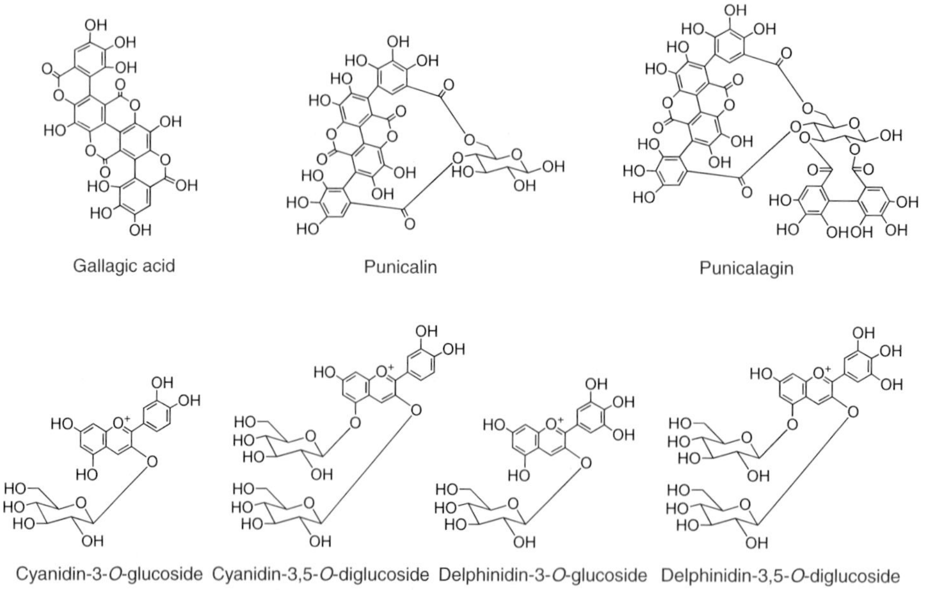

Figure 7.43 Pomegranate juice is a rich source of antioxidants containing gallagic acid, punicalin, punicalagin and anthocyanins.

tissue into compartments packed with sacs filled with sweetly acid, juicy, red, pink or whitish pulp or aril. In each sac there is one angular, soft or hard seed.

Commercial pomegranate juice is increasing in popularity. It has a high antioxidant capacity seemingly because industrial processing extracts hydrolysable tannins from the rind. The juice contains gallagic acid, an analogue of ellagic acid containing four gallic acid residues, and punicalin, the principal monomeric, hydrolysable tannin, in which gallagic acid is bound to glucose. Punicalagin is an additional hydrolysable tannin in which ellagic acid, as well as gallagic acid, is also linked to the glucose moiety. Juices also contain the 3-*O*-glucosides and 3,5-*O*-diglucosides of cyanidin and delphinidin (Figure 7.43) and several ellagic acid derivatives (Gil et al. 2000). Gallagic acid has restricted occurrence in plants, but it has been reported as a toxic principal in *Terminalia* spp. responsible for losses of browsing cattle and sheep. This has been discussed by Clifford and Scalbert (2000). There is no evidence of pomegranate toxicity in humans.

7.5 Herbs and spices

Herbs and spices are botanically heterogeneous but with major contributions from the Apiaceae and Lamiaceae. These commodities are also phytochemically complex and variable geographically within a species or taxon and only a brief overview is possible here, focusing on those commodities most commonly used in Europe. Frequently herbs and spices contain phytochemicals not found in other foodstuffs and may sometimes resemble herbal medicines. However, the quantity consumed in food suggests that any pharmacological effects will be limited, although often only qualitative composition data are available.

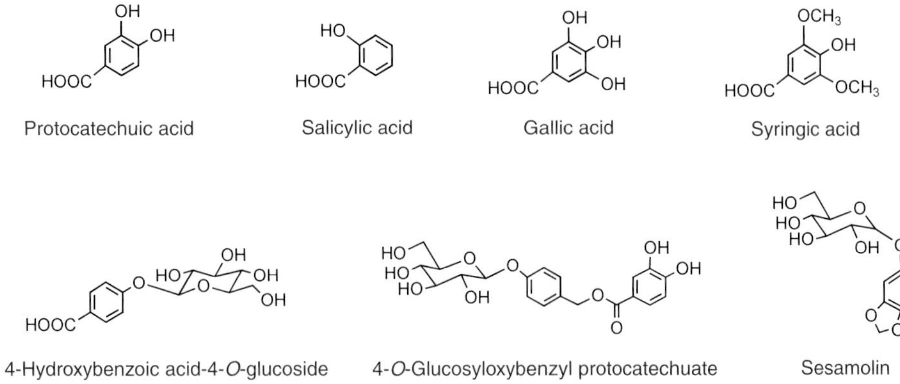

Figure 7.44 Herbs contain substantial quantities of a number of hydroxybenzoates derivatives.

It should be noted, however, that because herbs and spices impact strongly upon sensory properties and food palatability/acceptability their importance in the diet is out of proportion to their usage level, contributing significantly to the pleasure of eating. Many of the data that follow are taken from the NEODIET reviews (Lindsay and Clifford 2000), as well as Belitz and Grosch (1987) and Shan et al. (2005).

Hydroxybenzoic acids. Hydroxybenzoic acid glycosides are characteristic of some herbs and spices (Tomás-Barberán and Clifford 2000). After hydrolysis, protocatechuic acid is the dominant hydroxybenzoate in cinnamon bark (23–27 mg/kg) accompanied by salicylic acid (7 mg/kg) and syringic acid (8 mg/kg). Gallic acid occurs in clove buds (*Eugenia caryophyllata* Thunb.) (175 mg/kg) along with protocatechuic acid (~10 mg/kg) and syringic acid (8 mg/kg) (Figure 7.44). The fruit of anise (*Pimpinella anisum*) was reported to contain 730–1080 mg/kg 4-hydroxybenzoic acid-4-*O*-glucoside (Figure 7.44). The fruit of star anise (*Illicium verum*), dill (*Anethum graveolens*), fennel (*Foeniculum vulgare*), caraway (*Carum carvi*) and parsley (*Petroselinum crispum*) contain 730–840 mg/kg, 42–188 mg/kg, 30–106 mg/kg, 37–42 mg/kg and 165 mg/kg respectively. A glucosylated benzoate conjugate of protocatechuic acid has been isolated from oregano (Figure 7.44) (Kikuzaki and Nakatani 1989). Sesame (*Sesamum indicum*) seeds and oil contain 0.3-0.5% sesamolin, a glucoside of 1,2-methylenedioxy-4-hydroxybenzene (Figure 7.44). This compound should not be confused with the sesame lignans which are referred to as sesamolins.

Cinnamic acid derivatives. The Lamiaceae supplies many leafy herbs including basil (*Ocimum basilicum*), marjoram (*Origanum marjoram*), oregano, melissa (*Melissa officinalis*), peppermint (*Mentha* × *piperita*), rosemary (*Rosmarinus officinalis*), sage (*Salvia officinalis*), spearmint (*Mentha spicata*) and thyme (*Thymus* spp.). Herbs are the only dietary source of rosmarinic acid, the caffeic acid conjugate of α-hydroxyhydrocaffeic acid (Figure 7.45), at concentrations ranging from 10 to 20 g/kg dry basis (Shan et al. 2005). In some cases rosmarinic acid is accompanied by free cinnamic acids and some uncharacterized rosmarinic acid-like conjugates, especially in oregano (~13 g/kg) (Shan et al. 2005). These unknowns may include the previously reported 2-*O*-caffeoyl-3-[2'-(4"-hydroxybenzyl)-4',5'-dihydroxy]phenylpropionic acid, which is a 4-hydroxybenzyl derivative of rosmarinic acid (Figure 7.45) (Kikuzaki and Nakatani 1989). Cinnamic

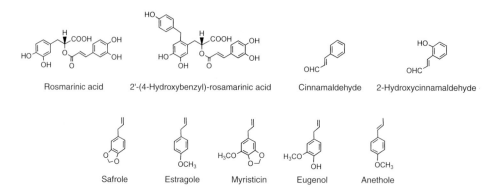

Figure 7.45 Herbs are the only dietary source of rosmarinic acid. They also contain a number of phenylpropanoids.

acid glycosides are found in sage (Lu and Foo 2000). A smaller quantity of rosmarinic acid (500 mg/kg dry weight) has been found in the botanically unrelated borage (*Borago officinalis*) (Clifford 1999).

The seeds of anise (*Pimpinella anisum*), fennel (*Foeniculum vulgare*), caraway (*Carun carvi*) and coriander (*Coriandrum sativum*) characteristically contain chlorogenic acids at concentrations up to 3 g/kg (Clifford 1999). Cinnamon (*Cinnamonum zeylanicum*) and cassia (*Cinnamonum cassia*) contain cinnamaldehyde and 2-hydroxycinnamaldehyde (Figure 7.45) at ~170 g/kg (Shan et al. 2005).

Phenylpropanoids of both subclasses (methyl-vinyl or allyl, R—CH$_2$—CH=CH$_2$) and propenyl (vinyl-methyl, R—CH=CH—CH$_3$) feature prominently, and some are of toxicological concern. Safrole (1-allyl-3,4-methylenedioxybenzene) (Figure 7.45), a major constituent of sassafras oil (*Sassafras albidum* Lauraceae), was shown to be carcinogenic in rodents and use of the oil and the compound for food flavouring was banned from 1960 (Singleton and Kratzer 1969), but transgressions leading to action by the regulatory authorities do still occur. This compound is quite widespread in other essential oils. It also occurs in black pepper (*Piper nigrum*), but as a relatively minor constituent, usually about 0.1%.

Estragole (1-allyl-4-methoxybenzene) (Figure 7.45), is found in the aerial parts of a number of culinary herbs including basil and fennel (*Foeniculum vulgare*) (Hussain et al. 1990). Chinese prickly ash (*Zanthoxylum bungeanum* Maxim.) is particularly rich at ~53 g/kg (Shan et al. 2005). Myristicin (1-allyl-3,4-methylenedioxy-5-methoxybenzene) (Figure 7.45) a characteristic constituent of nutmeg oil, nutmeg and mace from *Myristica fragrans* is a demonstrated hallucinogen. Ground nutmeg contains 1–3% myristicin and nutmeg oil 4% myristicin and 0.6% safrole. Both the oil and whole nutmegs have been associated with human fatalities (Singleton and Kratzer 1969; Fisher 1992) although such events are rare and extremely unlikely in conventional domestic usage. However, several intoxications have been reported after an ingestion of approximately 5 g of nutmeg, corresponding to 1–2 mg myristicin/kg body weight (Hallström and Thuvander 1997).

Eugenol (1-allyl-3-methoxy-4-hydroxybenzene) (Figure 7.45) is found in marjoram essential oil (10%) (Belitz and Grosch 1987), sweet basil, ground cinnamon (0.02–0.4%), ground cloves (1–20%), cinnamon oleoresin (2–6%) clove (*Syzygium aromaticum*)

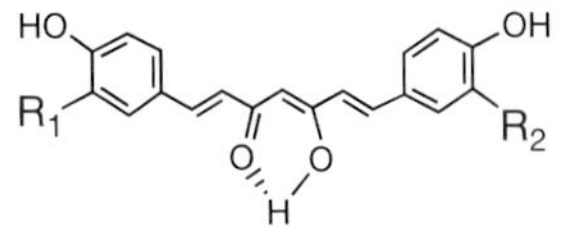

Curcumin (R_1, R_2 = OCH_3)
Demethoxycurcumin (R_1 = H, R_2 = OCH_3)
Bisdemethoxycurcumin (R_1, R_2 = H)

Figure 7.46 Curcumin and related compounds provide the colour and flavouring of tumeric.

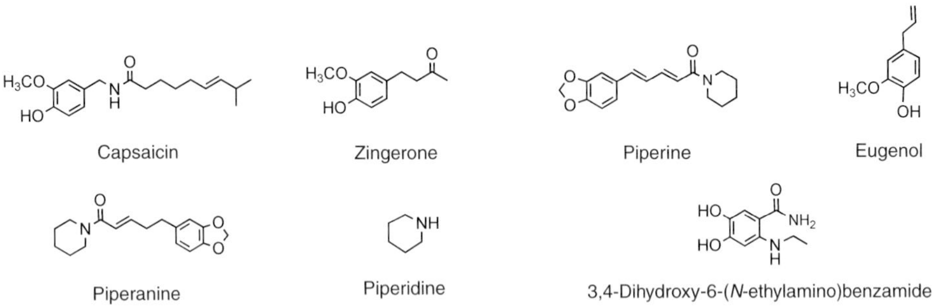

Capsaicin

Zingerone

Piperine

Eugenol

Piperanine

Piperidine

3,4-Dihydroxy-6-(N-ethylamino)benzamide

Figure 7.47 Capsaicins, piperines and other compounds that interact with the vanilloid receptor that has roles in taste, pain and analgesia.

oleoresin (60–90%) and in cinnamon leaf oil (70–90%) (Fisher 1992; Shan et al. 2005). Anethole (1-propenyl-4-methoxybenzene) (Figure 7.45) is found in star anise (*Illicium verum*) at ~5 g/kg (Shan et al. 2005).

Curcuminoids. Curcuminoids are cinnamoyl-methanes (diaryl-heptenoids) characteristic of ginger (*Zingiber officinale*), cardamon (*Elettara cardamonum*) and turmeric (*Curcuma longa*) used for their colouring and flavouring properties but more recently also for their putative antioxidant, anti-inflammatory and anti-carcinogenic properties. The main curcuminoids in tumeric are curcumin, demethoxycurcumin and bisdemethoxycurcumin (Figure 7.46) (Jayaprakasha et al. 2002). Ground turmeric typically contains 3–8% curcuminoids with some 30–40% in turmeric oleoresin (Clifford 2000; Kikuzaki et al. 2001).

Capsaicins and piperines. Capsaicins are compounds responsible for pungency. This is mediated through the vanilloid receptor that has roles in taste, pain and analgesia. Capsaicin has been identified as the chemical that gives the heat to chilli peppers (Figure 7.47). Other food-related ligands are zingerone from ginger, piperine and the phenylpropanoid, eugenol (see above). Typical dried ginger contains 1–4% pungent constituents with some 10–30% in ginger oleoresin (Fisher 1992). Oil of cloves has long been used to provide pain relief for teething infants, presumably via the interaction of eugenol (Figure 7.47) and the vanilloid receptor (Clifford 2000).

Green, black and white peppers are characterized by a series of phenolic amides. The major constituents, piperine and piperanine, are formed from C_6–C_5 phenolic acids (varying in side chain unsaturation) and piperidine with total concentrations in the range

3–6 g/kg (Shan et al. 2005) accompanied by 3,4-dihydroxy-6-(*N*-ethylamino)benzamide (Figure 7.47) (Bandyopadhyay et al. 1990).

Terpenes and terpenoid phenols. Monoterpenes such as borneol, bornylacetate, camphor, carvacrol, *p*-cymene, eucalyptol, (−)-menthol, (+)-α-pinene, (−)-β-pinene, γ-terpinene and thujone are widespread occurring, for example, in basil, mint (*Mentha rotundifolia*), oregano, juniper, rosemary, sage and thyme, and individual compounds may reach ~6 g/kg. These may be accompanied by phenolic terpenes such as thymol (4-isopropylphenol), carnosic acid, carnosol, epirosmanol, rosmanol, rosmariquinone and rosmaridiphenol (Figure 7.48) with individual compounds occurring at concentrations in the range 1–10 g/kg (Belitz and Grosch 1987; Fisher 1992; Clifford 2000; Shan et al. 2005).

Other monoterpenes of note include limonene (Section 7.4.5, Figure 7.29) which is the precursor of carvone; (+)-carvone provides the characteristic odour of caraway while

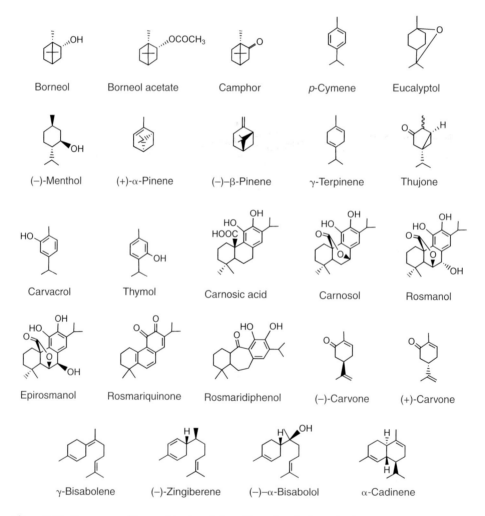

Figure 7.48 Terpenes and terpenoids phenols found in various herbs and spices.

its isomer (−)-carvone smells of spearmint. The C_{15} sesquiterpenes γ-bisabolene and (−)-zingiberene contribute to the aroma of ginger while α-bisabolol is a major component in chamomile (*Matricaria chamomilla*), dried flowers of which are used to make a herbal tea, and α-cadinene is one of many terpenoids found in juniper (*Juniperis communis*) berries, using in making gin. The structures of these compounds are illustrated in Figure 7.48.

Flavonoids. Most classes of flavonoids are found in herbs and spices. Frequently these include relatively uncommon aglycones and/or common aglycones with comparatively uncommon substitution patterns. Flavonoids do not generally exceed ∼0.2–0.4 g/kg in Lamiaceae herbs but reach ∼1.5–3 g/kg in Apiaceae herbs, ∼3.5 g/kg in cloves and ∼7 g/kg in bay leaf (*Laurus nobilis*) (Shan et al. 2005). Flavonol glycosides are found in basil (Baritaux et al. 1991) and ginger (Nakatani et al. 1991; Kawabata et al. 2003), glycosides and glucuronides of flavones in sage (Canigueral et al. 1989), glycosides of the isoflavone genistein (up to 100 mg/kg) in some samples of cumin (*Cuminum cymimum*) (Clarke et al. 2004) and glycosides of the relatively uncommon 6-hydroxyapigenin in marjoram and sage (Lu and Foo 2000; Miura et al. 2002; Kawabata et al. 2003). Lemongrass (*Cymbopogon citratus*) contains unusual C-glycosides of the flavones luteolin and chrysoeriol (Figure 7.49) as well as caffeic acid and chlorogenic acids (Cheel et al. 2005). Fennel (*Foeniculum vulgare*) also contains a diverse spectrum of flavone glycosides and phenolics (Figure 7.49) (Parejo et al. 2004).

Basil contains a complex mixture of acylated peonidin and cyanidin-based anthocyanins (Phippen and Simon 1998). Mint, sage and thyme contain lipophilic methylated flavone aglycones (Voirin and Bayet 1992; Lu, and Foo 2000; Miura et al. 2002) and flavanone glucosides (Guedon and Pasquier, 1994). Flavan-3-ols have been found in mint, basil, rosemary, sage and dill (*Anethum graveolens*) (1–2.5 g/kg) (Shan et al. 2005) and cinnamon bark is rich in proanthocyanidins, including A-type oligomers containing (epi)afzelchin units (Gu et al. 2003). Cinnamon contains in excess of 80 g/kg proanthocyanidins and curry powder some 700 mg/kg Gu et al. (2004).

Other compounds. Coriander contains photoactive furoisocoumarins, the principal component being coriandrin (Figure 7.50) (Ceska et al. 1988). Gallotannins have been

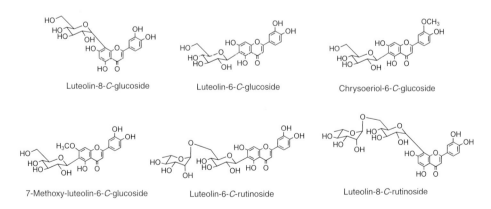

Figure 7.49 C-glycosides of luteolin and chrysoeriol are found in lemongrass. Fennel also contains a diverse array of flavonoids including luteolin-8-C-rutinoside.

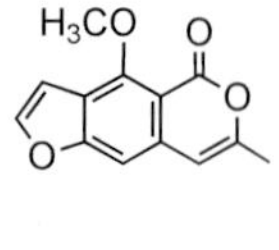

Coriandrin

Figure 7.50 Coriandrin is a photoactive furoisocoumarin from coriander.

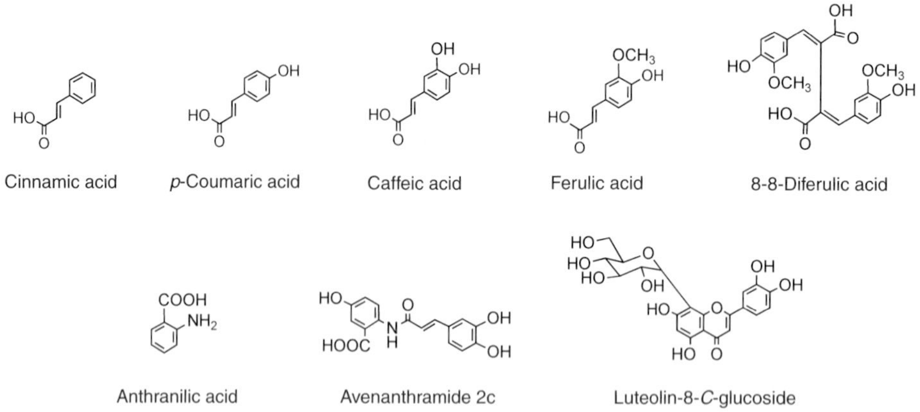

Cinnamic acid   p-Coumaric acid   Caffeic acid   Ferulic acid   8-8-Diferulic acid

Anthranilic acid   Avenanthramide 2c   Luteolin-8-C-glucoside

Figure 7.51 A range of conjugated phenylpropanoids, including avenanthramides, have been detected in cereals. Millet is rich in luteolin-8-C-glucoside.

reported to occur in cloves (Shan et al. 2005). The glucosinolates, sinalbin and sinigrin are characteristic of white and black mustard seed, respectively (Figure 7.11), while horseradish contains glucobrassicin (Figure 7.12) (Belitz and Grosch 1987) (see Section 7.3.3).

7.6 Cereals

Although cereals are staples, cereal brans and whole grains are viewed in the industrialized world more as health-promoting supplements. They contain some distinctive phytochemicals not encountered in other commodities.

Bran cell wall arabino-xylans contain arabinose residues esterified with cinnamic acids, especially ferulic acid (Figure 7.51). A portion of this total cinnamate exists as ferulate dimers linked in various ways including 5,5′ or 5,8′ carbon–carbon bonds. Barley (*Hordeum vulgare*) bran contains ∼50 mg/kg bound ferulic, ∼30 mg/kg bound p-coumaric and 3 mg/kg diferulic acid. The endosperm content is ∼3 mg/kg total and the aleurone is intermediate. Rice (*Oryza sativa*) endosperm cell walls contain 12 g/kg esterified cinnamic acids comprising ∼9 g/kg ferulic, ∼2.5 g/kg p-coumaric and ∼0.5 g/kg diferulic esters. While whole wheat (*Triticum vulgare*) contains some 20–30 mg/kg cinnamic acids esterified to polysaccharides the derived wheat bran contains some 4–7 g/kg and maize (*Zea mays*) bran as much as 30 g/kg.

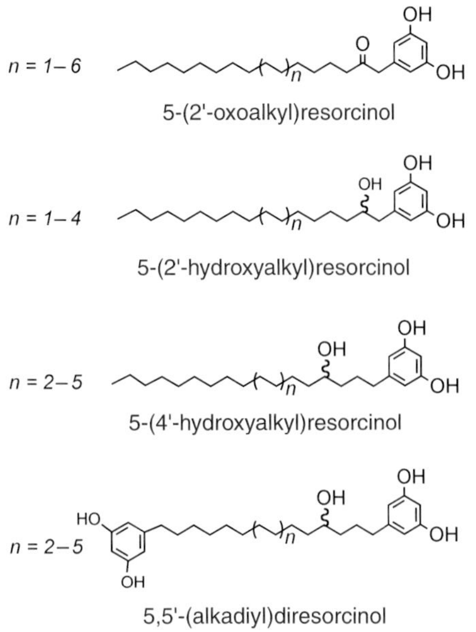

Figure 7.52 Skeletal structures of 5-alkyresorcinol-related analogues in rye.

Water chestnuts (*Eleocharis dulcis*, Cyperaceae) are botanically closer to the cereals than to other common fruits and vegetables and are characterized also by a significant content of cell wall-bound cinnamates (>7 g/kg ferulate and >4.5 g/kg diferulate, the majority of which is 8-O-4 linked).

Wheat, maize and rye (*Secale cereale*) contain ferulic and p-coumaric esters of sterols and stanols. Oats (*Avena sativa*) contain a series of 24 caffeic and ferulic esters of glycerol, long chain alkanols, alkandiols ($n = 22, 23, 24$) and ω-hydroxy acids ($n = 26, 28$). In addition, there is a large series of compounds (>25) that are esters of anthranilic acid or 5-hydroxyanthranilic acid with either p-coumaric, caffeic or ferulic acids (avenanthramides) (Figure 7.51) or with their ethylenic analogues (avenulamides). Oat meal has been reported to contain some 200–300 mg/kg of esterified ferulic acid.

Barley is the only common cereal with significant proanthocyanidins content (0.6–1.3 g/kg) (Santos-Buelga and Scalbert 2000). Millet (*Pennisetum americanum*) flour is comparatively rich in vitexin, the 8-C-glucoside of the flavone luteolin (Figure 7.51). High intakes have been associated with goitre in parts of west Africa (Gaitan et al. 1989; Akingbala 1991; Santos-Buelga and Scalbert 2000). Wheat, oat and rye bran contains some 5 mg/kg lignans (Cassidy et al. 2000) and whole flax (*Linum usitatissimum*) seed some 6–13 g/kg (Johnsson et al. 2000).

Several series of resorcinol derivatives have been found in wheat, rice, rye and triticale. These include 5-alkyl resorcinols and pairs of isomeric 5-alkenyl resorcinols ($n = 17, 19, 21, 23, 25$), accompanied by smaller amounts of 5-(2′-oxoalkyl)-, 5-(4′-hydroxyalkyl), 5-(2′-hydroxyalkenyl)-resorcinols and 5,5′-(alkadiyl)diresorcinols (Figure 7.52) (Suzuki et al. 1999). Whole wheat grains contain some 300–1200 mg/kg total resorcinols that are concentrated in the bran (22–26 g/kg for durum wheat). Comparatively low levels

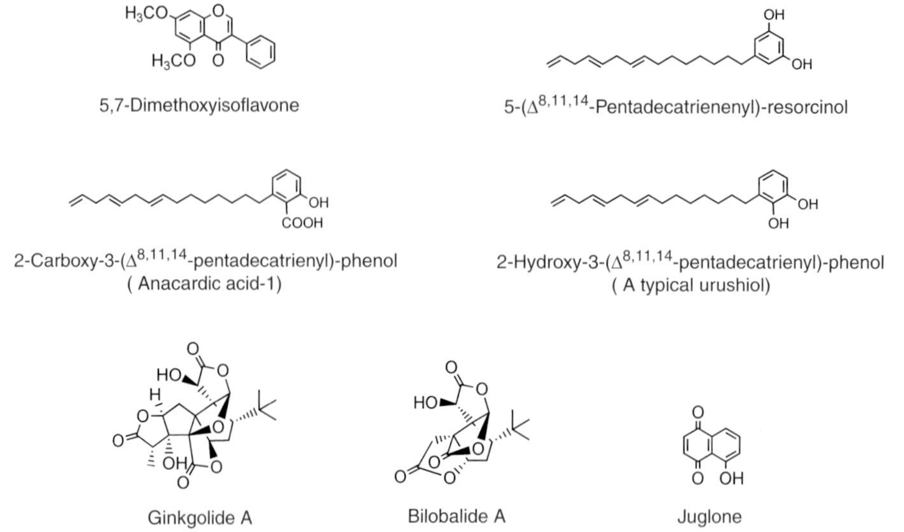

Figure 7.53 Structures of some of the secondary metabolites that occur in various types of nuts.

are found in other milling streams, and the contents decline on baking. Mullin and Emery (1992) reported that wheat bran, wheat-bran-enriched or whole wheat breakfast cereals contained 343–1455 mg/kg whereas breads or bran muffins contained only 61–217 mg/kg. Total alkylresorcinols in a typical serving of cereal-based foods therefore varied from 40 mg (wheat bran breakfast cereal) to 1 mg (one slice of seven-grain bread or rye bread). These compounds have been blamed for appetite suppression in domestic animals, although a small-scale study in rats failed to observe any effects on growth or nitrogen balance. Their metabolism and effects in humans are unknown (Clifford 2000).

7.7 Nuts

Nuts encompass a botanically diverse collection of fruits that contain an edible and usually rather hard and oily kernel within a hard or brittle outer shell. They are consumed raw or roasted as snack foods or decorative/comparatively minor ingredients in baked goods and confectionary. Data on the composition of the edible tissues are scarce, and the emphasis has clearly been on those nuts with potentially undesirable constituents. Much of what follows is taken from Shahidi and Naczk (1995) and the NEODIET series of reviews of Lindsay and Clifford (2000).

Cashew (*Anacardium occidentale*) kernels contain flavan-3-ols and proanthocyanidins and pecans (*Carya illinoensis*) a range of phenolic acids and flavan-3-ols (Shahidi and Naczk 1995). Isoflavones, such as 5,7-dimethoxyisoflavone (Figure 7.53), occur in peanuts (*Arachis hypogaea*) but at a much lower concentration than found in soya (Turner *et al.* 1975).

The proanthocyanidins of nuts have been characterized. Hazelnuts (*Corylus avellana*) and pecans are particularly rich with ~5 g/kg, whereas almonds (*Prunus dulcis*)

Table 7.4 Concentration of flavan-3-ol monomers, dimers and trimers and total proanthocyanidins in nuts (Gu et al. 2004)

Nut	Monomers	Dimers	Trimers	Total PAs	Type
Hazelnuts	98 ± 16	125 ± 38	136 ± 39	5007 ± 1520	PC, PD
Pecans	172 ± 25	421 ± 54	260 ± 20	4941 ± 862	PC, PD
Pistachios	109 ± 43	133 ± 18	105 ± 12	2373 ± 520	PC, PD
Almonds	78 ± 9	95 ± 16	88 ± 17	1840 ± 482	PC, PP
Walnuts	69 ± 34	56 ± 9	72 ± 12	673 ± 147	PC
Roasted peanuts	51 ± 10	41 ± 7	37 ± 5	156 ± 23	A, PC
Cashews	67 ± 29	20 ± 4	n.d.	87 ± 32	PC

Data expressed as mg/kg fresh weight ± standard deviation. n.d. – not detected; PA – proanthocyanidins; PC – procyanidins; PD – prodelphidins; PP – propelargonidins; A – indicates existence of A-type proanthocyanidins.

and pistachios contain 1.8–2.4 mg/kg, walnuts (*Juglans* spp.) ~0.67 g/kg, roasted peanuts ~0.16 g/kg and cashews contain only ~0.09 g/kg. All the foregoing have proanthocyanidins containing procyanidin units. Almonds also have propelargonidin units and hazelnuts, pecans and pistachios also have prodelphinidin units, with peanuts being the only one to have A-type units (Table 7.4). Cashews contain nothing larger than dimers, peanuts nothing larger than hexamers, and the others tannins containing more than 10 units (Gu et al. 2004). That nuts do not normally taste astringent must reflect the binding of the proanthocyanidins to matrix substances precluding their binding to and ready precipitation of salivary proteins when consumed (Clifford 1986).

Alkyl resorcinols (see also Section 7.6) have been reported in various members of the Anacardiaceae which are known for their ability to cause skin irritation. Cashew nuts contain several poly-unsaturated derivatives such as 5-($\Delta^{8,11,14}$-pentadecatrienyl)-resorcinol (Figure 7.53) and the shell oil with a greater concentration is notoriously irritating especially before roasting. This oil also contains many structurally related compounds, including anacardic acids (Figure 7.53) (derivatives of salicylic acid, i.e. 2-carboxy-3-alkylphenols) having saturated, mono-, di- and tri-unsaturated ($\Delta^{8,11,14}$) side chains of 13, 15 and 17 carbons which decarboxylate to cardols (3-alkylphenols) on roasting. Roasted cashew nuts contain ~0.65 g/kg anacardic acids (Trevisan et al. 2005). A related compound, 5-(1,2-heptadecenyl)-resorcinol has been identified as the contact allergen in mango latex (Section 7.4.8, Figure 7.32).

The botanically related Australian cashew (*Semecarpus australiensis*) has a substantial content (1.7%) of urushiols (3-pentadecylcatechols and 3-heptadecylcatechols) (Figure 7.52) having the same pattern of unsaturation (one, two or three double bonds) as those found at a much lower concentration (0.17%) in poison ivy (*Rhus toxicodendron*). No data could be found for the phenol composition of the related pistachio nut (*Pistachio vera*), although it has been reported to contain protein allergens that cross-react with those of cashew and mango (Clifford 2000). The leaves and nuts of the botanically unrelated *Ginkgo biloba* are of interest because of the increasing use of the leaves as a nutraceutical, contain anacardic acids (ginkgolic acids) having a different pattern of unsaturation to those discussed above (Δ^7-pentadecenyl) (Schotz 2004). The active ingredients in Ginkgo, believed to be responsible for improved peripheral and cerebrovascular circulation that delays decline in cognitive function and memory processes, are a mixture of terpenoid lactones comprising five ginkgolides and bilobalide (Figure 7.53).

Fruits of black walnut (*Juglans nigra*) and buttermilk walnut (*Juglans cinerea*) fruits contain 1,4,5-trihydroxynaphthalene-4-β-D-glucoside from which juglone (5-hydroxy-1,4-naphthoquinone) (Figure 7.53) is produced during ripening by hydrolysis and oxidation. This quinone is responsible for the yellow-brown staining and irritation of the hands that can occur after handling these nuts (Clifford 2000).

7.8 Algae

Marine algae are utilized to a limited extent for food and as a source of polysaccharides used as food additives but are increasingly being investigated for their novel, potentially bioactive components.

In the United Kingdom the red alga *Porphyra umbilicalis* is the basis of laver bread prepared traditionally in parts of Wales and Ireland. Similar products are prepared in the United States, where they are known by the Japanese term 'nori' which is used to garnish or wrap 'sushi'. Japanese nori is derived from *P. yessoensis* and *P. tenera* (Clifford 2000).

Red algae (Rhodophyceae) synthesize a substantial range of halogenated compounds including mono- and dihydroxy C_6–C_1, C_6–C_2 and C_6–C_3 phenols containing one or two bromine atoms (Fenical 1975). 2,4,6-Tri-bromophenol (Figure 7.54) predominates and the total bromophenols content ranges from 8 to 180 μg/kg.

Algal polysaccharides such as agar, obtained by aqueous extraction of red algae (*Gelidium* spp., *Pterocladia* spp. and *Gracilaria* spp.), alginates obtained by alkali extraction of brown algae (particularly *Macrocystis pyrifera*, but also *Laminaria* spp., *Ascophyllum* spp. and *Sargassum* spp.) and carrageenans obtained by mild alkali extraction of red algae (particularly *Chondus crispus*, but also *Eucheuma* spp., *Gigartina* spp., *Gloiopeltis* spp. and *Iridaea* spp.) have a widespread usage at low levels as emulsifiers, stabilizers and gelling agents in processed foods. The basic monomers of agar are β-D-galactose and 3,6-anhydro-α-L-galactose with alternate 1–3 and 1–4 linkages and a low level of sulphation. Alginates contain β-D-mannuronic acid and α-L-glucuronic acids linked 1–4. Carrageenans consist of β-D-galactose and 3,6-anhydro-β-D-galactose with extensive mono- and di-sulphation (Belitz and Grosch 1987).

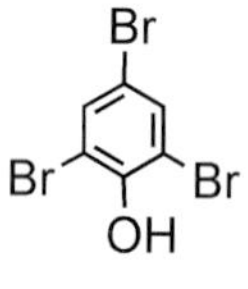

2,4,6-Tribromophenol

Figure 7.54 2,4,6-Tribromophenol is the main halogenated phenolic compound in red algae.

7.9 Beverages

7.9.1 Tea

Tea is one of the most widely consumed beverages in the world. Grown in about 30 countries, the botanical classification of *Camellia* spp. is complex and confused, with many forms of commercial tea that may or may not be distinct species (Kaundun and Matsumoto 2002). The main forms recognized are *C. sinensis* var. *sinensis* that originated on the northern slopes of the Himalayas, which has small leaves, a few centimetres in length, and *C. sinensis* var. *assamica*, with leaves 10–15 cm or more in length, that developed on the southern slopes of the Himalayas (Willson 1999). As examples of the variability, there is a large-leaved var. *sinensis* found in Yunnan that is rich in (−)-epicatechin gallate (Shao *et al.* 1995), a form of var. *sinensis* comparatively rich in methylated flavan-3-ols (Chiu and Lin 2005) and the so-called var. *assamica* × var. *sinensis* hybrids used for Japanese green tea production with a comparatively low flavan-3-ol content but rich in theanine (*N*-ethylglutamine) (Figure 7.55) and other amino acids (see Table 7.5) (Takeo 1992).

Tea is generally consumed in one of three forms, green, oolong or black, but there are many more variations which arise through differences in the nature of the leaf used, and the method of processing (Hampton 1992; Takeo 1992), including some that involve a microbial transformation stage (Shao *et al.* 1995). Approximately 3.2 million metric tons of dried tea are produced annually, 20% of which is green tea, 2% is oolong and the remainder is black tea. In all cases the raw material is young leaves, the tea flush, which are preferred as they have a higher flavan-3-ol content and elevated levels of active enzymes. The highest quality teas utilize 'two leaves and a bud', with progressively lower quality taking four or even five leaves (Willson, 1999).

There are basically two types of green tea (Takeo 1992). The Japanese type utilizes shade-grown hybrid leaf with comparatively low flavan-3-ol levels and high amino acids content, including theanine. After harvesting the leaf is steamed rapidly to inhibit polyphenol oxidase and other enzymes. Chinese green tea traditionally uses selected forms of var. *sinensis* and dry heat (firing) rather than steaming, giving a less efficient inhibition of the polyphenol oxidase activity and allowing some transformation of the flavan-3-ols.

In the production of black tea there are again two major processes (Hampton 1992). The so-called orthodox and the more recently introduced, but now well-established, cut–tear–curl process. In both processes the objective is to achieve efficient disruption of cellular compartmentation thus bringing phenolic compounds into contact with polyphenol oxidases and at the same time activating many other enzymes. The detailed preparation of the leaf, known as withering, time and temperature of the fermentation stage, and the method of arresting the fermentation to give a relatively stable product, all vary geographically across the black tea-producing areas. However, oxidation for 60–120 min at about 40°C before drying gives some idea of the conditions employed.

When harvested, the fresh tea leaf is unusually rich in polyphenols (∼30% dry weight) (Table 7.5) and this changes with processing even during the manufacture of commercial green tea, and progressively through semi-fermented teas to black teas and those with a microbial processing stage. Flavan-3-ols are the dominant polyphenols of fresh leaf. Usually (−)-epigallocatechin gallate dominates, occasionally taking

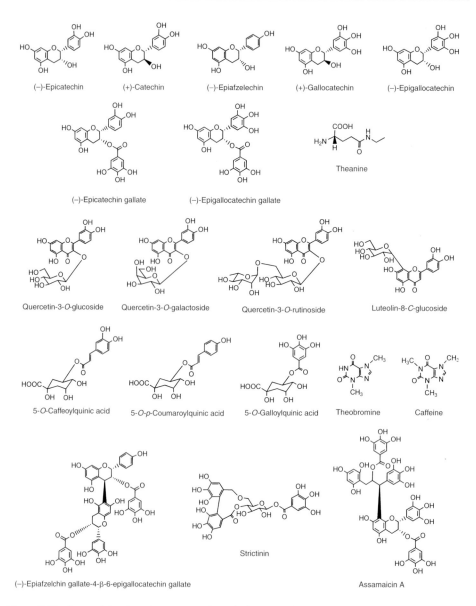

Figure 7.55 Some of the phenolic compounds, including proanthocyanidins and hydrolysable tannins, present in green tea (*Camellia sinensis*) infusions. The predominant purine alkaloid is caffeine with trace levels of theobromine.

second place to (−)-epicatechin gallate, together with smaller but still substantial amounts of (−)-epigallocatechin, (+)-gallocatechin, (+)-catechin, (−)-epicatechin and (−)-epiafzelchin (Figure 7.55). The minor flavan-3-ols also occur as gallates, and (−)-epigallocatechin may occur as a digallate, esterified with *p*-coumaric acid or caffeic acid, and with various levels of methylation (Hashimoto *et al.* 1992). There are at least

Table 7.5 Approximate composition of green tea shoots (% dry weight)

	Var. *assamica*	Small leafed var. *sinensis*	So-called hybrid of small leafed var. *sinensis* and var. *assamica*	Large leafed var. *sinensis*
Substances soluble in hot water				
Total polyphenols	25–30	14–23		32–33
Flavan-3-ols				
(−)-Epigallocatechin gallate	9–13	7–13	11–15	7–8
(−)-Epicatechin gallate	3–6	3–4	3–6	13–14
(−)-Epigallocatechin	3–6	2–4	4–6	1–2
(−)-Epicatechin	1–3	1–2	2–3	2–3
(+)-Catechin				4
Other flavan-3-ols	1–2			2
Flavonols and flavonol glycosides	1.5	1.5–1.7		1
Flavandiols	2–3			1
Phenolic acids and esters (despides)	5			
Caffeine	3–4	3		
Amino acids	2	4–5	2–3	
Theanine	2	2–5	2	
Simple carbohydrates (e.g. sugars)	4			
Organic acids	0.5			
Substances partially soluble in hot water				
Polysaccharides:				
Starch, pectic substances	1–2			
Pentosans, etc.	12			
Proteins	15			
Ash	5			
Substances insoluble in water				
Cellulose	7			
Lignin	6			
Lipids	3			
Pigments (chlorophyll, carotenoids, etc.)	0.5			
Volatile substances	0.01–0.02			

Note: Blanks in table indicate data are not available.

15 flavonol glycosides comprising mono-, di- and tri-glycosides based upon kaempferol, quercetin and myricetin and various permutations of glucose, galactose, rhamnose, arabinose and rutinose (Engelhardt et al. 1992; Finger et al. 1991; Price et al. 1998; Lakenbrink et al. 2000a), three apigenin-C-glycosides (Sakamoto 1967, 1969, 1970), several caffeoyl- and p-coumaroylquinic acids (chlorogenic acids) and galloylquinic acids (theogallins) and at least 27 proanthocyanidins including some with epiafzelchin units (Nonaka et al. 1983; Lakenbrink et al. 1999). In addition some forms have a significant content of hydrolysable tannins, such as strictinin (Nonaka et al. 1984), perhaps indicating an affinity with *C. japonica*, *C. sasanqua* and *C. oleifera* (Hatano et al. 1991; Han et al. 1994; Yoshida et al. 1994) whereas others contain chalcan–flavan dimers known as assamaicins (Hashimoto et al. 1989a). Relevant structures are illustrated in Figure 7.55 and data on the levels of some of these compounds in green tea shoots are presented in Table 7.5.

In green teas, especially Japanese production, most of these various polyphenols survive and can be found in the marketed product. In Chinese green teas and the semi-fermented teas such as oolong, some transformations occur, for example leading to the production of theasinensins (flavan-3-ol dimers linked $2 \rightarrow 2'$), oolong homo-bis-flavans (linked $8 \rightarrow 8'$ or $8 \rightarrow 6'$), oolongtheanin and $8C$-ascorbyl-epigallocatechin gallate (Figure 7.56) (Hashimoto et al. 1987, 1988, 1989a, 1989b). In black tea production the transformations are much more extensive with some 90% destruction of the flavan-3-ols in orthodox processing and even greater transformation in cut-tear-curl processing. Some

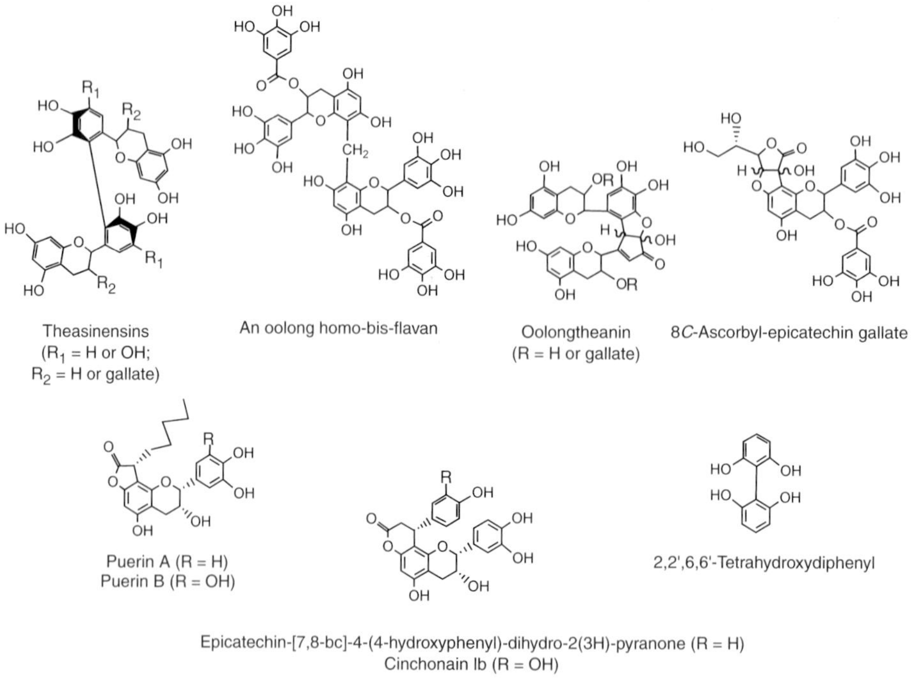

Figure 7.56 Transformation products found in Chinese green teas and semi-fermented teas such as oolong.

losses of galloylquinic acid, quercetin glycosides and especially myricetin glycosides have been noted, and recent studies on thearubigins suggest that theasinensins and possibly proanthocyanidins may also be transformed. Pu'er tea is produced by a microbial fermentation of black tea. Some novel compounds have been isolated and it is suggested that they form during the fermentation (Zhou et al. 2005). These include two new 8-C substituted flavan-3-ols, puerins A and B, and two known cinchonain-type phenols, epicatechin-[7,8-bc]-4-(4-hydroxyphenyl)-dihydro-2(3H)-pyranone and cinchonain Ib, and 2,2′,6,6′-tetrahydroxydiphenyl (Figure 7.56). However, various cinchonains have previously been reported in unfermented plant material (Nonaka and Nishioka 1982; Nonaka et al. 1982; Chen et al. 1993)

It is generally considered that polyphenol oxidase, which has at least three isoforms, is the key enzyme in the fermentation processes that produce black teas, but there is evidence also for an important contribution from peroxidases with the essential hydrogen peroxide being generated by polyphenol oxidase (Subramanian et al. 1999). The primary substrates for polyphenol oxidase are the flavan-3-ols which are converted to quinones. These quinones react further, and may be reduced back to phenols by oxidizing other phenols, such as gallic acid, flavonol glycosides and theaflavins, that are not direct substrates for polyphenol oxidase (Opie et al. 1993, 1995).

Many of the transformation products are still uncharacterized. The best known are the various theaflavins and theaflavin gallates (Figure 7.57), characterized by their bicyclic undecane benztropolone nucleus, reddish colour and solubility in ethyl acetate. These form through the Michael addition of a B-ring trihydroxy (epi)gallocatechin quinone to a B-ring dihydroxy (epi)catechin quinone prior to carbonyl addition across the ring and subsequent decarboxylation (Goodsall et al. 1996), but it is now accepted that the theasinensins (Figure 7.56) form more rapidly and may actually be theaflavin precursors (Hashimoto et al. 1992; Tanaka et al. 2002a). Theaflavonins and theogallinin, ($2 \rightarrow 2'$-linked theasinensin analogues) (Figure 7.56) formed from (−)-epigallocatechin/(−)-epigallocatechin gallate and isomyricetin-3-O-glucoside or 5-O-galloylquinic acid, respectively, have also been found in black tea (Hashimoto et al. 1992).

Coupled oxidation of free gallic acid or ester gallate produces quinones that can replace (epi)gallocatechin quinone leading to (epi)theaflavic acids and various theaflavates (Figure 7.57) (Wan et al. 1997). Interaction between two quinones derived from trihydroxy precursors can produce benztropolone-containing theaflagallins (Hashimoto et al. 1986) or yellowish theacitrins that have a tricyclic dodecane nucleus (Davis et al. 1997). Mono- or di-gallated analogues are similarly formed from the appropriate gallated precursors and in the case of theaflavins coupled oxidation of benztropolone gallates can lead to theadibenztropolones (and higher homologues at least in model systems). Oxidative degallation of (−)-epigallocatechin gallate produces the pinkish-red desoxyanthocyanidin, tricetanidin (Figure 7.57) (Coggon et al. 1973).

The brownish water-soluble thearubigins are the major phenolic fraction of black tea and these have been only partially characterized. Masses certainly extend to $\sim$2000 daltons. Early reports that these were polymeric proanthocyanidins (Brown et al. 1969) probably arose through detection of proanthocyanidins that had passed through from the fresh leaf unchanged. The few structures that have been identified include dibenztropolones (Figure 7.57) where the 'chain extension' has involved coupled oxidation of ester gallate (Sang et al. 2002, 2004), theanaphthoquinones formed when a bicylco-undecane

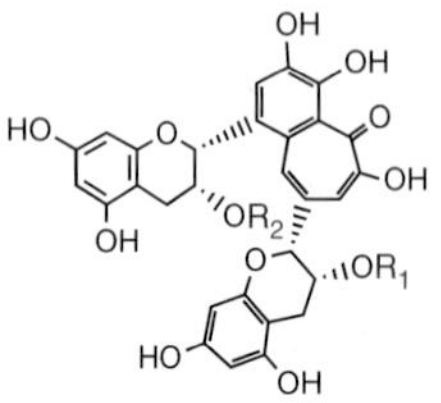

Theaflavins
(Theaflavin R_1 and R_2 = H)
(Theaflavin-3-gallate R_1 = H, R_2 = gallate)
(Theaflavin-3'-gallate R_1 = gallate, R_2 = H)
Theaflavin-3,3'-digallate R_1 and R_2 = gallate)

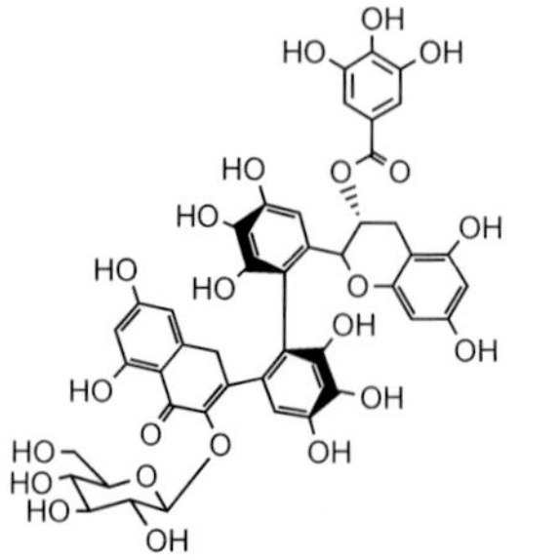

Theaflavonin

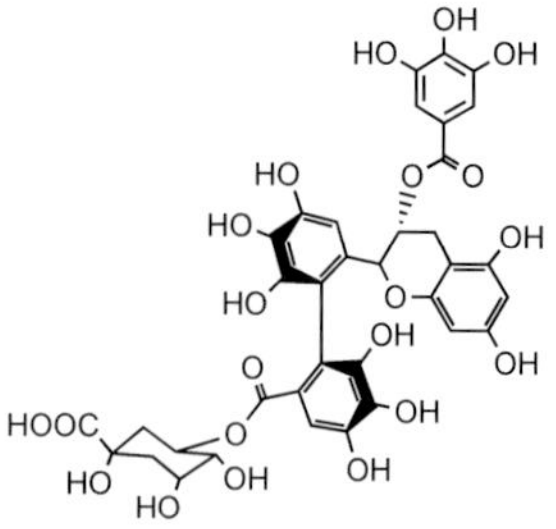

Theogallinin

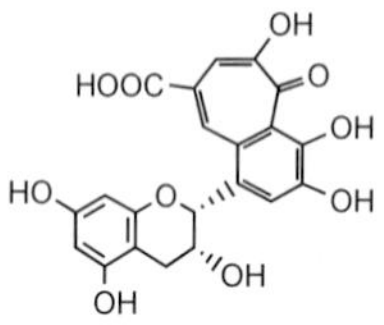

Epitheaflavic acid

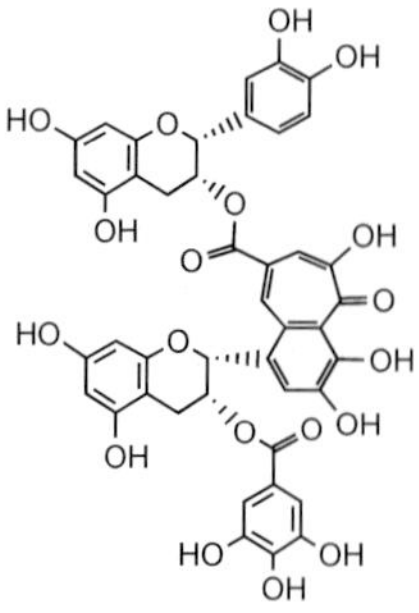

Theaflavate A

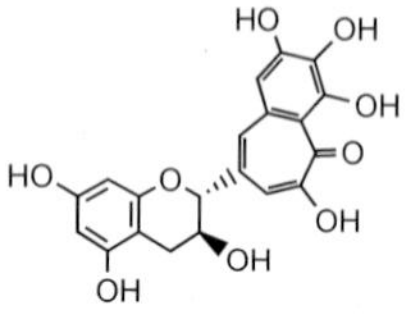

Theaflagallin

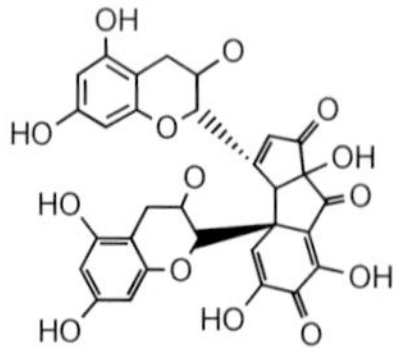

Theacitrins
(Theacitrin R_1 and R_2 = H)
(Theacitrin-3-gallate R_1 = H, R_2 = gallate)
(Theacitrin-3'-gallate R_1 = gallate, R_2 = H)
Theacitrin-3,3'-digallate R_1 and R_2 = gallate)

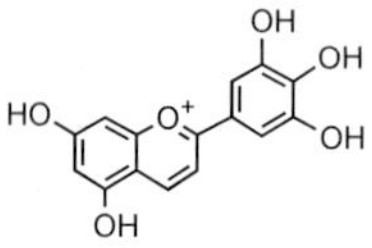

Tricetanidin

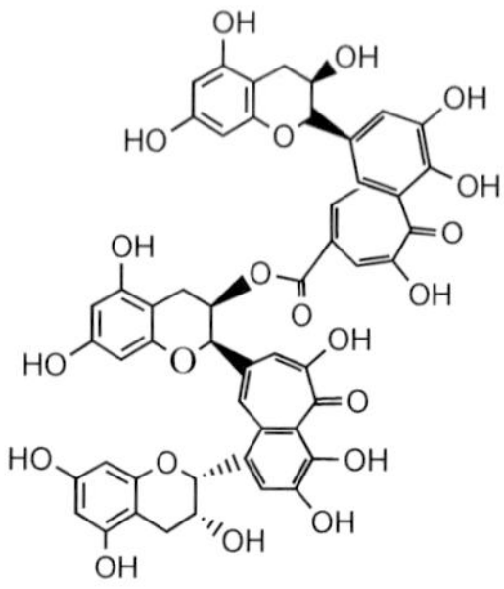

Theadibenztropolone A

Figure 7.57 Fermentation transformation products that have been detected in black tea.

benztropolone nucleus collapses back to a bicyclo-decane nucleus (Tanaka et al. 2000, 2001), and dehydrotheasinensins (Figure 7.58) (Tanaka et al. 2005a). Production of higher mass thearubigins could involve coupled oxidation of gallate esters yielding tribenztropolones, etc., coupled oxidation of large mass precursors such as proanthocyanidin gallates or theasinensin gallates rather than flavan-3-ol gallates (Menet et al. 2004), or interaction

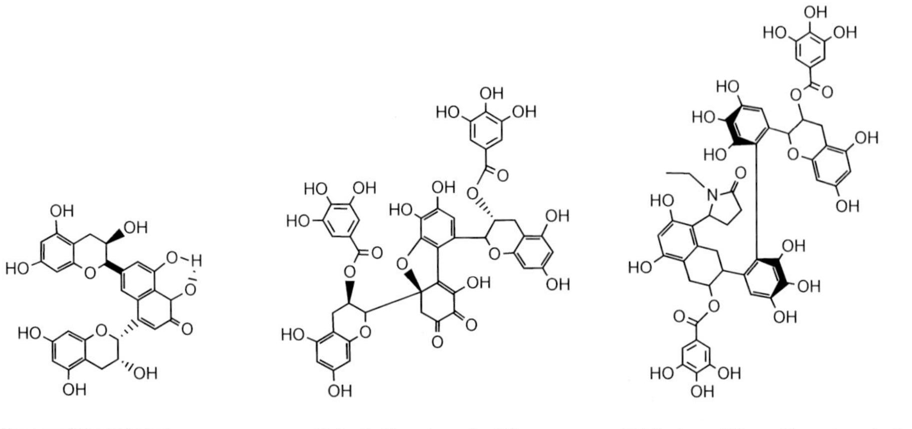

Theanaphthoquinone   **Dehydrotheasinensin AQ**   **8′-Ethylpyrrolidinonyltheasinensin A**

Figure 7.58 Polymeric phenolics from black tea that have been associated with the production of thearubigins.

of quinones with peptides and proteins. Though long anticipated, 8′-ethylpyrrolidinonyl-theasinensin A, (Figure 7.58) the first such product containing an N-ethyl-2-pyrrolidinone moiety, was only isolated from black tea in 2005 (Tanaka et al. 2005b). It is probably formed from a theasinensin and the quinone-driven Strecker aldehyde produced by decarboxylation of theanine.

Model system studies have led to the characterization of some additional structures, but since their relevance to commercial black tea is currently unclear they are not discussed further in this chapter (Tanaka et al. 2001, 2002b,c, 2003). Much remains to be done in this area, and it is interesting to note, that for consumers of black tea, consumption of these uncharacterized derived polyphenols at ~100 mg per cup greatly exceeds their consumption of chemically-defined polyphenols such as flavonoids (Gosnay et al. 2002; Woods et al. 2003).

Aqueous infusions of tea leaves contain the purine alkaloid caffeine and traces of theobromine. Green and semi-fermented teas retain substantial amounts of the flavan-3-ols but decline progressively with increased fermentation and are lowest in cut–tear–curl black teas. Beverages from green, semi-fermented and black teas also have significant contents of flavonol glycosides and smaller amounts of chlorogenic acids, flavone-C-glycosides and theogallin (Figure 7.55) which are less affected by processing but may vary more markedly with the origin of the fresh leaf (Engelhardt et al. 1992; Shao et al. 1995; Lin et al. 1998; Price et al. 1998; Lakenbrink et al. 2000b; Luximon-Ramma et al. 2005). Black tea beverage uniquely contains theaflavins and to a greater extent the high molecular weight thearubigins which are responsible for the astringent taste of black tea and the characteristic red-brown colour. Thearubigins are difficult to analyse, since they either do not elute from or are not resolved on reverse phase HPLC columns. Indirect estimates indicate that they comprise around 80% of the phenolic components in black tea infusions (Stewart et al. 2005). Details of how some of the phenolic compounds in green tea are modified by fermentation to produce black tea are presented in Table 7.6.

Table 7.6 Concentration of the major phenolics in infusions of green and black tea manufactured from the same batch of *Camellia sinensis* leaves (Del Rio et al. 2004)

Compound	Green tea	Black tea	Black tea content as a percentage of green tea content
Gallic acid	6.0 ± 0.1	125 ± 7.5	2083
5-*O*-Galloylquinic acid	122 ± 1.4	148 ± 0.8	121
Total gallic acid derivatives	**128**	**273**	**213**
(+)-Gallocatechin	383 ± 3.1	n.d.	0
(−)-Epigallocatechin	1565 ± 18	33 ± 0.8	2.1
(+)-Catechin	270 ± 9.5	12 ± 0.1	4.4
(−)-Epicatechin	738 ± 17	11 ± 0.2	1.5
(−)-Epigallocatechin gallate	1255 ± 63	19 ± 0.0	1.5
(−)-Epicatechin gallate	361 ± 12	26 ± 0.1	7.2
Total flavan-3-ols	**4572**	**101**	**2.2**
3-*O*-Caffeoylquinic acid	60 ± 0.2	10 ± 0.2	17
5-*O*-Caffeoylquinic acid	231 ± 1.0	62 ± 0.2	27
4-*O*-*p*-Coumaroylquinic acid	160 ± 3.4	143 ± 0.2	89
Total hydroxycinnamoyl quinic acids	**451**	**215**	**48**
Quercetin-rhamnosylgalactoside	15 ± 0.6	12 ± 0.2	80
Quercetin-3-*O*-rutinoside	131 ± 1.9	98 ± 1.4	75
Quercetin-3-*O*-galactoside	119 ± 0.9	75 ± 1.1	63
Quercetin-rhamnose-hexose-rhamnose	30 ± 0.4	25 ± 0.1	83
Quercetin-3-*O*-glucoside	185 ± 1.6	119 ± 0.1	64
Kaempferol-rhamnose-hexose-rhamnose	32 ± 0.2	30 ± 0.3	94
Kaempferol-galactoside	42 ± 0.6	29 ± 0.1	69
Kaempferol-rutinoside	69 ± 1.4	60 ± 0.4	87
Kaempferol-3-*O*-glucoside	102 ± 0.4	69 ± 0.9	68
Kaempferol-arabinoside	4.4 ± 0.3	n.d.	0
Unknown quercetin conjugate	4 ± 0.1	4.3 ± 0.5	108
Unknown quercetin conjugate	33 ± 0.1	24 ± 0.9	73
Unknown kaempferol conjugate	9.5 ± 0.2	n.d.	0
Unknown kaempferol conjugate	1.9 ± 0.0	1.4 ± 0.0	74
Total flavonols	**778**	**570**	**73**
Theaflavin	n.d.	64 ± 0.2	∞
Theaflavin-3-gallate	n.d.	63 ± 0.6	∞
Theaflavin-3′-gallate	n.d.	35 ± 0.8	∞
Theaflavin-3,3′-digallate	n.d.	62 ± 0.1	∞
Total theaflavins	**n.d.**	**224**	∞

Data expressed as mg L^{-1} ± standard error ($n = 3$). n.d. – not detected. Green and black teas prepared by infusing 3 g of leaves with 300 mL of boiling water for 3 min.

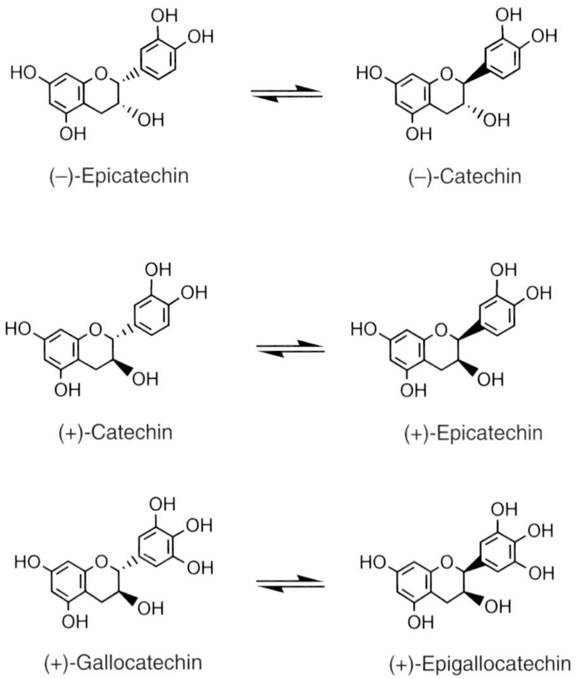

Figure 7.59 During brewing and production of instant tea beverages flavan-3-ols such as (+)-catechin, (−)-epicatechin and (+)-gallocatechin may epimerize.

Further changes may occur during the domestic brewing process and production of instant tea beverages. The flavan-3-ols may epimerize, producing for example (+)-epicatechin and (−)-catechin, (+)-epigallocatechin gallate, etc. (Figure 7.59) (Wang and Helliwell 2000; Ito et al. 2003). Black tea brew can form either scum or cream as it cools. Scum formation requires temporary hard water containing calcium bicarbonate that facilitates oxidation of brew phenols at the air-water interface (Spiro and Jaganyi 1993). Tea cream is a precipitate formed as black tea cools, being more obvious in strong infusions, and involves an association between theaflavins, some thearubigins and caffeine, exacerbated by calcium present in hard water (Jöbstl et al. 2005).

7.9.2 Maté

Maté is a herbal tea prepared from the dried leaves of *Ilex paraguariensis* which contain both caffeine and theobromine (Clifford and Ramìrez-Martìnez 1990). Originally, the drink was consumed by Guarani Indians in the forests of Paraguay and the habit was adopted by settlers in rural areas of South America, such as the Brazilian Panthanal and the Pampas in Argentina, where there was a belief that it ensured health, vitality and longevity. Its consumption is now becoming more widespread, perhaps aided by articles in the popular press claiming that it has Viagra-like qualities (Veash 1998).

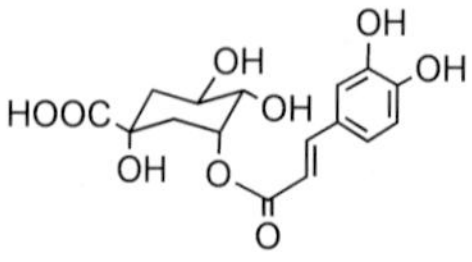

3-*O*-Caffeoylquinic acid

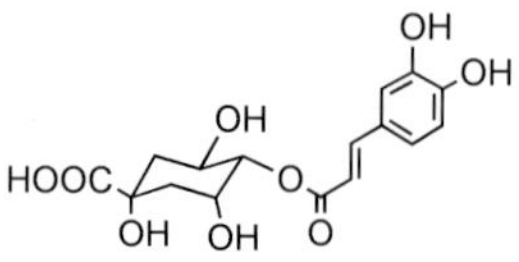

4-*O*-Caffeoylquinic acid

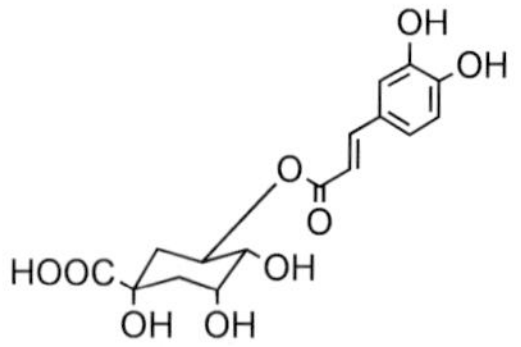

5-*O*-Caffeoylquinic acid

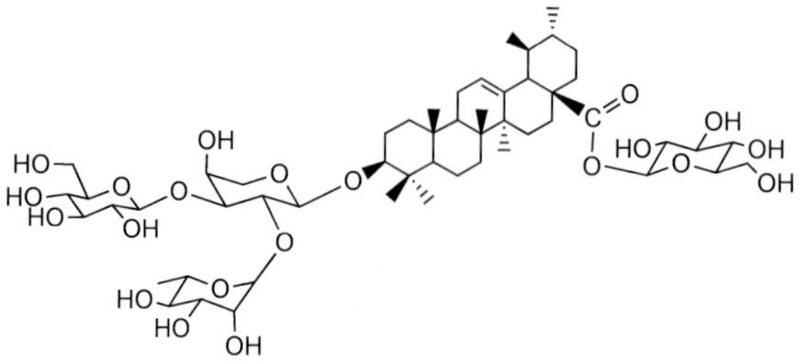

Matesaponin 2

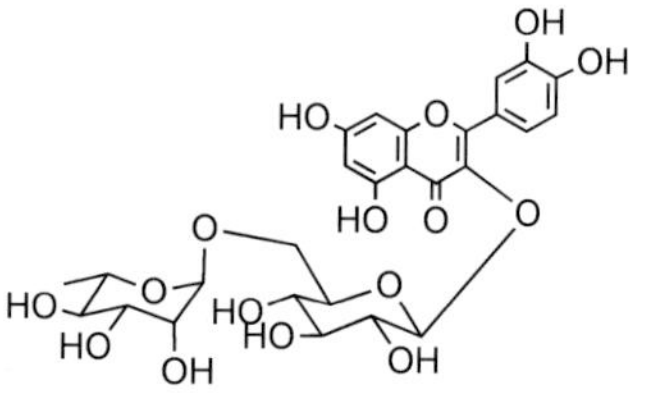

Quercetin-3-*O*-rutinoside

Figure 7.60 Maté is a herbal tea containing chlorogenic acids, saponins and trace amounts of querctin-3-*O*-rutinoside.

Maté, unlike *Camellia sinensis*, is a rich source of chlorogenic acids, containing substantial amounts of 3-, 4- and 5-*O*-caffeoylquininic acid (Figure 7.60) and three isomeric dicaffeoylquinic acids (Clifford and Ramìrez-Martìnez 1990). Maté leaves have yielded three new saponins named matesaponins 2, 3, and 4, which have been characterized by chemical and nmr methods as ursolic acid 3-*O*-[β-D-glucopyranosyl-(1 → 3)-[α-L-rhamnopyranosyl-(1 → 2)]]-α-L-arabinopyranosyl]-(28 → 1)-β-D-glucopyranosyl ester (Figure 7.60), ursolic acid 3-*O*-[β-D-glucopyranosyl-(1 → 3)-α-L-arabinopyranosyl]-(28 → 1)-[β-D-glucopyranosyl-(1 → 6)-β-D-glucopyranosyl]ester, and ursolic acid 3-*O*-[β-D-glucopyranosyl-(1 → 3)-[α-L-rhamnopyranosyl-(1 → 2)]]-α-L-arabinopyranosyl]-(28 → 1)-[β-D-glucopyranosyl-(1 → 6)-β-D-glucopyranosyl] ester, respectively (Gosmann *et al.* 1995). Maté infusions also contain quercetin-3-*O*-rutinoside (Figure 7.60) and glycosylated derivatives of luteolin and caffeic acid (Carini *et al.* 1998). Despite claims to the contrary (Morton 1989), neither condensed tannins nor hydrolysable tannins are present (Clifford and Ramirez-Martin 1990).

The traditional method of brewing and consumption is to add the dry, sometimes roasted, leaf to water boiling vigorously in a gourd or metal vessel. To avoid the boiling liquid burning the lips, it is drawn into the mouth using a straw and deposited at the back of the throat where there are comparatively few pain receptors. Although this practice avoids much of the discomfort otherwise associated with the consumption of a hot beverage, it still damages the oesophagus. Exposure of the damaged tissue during the error-prone stage of tissue repair makes such consumers unusually susceptible to oesophageal cancer (IARC 1991). While there seems little doubt that the primary causative agent is boiling hot water, the involvement of other substances is less certain. As a consequence of the repeated and ultimately severe tissue damage many substances will gain access to the tissue in an unmetabolized form, something that would not happen in healthy tissue. Redox cycling of dihydroxyphenols such as the caffeoylquinic acids may have a role to play, as may carcinogens from tobacco or alcohol (Castelletto *et al.* 1994). It is reassuring, that consumption of maté beverage, in a manner more closely resembling tea drinking, is not associated with an increased risk of oesophageal cancer, even in south American populations where this phenomenon was first observed (Rolon *et al.* 1995; Sewram *et al.* 2003).

7.9.3 Coffee

In economic terms, coffee is the most valuable agricultural product exported by third world and developing counties, amounting to about six million metric tonnes (Willson 1999). The green coffee bean is the processed, generally non-viable, seed of the coffee cherry. Commercial production exploits the seeds of *Coffea arabica* (so-called arabica coffees) accounting for ~70% of the world market and *C. canephora* (so-called robusta coffees) accounting for ~30%. There are many wild species some of which are virtually caffeine-free (e.g. *C. pseudozanguebariae*) (Clifford *et al.* 1989) but not suitable for commercial exploitation. Arabicas originated from the highlands of Ethiopia whereas the robustas originated at lower altitudes across Côte d'Ivoire, Congo and Uganda. There are many varieties of arabicas (*C. arabica* var. *arabica* accounting for the majority of commercial production) and robustas, and there has been limited commercial exploitation of various arabica × robusta hybrids and local use of *Coffea liberica*, *Coffea racemosa* and *Coffea dewevrei* in some African countries (Willson 1999). The establishment of coffee plantations in other parts of the world and the adoption of the beverage is discussed in several books (including Smith 1985).

Of the 57 recognized producing countries, the major producers in 1996 were Brazil, Colombia, Indonesia, Mexico and Uganda (Willson 1999). Most coffee is exported from these countries as green beans, but there are also significant exports of instant coffee powder produced in the coffee growing countries.

Increasingly mechanized harvesting is used. There are basically two methods for releasing the seeds from the harvested fruit, the wet process which requires copious supplies of good quality water, used mainly for arabicas, and the dry process used mainly for robustas. In the dry process the freshly picked or mechanically harvested cherries are sun dried for 2–3 weeks, followed by mechanical removal of the dried husk. In the wet process the cherries are soaked and fermented in water to remove the pulp prior to drying. After roasting most Robustas are blended as they are generally considered inferior to arabicas. Robustas are, however, preferred for instant coffee production as they give a higher yield of extractables and a less 'thin' liquor. With improved quality assurance robustas are now generally of very good quality but nevertheless subtly different from arabicas (Clarke 1985a; Willson 1999).

The commercial beans are roasted at air temperatures as high as 230°C for a few minutes, or at 180°C for up to ~20 min. There is a substantial and exothermic pyrolysis and a myriad of chemical reactions occur as a consequence of the internal temperature and internal pressure (5–7 atmospheres) achieved. Pyrolysis loss (as dry matter) ranges from 3 to 5% for a light roast to 5–8% for a medium roast and 8–14% for a dark roast. The pressure is largely due to entrapped carbon dioxide which effectively ensures an inert atmosphere within roasted whole beans. Grinding releases this, and subsequent extraction produces the beverage ready to drink, or on a commercial scale, a concentrated liquor that is converted to powder, either by spray drying or freeze drying. Domestic brewing extracts some 25–32% of solids, being highest in espresso, but varying with equipment, coffee particle size and the charge of coffee relative to water. Commercial instantization with water at up to 14 atmospheres, extracts some 50% from the roasted bean. Within the European Economic Community the yield of solubles is controlled by legislation at not more than 1 kg from 2.3 kg of green beans (Clarke 1985b). A cup of coffee contains some 1–2% solids by weight. Green beans may be decaffeinated either by the use of supercritical carbon dioxide or organic solvent. European Economic Community legislation requires less than 0.1% caffeine in the decaffeinated green bean which corresponds to less than 0.3% in instant powder (Clarke 1985b).

Green coffee beans are one of the richest dietary sources of chlorogenic acids which comprise 6–10% on a dry weight basis. 5-O-Caffeoylquinic acid is by far the dominant chlorogenic acid accounting for some 50% of the total. This is accompanied by significant amounts of 3-O- and 4-O-caffeoylquinic acid, the three analogous feruloylquinic acids and 3,4-O-, 3,5-O- and 4,5-O-dicaffeoylquinic acids (Figure 7.61) (Clifford 1999). Recently

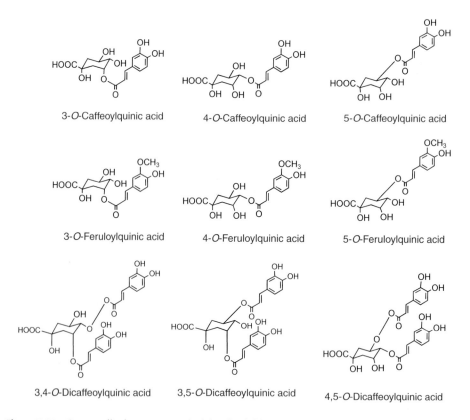

Figure 7.61 Green coffee beans contain high levels of chlorogenic acids which decline during roasting.

six caffeoylferuloylquinic acids have also been reported (Clifford et al. 2003) along with a series of amino acid conjugates (Clifford and Knight 2004) and novel mono- and diacyl chlorogenic acids incorporating 3,4-dimethoxycinnamic acid (Clifford et al. 2006). Robustas with the possible exception of those from Angola have a significantly greater content of chlorogenic acids than arabicas (Clifford and Jarvis 1988).

During roasting there is a progressive destruction and transformation of chlorogenic acids with some 8–10% being lost for every 1% loss of dry matter. Nonetheless substantial amounts survive to be extracted into domestic brews and commercial soluble coffee powders, and for many consumers coffee beverage must be the major dietary source of chlorogenic acids (Clifford 1999). Regular coffee drinkers will almost certainly have a greater intake of chlorogenic acids than flavonoids (Gosnay et al. 2002; Woods et al. 2003). While a portion of the green bean chlorogenic acids is completely destroyed, some is transformed during roasting. Early in roasting when there is still adequate water content, isomerization (acyl migration) occurs accompanied by some hydrolysis releasing the cinnamic acids and quinic acid. Later in roasting the free quinic acid epimerizes and lactonizes, and several chlorogenic lactones including caffeoyl quinides (Figure 7.62), also form (Scholz and Maier 1990; Bennat et al. 1994; Schrader et al. 1996; Farah et al. 2005). The cinnamic acids may be decarboxylated and transformed to a number of simple phenols and range of phenylindans probably via decarboxylation and cyclization of the vinylcatechol intermediate (Stadler et al. 1996). Two of these rather unstable compounds (Figure 7.62) have been found in roasted and instant coffee at 10–15 mg/kg.

The coffee bean also contains many other phytochemicals, many of which enter the beverage, and some that are not found elsewhere in the diet. The best known is caffeine present in green arabicas at ~1% and robustas at ~2%. There is some loss by sublimation during roasting but then near quantitative transfer to the brew or instant powder. A survey

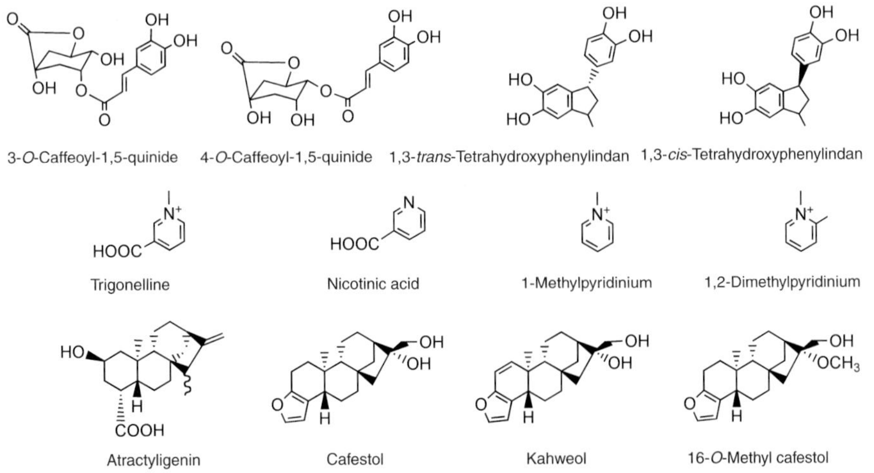

Figure 7.62 During roasting of coffee beans chlorogenic acids are transformed resulting in the appearance of lactones (caffeoyl quinides) and phenylindans. In addition, trigonelline is converted to nicotinic acid (vitamin B_3) and methylpyridiniums. Atractyligenins contribute to the bitter taste of roasted coffee. Cafestol, kahweol and 16-O-methyl cafestol occur as fatty acyl esters in unfiltered coffee, consumption of which can result in elevated plasma LDL cholesterol.

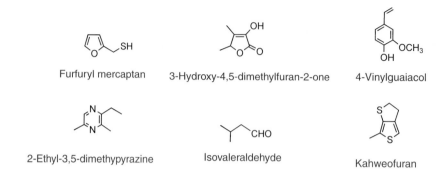

Figure 7.63 Volatile compounds from coffee having a major impact on aroma. Kahweofuran is reputed to have a coffee-like odour even in isolation.

of twelve instant coffees on the United Kingdom market in 1983 reported 2.83–4.83% caffeine (Clifford 1985).

Coffee contains fatty acyl serotonin (C_{18}; C_{20}; C_{22}; C_{24}) derivatives originating from the wax. They are thought by many to be undesirable irritants, and steaming processes have been developed for their removal (Clifford 1985). Trigonelline, the N-methyl betaine of nicotinic acid is present in green beans at about 1% and during roasting is converted partially to niacin (nicotinic acid) (Figure 7.62) making coffee beverage a potentially important source of vitamin B_3 (Clifford 1985). Trigonelline also yields 1-methylpyridinium (up to 0.25%) and 1,2-dimethylpyridinium (Figure 7.62) (up to 25 mg/kg) in proportion to the severity of roasting (Stadler *et al.* 2002). A range of diterpenes are present, some of which are glycosides, for example atractyligenin (Figure 7.62), thought to give a bitter taste to the beverage (Clifford 1985). Cafestol, kahweol and 16-*O*-methyl cafestol (Figure 7.61) occur as fatty acyl esters and enter the brew when coffee is prepared by extended boiling and the beverage is not filtered through paper or a bed of coffee grounds (Speer and Kölling-Speer 2001). These substances are responsible for the observed reversible elevation of plasma LDL cholesterol seen in some populations, notably in Scandinavia and Italy (Urgert *et al.* 1996, 1997; Urgert and Katan 1997). Education and encouragement of alternative brewing procedures has seen a lowering of plasma LDL cholesterol in some Scandinavian populations. Instant coffee powder also contains ~1% mannose and ~2% galactose. These sugars, which rarely occur free in foods, are formed by the hydrolysis of structural arabino-galactan and storage mannan polysaccharides during the high temperature commercial extraction process (Clifford 1985).

Roasted coffee is prized by many for its unique and very attractive aroma. The odour complex contains in excess of 800 known substances, many of which are heterocycles, and many of which will not be found significantly elsewhere in the diet. Of these, fourteen, 2-furfuryl-thiol, 4-vinylguaiacol, three alkylpyrazines, four furanones and five aliphatic aldehydes, have been identified as particularly important determinants of odour (Grosch 2001). Kahweofuran is another important volatile as it is reputed to have a coffee-like aroma. Selected structures are shown in Figure 7.63. These volatiles, present in roasted coffee at concentrations in the μg/kg to mg/kg range are produced during the high temperature roasting process by a complex series of reactions referred to as the Maillard reaction

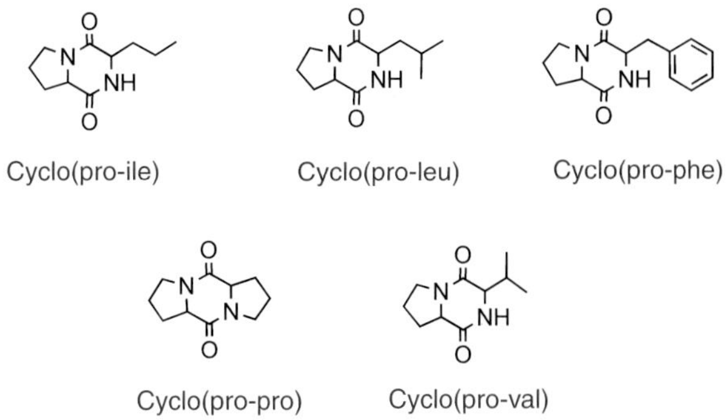

Cyclo(pro-ile)   Cyclo(pro-leu)   Cyclo(pro-phe)

Cyclo(pro-pro)   Cyclo(pro-val)

Figure 7.64 Cyclic diketopiperazines formed from proline and other amino acids contribute to the bitter taste of coffee.

in which sugars and amino acids are key reactants (Nursten 2005). Coffee bitterness is only partially due to caffeine, and it is thought that Maillard products, including cyclic diketopiperazines formed from proline and other amino acids are important (Figure 7.64) (Ginz and Engelhardt 2000).

7.9.4 Cocoa

Cocoa (*Theobroma cacao*) is a tree which originated in the tropical regions of South America. There are two forms sufficiently distinct as to be considered subspecies. Criollo developed north of the Panama isthmus and Forastero in the Amazon basin. A so-called hybrid, Trinitario, developed in Trinidad (Willson 1999). The cocoa plant is now cultivated worldwide, major producers being the Ivory Coast, Ghana, Nigeria, Indonesia, Brazil and Cameroon. The main cultivated form is *Theobroma cacao* var. *forastero* which accounts for more than 90% of the world's usage. Criollo and trinitario are also grown, and some regard these as providing better flavour qualities to cocoa-based products (Leung and Foster 1996).

Ripe cocoa pods contain about 30–40 seeds which are embedded in a sweet mucilaginous pulp comprised mainly of sugars. The pods are harvested and broken open and the pulp and seeds are formed into large mounds and covered with leaves. The pulp is fermented for 6–8 days. During this period sucrose is converted to glucose and fructose by invertase and the glucose is subsequently utilized in fermentation yielding ethanol which is metabolized to acetic acid. As the tissues of the beans loose cellular integrity and die, storage proteins are hydrolysed to peptides and amino acids while polyphenol oxidase converts phenolic components to quinones which polymerize yielding brown, highly insoluble compounds that give chocolate its characteristic colour (Haslam 1998). After fermentation, the seeds are dried in the sun, reducing the moisture content from 55% to 7.5%. The resulting cocoa beans are then packed for the wholesale trade.

Cocoa beans are used extensively in the manufacture of chocolate, but this chapter is confined to the use of cocoa as a beverage. To produce the cocoa powder used in the beverage, the beans are roasted at 150°C and the shell (hull) and meat of the bean (nib) are mechanically separated. The nibs, which contain about 55% cocoa butter, are then finely ground while hot to produce a liquid 'mass' or 'liquor'. This sets on cooling and is then pressed to express the 'butter' which is used in the manufacture of chocolate. The residual cake is pulverized to produce the cocoa powder traditionally used as a beverage. An alkalization process is also often employed to modify the dispersability, colour and flavour of cocoa powders. This involves the exposure of the nibs prior to processing to a warm solution of caustic soda (Bixler and Morgan 1999).

The dominant polyphenols in cocoa are flavan-3-ol derivatives. The principal components in fresh beans are (−)-epicatechin, (+)-catechin and oligomeric procyanidins ranging from dimers to decamers. Trace quantities of quercetin-3-O-glucoside and quercetin-3-O-arabinoside also occur (Hammerstone *et al.* 1999). Individual procyanidins that have been identified include the B_2 and B_5 dimers and the trimer C_1 (Figure 7.65) (Haslam 1999). N-Caffeoyl-3-O-hydroxytyrosine (clovamide) and N-p-coumaroyl-tyrosine (deoxyclovamide) are also present (Sanbongi *et al.* 1998). These compounds along with the proanthocyanidins contribute to the astringent taste of unfermented cocoa beans and roasted cocoa nibs but not to the same degree as other amides, in particular cinnamoyl-L-aspartic acid and caffeoyl-L-glutamic acid (Figure 7.65) (Stark and Hofman (2005). The main purine alkaloid is theobromine which is present in much higher amounts than caffeine (Ashihara and Crozier 1999). During fermentation, the conversion of many of the phenolic components to insoluble brown polymeric compounds takes place and, as a consequence, the level of soluble polyphenols falls by ∼90%. An 'average' home-made serving of hot cocoa has approximately 200 mg of flavan-3-ol type polyphenols (Vinson *et al.* 1999).

7.9.5 Wines

Wine is basically fermented grape juice with a minimal alcohol level of 8.5% by volume. The wild grapevine originated in the Far East (Mesopotamia) and Egypt and evidence for wine production dates from Neolithic times. Wine was consumed by many ancient civilizations including the Mesopotamians, Egyptians, Greeks and Romans. Once the floods had receded, Noah appears to have over indulged in wine (Genesis, Chapter IX, Verse 21) and St. Paul apparently recommended the consumption of wine on health grounds (Watkins 1997). More recently, Galileo made and drank his own red wines right up to his death at the age of 78 in 1642 (Sobel 1999). Today, wines are produced from numerous varieties of grapes, including Cabernet Sauvignon, Merlot, Pinot Noir, Syrah, Cinsault, Rondinella, Sangiovese, Nebiolo, Grenache, Tempranillo and Carignan. The main commercial producers are located in California in the United States, and in France, Italy, Australia, New Zealand, Spain, Chile, Argentina, South Africa, as well as Bulgaria, Romania, Southern Brazil, and more recently China and India.

A wide variety of processes are used in the making of red wine. Typically, however, black grapes are pressed and the juice (must), together with the crushed grapes, undergo alcoholic fermentation for 5–10 days at ∼25–28°C. The solids are removed and the young wine subjected to a secondary or malo-lactic fermentation during which malic acid is converted to lactic acid and carbon dioxide. This softens the acidity of the wine and adds to its complexity and stability. The red wine is then matured in stainless steel vats, or in the case of higher quality vintages in oak barrels, for varying periods before being filtered and bottled.

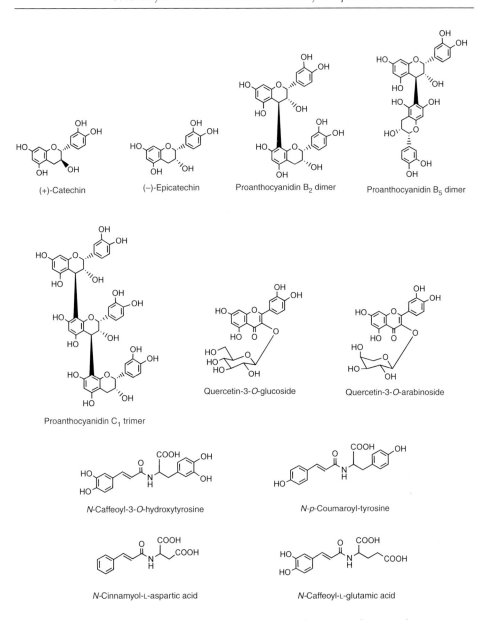

Figure 7.65 Monomeric flavan-3-ols and the proanthocyanidin B_2, B_5 dimers and C_1 trimer are found in fresh cocoa beans along with small amounts of quercetin-3-O-glucoside and quercetin-3-O-arabinoside. Along with the proanthocyanidins, the amides N-caffeoyl-3-O-hydroxytyrosine (clovamide) and N-p-coumaroyl-tyrosine (deoxyclovamide), and in particular N-cinnamoyl-L-aspartic acid and N-caffeoyl-L-glutamic acid, contribute to the astringency of cocoa.

White wines are produced from both black and, more traditionally, white varieties of grapes. The berries are crushed gently rather than pressed to prevent breaking of stems and seeds. Solid material is removed and the clarified juice fermented typically between 16 and 20°C for 5 days. The resultant must then undergoes malo-lactic fermentation, before maturation, filtration and bottling.

Wines are produced from an assortment of grape cultivars grown under climatic conditions that can vary substantially not only in different geographical regions but also locally on a year-to-year basis. To complicate matters further, grapes at different stages of maturity are used and vinification and ageing procedures are far from uniform. It is hardly surprising, therefore, that wines are extremely heterogeneous in terms of their colour, flavour, appearance, taste and chemical composition. (Singleton 1982; Haslam 1998). In general, however, red wines, and to a much lesser extent white wines, are an extremely rich source of a variety of phenolic and polyphenolic compounds.

In the making of red wine, with prolonged extraction, the fermented must can contain up to 40–60% of the phenolics originally present in the grapes. Subtle changes in these grape-derived phenolic components occur during the ageing of the wines especially when carried out in oak barrels or, as in recent years, during exposure to chips of oak wood. Consequently, there is a wide range in the level of phenolics between different red wines, the concentration of flavonols, for instance, varying by more than 10-fold and the overall level of phenolics by almost five-fold (Table 7.7) (Burns et al. 2000). Information on variations in the levels of a number of phenolic compounds in comprehensive range of French red wines have been published by Carando et al. (1999) and Landrault et al. (2001).

The phenolics in red wines are the hydroxycinnamates, coutaric, caftaric and fertaric acids, and malvidin-3-O-glucoside and other anthocyanins with lower levels of gallic acid, stilbenes and flavonols. From the data presented in Table 7.8, which are based on a study by Burns et al. (2000), it is evident that the levels of the flavan-3-ol monomers (+)-catechin and (−)-epicatechin are not high, and that there is a large discrepancy between the levels of phenolics measured by HPLC and the total phenolics determined by the Folin-Ciocalteau

Table 7.7 Range of concentrations of phenolic compounds in 15 red wines of different geographical origin (after Burns et al. 2000)[a]

Phenolic	Range (mg/L)
Total flavonols	5–55
Total stilbenes	1–18
Gallic acid	8–71
Total hydroxycinnmates	66–124
(+)-Catechin and (−)-epicatechin	8–60
Free and polymeric anthocyanins	41–150
Total phenols	824–4059

[a]Total phenols measured by colorimetric Folin-Ciocalteau assay, other estimates based on HPLC analyses that did not detect proanthocyanidins.

assay. Among the 'missing ingredients', that were not measured by HPLC, are proanthocyanidin B_{1-4} dimers, the C_1 and C_2 trimers (Ricardo da Silva *et al.* 1990) and oligomeric and polymeric forms with, respective, mean degrees of polymerization of 4.8 and 22.1 (Sun *et al.* 1998). The equivalent mean degrees of polymerization of proanthocyanidins in grapes were 9.8 and 31.5 indicating that substantial changes in flavan-3-ol composition occur during fermentation and aging of the wines. Among the processes involved is the formation of compounds corresponding to malvidin-3-*O*-glucoside linked through a vinyl bond to either (+)-catechin, (−)-epicatechin or the procyanidin dimer B_3 (Mateus *et al.* 2002). Similar blue coloured compounds with the flavan-3-ols linked to malvidin-3-*O*-(6″-*O*-*p*-coumaroyl)glucoside have also been detected in red wines (Mateus *et al.* 2003). The production of pyruvate and acetaldehyde by yeast during fermentation of Tempranillo grapes has been associated with the formation of malvidin-3-*O*-glucoside-pyruvic acid (vitisin A) and malvidin-3-*O*-glucoside-4-vinyl (vitisin B) (Morata *et al.* 2003) which are members of a group of red wine-derived compounds referred to as pyranoanthocyanidins. The structures involved are illustrated in Figure 7.66.

The production of white wine results in either low levels or an absence of skin- and seed-derived phenolics, so the overall level of phenolics can be much lower than that found in many red wines (Waterhouse and Teissedre 1997). This observation is reflected in a more detailed comparison of the constituents of French red wines, dry white wines and sweet white wines carried out by Landrault *et al.* (2001). A summary of the data obtained in this study is presented in Table 7.8.

7.9.6 Beer

There is evidence that the Sumerians were making ale 8000 years ago although it would not be recognizable as beer today. The Sumerians used two grains for fermentation, barley and wheat. There is evidence of eight kinds of ale from barley, another eight from wheat and three from mixed grains. Sumerians, as well as drinking beer, also used it as a form of currency and to pay salaries. The Egyptians were also keen brewers. Beer in Egypt 3000 years ago, based on remains in urns, was made from a mixture of malted barley and an ancient wheat called emmer. Prior to hops (*Humulus lupulus*) a variety of flavourings were used including mandrake (*Mandragora officinarum*) which tasted like leeks! Later, Europeans employed rosemary, yarrow (*Achillea millefolium*), coriander and bog myrtle (*Myrica gale*) as flavourings. Hops were used around 200 BC in Babylon and are first mentioned in Europe in 736 AD. In 1519 hops were condemned by the English as a 'wicked and pernicious weed' but hop growing began in Kent in 1524. Because hops are a natural antiseptic they gave the advantage of prolonged storage and allowed brewers to thin out the drink and reduce the sugar content. It took until the sixteenth century before the manufacture of beer rather than ale was common throughout Europe.

Beer is an alcoholic beverage made from malted grains (usually barley or wheat), hops, yeast and water. Originally the terms 'beer' and 'ale' referred to different beverages, ale by tradition being made without hops. However, most commercial products contain hops; the term beer now encompasses two broad categories: ales and lagers. Ales are brewed with 'top-fermenting' yeasts such as *Saccharomyces cerevisiae* at around 15–20°C. The term ale includes a broad range of beer styles including bitters, pale ales, porters and stouts. Lagers are brewed with 'bottom-feeding' yeasts such as *S. carlsbergensis* (*uvarum*) at colder

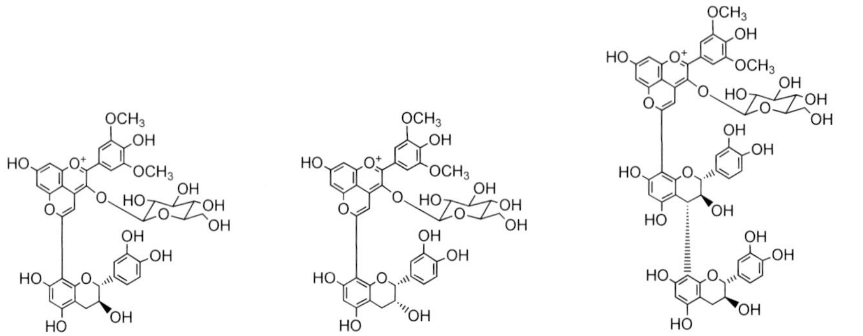

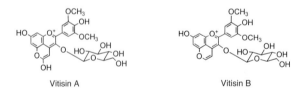

Malvidin-3-O-glucoside linked through a vinyl bond to (+)-catechin (I), (−)-epicatechin (II) and procyanidin dimer B₃ (III)

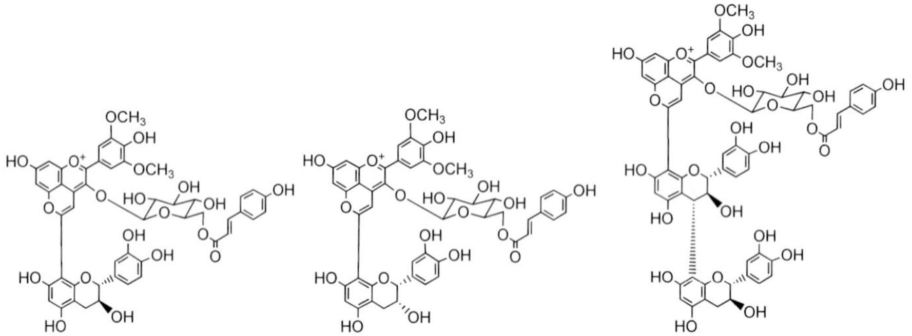

Malvidin-3-O-(6″-O-p-coumaroyl)glucose linked through a vinyl bond to (+)-catechin (IV), (−)-epicatechin (V) and procyanidin dimer B₃ (VI)

Vitisin A

Vitisin B

Figure 7.66 Red wines contain many phenolic compounds. Some are found in red grapes (see Figure 7.39) others including those illustrated in this figure, are formed during fermentation and ageing.

temperatures (6–10°C) and are matured (or lagered – from the German verb 'to store') over much longer periods of time (months).

The first stage in beer manufacture allows the cereal to germinate in a warm atmosphere to activate the amylolytic diastases in the grain, and initiate the enzymatic hydrolysis of starches. The germinated seed is gently dried (kilned) to preserve the enzyme activity and produce 'malt'. In brewing, the malt is milled and then mixed with warm water in a porridge-like consistency to allow the enzymes to degrade the starch and proteins – a process called mashing. The liquid extract from the mash, called 'wort', is then boiled

Table 7.8 Average concentrations of antioxidants and phenolic compounds in 34 red, 11 dry white and 7 sweet French wines (after Landrault et al. 2001)

	Red wine	Dry white wine	Sweet white wine
Antioxidant capacity	18.9 ± 0.7	3.1 ± 1.1	3.2 ± 0.2
Total phenol content	2155 ± 78	414 ± 102	657 ± 33
Flavan-3-ols			
(+)-Catechin	41 ± 6	15 ± 8	4.2 ± 0.7
(−)-Epicatechin	29 ± 3	12 ± 9	1.4 ± 0.3
Procyanidin B_1	15 ± 2	5.1 ± 2.3	3.4 ± 0.5
Procyanidin B_2	27 ± 5	8.9 ± 4.9	3.0 ± 0.5
Procyanidin B_3	59 ± 7	13 ± 5	10 ± 2
Procyanidin B_4	5.2 ± 1.0	4.0 ± 2.5	2.0 ± 1.1
Total flavan-3-ols	177 ± 22	59 ± 31	24 ± 1
Gallic acid	30 ± 2	4.0 ± 2.1	5.8 ± 1.1
Caffeic acid	11 ± 1	3.4 ± 0.5	1.6 ± 0.3
Caftaric acid	51 ± 4	33 ± 6	14 ± 3
Anthocyanins			
Malvidin-3-*O*-glucoside	20 ± 19	n.d.	n.d.
Peonidin-3-*O*-glucoside	1.8 ± 2.5	n.d.	n.d.
Cyanidin-3-*O*-glucoside	0.3 ± 0.4	n.d.	n.d.

Antioxidant capacity expressed as mM Trolox equivalents and all other values in mg/L. All figues are mean values ± standard error. n.d. – not detected.

with the addition of hops, which are the dried cones of the female hop plant. These cones contain the bitter compounds that serve to add aroma and flavour as well as acting as a preservative. The cooled 'wort' is then inoculated with yeast to begin the fermentation process. After fermentation, typically for 5–7 days, the fermented wort is allowed to stand to flocculate the yeast so that it can be removed. The beer is then matured for weeks to months, depending on type of beer, prior to being filtered, pasteurized and either bottled or canned. Some beers receive a secondary fermentation in the cask or bottle by addition of fresh yeast.

One of the compounds that contributes to the characteristic aroma of hops is the volatile monoterpene β-myrcene (Figure 7.67). Hops also contain a range of unique bitter substances humulone, cohumulone and adhupulone, which are referred to as α-acids, and lupulone, colupulone and adlupulone, known as the β-acids (Figure 7.67) (Belitz and Grosch 1987; Hofte et al. 1998; De Keukeleire 2000). During drying, storage and processing of the hops these compounds, of which humulone is the major component, are transformed by a series of isomerizations, oxidations and polymerizations into a range of incompletely characterized secondary products. When the hops or hop extracts are added to the wort, boiling extracts the bitter principles and further transforms them. Humulones are converted to the more bitter and less soluble *cis*- and *trans*-isohumulones, and these are further transformed to the less bitter humulinic acids. The lupulones are, likewise, converted to the hulupones and luputriones (Figure 7.68). Hululone is reported to inhibit

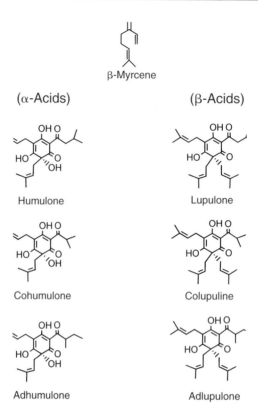

Figure 7.67 The monoterpene β-myrcene contributes to the characteristic aroma of hops while the α- and β-acids have a bitter taste.

angiogenesis and is, therefore, a potential tool for the therapy of various angiogenic diseases involving solid tumour growth and metastasis (Shimamura *et al.* 2001).

Beer contains a range of phenolic and polyphenolic compounds, which come partly from the barley (70%) and partly from the hops (30%). Flavan-3-ols are found equally in hops and malt. These include monomers such as (+)-catechin and (−)-epicatechin, and the dimers prodelphinidin B_3 and procyanidin B_3 (Figure 7.69). Trimers also occur although a more recent study, albeit with one unnamed American beer, reported an absence of high molecular weight polymeric proanthocyanidins and an average degree of polymerization of only 2.1 (Gu *et al.* 2003). The malt contributes most of the simple phenolics such as protocatechuic acid, ferulic acid and caffeic acid, small amounts of these compounds are also found in hops. If brewing with dark coloured malt, the antioxidant levels will be higher.

Hops contain quercetin conjugates and the prenylflavonoid xanthohumol which during the brewing process undergoes substantial conversion to the flavanone isoxanthohumol, which predominates in most beers. Other prenylflavonoids found in beers include 6- and 8-prenylnaringenin and 6-geranylnaringenin (Figure 7.69) (Stevens *et al.* 1999). The quantity of phenolics in beer has not been widely studied but in general low and sub-mg quantities per litre are present. De Pascual Teresa *et al.* (2000) determined the flavan-3-ol content of

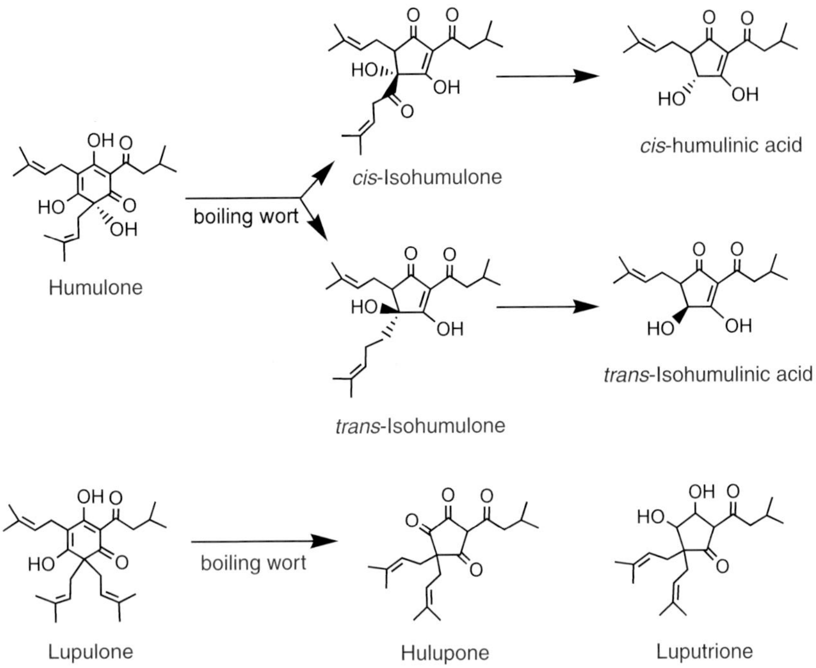

Figure 7.68 Transformations of humulone and lupulone that occur during the drying of hops and the boiling of wort in the production of beer.

a red wine and a beer and found 17.8 and 7.3 mg/L respectively. Considering serving size, the potential flavan-3-ol intake from these two sources is broadly comparable.

7.9.7 Cider

Cider is an alcoholic beverage produced from apples, which can either be made from specific cider varieties or dessert apples. Cider making in the Mediterranean basin was described in the works of Roman writer Pliny during the first century AD. Cider making then moved north and was well establish in France by the ninth century. It is thought that cider making was introduced to England from Normandy (Lea 1995).

The juice is extracted from the apples by milling and pressing the fruit. Cultured yeast in prime condition is now used to carry out fermentation. The yeast is added 24 h after the sulphur dioxide has been added. The fermentation can be carried out in traditional wooden storage vessels or modern sterilizable tanks. The process is monitored by comparing sample results to predefined set specifications. The cider is transferred to a maturation vessel once it has fermented. Cider can either be made from a blend of several varieties or from one single variety of apple. The majority of commercial ciders are blended ciders. The finished product is then passed through a series of filters and some are pasteurized. The stable finished product is then ready for packaging.

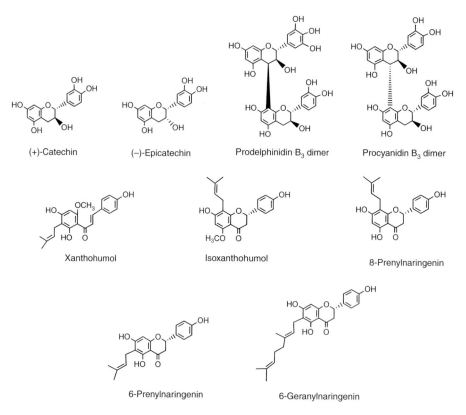

Figure 7.69 (+)-Catechin, (−)-epicatechin, prodelphinidin B_3 and procyanidin B_3 occur in beer together with xanthohumol, isoxanthohumol, desmethylxanthohumol, 6- and 8-prenylnaringenin and 6-geranylnaringenin.

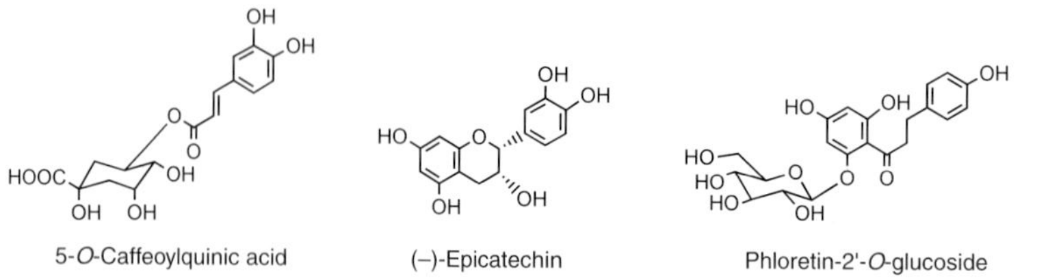

Figure 7.70 Apple-derived 5-O-caffeoylquinic acid, (−)-epicatechin and phloretin-2′-glucosides are present in cider.

The major phenolics in cider have been shown to be 5-O-caffeoylquinic acid, (−)-epicatechin and phloretin-2′-O-glucoside (Figure 7.70) (Slaiding et al. 2004). However, this report was unable to quantify the presence of procyanidins, which are thought to be a major component of cider (Lea 1990).

Hydroxycinnamic acid derivatives and flavan-3-ols are the two classes that are important in the cider industry due to their physiochemical properties. 5-O-Caffeoylquinic acid is one of the most important substrates for the endogenous enzymes, polyphenol oxidases, further reactions from the products formed give cider its yellow-brown colouring (Guyot et al. 1998).

Phenolics of apples are implicated in cider quality. They are involved in astringent and bitter tastes. The degree of polymerization of procyanidins is directly involved in the balance of bitterness and astringency. Bitterness is due to presence of oligomeric procyanidins with a degree of polymerization 2–5, whereas procyanidins with a degree of polymerization of structures 6–10 are more involved in astringency (Lea and Arnold 1978).

The method of production has been shown to affect the phenolic content, with modern techniques of pneumatically pressed cider fermented in stainless steel vats decreasing levels slower than the more traditional methods of pressing and fermenting in wooden barrels (del Campo et al. 2003). Oxidation which occurs during the juice extraction has also been linked with a reduction in the level of polymeric procyanidins (Lea et al. 1978). Fining to clarify the cider has been shown to decrease the procyanidin content (Lea 1990).

7.9.8 Scotch whisky

Whisky is believed to have been first produced by the Ancient Celts. Over the years 'uisge beatha' ('water of life') has been ascribed many medicinal and health promoting properties including the relief of colic, palsy and even smallpox. Both malt and grain whiskies are produced in Scotland. Scotch malt whisky is made from malted barley, water and yeast. The barley is steeped in tanks of water for 2–3 days before being spread out on the floors of the malting house to germinate. The plant hormone, gibberellic acid is added to the germinating seed to increase α-amylase synthesis which speeds up the hydrolysis of starch and the accompanying accumulation of sugars. To arrest germination when sugar levels are high, the malted barley is dried in a kiln often over a peat-fuelled fire, the smoke ('peat reek') imparting a distinctive aroma to the final spirit. Subsequent mashing and mixing of the malted barley produces a wort which is transferred to a fermenting vat, where added yeast converts the sugar to alcohol. The resulting 'wash' containing about 10% alcohol is then distilled twice in copper stills and the distillate containing about 65–70% alcohol is aged in sherry casks for at least three years (Piggott et al. 1993).

Scotch grain whisky is made from unmalted wheat, rye, oats or maize which is first gelatinized by cooking under pressure. This makes the starch more accessible to the amylases resulting in the production of fermentable sugars. A proportion of malted barley is then added to the sugar-rich 'wort'. Following fermentation, distillation is carried out in a continuously operating, two-columned Coffey still and the distillate is then aged in oak casks. Although unblended single malts are commonly drunk, most Scotch whisky bought in shops is a blend of malt and grain whiskies. A blended whisky can be a combination of anything from 15 to 50 single whiskies of varying ages. Producers tend to have their own secret formula for the blending process so each whisky brand will differ in phytochemical composition. Consequently, data in the following sections are approximate.

Scotch malt whiskies contain complex mixtures of phenolic compounds which are extracted from the wooden casks in which the maturation process takes place. The phenolic profile is influenced by several factors including the length of maturation, the species of oak from which the casks are made, the pre-treatment of the cask by charring of the wood, prior

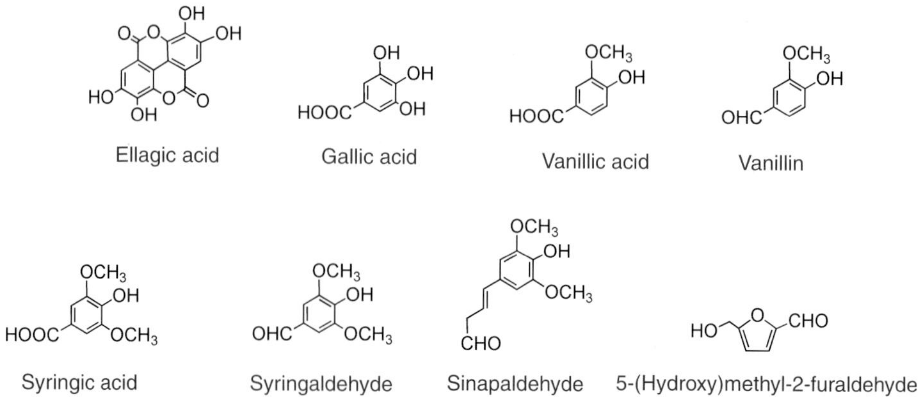

Ellagic acid · **Gallic acid** · **Vanillic acid** · **Vanillin**

Syringic acid · **Syringaldehyde** · **Sinapaldehyde** · **5-(Hydroxy)methyl-2-furaldehyde**

Figure 7.71 Scotch whisky contains a number of phenolic compounds most of which have their origins in the oak casks in which the whisky matures.

usage of the cask for bourbon or sherry storage and the number of times which the cask has been used for maturation (Rous and Aldersen 1983; Singleton, 1995). Simple phenolics in whisky may arise from the thermal degradation of benzoic acid derivatives from malt and peat smoke. Phenolic aldehydes such as vanillin, syringaldehyde, coniferaldehyde and sinapaldehyde are formed from the breakdown of wood lignin during cask charring and maturation. Ellagic acid is generally present in high concentrations. Heterocyclic oxygen compounds, such as furaldehydes (Figure 7.71) and lactones are also present, formed from hexoses during mashing and distillation.

7.10 Databases

Information on the occurrence and levels of various flavonoids in fruits, vegetables, beverages and foods can be found in on-line databases prepared by the US Department of Agriculture, Agricultural Research Service (2002, 2003, 2004). Other reports relate to the flavonol (Hertog et al. 1992, 1993) and flavan-3-ol (Arts et al. 2000a,b) content of Dutch produce and the flavan-3-ol content of Spanish foodstuffs and beverages (de Pascual-Teresa et al. 2000). Gu et al. (2003) have produced a report on the proanthocyanidin content of 88 different foods obtained in the United States.

References

Adams, H., Vaughan, J.G. and Fenwick, G.R. (1989) The use of glucosinolates for cultivar identification in swede, *Brassica napus* L. var. *napobrassica* L. petem. *J. Sci. Food Agric.*, **46**, 319–324.

Adão, R.C. and Glória, M.B.A. (2005) Bioactive amines and carbohydrate changes during ripening of 'Prata' banana (*Musa acuminata* × *M. balbisiana*). *Food Chem.*, **90**, 705–711.

Akingbala, J.O. (1991) Effect of processing on flavonoids in millet (*Pennisetum americanum*) flour. *Cereal Chem.*, **68**, 180–183.

Alasalvar, C., Grigorr, J.M. and Zhang, D. (2001) Comparison of volatiles, phenolics, sugars, antioxidant vitamins and sensory quality of different coloured carrot varieties. *J. Agric. Food Chem.*, **49**, 1410–1416.

Al-Farsi, M., Alasalvar, C., Morris, A. *et al.* (2005) Comparison of antioxidant activity, anthocyanins, carotenoids, and phenolics of three native fresh and sun dried date (*Phoenix dactylifera* L.) varieties grown in Oman. *J. Agric. Food Chem.*, **53**, 7592–7599.

Amakura, Y., Okada, M., Tsuji, S. and Tonagai, Y. (2000) High performance liquid chromatographic determination with photo diode array detection of ellagic acid in fresh and processed fruits. *J. Chromatogr. A*, **896**, 87–93.

Amiot, M.J., Fleuriet, A. and Macheix, J.J. (1986) Importance and evolution of phenolic compounds in olive during growth and maturation. *J. Agric. Food. Chem.*, **34**, 823–826.

Arts, I.C.W., van de Putte, B. and Hollman, P.C.H. (2000a) Catechin content of foods commonly consumed in the Netherlands. 1. Fruits, vegetables, staple foods and processed foods. *J. Agric. Food Chem.*, **48**, 1746–1751.

Arts, I.C.W., van de Putte, B. and Hollman, P.C.H. (2000b) Catechin content of foods commonly consumed in the Netherlands. 2. Tea, wine, fruit juices and milk chocolate. *J. Agric. Food Chem.*, **48**, 1752–1757.

Ashihara, H. and Crozier, A. (1999) Biosynthesis and metabolism of caffeine and related purine alkaloids in plants. *Adv. Bot. Res.*, **30**, 117–205.

Anon (1897) *The Gardeners Chronicle*, ii, 19.

Bandyopadhyay, C., Narayan, V.S. and Variyar, P.S. (1990) Phenolics of green pepper berries (*Piper nigrum* L.). *J. Agric. Food Chem.*, **38**, 1696–1699.

Barnes, S., Kirk, M. and Coward, L. (1994) Isoflavones and their conjugates in soy foods: extraction conditions and analysis by HPLC-mass spectrometry. *J. Agric. Food Chem.*, **42**, 2466–2474.

Bandoniene, D. and Murkovic, M. (2000) On-line HPLC-DPPH screening method for evaluation of radical screening phenols extracted from apples (*Malus domestica* L.). *J. Agric. Food Chem.*, **50**, 2482–2487.

Barillari, J., Canistro, D., Paolini, M. *et al.* (2005) Direct antioxidant activity of purified glucoerucin, the dietary secondary metabolite contained in rocket (*Eruca sativa* Mill.) seeds and sprouts. *J. Agric. Food Chem.*, **53**, 2475–2482.

Baritaux, O., Amiot, M.J., Richard, H. *et al.* (1991) Enzymatic browning of basil (*Ocimum basilicum* L.). Studies on phenolic compounds and polyphenol oxidase. *Sci. Aliments*, **11**, 49–62.

Beatti, J., Crozier, A. and Duthie, G.G. (2005) Potential health benefits of berries. *Curr. Nutr. Food Sci.*, **1**, 71–86.

Beauchamp, G.K., Keasting, R.S.J., Morel, D. *et al.* (2005) Ibuprofen-like activity in extra-virgin olive oil. *Nature*, **437**, 45–46.

Belitz, H.-D. and Grosch, W. (1987) *Food Chemistry*. Springer Verlag, Berlin.

Bennat, C., Engelhardt, U.H., Kiehne, A. *et al.* (1994) HPLC Analysis of chlorogenic acid lactones in roasted coffee. *Z. Lebens. Unters. Forschung*, **199**, 17–21.

Berardini, N., Carle, R. and Schieber, A. (2004) Characterization of gallotannins and benzophenone derivatives from mango (*Mangifera indica* L. cv. 'Tommy Atkins') peels, pulp and kernels by high-performance liquid chromatography/electrospray ionization mass spectrometry. *Rapid Commun. Mass Spectrom*, **18**, 2208–2216.

Brown, A.G., Eyton, W.B., Holmes, A. *et al.* (1969) Identification of the thearubigins as polymeric proanthocyanidins. *Nature*, **221**, 742–744.

Bixler, R.G. and Morgan, J.N. (1999) Cocoa bean and chocolate processing. In I. Knight (ed.), *Chocolate and Cocoa: Health and Nutrition*. Blackwells, Oxford, pp. 43–60.

Burns, J., Gardner, P.T., McPhail, D.B. *et al.* (2000). Relationship among antioxidant activity, vasodilation capacity and phenolic content of red wines. *J. Agric. Food Chem.*, **48**, 220–230.

Burns, J., Gardner, P.T., Matthews, D. *et al.* (2001) Extraction of phenolics and changes in antioxidant activity of red wines during vinification. *J. Agric. Food Chem.*, **49**, 5797–5808.

Burns, J.A., Mullen, W., Landrault, N. *et al.* (2002a) Variations in the profile and content of anthocyanins in wines made from Cabernet Sauvignon and hybrid grapes. *J. Agric. Food Chem.*, **50**, 4096–4102.

Burns, J.A., Yokota, T., Ashihara, H. *et al.* (2002b) Plant foods and herbal sources of resveratrol. *J. Agric. Food Chem.*, **50**, 3337–3340.

Canigueral, S., Iglesias, J., Hamburger, M. *et al.* (1989) Phenolic constituents of *Salvia lavandulifolia* ssp. *lavandulifolia*. *Planta Medica*, **55**, 92.

Carando, S., Teissedre, P.-L., Pascual-Martinez, L. *et al.* (1999). Levels of flavan-3-ols in French wines. *J. Agric. Food Chem.*, **47**, 4161–4166.

Cassidy, A., Hanley, B. and Lamuela-Raventós, R.M. (2000) Isoflavones, lignans and stilbenes – origins, metabolism and potential importance to human health. *J. Sci. Food Agric.*, **80**, 1044–1062.

Carini, M., Facino, R.M., Aldini, G. *et al.* (1998) Characterization of phenolic antioxidants from maté (*Ilex paraguayensis*) by liquid chromatography/mass spectrometry and liquid chromatography/tandem mass spectrometry. *Rapid Commun. Mass Spectrom.*, **12**, 1813–1819.

Castelletto, R., Castellsague, X., Munoz, N. *et al.* (1994) Alcohol, tobacco, diet, maté drinking, and esophageal cancer in Argentina. *Cancer Epidemiol. Biomarkers Prev.*, **3**, 557–564.

Cerdá, B., Tomás-Barberán, F.A. and Espín, J.C. (2005) Metabolism of antioxidant and chemopreventive ellagitannins from strawberries, raspberries, walnuts and oak-aged wine in humans. identification of biomarkers and individual variability. *J. Agric. Food Chem.*, **53**, 227–235.

Ceska, O., Chaudhary, S.K., Warrington, P. *et al.* (1988) Coriandrin, a novel highly photoactive compound isolated from *Coriandrum sativum*. *Phytochemistry*, **27**, 2083–2087.

Cheel, J., Theoduloz, C., Rodrígues *et al.* (2005) Free radical scavengers and antioxidants from lemongrass (*Cymbopogon citrates* (DC.) Stapf). *J. Agric. Food Chem.*, **53**, 2511–2517.

Chen, H.-F., Tanaka, T., Nonaka, G.-I. *et al.* (1993) Phenylpropanoid-substituted catechins from *Castanopsis hystrix* and structure revision of cinchonains. *Phytochemistry*, **33**, 183–187.

Chiu, F.L. and Lin, J.K. (2005) HPLC analysis of naturally occurring methylated catechins, $3''$- and $4''$-methyl-epigallocatechin gallate, in various fresh tea leaves and commercial teas and their potent inhibitory effects on inducible nitric oxide synthase in macrophages. *J. Agric. Food Chem.*, **53**, 7035–7042.

Cho, MJ., Howard, L.R. and Prior, R.L. (2004) Flavonoid glycosides and antioxidant capacity of various blackberry, blueberry and red grape genotypes determined by high-performance liquid chromatography/mass spectrometry. *J. Sci. Food Agric.*, **84**, 1771–1782.

Chu, Y.-H., Chang, C.-L. and Hsu, H.-F. (2000) Flavonoid content of several vegetables and their antioxidant activity. *J. Sci. Food Agric.*, **80**, 561–566.

Clarke, D.B., Barnes, K.A. and Lloyd, A.S. (2004) Determination of unusual soya and non-soya phytoestrogen sources in beer, fish products and other foods. *Food Addit. Contam.*, **21**, 949–962.

Clarke, R.J. (1985a) Green coffee processing. In M.N. Clifford and K.C. Willson (eds), *Coffee: Botany, Biochemistry and Production of Beans and Beverage*. Croom Helm, London, pp. 230–250.

Clarke, R.J. (1985b) The technology of converting green coffee into the beverage. In M.N. Clifford and K.C. Willson (eds), *Coffee: Botany, Biochemistry and Production of Beans and Beverage*. Croom Helm, London, pp. 375–393.

Clifford, M.N. (1985) Chemical and physical aspects of green coffee and coffee products. In M.N. Clifford and K.C. Willson (eds), *Coffee: Botany, Biochemistry and Production of Beans and Beverage*. Croom Helm, London, pp. 305–374.

Clifford, M.N. (1986) Phenol–protein interactions and their possible significance for astringency. In G.G. Birch and M.G. Lindley (eds), *Interactions of Food Components*. Elsevier Applied Science, London, pp. 143–164.

Clifford, M.N. (1999) Chlorogenic acids and other cinnamates – nature, occurrence and dietary burden. *J. Sci. Food Agric.*, **79**, 362–372.

Clifford, M.N. (2000a) Anthocyanins – nature, occurrence and dietary burden. *J. Sci. Food Agric.*, **80**, 1063–1072.

Clifford, M.N. (2000b) Miscellaneous phenols in foods and beverages – nature, occurrence and dietary burden. *J. Sci. Food Agric.*, **80**, 1126–1137.

Clifford, M.N. (2003) The analysis and characterization of chlorogenic acids and other cinnamates. In G. Williamson, and S. Santos-Buelga (eds), *Methods in Polyphenol Analysis*. Royal Society of Chemistry, London, pp. 314–337.

Clifford, M.N. and Jarvis, T. (1988) The chlorogenic acids content of green robusta coffee beans as a possible index of geographic origin. *Food Chem.*, **9**, 291–298.

Clifford, M.N., Johnston, K.L., Knight, S. *et al.* (2003) A hierarchical scheme for LC-MS^n identification of chlorogenic acids. *J. Agric. Food Chem.*, **51**, 2900–2911.

Clifford, M.N. and Knight, S. (2004) The cinnamoyl–amino acid conjugates of green robusta coffee beans. *Food Chem.*, **87**, 457–463.

Clifford, M.N., Knight, S., Surucu, B. *et al.* (2006) Characterization by LC-MSn of four new classes of chlorogenic acids in green coffee beans: dimethoxy-cinnamoylquinic acids, diferuloylquinic acids, caffeoyl-dimethoxycinnamoylquinic acids, and feruloyl-dimethoxycinnamoylquinic acids. *J. Agric. Food Chem.*, **54**, 1957–1969.

Clifford, M.N. and Ramìrez-Martìnez, J.R. (1990) Chlorogenic acids and purine alkaloid content of maté (*Ilex paraguariensis*) leaf and beverage. *Food Chem.*, **35**, 13–21.

Clifford, M.N. and Scalbert, A. (2000) Ellagitannins – nature, occurrence and dietary burden. *J. Sci. Food Agric.*, **80**, 1118–1125.

Clifford, M.N., Williams, T. and Bridson, D. (1989) Chlorogenic acids and caffeine as possible taxonomic criteria in *Coffea* and *Psilanthus*. *Phytochemistry*, **28**, 829–838.

Coggon, P., Moss, G.A., Graham, H.N. *et al.* (1973) The biochemistry of tea fermentation: oxidative degallation and epimerization of the tea flavanol gallates. *J. Agric. Food Chem.*, **21**, 727–733.

Cojocaru, M., Droby, S., Glotter, A. *et al.* (1986) 5-(12-Heptadecenyl)-resorcinol, the major component of the antifungal activity in the peel of mango fruit. *Phytochemistry*, **25**, 1093–1095.

Cole, R.A. (1985) Relationship between the concentration of chlorogenic acid in carrot root and the incidence of carrot fly larval damage. *Ann. Appl. Biol.*, **106**, 211–217.

Collin, B.H., Horska, A., Hotten, P.M., Riddoch, C. and Collins, A.R. (2001) Kiwi fruit protects against oxidative DNA damage in human cells and in vitro. *Nutr. Cancer*, **39**, 148–153.

Croteau, R., Kutchan, M. and Lewis, N.G. (2000) Natural products (secondary metabolites). In B.B. Buchanan, W. Gruissem and R.L. Jones (eds) *Biochemistry and Molecular Biology of Plants*. American Society of Plant Physiologists, Rockville, MD, pp. 1250–1318.

Crozier, A., McDonald, M.S., Lean, M.E.J. *et al.* (1997) Quantitative analysis of the flavonoid content of tomatoes, onions, lettuce and celery. *J. Agric. Food Chem.*, **45**, 590–595.

Crozier, A., Burns, J., Aziz, A.A. *et al.* (2000) Antioxidant flavonols from fruits and vegetables: measurements and bioavailability. *Biol. Res.*, **33**, 78–88.

Davis, A.L., Lewis, J.R., Cai, Y. *et al.* (1997) A polyphenolic pigment from black tea. *Phytochemistry*, **46**, 1397–1402.

Dégenéve, A. (2004) *Antioxidants in Fruits and Vegetables*. MSc thesis, University of Glasgow.

De Keukeleire, D. (2000) Fundamentals of beer and hop chemistry. *Quìmca Nova*, **23**, 108–112.

de Pascual-Teresa, S., Santos-Buelga, C. and Rivas-Gonzalo, J.C. (2000) Quantitative analysis of flavan-3-ols in Spanish foodstuffs and beverages. *J. Agric. Food Chem.*, **48**, 5331–5337.

del Campo, G., Santos, J.I., Berregi, I. *et al.* (2003) Ciders produced by two types of presses and fermented in stainless steel and wooden vats. *J. Inst. Brew.*, **109**, 342–348.

Donner, H., Gao, L. and Mazza, G. (1997) Separation and characterization of simple and malonylated anthocyanins in red onions. *Allium cepa* L. *Food Res. Int.*, **30**, 637–643.

Duester, K.C. (2001) Avocado fruit is a rich source of β-sitosterol. *J. Am. Diet. Assoc.*, **101**, 404–405.

Echeverri, F., Torres, F., Quiñones, W. *et al.* (1997) Danielone, a phytoalexin from papaya fruit. *Phytochemistry*, **44**, 255–256.

Edenhardner, R., Keller, G., Platt, K.L. *et al.* (2001) Isolation and characterization of structurally novel antimutagenic flavonoids from spinach (*Spinacia oleracea*). *J. Agric. Food. Chem.*, **49**, 2767–2773.

Engelhardt, U.H., Finger, A., Herzig, B. *et al.* (1992) Determination of flavonol glycosides in black tea. *Deusche Lebensmittel.*, **88**, 69–73.

Fahey, J.W., Zhang, Y. and Talalay, P. (1997) Broccoli sprouts: an exceptionally rich source of inducers of enzymes that protect against chemical carcinogens. *Proc. Natl. Acad. Sci. USA*, **94**, 10367–10373.

Fang, N., Yu, S. and Prior, R.L. (2002) LC/MS/MS characterization of phenolic compounds in dried plums. *J. Agric. Food Chem.*, **50**, 3579–3585.

Farah, A., de Paulis, T. and Trugo, L.C. (2005) Effect of roasting on the formation of chlorogenic acid lactones in coffee. *J. Agric. Food Chem.*, **53**, 1505–1513.

Fenical, W. (1975) Halogenation in the Rhodophyta. A review. *J. Phycol.*, **11**, 245–259.

Fenwick, G.R., Heaney, R.K. and Mullin, W.J. (1983) Glucosinolates and their breakdown products in food and food plants. *Crit. Rev. Food Sci. Nutr.*, **18**, 123–201.

Ferreira, D., Guyot, S., Marnet, N. *et al.* (2002) Composition of phenolic compounds in a Portuguese pear (*Pyrus communis* L. var. Bartolomeu) and changes after sun-drying. *J. Agric. Food Chem.*, **50**, 4537–4544.

Ferreres, F., Gil, M.I., Castañer, M. *et al.* (1997) Phenolic metabolites in red pigmented lettuce (*Lactuca sativa*). Changes with minimal processing and cold storage. *J. Agric. Food Chem.*, **45**, 4249–4254.

Finger, A., Engelhardt, U.H. and Wray, V. (1991) Flavonol triglycosides containing galactose in tea. *Phytochemistry*, **30**, 2057–2060.

Fisher, C. (1992) Phenolic compounds in spices. In C.-T. Ho, C.Y. Lee and M.-T. Huang (eds), *Phenolic Compounds in Foods and Their Effects on Health. 1. Analysis, Occurrence and Chemistry*. American Chemical Society Symposium Series Vol. 506. American Chemical Society, Washington, DC, pp. 118–129.

Fletcher, R.J. (2003) Food sources of phyto-oestrogens and their precursors in Europe. *Brit. J. Nutr.*, **89**, S30–S43.

Gaitan, E., Lindsay, R.H., Reichert, R. D. *et al.* (1989) Antithyroid and goitrogenic effects of millet: role of C-glycosylflavones. *J. Clin. Endocrinol. Metab.*, **68**, 707–714.

Gil, M.I., Ferreres, F. and Tomás-Barberán, F.A. (1999) Effect of postharvest storage and processing on the antioxidant constituents (flavonoids and vitamin C) of fresh cut spinach. *J. Agric. Food Chem.*, **47**, 2213–2217.

Gil, M.I., Tomás-Barberán, F.A., Hess-Pierce, B. *et al.* (2000) Antioxidant activity of pomegranate juice and its relationship with phenolic composition and processing. *J. Agric. Biol. Chem.*, **48**, 4581–4589.

Ginz, M. and Engelhardt, U.H. (2000) Identification of proline-based diketopiperazines in roasted coffee. *J. Agric. Food Chem.*, **48**, 3528–3532.

Galvis-Sánchez, A.C., Gil-Izquierdo, A. and Gil, M.I. (2003) Comparative study of six pear cultivars in terms of their phenolic and vitamin C contents and antioxidant capacity. *J. Sci. Food Agric.*, **83**, 995–1003.

Goodsall, C.W., Parry, A.D. and Safford, D.S. (1996) Investigations into the mechanism of theaflavin formation during black tea manufacture. In J. Vercauteren *et al.* (eds), *Polyphenols 96*. XVIIIe Journeés Internationales Groupe Polyphénols, Vol. 2., INRA Editions, Paris, pp. 287–288.

Gosmann, G., Guillaume, D., Taketa, A.T. *et al.* (1995) Triterpenoid saponins from *Ilex paraguariensis*. *J. Nat. Prod.*, **58**, 438–441.

Gosnay, S.L., Bishop, J.A., New, S.A. et al. (2002) Estimation of the mean intakes of fourteen classes of dietary phenolics in a population of young British women aged 20–30 years. *Proc. Nutr. Soc.*, **61**, 125A.

Grolier, P., Bartholin, G., Broers, L. et al. (1999) Composition of tomatoes and tomato products in antioxidants. In R. Bilton, M. Gerber, P. Grolier et al. (eds), *The White Book on Antioxidants in Tomatoes and Tomato Products and Their Health Benefits*. Tomato News, CMITI, Avignon, pp. 1–104.

Grolier, P., Bartholin, G., Broers, L. et al. (2001) Composition of tomatoes and tomato products in antioxidants. In R. Bilton, M. Gerger, P. Grolier and C. Leoni (eds), *The White Book on Antioxidants in Tomatoes and Tomato Products and Their Health Benefits*. Tomato News, CMITI, Avignon, pp. 1–104.

Grosch, W. (2001) Chemistry III. Volatile compounds. In R.J. Clarke and O.G. Vitzthum (eds), *Coffee: Recent Developments*. Blackwell Science, Oxford, pp. 68–89.

Gu, L., Kelm, M.A., Hammerstone, J.F. et al. (2003) Screening of foods containing proanthocyanidins and their structural characterization using LC-MS/MS and thiolytic degradation. *J. Agric. Chem.*, **51**, 7513–7521.

Gu, L., Kelm, M.A., Hammerstone, J.F. et al. (2004) Concentration of proanthocyanidins in common foods and estimates of normal consumption. *J. Nutr.*, **134**, 613–617.

Guedon, D.J. and Pasquier, B.P. (1994) Analysis and distribution of flavonoid glycosides and rosmarinic acid in 40 *Mentha* × *piperita* clones. *J. Agric. Food Chem.*, **42**, 679–684.

Guyot, S., Marnet, N., Laraba, D. et al. (1998) Reversed-phase HPLC following thiolysis for quantitative estimation and characterization of the four main classes of phenolic compounds in different tissue zones of a French cider apple variety (*Malus domestica* var. Kermerrien). *J. Agric. Food Chem.*, **46**, 1698–1705.

Guyot, S., Marnet, N., Sanoner, P. et al. (2003) Variability of the polyphenolic composition of cider apple (*Malus domestica*) fruits and juices. *J. Agric. Food Chem.*, **51**, 6240–6247.

Häkkinen, S.H., Karenlampi, S.O., Heinonen, I.M. et al. (1999) Content of flavonols quercetin, myricetin and kaempferol in 25 edible berries. *J. Agric. Food Chem.*, **47**, 2274–2279.

Hallström, H. and Thuvander, A. (1997) Toxicological evaluation of myristicin. *Nat. Toxins*, **5**, 186–192.

Hammerstone, J.F., Lazarus, S.A., Mitchell, A.E. et al. (1999) Identification of procyanidins in cocoa (*Theobroma cacao*) and chocolate using high-performance liquid chromatography/mass spectrometry. *J. Agric. Food Chem.*, **47**, 490–496.

Hampton, M.G. (1992) Production of black tea. In K.C. Willson, and M.N. Clifford (eds), *Tea: Cultivation to Consumption*. Chapman and Hall, London, pp. 459–512.

Han, L., Hatano, T., Yoshida, T. et al. (1994) Tannins of theaceous plants. V. Camelliatannins F, G and H, three new tannins from *Camellia japonica* L. *Chem. Pharm. Bull.*, **42**, 1399–1409.

Harborne, J.B., Williams, C.A., Greenham, J. et al. (1974) Distribution of charged flavones and caffeoylshikimic acid in *Palmae*. *Phytochemistry*, **13**, 1557–1559.

Hasegawa, S., Lam, L.K.T. and Miller, E.G. (2000) Citrus flavonoids: biochemistry and possible importance in human nutrition. In F. Shahidi and C.-T. Ho (eds), *Phytochemicals and Phytopharmaceuticals*. AOCS Press, Champaign, IL, pp. 95–105.

Hashimoto, F., Nonaka, G.-I. and Nishioka, I. (1986) Tannins and related compounds XXXVI. Isolation and structures of theaflagallins, new red pigments from black tea. *Chem. Pharm. Bull.*, **34**, 61–65.

Hashimoto, F., Nonaka, G.-I. and Nishioka, I. (1987) Tannins and related compounds LVI. Isolation of four new acylated flavan-3-ols from oolong tea. *Chem. Pharm. Bull.*, **35**, 611–616.

Hashimoto, F., Nonaka, G.-I. and Nishioka, I. (1988) Tannins and related compounds LXIX. Isolation and structure elucidation of B,B′-linked bisflavanoids, theasinensins D–G and oolongtheanin from oolong tea. *Chem. Pharm. Bull.*, **36**, 1676–1684.

Hashimoto, F., Nonaka, G.-I. and Nishioka, I. (1989a) Tannins and related compounds LXXVII. Novel chalcan–flavan dimers, assamaicins A, B and C, and a new flavan-3-ol and proanthocyanidins from the fresh leaves of *Camellia sinensis* L. var. *assamica* Kitamura. *Chem. Pharm. Bull.*, 37, 77–85.

Hashimoto, F., Nonaka, G.-I. and Nishioka, I. (1989b) Tannins and related compounds XC. 8C-ascorbyl-(−)-epigallocatechin-3-gallate and novel dimeric flavan-3-ols, oolonghomobisflavans A and B from oolong tea. *Chem. Pharm. Bull.*, 37, 3255–3263.

Hashimoto, F., Nonaka, G.-I. and Nishioka, I. (1992) Tannins and related compounds. CXIV. Structures of novel fermentation products, theogallinin, theaflavonin and desgalloyltheaflavonin from black tea, and changes of tea leaf polyphenols during fermentation. *Chem. Pharm. Bull.*, 40, 1383–1389.

Haslam, E. (1998) *Practical Polyphenols – From Structure to Molecular Recognition and Physiological Action*. Cambridge University Press, Cambridge.

Hatano, T., Shida, S., Han, L. *et al.* (1991) Tannins of Theaceous Plants: III. Camelliatannins A and B, two new complex tannins from *Camellia japonica* L. *Chem. Pharm. Bull.*, 39, 876–880.

Hemple, J. and Bohm, H. (1996) Quality and quantity of prevailing flavonoid glycosides in yellow and green French beans (*Phaseolus vulgaris* L.). *J. Agric. Food Chem.*, 44, 2114–2116.

Henderson, L., Gregory, J. and Swan, G. (2002) *The National Diet and Nutrition Survey: Adults Aged 19 to 64 Years. Volume 1 Types and Quantities of Food Consumed*. HMSO, London.

Herrmann, K. (1976) Flavonols and flavones in food plants: a review. *J. Food Technol.*, 11, 433–448.

Herrmann, K. (1988) On the occurrence of flavonol and flavone glycosides in vegetables. *Z. Lebensm. Unters. Forsch.*, 186, 1–5.

Hertog, M.G.L., Hollman, P.C.H. and Venema, D.P. (1992) Optimization of quantitative HPLC determination of potentially anticarcinogenic flavonoids in fruit and vegetables. *J. Agric. Food. Chem.*, 40, 1591–1598.

Hertog, M.G.L., Hollman, P.C.H. and van de Putte, B. (1993) Content of potentially anticarcinogenic flavonoids of tea infusions, wines and fruit juices. *J. Agric. Food Chem.*, 41, 1242–1246.

Hofte, A.J.P., van der Hoewen, R.A.M., Fung, S.Y. *et al.* (1998) Characterization of hop acids by liquid chromatography with negative ionization electrospray mass spectrometry. *J. Am. Soc. Brew. Chem.*, 56, 118–122.

Hong, Y.-J., Barrett, D.M. and Mitchell, A.E. (2004) Liquid chromatography/mass spectrometry investigation of the impact of thermal processing and storage on peach procyanidins. *J. Agric. Food Chem.*, 52, 2366–2371.

Hong, Y.J., Tomas-barberan, F.A., Kader, A.A. *et al.* (2006) The flavonoid glycoside and procyanidin composition of Deglet Noor dates (*Phoenix dactylifera*). *J. Agric. Food Chem.*, 54, 2405–2411.

Hussain, R.A., Poveda, L.J., Pezzuto, J.M. *et al.* (1990) Sweetening agents of plant origin: phenylpropanoid constituents of seven sweet-tasting plants. *Econ. Bot.*, 44, 174–182.

IARC (1991) *IARC Monographs on the Evaluation of Carcinogenic Risks to Humans. Volume 51: Coffee, Tea, Maté, Methylxanthines and Glyoxal*. IARC, Lyon.

Ichikawa, M., Ryu, K., Yoshida, J. *et al.* (2003) Identification of six phenylpropanoids from garlic skins as major antioxidants. *J. Agric. Food Chem.*, 51, 7313–7317.

Ito, R., Yamamoto, A., Kodama, S. *et al.* (2003) A study on the change of enantiomeric purity of catechins in green tea infusion. *Food Chem.*, 83, 563–568.

Janssen, P.L., Hollman, P.C., Venema, D.P. *et al.* (1996) Salicylates in foods. *Nutr. Rev.*, 54, 357–359.

Jayaprakasha, G.K., Rao, L.J.M. and Sakariah, K.K (2002) Improved HPLC method for the determination of curcumin, demethoxycurcumin and bisdemethoxycurcumin. *J. Agric. Chem.*, 50, 3668–3672.

Jenner, A.M., Rafter, J. and Halliwell, B. (2005) Human fecal water content of phenolics: the extent of colonic exposure to aromatic compounds. *Free Rad. Biol. Med.*, 38, 763–772.

Johnsson, P., Kamal-Eldin, A., Lundgren, L.N. *et al.* (2000) HPLC method for analysis of secoisolariciresinol diglucoside in flaxseeds. *J. Agric. Food Chem.*, **48**, 5216–5219.

Jöbstl, E., Fairclough, J.P.A., Davies, A.P. *et al.* (2005) Creaming in black tea. *J. Agric. Food Chem.*, **53**, 7997–8002.

Kahle, K., Kraus, M., Richling, E. (2005) Polyphenol profiles of apple juices. *Mol. Nutr. Food Res.*, **49**, 797–806.

Kanazawa, K. and Sakakibara, H. (2000) High content of dopamine, a strong antioxidant, in Cavendish banana. *J. Agric. Food Chem.*, **48**, 844–848.

Karadeniz, F., Durst, R.W. and Wrolstad, R.E. (2000) Polyphenolic composition of raisins. *J. Agric. Food Chem.*, **48**, 5343–5350.

Karlsson, P.C., Huss, U., Jenner, A. *et al.* (2005) Human fecal water inhibits COX-2 in colonic HT-29 cells: role of phenolic compounds. *J. Nutr.*, **135**, 2343–2349.

Kashiwada, Y., Nonaka, G. and Nishioka, I. (1986) Tannins and related compounds. XLVIII. Rhubarb. Isolation and characterization of new dimeric and trimeric procyanidins. *Chem. Pharm. Bull. (Tokyo)*, **34**, 4083–4091.

Kaundun, S.S. and Matsumoto, S. (2002) Heterologous nuclear and chloroplast microsatellite amplification and variation in tea, *Camellia sinensis*. *Genome*, **45**, 1041–1048.

Kawabata, J., Mizuhata, K., Sato, E. *et al.* (2003) 6-Hydroxyflavonoids as α-glucosidase inhibitors from marjoram (*Origanum majorana*) leaves. *Biosci. Biotechnol. Biochem.*, **67**, 445–447.

Keys, A. (1980) Wine, garlic and CHD in seven countries. *Lancet*, **i**, 145–146.

Kidmose, U., Knuthsen, P., Edelenbos, M. *et al.* (2001) Carotenoids and flavonoids in organically grown spinach (*Spinacia oleracea* L.) genotypes after deep frozen storage. *J. Sci. Food Agric.*, **81**, 918–923.

Kikuzaki, H. and Nakatani, N. (1989) Structure of a new oxidative phenolic acid from oregano (*Origanum vulgare* L.). *Agric. Biol. Chem.*, **53**, 519–524.

Kikuzaki, H., Kawai, Y. and Nakatani, N. (2001) 1,1-Diphenyl-2-picrylhydrazyl radical-scavenging active compounds from greater cardamom (*Amomum subulatum* Roxb.). *J. Nutr. Sci. Vitaminol., (Tokyo)*, **47**, 167–171.

Kim, O.K., Murakami, A. and Nakamura, Y. (2000) Novel nitric oxide and superoxide generation inhibitors, persenone A and B, from avocado fruit. *J. Agric. Food Chem.*, **48**, 1557–1563.

Kleijnen, J., Knipschild, P. and ter Reit, G. (1989) Garlic, onions and cardiovascular risk factors. A review of the evidence from human experiments with emphasis on commercially available preparations. *Br. J. Clin. Pharmacol.*, **28**, 535–544.

Kobaek-Larsen, M., Christensen, L.P., Vach, W. *et al.* (2005) Inhibitory effects of feeding with carrots or (−)-falcarinol on development of azomethane-induced preneoplastic lesions in the rat colon. *J. Agric. Food Chem.*, **53**, 1823–1827.

Lakenbrink, C., Engelhardt, U.H. and Wray, V. (1999) Identification of two novel proanthocyanidins in green tea. *J. Agric. Food Chem.*, **47**, 4621–4624.

Lakenbrink, C., Lam, T.M.L., Engelhardt, U.H. *et al.* (2000a) New flavonol triglycosides from tea (*Camellia sinensis*). *Nat. Prod. Lett.*, **14**, 233–238.

Lakenbrink, C., Lapczynski, S., Maiwald, B. *et al.* (2000b) Flavonoids and other polyphenols in consumer brews of tea and other caffeinated beverages. *J. Agric. Food Chem.*, **48**, 2848–2852.

Landrault, N., Larronde, F., Delaunay, J.C. *et al.* (2002) Levels of stilbene oligomers and astilbin in French varietal wines and in grapes during noble rot development. *J. Agric. Food Chem.*, **50**, 2046–2052.

Landrault, N., Poucheret, P., Ravel, P. *et al.* (2001) Antioxidant capacities and phenolics levels of French wines from different varieties and vintages. *J. Agric. Food Chembba.*, **49**, 3341–3348.

Lea, A.G.H. (1990) Bitterness and astringency: the procyanidins of fermented apple ciders. In R.L. Rouseff (ed.), *Bitterness in Foods and Beverages*. Elsevier Science, Amsterdam, pp. 123–144.

Lea, A.G.H. (1995) Cidermaking. In A.G.H. Lea and J.R. Piggot (eds), *Fermented Beverage Production*. Chapman & Hall, London, pp. 66–96.

Lea, A.G.H. and Arnold, G.M. (1978) The phenolics of cider: bitterness and astringency. *J. Sci. Food Agric.*, **29**, 478–483.

Leung, A.Y. and Foster, S. (1996) *Encyclopedia of Common Natural Ingredients Used in Food, Drugs and Cosmetics*, 2nd edn. Wiley, New York.

Liggins, J., Bluck, L.J.C., Runswick, S. *et al.* (2000) Daidzein and genistein contents of vegetables. *Br. J. Nutr.*, **84**, 717–725.

Lin, J.-K., Lin, C.-L., Liang, Y.-C. *et al.* (1998) Survey of catechins, gallic acid, and methylxanthines in green, oolong, pu'erh, and black teas. *J. Agric. Food Chem.*, **46**, 3635–3642.

Lindsay, D.G. and Clifford M.N. (eds) (2000) Special issue devoted to critical reviews produced within the EU Concerted Action 'Nutritional Enhancement of Plant-based Food in European Trade' (NEODIET). *J. Sci. Food Agric.*, **80**, 793–1137.

Lister, C.A. and Podivinsky, E.P. (1998) Antioxidants in New Zealand fruit and vegetables. *Polyphenol Commun.*, 273.

Llorach, R., Gil-Izquierdo, A., Ferreres, F. *et al.* (2003) HPLC-DAD-MS/MS ESI characterization of unusual highly glycosylated flavonoids from cauliflower (*Brassica oleracea* L. var. *botrytis*) agroindustrial byproducts. *J. Agric. Food Chem.*, **51**, 3895–3899.

Lu, Y. and Foo, L.Y. (2000) Flavonoid and phenolic glycosides from *Salvia officinalis*. *Phytochemistry*, **55**, 263–267.

Lu, Q.Y., Arteaga, J.R., Zhang, Q.S. *et al.* (2005) Inhibition of prostate cancer cell growth by an avocado extract: role of lipid-soluble bioactive substances. *J. Nutr. Biochem.*, **16**, 23–30.

Luximon-Ramma, A., Bahorun, T., Crozier, A. *et al.* (2005) Characterization of the antioxidant functions of flavonoids and proanthocyanidins in Mauritian black teas. *Food Res. Int.*, **38**, 357–367.

Määttä, K.R., Kamal-Eldin, A. and Törrönen, A.R. (2003) High-performance liquid chromatography (HPLC) analysis of phenolic compounds in berries with diode array and electrospray ionization mass spectrometric (MS) detection: *Ribes* species. *J. Agric. Food Chem.*, **51**, 6736–6744.

Makris, D.P. and Rossiter, J.T. (2001) Domestic processing of onion bulbs (*Allium cepa*) and asparagus spears (*Asparagus officinalis*): effect on flavonol content and antioxidant status. *J. Agric. Food Chem.*, **49**, 3216–3222.

Marin, A., Ferreres, F., Tomas-Barberán, F.A. *et al.* (2004) Characterization and quantification of antioxidant constituents of sweet pepper (*Capsicum annum* L.). *J. Agric. Food Chem.*, **52**, 3861–3869.

Menet, M.C., Sang, S., Yang, C.S. *et al.* (2004) Analysis of theaflavins and thearubigins from black tea extract by MALDI-TOF mass spectrometry. *J. Agric. Food Chem.*, **52**, 2455–2461.

Mateus, N., Carvalho, E., Carvalho, A.R.F. *et al.* (2003) Isolation and structural characterization of new acylated anthocyanin-vinyl-flavanol pigments occurring in aging red wines. *J. Agric. Food Chem.*, **51**, 277–282.

Mateus, N., Silva, A.M.S., Santos-Buelga, C. *et al.* (2002) Identification of anthocyanin-flavanol pigments in red wines by NMR and mass spectrometry. *J. Agric. Food Chem.*, **50**, 2110–2116.

McCance, R.A. and Widdowson, E.M. (1991) *The Composition of Foods*, 5th edn. Royal Society of Chemistry, London.

McGhie, T.K., Aingie, G.D., Barnet, L.E. *et al.* (2003) Anthocyanin glycosides from berry fruit are absorbed and excreted unmetabolized by both human and rats. *J. Agric. Food Chem.*, **51**, 4539–4548.

Miura, K., Kikuzaki, H. and Nakatani, N. (2002) Antioxidant activity of chemical components from sage (*Salvia officinalis* L.) and thyme (*Thymus vulgaris* L.) measured by the oil stability index method. *J. Agric. Food Chem.*, **50**, 1845–1851.

Morata, A., Gómes-Cordovés, M.C., Colomo, B. *et al.* (2003) Pyruvic acid and acetaldehyde production by different strains of *Saccharomyces cerevisiae*: relationship to vitisin A and B formation in red wines. *J. Agric. Food Chem.*, **51**, 7402–7409.

Morton, J.F. (1989) Tannins as a carcinogen in bush tea, tea, maté and khat. In J.J. Hemmingway and J.J. Karchesy (eds), *Chemistry and Significance of Condensed Tannins*. Plenum Press, New York, pp. 403–415.

Mozetic, B., Trebse, P. and Hribar, J. (2002) Determination and quantitation of anthocyanins and hydroxycinnamic acids in different cultivars of sweet cherries (*Prunus avium* L.) from Nova Gorica region (Slovenia). *Food Technol. Biotechnol.*, **40**, 207–212.

Mullen, W., McGinn, J., Lean, M.E.J. *et al.* (2002a) Ellagitannins, flavonoids, and other phenolics in red raspberries and their contribution to antioxidant capacity and vasorelaxation properties, *J. Agric. Food Chem.*, **50**, 6902–6909.

Mullen, W., Stewart, A.J., Lean, M.E.J. *et al.* (2002b) Effect of freezing and storage on the phenolics, ellagitannins, flavonoids, and antioxidant capacity of red raspberries. *J. Agric. Food Chem.*, **50**, 5197–5201.

Mullen, W., Boitier, A., Stewart, A.J. *et al.* (2004) Flavonoid metabolites in human plasma and urine after consumption of red onions: analysis by HPLC with photodiode array and full scan tandem mass spectrometric detection. *J. Chromatogr. A*, **1058**, 163–168.

Mullin W.J. and Emery J.P.H. (1992) Determination of alkylresorcinols in cereal-based foods. *J. Agric. Food Chem.*, **40**, 2127–2130

Nagel, C.W., Baranowski, J.D., Wulf, L.W. *et al.* (1979) The hydroxycinnamic acid tartaric acid ester content of musts and grape varieties grown in the Pacific Northwest. *Am. J. Enol. Vitic.* **30**, 198–201.

Nakatani, N., Jitoe, A., Masuda, T. *et al.* (1991) Flavonoid constituents of *Zingiber zerumbet* Smith. *Agric. Biol. Chem.*, **55**, 455–460.

Nonaka, G.-I., Kawahara, O. and Nishioka, I. (1982) Tannins and related compounds. VIII. A new type of proanthocyanidin, Cinchonains IIa and IIb from *Cinchona succirubra*. *Chem. Pharm. Bull.*, **30**, 4277–4282.

Nonaka, G.-I., Kawahara, O. and Nishioka, I. (1983) Tannins and related compounds. XV. A new class of dimeric flavan-3-ol gallates, theasinensins A and B, and proanthocyanidin gallates from green tea. *Chem. Pharm. Bull.*, **31**, 3906–3914.

Nonaka, G.-I. and Nishioka, I. (1982) Tannins and related compounds. VII. Phenylpropanoid-substituted epicatechins, cinchonains from *Cinchona succirubra*. *Chem. Pharm. Bull.*, **30**, 4268–4276.

Nonaka, G.-I., Sakai, R. and Nishioka, I. (1984) Hydrolysable tannins and proanthocyanidins from green tea. *Phytochemistry*, **23**, 1753–1755.

Núñez-Sellés, A.J., Vélez Castro, H.T., Agüero-Agüero, J. *et al.* (2002) Isolation and quantitative analysis of phenolic antioxidants, free sugars, and polyols from mango (*Mangifera indica* L.) stem bark aqueous decoction used in Cuba as a nutritional supplement. *J. Agric. Food Chem.*, **50**, 762–766.

Nursten, H.E. (2005) *The Maillard Reaction: Chemistry, Biology and Implications*. Springer, Heidelberg.

Opie, S.C., Clifford M.N. and Robertson, A. (1993) The role of (−)-epicatechin and polyphenol oxidase in the coupled oxidative breakdown of theaflavins. *J. Sci. Food Agric.*, **63**, 435–438.

Opie, S.C., Clifford, M.N. and Robertson, A. (1995) The formation of thearubigin-like substances by *in vitro* polyphenol oxidase-mediated fermentation of individual flavan-3-ols. *J. Sci. Food Agric.*, **67**, 501–505.

Paganga, G., Miller, N.G. and Rice-Evans, C.A. (1999) The phenolic content of fruits and vegetables and their antioxidant activities. What does a serving constitute? *FEBS Lett.*, **401**, 78–82.

Pan, G.G., Kilmartin, P.A., Smith, B.G. *et al.* (2002) Detection of orange juice adulteration by tangelo juice using multivariate analysis of polymethoxylated flavones and carotenoids. *J. Sci. Food Agric.*, **82**, 421–427.

Parejo, I., Jauregui, O., Sánchez-Rabaneda, F. *et al.* (2004) Separation and characterization of phenolic compounds in fennel (*Foeniculum vulgare*) using liquid chromatography-negative ion electrospray ionization tandem mass spectrometry. *J. Agric. Food Chem.*, **52**, 3679–3687.

Pavoro, A. (1999) *The New Kitchen Garden*. Dorling Kindersley, London.

Phippen, W.B. and Simon, J.E. (1998) Anthocyanins in basil (*Ocimum basilicum* L.). *J. Agric. Food Chem.*, **46**, 1734–1738.

Piggott, J.R., Conner, J.M., Paterson, A. *et al.* (1993) Effects on Scotch whisky composition and flavour of maturation in oak casks with varying histories. *Int. J. Food Sci. Technol.*, **28**, 303–318.

Piironen, V., Lindsay, D.G., Miettinen, T.A. *et al.* (2000) Plant sterols: biosynthesis, biological function and their importance in human nutrition. *J. Sci. Food Agric.*, **80**, 936–966.

Plumb, G.W., Price, K.R., Rhodes, M.J.C. *et al.* (1997) Antioxidant properties of the major phenolic compounds in broccoli. *Free Rad. Res.*, **27**, 429–435.

Podsedek, A., Wilskajeszka, J. and Anders, B. (1998) Characterization of proanthocyanidins of apple, quince and strawberry fruits. *Polyphenol Commun.*, 319–320.

Price, K. R., Rhodes, M.J.C. and Barnes, K.A. (1998) Flavonol glycoside content and composition of tea infusions made from commercially available teas and tea products. *J. Agric. Food Chem.*, **46**, 2517–2522.

Price, S.F., Breen, P.J., Valladao, M. *et al.* (1995) Cluster sun exposure and quercetin in Pinot-Noir grapes and wine. *Am. J. Enol. Vitic.*, **46**, 187–194.

Prior, R.L., Lazarus, S.A., Cao, G. *et al.* (2001) Identification of procyanidins and anthocyanidins in blueberries and cranberries (*Vaccinium* spp) using high performance liquid chromatography-mass spectroscopy. *J. Agric. Food Chem.*, **49**, 1270–1276.

Readers Digest Association Ltd. (1996) *The Readers Digest Guide to Vitamins, Minerals and Supplements.* Readers Digest Association Ltd., London.

Rhodes, M.J.C. and Price, K.R. (1998) Phytochemicals: classification and occurrence. In M.J. Sadler, J.J. Strain and B. Caballero (eds), *Encyclopaedia of Human Nutrition*, Vol. 3. Academic Press, New York, pp. 1539–1549.

Ricardo da Silva, J.M., Rosec, J.-P., Bourzeic, M. *et al.* (1990) Separation an quantitative determination of grape and wine proanthocyanidins by high performance liquid chromatography. *J. Sci. Food Agric.*, **53**, 85–92.

Risch, B. and Herrmann, K. (1988) Contents of hydroxycinnamic acids derivatives in citrus fruits. *Z. Leben. Unters. Forschung*, **187**, 530–34.

Rolon, P.A., Castellsague, X., Benz, M. *et al.* (1995) Hot and cold maté drinking and esophageal cancer in Paraguay. *Cancer Epidemiol. Biomarkers Prev.*, **4**, 595–605.

Rous, C. and Aldersen, B. (1983) Phenolic extraction curves for white wine aged in French and American oak barrels. *Am. J. Enol. Vitic.*, **34**, 211–215.

Rubnov, S., Kashman, Y., Rabinowitz, R., Schlesinger, M. and Mechoulam, R. (2001) Suppressers of cancer cell proliferation from fig (*Ficus caria*) resin: isolation and structure elucidation. *J. Nat. Prod.*, **64**, 993–996.

Sakamoto, Y. (1967) Flavones in green tea. Part I. Isolation and structure of flavones occurring in green tea infusion. *Agric. Biol. Chem.*, **31**, 1029–1034.

Sakamoto, Y. (1969) Flavones in green tea. Part II. Identification of isovitexin and saponarin. *Agric. Biol. Chem.*, **33**, 959–961.

Sakamoto, Y. (1970) Flavones in green tea. Part III. Structure of pigments IIIa and IIIb. *Agric. Biol. Chem.*, **34**, 919–925.

Sanbongi, C., Osakabe, N., Natsume, M. *et al.* (1998) Antioxidative polyphenols isolated from *Theobroma cacao*. *J. Agric. Food Chem.*, **46**, 454–457.

Sang, S., Tian, W., Meng, X. *et al.* (2002) Theadibenztropolone A, a new type pigment from enzymatic oxidation of (−)-epicatechin and (−)-epigallocatechin gallate and characterized from black tea using LC/MS/MS. *Tetrahedron Lett.*, **43**, 7129–7133.

Sang, S., Tian, S., Stark, R.E. *et al.* (2004) New dibenzotropolone derivatives characterized from black tea using LC/MS/MS. *Bioorg. Med. Chem.*, **12**, 3009–3017.

Sanoner, P., Guyot, S. and Drilleau, J.F. (1998) Characterization of the phenolic classes in four cider apple cultivars by thiolysis and HPLC. *Polyphenol Commun.*, 401–403.

Sanoner, P., Guyot, S., Marnet, N. *et al.* (1999). Polyphenol profiles of French cider apple varieties (*Malus domestica* sp.). *J. Agric. Food Chem.*, **47**, 4847–4853.

Santos-Buelga, C. and Scalbert, A. (2000) Proanthocyanidins and tannin-like compounds – nature, occurrence, dietary intake and effects on nutrition and health. *J. Sci. Food Agric.*, **80**, 1094–1117.

Schenker, S. (2001) Cranberries. *Nutr. Bull.*, **26**, 115–116.

Scholz, B.M. and Maier, H.G. (1990) Isomers of quinic acid and quinide in roasted coffee. *Z. Lebens.-Unters. Forschung*, **190**, 132–134.

Schotz, K. (2004) Quantification of allergenic urushiols in extracts of *Ginkgo biloba* leaves, in simple one-step extracts and refined manufactured material (EGb 761). *Phytochem. Anal.*, **15**, 1–8.

Schrader, K., Kiehne, A., Engelhardt, U.H. *et al.* (1996) Determination of chlorogenic acids with lactones in roasted coffee. *J. Sci. Food Agric.*, **71**, 392–398.

Schuster, B. and Herrmann, K. (1985) Hydroxybenzoic acid and hydroxycinnamic acid derivatives in soft fruits. *Phytochemistry*, **24**, 2761–2764.

Seo, J.S., Burri, B.J., Quan, Z. *et al.* (2005) Extraction and chromatography of carotenoids from pumpkin. *J. Chromatogr. A*, **1073**, 371–375.

Sewram, V., De Stefani, E., Brennan, P. *et al.* (2003) Maté consumption and the risk of squamous cell esophageal cancer in Uruguay. *Cancer Epidemiol. Biomarkers Prev.*, **2**, 08–513.

Shahidi, F. and Naczk, M. (1995) *Food Phenolics: Sources, Chemistry, Effects and Applications*. Technomic Publishing, Inc., Lancaster, PA.

Shan, B., Cai, Y.Z., Sun, M. *et al.* (2005) Antioxidant capacity of 26 spice extracts and characterization of their phenolic constituents. *J. Agric. Food Chem.*, **53**, 7749–7759.

Shao, W., Powell, C. and Clifford M.N. (1995) The analysis by HPLC of green, black and Pu'er teas produced in Yunnan. *J. Sci. Food Agric.*, **69**, 535–540.

Shimamura, M., Hazato, T., Ashino, H. *et al.* (2001) Inhibition of angiogenesis by humulone, a bitter acid from beer hops. *Biochem. Biophys. Res. Commun.*, **289**, 220–224.

Simpson, B.B. and Conner-Ogorzaly M. (1986) *Economic Botany: Plants of Our World*. McGraw-Hill International, New York.

Singleton, V.L. (1982) Grape and wine phenolics: background and prospects. In A.D. Webb (ed.), *Proceedings of the University of California, Davis, Grape and Wine Centennial Symposium*. University of California Press, Davis, USA, pp. 215–227.

Singleton, V.L. (1995) Maturation of wines and spirits – comparisons, facts, and hypotheses. *Am. J. Enol. Vitic.*, **46**, 98–115.

Singleton, V.L. and Kratzer, F.H. (1969) Toxicity and related physiological activity of phenolic substances of plant origin. *J. Agric. Food Chem.*, **17**, 497–511.

Slaiding, I., Junquera, M. and Walker, C. (2004) Antioxidants – Final Report 2 (Rep. No. CS 2300). Brewing Research International, Redhill, UK.

Smith, R.F. (1985) A history of coffee. In M.N. Clifford and K.C. Willson (eds), *Coffee: Botany, Biochemistry and Production of Beans and Beverage*. Croom Helm, London, pp. 1–12.

Sobel, A. (1999) *Galileo's Daughter. A Drama of Science, Fate and Love*. Fourth Estate Ltd, London.

Soler-Rivas, C., Espin, J.C. and Wichers, H.J. (2000) Oleuropein and related compounds. *J. Sci. Food Agric.*, **80**, 1013–1023.

Spanos, G.A. and Wrolstad, R.E. (1992) Phenolics of apple, pear, and white grape juices and their changes with processing and storage – a review. *J. Agric. Food Chem.*, **40**, 1478–1487.

Speer, K. and Kölling-Speer, I. (2001) Chemistry IC. Lipids. In R.J. Clarke and O.G. Vitzthum (eds), *Coffee: Recent Developments*. Blackwell Science, Oxford, pp. 33–49.

Spiro, M. and Jaganyi, D. (1993) What causes scum on tea. *Nature*, **364**, 581.

Stadler, R.H., Varga, N., Milo, C. et al. (2002) Alkylpyridiniums. 2. Isolation and quantification in roasted and ground coffees. *J. Agric. Food Chem.*, **50**, 1200–1206.

Stadler, R.H., Welti, D.H., Stämpfli, A.A. et al. (1996) Thermal decomposition of caffeic acid in model systems. Identification of novel tetraoxygenated phenylindan isomers and their stability in aqueous solution. *J. Agric. Food Chem.*, **44**, 898–905.

Stark, T. and Hofmann, T. (2005) Isolation, structure determination, synthesis, and sensory activity of N-phenylpropenoyl-L-amino acids from cocoa (*Theobroma cocao*). *J. Agric. Food Chem.*, **53**, 5419–5428.

Steiner, M., Khan, A.H., Holbert, D. et al. (1996) A double blind crossover study in moderately hypercholesterolemic men that compared the effect of aged garlic extract and placebo administration on blood lipids. *Am. J. Clin. Nutr.*, **64**, 866–870.

Steinmetz, K.A., Kushi, I.H., Bostick, R.M. et al. (1994) Vegetables, fruit and colon cancer in the Iowa Women's Health Study. *Am. J. Epidemiol.*, **139**, 1–8.

Steven, J.F., Taylor, A.W. and Deinzer, M.L. (1999) Quantitative analysis of xanthohumol and related prenylflavonoids in hops and beer by liquid chromatography-mass spectrometry. *J. Chromatogr. A*, **832**, 97–107.

Stewart, A.J., Bozonnet, S., Mullen, W. et al. (2000) Occurrence of flavonols in tomato fruits and tomato-based products. *J. Agric. Food Chem.*, **48**, 2663–2669.

Stewart, A.J., Mullen, W. and Crozier, A. (2005) On-line high-performance liquid chromatography analysis of the antioxidant activity of phenolic compounds in green and black tea. *Mol. Nutr. Food Res.*, **49**, 52–60.

Subramanian, R., Venkatesh, P., Ganguli, S. et al. (1999) Role of polyphenol oxidase and peroxidase in the generation of black tea theaflavins. *J. Agric. Food Chem.*, **47**, 2571–2578.

Sun, B., Leandro, C., Ricardo da Silva, J.M. et al. (1998) Separation of grape and wine proanthocyanidins according to their degree of polymerization. *J. Agric. Food Chem.*, **46**, 1390–1396.

Suzuki, Y., Esumi, Y. and Yamaguchi, I. (1999) Structures of 5-alkylresorcinol-related compounds in rye. *Phytochemistry*, **52**, 281–289.

Takata, R.H. and Scheuer, P.J. (1976) Isolation of glyceryl esters of caffeic and p-coumaric acids from pineapple stems. *Lloydia*, **39**, 409–411.

Takeo, T. (1992) Green and semi-fermented teas. In K.C. Willson and M.N. Clifford (eds), *Tea: Cultivation to Consumption*. Chapman & Hall, London, pp. 413–457.

Tanaka, T., Betsumiya, Y., Mine, C. et al. (2000) Theanaphthoquinone, a novel pigment oxidatively derived from theaflavin during tea fermentation. *Chem. Commun.*, 1365–1366.

Tanaka, T., Inoue, K., Betsumiya, Y. et al. (2001) Two types of oxidative dimerization of the black tea polyphenol theaflavin. *J. Agric. Food Chem.*, **49**, 5785–5789.

Tanaka, T., Matsuo, Y., Kouno, I. (2005a) A novel black tea pigment and two new oxidation products of epigallocatechin-3-O-gallate. *J. Agric. Food Chem.*, **53**, 7571–7578.

Tanaka, T., Mine, C., Inoue, K. et al. (2002b) Synthesis of theaflavin from epicatechin and epigallocatechin by plant homogenates and role of epicatechin quinone in the synthesis and degradation of theaflavin. *J. Agric. Food Chem.*, **50**, 2142–2148.

Tanaka, T., Mine, C. and Kuono, I. (2002c) Structures of two new oxidation products of green tea polyphenols generated by model tea fermentation. *Tetrahedron*, **58**, 8851–8856.

Tanaka, T., Mine, C., Watarumi, S. et al. (2002a) Accumulation of epigallocatechin quinone dimers during tea fermentation and formation of theasinensins. *J. Nat. Prod.*, **65**, 1582–1587.

Tanaka, T., Watarumi, S., Fujieda, M. *et al.* (2005b) New black tea polyphenol having *N*-ethyl-2-pyrrolidinone moiety derived from tea amino acid theanine: isolation, characterization and partial synthesis. *Food Chem.*, **93**, 81–87.

Tanaka, T., Watarumi, S., Matsuo, Y. *et al.* (2003) Production of theasinensins A and D, epigallocatechin gallate dimers of black tea, by oxidation-reduction dismutation of dehydrotheasinensin A. *Tetrahedron*, **59**, 7939–7947.

Tannahill, R. (1988) *Food in History*. Penguin, London.

Tomás-Barberán, F.A. and Clifford M.N. (2000) Dietary hydroxybenzoic acid derivatives – nature, occurrence and dietary burden. *J. Sci. Food Agric.*, **80**, 1024–1032.

Tomás-Barberán, F.A., Gil, M.I., Cremin, P. *et al.* (2001) HPLC-DAD-ESIMS analysis of phenolic compounds in nectarines, peaches and plums. *J. Agric. Food Chem.*, **49**, 4748–4760.

Trevisan, M.T., Pfundstein, B., Haubner, R. *et al.* (2005) Characterization of alkyl phenols in cashew (*Anacardium occidentale*) products and assay of their antioxidant capacity. *Food Chem. Toxicol.*, **44**, 189–197.

Turner, R.B., Lindsey, D.L., Davis, D.D., *et al.* (1975) Isolation and identification of 5,7-dimethoxyisoflavone, an inhibitor of *Aspergillus flavus* from peanuts. *Mycopathologia*, **57**, 39–40.

Urgert, R. and Katan, M.B. (1997) The cholesterol-raising factor from coffee beans. *Ann. Rev. Nutr.*, **17**, 305–324.

Urgert, R., Meyboom, S., Kuilman, M. *et al.* (1996) Comparison of effect of cafetiere and filtered coffee on serum concentrations of liver aminotransferases and lipids: six month randomized controlled trial. *Br. Med. J.*, **313**, 1362–1366.

Urgert, R., Weusten-van der Wouw, M.P., Hovenier, R. *et al.* (1997) Diterpenes from coffee beans decrease serum levels of lipoprotein(a) in humans: results from four randomized controlled trials. *Eur. J. Clin. Nutr.*, **51**, 431–436.

US Department of Agriculture, Agricultural Research Service (2002) USDA-Iowa State University Database on the Isoflavone Content of Foods. Nutrient Data Laboratory Web Site (http://www.nal.usda.gov/fnic/foodcomp/Data/isoflav/isoflav.html).

US Department of Agriculture, Agricultural Research Service (2003) USDA Database on the Flavonoid Content of Selected Foods. Nutrient Data Laboratory Web Site (http://www.nal.usda.gov/fnic/foodcomp/Data/flav/flav.html).

US Department of Agriculture, Agricultural Research Service (2004) USDA Database on the Procyanidin Content of Selected Foods. Nutrient Data Laboratory Web Site (http://www.nal.usda.gov/fnic/foodcomp/Data/PA/PA.html).

Vallejo, F. and Tomás-Barberán, F.A. (2004) Characterization of flavonols in broccoli (*Brassica oleracea* L. var. italica) by liquid chromatography-UV diode array detection-electrospray ionization mass spectrometry. *J. Chromatogr. A*, **1054**, 181–193.

Vallejo, F., Tomás-Barberán, F.A. and García-Viguera, C. (2003) Health-promoting compounds in broccoli as influenced by refrigerated transport and retail sale period. *J. Agric. Food Chem.*, **51**, 3029–3034.

van de Berg, H., Faulks, R., Fernando Granado, H. *et al.* (2000) The potential for the improvement of carotenoids levels in foods and the likely systemic effects. *J. Sci. Food Agric.*, **80**, 880–912.

Veash, N. (1998) Coming soon: the poor man's Viagra. *The Observer*, News Section, 13 December, p. 8.

Venema, D.P., Hollman, P.C.H., Janssen, K.P.L.T.M. *et al.* (1996) Determination of acetylsalicylic acid and salicylic acid in foods using HPLC with fluorescence detection. *J. Agr. Food Chem.*, **44**, 1762–1767.

Vinson, J.A., Proch, J. and Zubick, L. (1999) Phenol antioxidant quantity and quality in foods: cocoa, dark chocolate and milk chocolate. *J. Agric. Food Chem.*, **47**, 4822–4824.

Voirin, B. and Bayet, C. (1992) Developmental variations in leaf flavonoid aglycones of *Mentha* × *piperita*. *Phytochemistry*, **31**, 2299–2304.

Wan, X. C., Nursten, H.E., Cai, Y. *et al.* (1997) A new type of tea pigment from the chemical oxidation of epicatechin gallate and isolated from tea. *J. Sci. Food Agric.*, **74**, 401–408.

Wang, H. and Helliwell, K. (2000) Epimerization of catechins in green tea infusions. *Food Chem.*, **70**, 337–341.

Wang, M., Simon, J.E., Aviles, I.F. *et al.* (2003a) Analysis of antioxidative phenolic compounds in artichoke (*Cynara scolymus* L.). *J. Agric Food Chem.*, **51**, 601–608.

Wang, M., Tadmore, Y., Qing-Li, W. *et al.* (2003b) Quantification of protodioscin and rutin in asparagus shoots by LC/MS and HPLC methods. *J. Agric. Food Chem.*, **51**, 6132–6136.

Waterhouse, A. and Teissedre, P.-L. (1997) Levels of phenolics in California varietal wines. In T.R. Watkins (ed.), *Wine: Nutritional and Therapeutic Benefits*. ACS Symposium Series 661, American Chemical Society, Washington, DC, pp. 12–23.

Watkins, T.R. (1997) *Wine: Nutritional and Therapeutic Benefits*. ACS Symposium Series 661. American Chemical Society, Washington, DC, USA.

Wen, L., Wrolstad, R.E. and Hsu, V.L. (1999) Characterization of sinapyl derivatives in pineapple (*Ananas comosus* [L.] Merill) juice. *J. Agric. Food Chem.*, **47**, 850–853.

Whiteman, K. (1999) *The Definitive Guide to the Fruits of the World*. Hermes House, London.

Williams, C.L. (1995) Healthy eating: clarifying advice about fruits and vegetables. *Brit. Med. J.*, **310**, 1453–1455.

Willson, K.C. (1999) *Coffee, Cocoa and Tea*. CABI Publishing, Wallingford, UK.

Winter, M. and Herrmann, K. (1986) Esters and glucosides of hydroxycinnamic acids in vegetables. *J. Agric. Food Chem.*, **34**, 616–620.

Woods, E., Clifford, M.N., Gibbs, M. *et al.* (2003) Estimation of mean intakes of 14 classes of dietary phenols in a population of male shift workers. *Proc. Nutr. Soc.*, **62**, 60A.

Wu, X., Gu, L., Prior, R.L. *et al.* (2004) Characterization of anthocyanins and proanthocyanidins in some cultivars of *Ribes, Aronia*, and *Sambucus* and their antioxidant capacity. *J. Agric. Food Chem.*, **52**, 7846–7856.

Wu, X. and Prior, R.L. (2005) Systematic identification and characterization of anthocyanins by HPLC-ESI-MS/MS in common foods in the United States: fruits and berries. *J. Agric. Food Chem.*, **53**, 2589–2599.

Yoshida, T., Nakazawa, T., Hatano, T. *et al.* (1994) Tannins from theaceous plants. 6. A dimeric hydrolyzable tannin from *Camellia oleifera*. *Phytochemistry*, **37**, 241–244.

Yoshimi, N., Matsunaga, K., Katayama, M. *et al.* (2001) The inhibitory effects of mangiferin, a naturally occurring glucosylxanthone, in bowel carcinogenesis of male F344 rats. *Cancer Lett.*, **163**, 163–170.

Zane, E. and Wender, S.H. (1961) Flavonols in spinach leaves. *J. Org. Chem.*, **26**, 4718–4719.

Zhou, Z.H., Zhang, Y.J., Xu, M. *et al.* (2005) Puerins A and B, two new 8-C-substituted flavan-3-ols from pu'er tea. *J. Agric. Food Chem.*, **53**, 8614–8617.

Zidorn, C., Johrer, K., Ganzera, M. *et al.* (2005) Polyacetylenes from the Apiaceae vegetables carrot, celery, fennel, parsley, and parsnips and their cytotoxic activities. *J. Agric. Food Chem.*, **53**, 2518–2523.

Chapter 8
Absorption and Metabolism of Dietary Plant Secondary Metabolites

Jennifer L. Donovan, Claudine Manach, Richard M. Faulks and Paul A. Kroon

8.1 Introduction

The food supply contains numerous classes of plant secondary metabolites that may possess biological activity. The potential beneficial heath effects of these food constituents will be highly dependent upon their uptake from foods, their metabolism and their disposition in target tissues and cells. This chapter presents an overview of the mechanisms that regulate the absorption and metabolism of the most commonly consumed secondary metabolites present in the food supply including flavonoids, hydroxycinnamates, phenolic acids, dihydrochalcones, betalains, glucosinolates and carotenoids. Structures of representative compounds from each of these classes are shown below in Figure 8.1.

8.2 Flavonoids

Flavonoids are a class of plant secondary metabolites that are abundant components of fruit and vegetables. They are divided into five major subclasses: flavonols, flavan-3-ols (monomers and proanthocyanidins), flavones, flavanones and anthocyanins (Manach *et al.* 2004). Isoflavones are a class of dietary polyphenols present in soya-based foods. The structures of representative compounds from each of these polyphenol subclasses are shown in Figure 8.2 (also see Chapter 1). Flavonoids and isoflavones have received attention due to their potent antioxidant activity and possible role in the prevention of cancer, cardiovascular, neurodegenerative and infectious diseases as well as osteoporosis.

The metabolism and pharmacokinetics of flavonoids has been an area of active research in the last decade. Although the majority of flavonoids are absorbed in some form after consumption from foods and beverages, most flavonoids are present in blood and tissues as glucuronidated, sulphated and methylated conjugates of the aglycones in vivo, and not in the forms that are present in foods. Plasma levels of flavonoids and metabolites along with other pharmacokinetic properties have been documented in humans after short-term feeding studies in young, healthy volunteers. Data from animal models and in vitro experiments have increased our understanding of the mechanisms that regulate the bioavailability of flavonoids. The focus of this chapter is to review the current knowledge regarding the mechanisms that regulate the bioavailability and metabolism of common

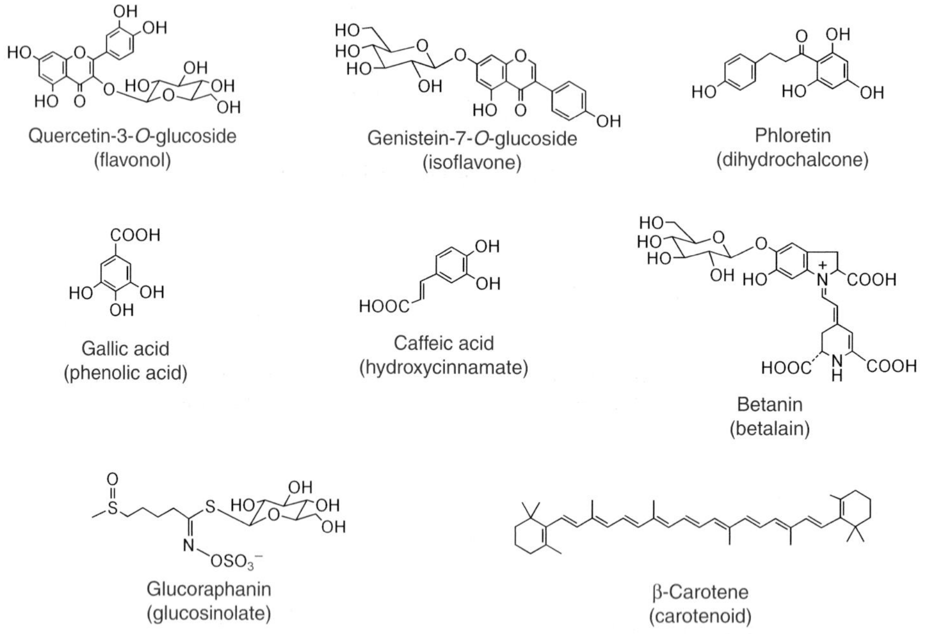

Figure 8.1 Representative structures of compounds in each of the main classes of dietary plant secondary metabolites.

dietary flavonoids, to present the specific flavonoid metabolites that have been identified in vivo thus far and to provide a summary of the pharmacokinetics of flavonoids and metabolites after consumption of typical flavonoid-rich food sources. Although not the primary focus of this chapter, data regarding the bioavailability of isoflavones is also presented as their bioavailability is probably the best understood of all the classes of dietary polyphenols.

8.2.1 Mechanisms regulating the bioavailability of flavonoids

8.2.1.1 Absorption

Multiple factors appear to influence the rate and extent of intestinal absorption of flavonoids. The majority of flavonoids are present as various glycosides in foods and in the diet (see Chapter 7). The hydrolysis of the glycoside moiety is a requisite step for absorption. The type of sugar attached to the flavonoid is the most important determinant of the site and extent of absorption, but the position of the sugar affects the mechanisms involved in intestinal uptake. Polymerization and galloylation are common in the flavan-3-ol subclass and this will also significantly affect the intestinal absorption of this group of compounds. Members of the adenosine triphosphate (ATP)-binding cassette (ABC) superfamily of transporters including multidrug resistance protein (MRP) and P-glycoprotein (P-gp) are involved in regulating the intestinal efflux of some flavonoids and ultimately influence the net amount that is absorbed into systemic circulation.

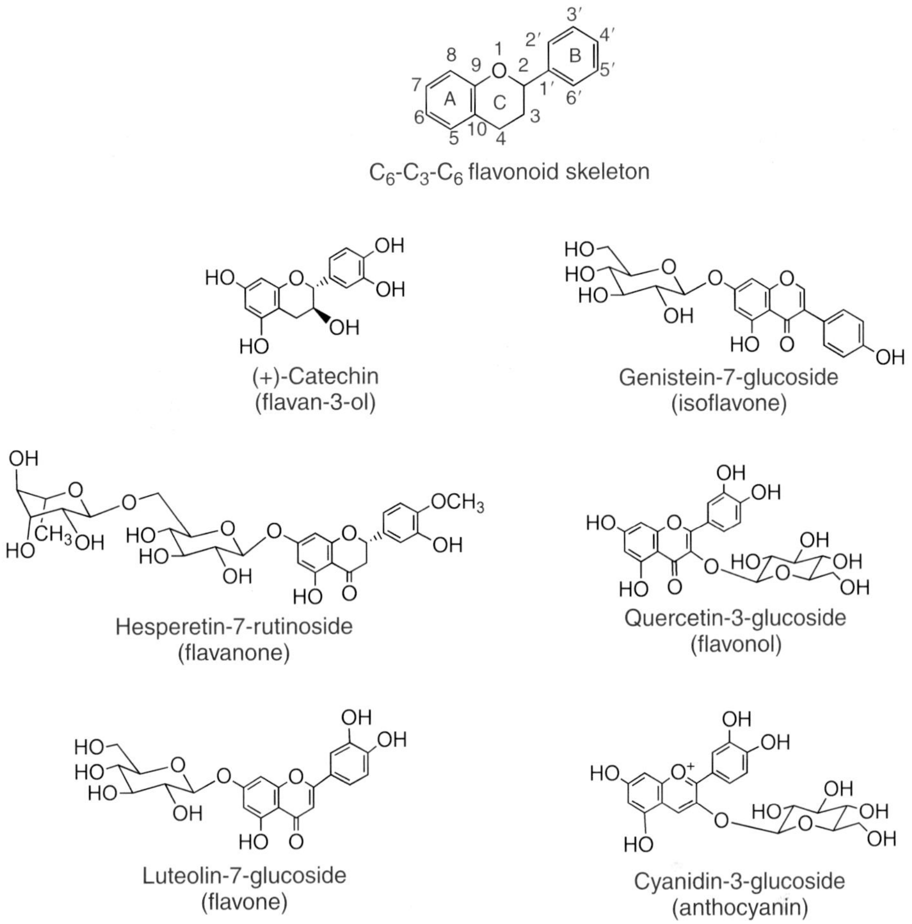

Figure 8.2 Numbering of the C_6-C_3-C_6 flavonoid skeleton and the chemical structures of the main subclasses of flavonoids and isoflavonoids present in the diet.

Glycosylated flavonoids
Almost all flavonoids, with the exception of flavan-3-ols, are present in the diet as various glycosides. These glycosides must be hydrolysed before significant absorption by the small intestine can occur. Of the flavonoid glycosides, the flavonol quercetin has been the most extensively studied, but it appears that other flavonoid subclasses follow similar mechanisms. Quercetin glycosides are hydrolysed by endogenous mammalian β-glucosidases present in the small intestine. Lactase phloridzin hydrolase (LPH) is present on the brush-border of small intestine epithelial cells (Day et al. 2000). This enzyme has a broad specificity for β-glucosides conjugated to flavonoids and will hydrolyse the glucoside to the aglycone prior to absorption. The resulting aglycone may then enter epithelial cells by passive diffusion due to its increased hydrophobicity, and this process is possibly enhanced by proximity to the cellular membrane. Alternatively, a broad specificity cytosolic β-glucosidase (CBG) has been identified in epithelial cells (Day et al. 1998). For hydrolysis

to occur by CBG, the polar glucosides would need to be transported into epithelial cells, possibly by the active sodium-dependent glucose transporter SGLT1 (Gee et al. 2000).

The relative contributions of 'LPH/diffusion' and 'transport/CBG' depend on the position of glycosylation. Quercetin-3-glucoside is not a substrate for CBG (Day et al. 1998) and larger molecules such as quercetin-3,4'-O-diglucoside interact poorly with the sugar transporter whereas quercetin-4'-O-glucoside is a good substrate for CBG (Gee et al. 2000). The absorption of quercetin-4'-O-glucoside appears to follow both 'LPH/diffusion' and 'transport/CBG' pathways, whereas quercetin-3-O-glucoside follows only the 'LPH/diffusion' pathway (Day et al. 2003).

Quercetin rhamnoglucosides (e.g. rutin) are substrates for neither CBG nor LPH and, before absorption, must be deglycosylated by microfloral rhamnosidases and β-glucosidases present in the colon. Some quercetin appears to be degraded during this process. Although the product of hydrolysis that ultimately enters epithelial cells is the same as for the glucosides in the small intestine (i.e. quercetin aglycone), absorption of rhamnoglucosides is delayed and appears to be ultimately less efficient (Hollman et al. 1999; Jaganath et al. 2006).

Until recently, it appeared that anthocyanins differed from other glycosylated flavonoids as they were present in their native, glycosylated forms in plasma rather than as glucuronate and sulphate conjugates (Lapidot et al. 1998; Cao et al. 2001). However, glucuronate and sulphate conjugates of anthocyanins have now been identified as the major metabolites of anthocyanins in human urine using high performance liquid chromatography-tandem mass spectrometry (HPLC-MS-MS) analysis (Wu et al. 2002; Felgines et al. 2003). Thus, they also appear to be deglycosylated during absorption. The mechanism of deglycosylation has not been well studied although anthocyanins were not substrates for CBG (Day et al. 1998).

The mechanisms involved in absorption of the other glycosylated flavonoid classes have not been studied as extensively. Fuhr and Kummert (1995) first suggested that naringenin-7-neohesperidoside (naringin) cleavage was required for absorption. More recently, it was demonstrated that the flavanones hesperetin-7-O-rutinoside (hesperidin) and naringenin-7-O-rutinoside (narirutin) were present in plasma exclusively in conjugated form also indicating the hydrolysis of the glycoside moiety prior to absorption. The time to reach the maximum plasma concentration (T_{max}) of these compounds was between 5 and 6 h indicating that the rutinose moiety was hydrolysed in the distal part of the intestine by microflora rather than by an endogenous enzyme (Manach et al. 2003). Studies using the rat everted small intestine demonstrate that luteolin-7-O-glucoside, a flavone, is absorbed after hydrolysis to luteolin aglycone (Shimoi et al. 1998). The isoflavones are similar to the flavonoids in that they undergo hydrolysis prior to intestinal absorption (Setchell et al. 2002). Thus, evidence exists that representative compounds from all of the classes of flavonoid glycosides are hydrolysed to the aglycones prior to or during intestinal absorption.

Non-Glycosylated Flavonoids: The Flavan-3-ols
Flavan-3-ols are the only subclass of flavonoids that are not present as glycosides in the diet. The monomers are (+)-catechin, (−)-epicatechin and (−)-epiga llocatechin and the gallate esters (−)-epicatechin gallate and (−)-epigallocatechin gallate. Although the stereochemistry of flavan-3-ol monomers in raw plant foods is well defined (see Chapter 1) there is evidence that epimerisation at C2 may occur during metabolism (Yang et al. 2000). These epimers if present would not be resolved by routine chromatographic procedures.

Mammalian metabolites are usually assumed to have the same stereochemistry as the substance fed, but this can only be determined following isolation of the metabolite. Even when a synthetic and stereochemically-defined standard is available it does not define the stereochemistry of a substance in plasma or urine. Because of this uncertainty, in this chapter we define a large portion of flavan-3-ols are present in foods in the form of oligomers and polymers (proanthocyanidins). Galloylation and polymerization appear to have dramatic effects on the extent of intestinal absorption.

Monomers. The T_{max} values after feeding (+)-catechin, (−)-epicatechin and (−)-epigallocatechin gallate observed in human studies indicate that these compounds are absorbed mainly in the small intestine (Donovan *et al.* 1999; Richelle *et al.* 1999; Lee *et al.* 2002). Perfusion studies of the rat small intestine demonstrate that both (+)-catechin and (−)-epicatechin were efficiently absorbed by the jejunum and ileum (Kuhnle *et al.* 2000; Donovan *et al.* 2001). The possibility of hydrolysis of the gallate moiety prior to absorption of the flavanol gallate esters has been suggested. Cleavage of (−)-epigallocatechin gallate can occur in the mouth and the esterase responsible, identified in saliva, is thought to be derived from human oral epithelial cells. This esterase was not reported to be present in human plasma or liver (Yang *et al.* 1999). Amelsvoort *et al.* (2001) showed that only 3% of (−)-epicatechin gallate and 5% of (−)-epigallocatechin gallate were de-galloylated in plasma after the administration of purified green tea flavan-3-ols to humans.

The effect of dose on catechin absorption was also studied using an in situ intestinal perfusion in the rat (Donovan *et al.* 2001). Approximately one-third of the catechin dose was absorbed at all concentrations ranging from 1 to 100 μM, suggesting that absorption of (+)-catechin by the small intestine is directly proportional to the dose over concentrations expected to occur in the small intestine after customary dietary intakes. The data suggested that (+)-catechin enters intestinal epithelial cells by passive diffusion, a mechanism that is generally proportional to the dose.

Oligomers and polymers. Oligomeric and polymeric proanthocyanidins do not appear to be significantly absorbed in the small intestine although procyanidin B_1 (epicatechin-(4β-8)-catechin) and B_2 (epicatechin-(4β-8)-epicatechin) have both been detected at low levels in human plasma (Holt *et al.* 2002; Sano *et al.* 2003). The dimer B_5 (epicatechin-(4β-6)-epicatechin) could not be detected in plasma after chocolate consumption (Holt *et al.* 2002). No studies have detected proanthocyanidins with degrees of polymerization larger than the dimers in plasma. The absorption of proanthocyanidins with different degrees of polymerization was investigated with colonic carcinoma (caco-2) cells, a commonly used model of intestinal absorption. Monomeric (+)-catechin and low molecular weight proanthocyanidins could be absorbed by the small intestine but larger molecular weight proanthocyanidins were not absorbed (Déprez *et al.* 2001).

It has been suggested that proanthocyanidins may be broken down into bioavailable monomers by acid present in the stomach (Spencer *et al.* 2000). This depolymerization would in turn release significant quantities of the monomer for absorption. The degree of polymerization of proanthocyanidins has been reported to decrease in the rat intestine (Abia and Fry 2001), and an energy dependent cleavage of dimers was described in the rat small intestine (Spencer *et al.* 2001). However, in vivo experiments have failed to support the theory that proanthocyanidins are cleaved to bioavailable monomers. When rats were fed purified procyanidin B_3 (catechin-(4α-8)-catechin), no catechin or metabolites could be detected in plasma or urine (Donovan *et al.* 2002b). Two studies that investigated the absorption of (+)-catechin, (−)-epicatechin and proanthocyanidins derived from a grape seed extract demonstrated that plasma levels of catechin or epicatechin were not different when these monomers were consumed alone or along with oligomeric procyanidins (Donovan *et al.* 2002b; Nakamura and Tonogai 2003). These data show that oligomeric procyanidins were not cleaved into bioavailable monomers at any point during the digestive process in the rat, a conclusion supported by the results of Tsang *et al.* (2005). A recent clinical study demonstrated that when subjects consumed a cocoa beverage containing proanthocyanidins, no change in the relative amounts of monomers and proanthocyanidins was observed in the stomach contents. These data demonstrate that procyanidins were stable in the human stomach (Rios *et al.* 2002).

The conflicting results obtained in vitro versus in vivo may be due to a buffering effect of the food bolus making the condition in the stomach less acidic than required for proanthocyanidin depolymerization (Rios *et al.* 2002). However, proanthocyanidins were not cleaved in the rat stomach even when a grape seed extract was administered by gavage without food (Nakamura and Tonogai 2003). Whether acid-catalysed proanthocyanidin depolymerization occurs in humans after consumption of proanthocyanidins without food (e.g. after consumption from dietary supplements) remains to be determined.

8.2.1.2 Intestinal efflux of absorbed flavonoids

Intestinal excretion is an important mechanism that limits the absorption of certain flavonoids (Crespy *et al.* 1999; Walle *et al.* 1999; Walgren *et al.* 2000). Conjugated metabolites formed in the small intestine are actively effluxed back into the intestinal lumen by interaction with membrane-bound transporters in the ABC family. The efflux of quercetin and epicatechin metabolites is thought to occur by MRP2, located on the luminal side of epithelial cells (Walgren *et al.* 2000; Vaidyanathan and Walle 2001). Studies using specific inhibitors with Caco-2 cells, Chinese hamster ovary cells and Madin-Darby canine kidney cells indicate that the monocarboxylate transporter P-gp and MRP(1/2) all play important roles in regulating the cellular uptake of (−)-epicatechin gallate (Vaidyanathan and Walle 2003). In the Loc-I-Gut® model in which the human jejunum is temporarily isolated by the insertion of inflatable balloons, quercetin-3′-*O*-glucuronide was selectively excreted into the lumen, presumably leaving other metabolites such as the 3- and 7-*O*-glucuronides available for systemic circulation (Petri *et al.* 2003). Thus, ABC transport at the intestinal level not only limits the net absorption of quercetin but may affect the pattern of conjugates present in vivo.

The amount of active intestinal efflux of flavonoids representing the major subclasses of flavonoids was studied after in situ intestinal perfusion (Crespy *et al.* 2003). For quercetin, 52% of the perfused dose was re-excreted back into the lumen, whereas only 10–20% of the dose was re-excreted for luteolin, eriodicytol and kaempferol (Crespy *et al.* 2003). (+)-Catechin, alternatively, did not appear to be a substrate for these efflux transport proteins (Donovan *et al.* 2001). Some studies suggest significant differences exist in the extent of absorption between (+)-catechin and (−)-epicatechin (Holt *et al.* 2002). Other studies indicate that although epigallocatechin gallate was present in plasma after green tea consumption, epicatechin gallate was not detected in plasma. The reasons for the apparent differences in the net absorption of some of these flavonoids may be more dependent upon interaction with efflux transporters than on the amount of intestinal absorption.

ABC transport proteins are expressed in many tissues besides the intestine including the liver, kidney and at the luminal membranes of the blood-brain barrier. Thus, the interaction between flavonoids and ABC transporters will not only affect the extent of intestinal absorption and the pattern of metabolites entering circulation but will also play a role in the distribution of flavonoids and metabolites to some of their target sites of action.

8.2.1.3 Metabolism

Flavonoids undergo extensive metabolism prior to entry into systemic circulation. Most flavonoids exist predominantly, or even exclusively, as metabolites conjugated with combinations of glucuronate, sulphate or methyl groups after consumption in common foods (Lee *et al.* 1995; Donovan *et al.* 1999). The conjugation reactions occur within various tissues and cells. The intestine and the liver appear to be the most important organs involved in flavonoid metabolism, although other organs such as the kidney may also contribute. Glucuronidation occurs on the luminal side of the endoplasmic reticulum by uridine-5′-diphosphate glucuronosyltransferases (UGTs), a superfamily of enzymes. Sulphation and methylation both occur in the cytosol by sulphotransferases (SULT) and catechol-O-methyltransferases. The specific isoforms involved have not been identified. The UGT1A family is thought to be responsible for glucuronidation of flavonoids (Cheng *et al.* 1999). SULT1A1 and SULT1A2 are implicated in the sulphation of phenol-type substrates and SULT1A1 and SULT1A3 were determined to be responsible for (−)-epicatechin sulphation (Ghazali and Waring 1999; Vaidyanathan and Walle 2002).

Small intestinal metabolism
All flavonoids when consumed in the diet will first be exposed to the small intestine. The small intestine is thought to be the major organ of glucuronidation of many flavonoids. Crespy and colleagues demonstrated that quercetin could be glucuronidated by microsomal preparations of the rat small intestine and that the jejunum had a higher metabolic capacity for glucuronidation than the ileum (Crespy *et al.* 1999). Quercetin was also extensively glucuronidated by the rat small intestine (Crespy *et al.* 1999 ; Gee *et al.* 2000). Petri *et al.* (2003) showed extensive glucuronidation using a human in vivo intestinal perfusion technique, the Loc-I-Gut® human in situ intestinal perfusion model (Petri *et al.* 2003). In situ perfusion studies in the rat also indicate that the small intestine is the most important organ of glucuronidation of (+)-catechin and (−)-epicatechin. Those studies also demonstrated that methylation occurs in the small intestine, although to a lesser extent than glucuronidation (Kuhnle *et al.* 2000; Donovan *et al.* 2001). Piskula and Terao (1998) measured the activity of microsomal preparations of kidney, lung, intestine and plasma using (−)-epicatechin as a substrate and showed that the intestine had the highest capacity for glucuronidation, with approximately ten times the activity of the liver. Studies using the rat everted small intestine demonstrated that luteolin is also glucuronidated during the absorption process (Shimoi *et al.* 1998). There is little evidence of sulphation by the small intestine as it does not occur significantly in the rat small intestine. SULT activity is characteristically much higher in humans than in rats (Pacifici *et al.* 1988; Dunn and Klaassen 1998). In addition, sulphate conjugates of (−)-epicatechin were formed using Caco-2

cell model indicating their possible formation in the human intestine (Vaidyanathan and Walle 2001).

Hepatic metabolism
After absorption and intestinal metabolism, the major products in the hepatic portal vein are most certainly glucuronides and perhaps methylated glucuronides. There is now strong evidence that these polar conjugates gain access to hepatocytes and are further modified therein. Studies in HepG2 cells, an established model of human hepatic metabolism, demonstrated that quercetin glucuronides can be taken up and methylated on the catechol ring (O'Leary et al. 2003). In situ perfusion studies demonstrated that catechin glucuronides formed in the rat small intestine were subsequently sulphated, as well as methylated, in the liver (Kuhnle et al. 2000; Donovan et al. 2001).

Studies in HepG2 cells also showed that certain glucuronides can be hydrolysed and then re-glucuronidated at a different position or conjugated with sulphate. Hydrolysis has been attributed to β-glucuronidase activity within hepatocytes (O'Leary et al. 2003). It is unknown whether β-glucuronidases are also active on flavan-3-ol glucuronides. However, in spite of the high capacity for glucuronidation of (+)-catechin by the rat small intestine, a large proportion of catechin is unconjugated in liver tissue after feeding rats (+)-catechin (Manach et al. 1999). The extent to which mammalian β-glucuronidases mediate flavonoid metabolism, as well as a transient exposure of aglycone within cells, deserves further exploration.

Although quercetin and flavan-3-ol metabolites are clearly able to enter hepatocytes, the mechanism of uptake into hepatocytes is unknown. The organic anionic transport polypedtide-2 (OATP2) has been implicated in the uptake of glucuronides into the liver. OATP2 is not present in HepG2 cells, but inhibitors of OATP2 reduced uptake in this model indicating the presence of a similar but as yet unidentified transporter (O'Leary et al. 2003). Within the liver, SULT activity is predominant but further methylation may also occur. Catechin glucuronides were subsequently sulphated, as well as methylated in rat liver after in situ perfusion (Donovan et al. 2001). Piskula and Terao (1998) also reported that SULT activity for (−)-epicatechin was present exclusively in liver.

8.2.1.4 Elimination

After consumption of flavonoids only a very small fraction of the dose is typically recovered in urine as forms containing the intact flavonoid ring. Indirect evidence of elimination by bile in humans, along with animal models, supports the theory that elimination in bile is quantitatively the most important route of elimination for some or most flavonoids. Crespy and colleagues (2003) have compared the biliary excretion of flavonoids representing the major subclasses using the in situ perfusion of the rat small intestine. In this model the bile duct is cannulated prior to perfusion and biliary secretion can be determined. The authors found that the flavanone eriodicytol had the highest elimination in bile followed by luteolin, kaempferol, quercetin and then (+)-catechin which had minor elimination by this route. The flavonoids eliminated in bile were always present as conjugated metabolites. Studies in rats demonstrated that (+)-catechin was eliminated in bile mainly as glucuronide conjugates of 3'-*O*-methylcatechin, although glucurono-sulpho-conjugates have also been detected (Shaw and Griffiths, 1980; Donovan et al. 2001). Finally, a clinical study showed that after an intravenous dose of ^{14}C-labelled quercetin, a substantial

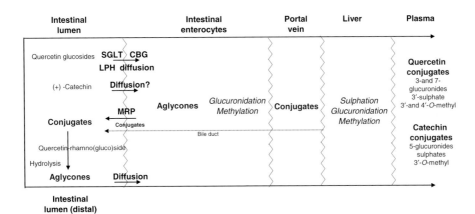

Figure 8.3 A schematic representation of the general mechanisms of flavonoid absorption, metabolism, and elimination using quercetin glycosides and (+)-catechin as examples. MRP, multidrug resistance protein; CBG, cytosolic β-glucosidase; LPH, lactase phloridzin hydrolase; SGLT, sodium-dependent glucose transporter.

proportion of the dose was later metabolized into ^{14}C-carbon dioxide, presumably by microflora in the large intestine (Walle *et al.* 2001b).

8.2.2 Overview of mechanisms that regulate the bioavailability of flavonoids

A schematic representation of the mechanisms of flavonoid absorption, metabolism and elimination using (+)-catechin and quercetin as examples is shown in Figure 8.3. Glycosylated flavonoids are either hydrolysed by LPH and are then absorbed by passive diffusion, or are transported into epithelial cells and hydrolysed therein by CBG. Certain glycosides are not hydrolysed until reaching the large intestine and after exposure to microfloral enzymes. The flavonoids that are absorbed by the intestine are extensively glucuronidated and sometimes methylated therein. They may be effluxed back into the intestinal lumen or delivered to the portal blood. After delivery to the portal blood, metabolites are able to enter into hepatocytes and are further metabolized before entry into circulation or elimination in bile. The unabsorbed flavonoids along with those actively excreted by the small intestine and by bile will reach the distal portion of the intestine where they become available for metabolism by microflora or enterohepatic recirculation.

8.2.3 Flavonoid metabolites identified in vivo and their biological activities

It is clear that dietary flavonoids are partially absorbed in humans, as they can be detected in the blood and urine of volunteers fed flavonoid-containing foods or supplements. Therefore, dietary flavonoids have the potential to exert biological effects in humans. But, during first-pass metabolism, flavonoids are modified, and the most important modifications that occur involve conjugation of the aglycone or methylated derivatives with glucuronate or sulphate groups. For most flavonoids, even at supranormal oral doses, flavonoid glycosides (as present in plant-derived foods) and aglycones are either absent from plasma or present only as a small fraction of the total flavonoid pool (Kroon *et al.* 2004). A notable exception are the flavan-3-ols typical of green tea for which a proportion has been reported to be present in plasma in the unconjugated forms found in tea (Nakagawa *et al.* 1997), but note that as discussed previously there may be some change in stereochemistry.

The chemico-physical properties of flavonoids are therefore altered during first-pass metabolism. Conjugation affects properties such as size/mass, charge and hydrophobicity, which may impact on their solubility and ability to cross biological membranes. It is also likely to affect their rate of excretion (via the kidneys or liver) and therefore the half-life in plasma. Conjugations will effectively reduce the number of free hydroxyl groups, which is likely to impact on the antioxidant properties and possibly the ability to interact with important functional cellular proteins including enzymes, receptors and transporters (see Chapter 1, and Clifford and Brown, 2006). It is therefore important to determine the likely impact of these conjugates/metabolites on relevant tissues, cells and proteins in order to provide mechanistic insight regarding the role of flavonoids in protecting against age-related diseases and maintaining optimal health.

In order to be able to investigate the effects of physiological metabolites in relation to processes underlying the initiation or progression of disease, several key steps are required: (1) analytical methods with appropriate sensitivity and selectivity to facilitate identification of the individual metabolites/conjugates in human fluids, (2) authentic samples of the individual metabolites/conjugates to facilitate measurement of the levels achieved in vivo and for in vitro studies concerned with biological impact and (3) appropriate models for measuring the biological response. The following sections are concerned with the flavonoid conjugate content/composition of plasma (and urine) of humans and the biological activities of those metabolites/conjugates in various biological systems used to assess cellular and tissue responses.

8.2.3.1 Approaches to the identification of flavonoid conjugates in plasma and urine

Early studies concerned with measuring absorption of flavonoids in humans, and all those up until the mid 1990s, applied chemical or enzymatic hydrolysis to plasma and urine samples in order to convert all the flavonoid conjugates (sulphates, glucuronides) to aglycones, thereby increasing sensitivity and simplifying chromatograms. This approach is useful for estimating the bioavailability (amount reaching plasma) of particular flavonoids from an oral dose and is still used extensively for this purpose today. However, the application of advanced analytical methods such as the combination of high resolution chromatography systems (especially reversed-phase HPLC) with detection systems such as MS, coularray electrochemical and photodiode array have enabled researchers to obtain information regarding the structure of the flavonoid metabolites/conjugates as they are found in vivo. There are now several published reports describing the flavonoid conjugate composition of human plasma or urine samples following ingestion of a flavonoid-rich meal (reviewed in Kroon *et al.* 2004).

Various approaches for sample and data analysis have been taken, each providing different levels of information. Although not central to the content of this chapter, some methodological aspects will be covered here since they are of relevance when interpreting published data. The reader is referred to a book entitled 'Methods in Polyphenol Analysis'

(Santos-Buelga and Williamson 2003) which covers most of the analytical methods that have been used and which provides an excellent technical summary for those in the field.

The simplest approach that provides information on metabolite/conjugate structure involves comparing the levels of flavonoid peaks from plasma samples that have been deconjugated (either by treatment with β-glucuronidase and/or sulphatase enzymes or by chemical means) with samples that have not been treated. The information provided by this approach is limited but can be useful. Potentially, the extent of conjugation can be calculated and the amounts of different types of metabolites determined. To perform this type of analysis, it is necessary to determine whether peaks on chromatograms are authentic conjugated derivatives of the flavonoid(s) of interest or merely contaminating compounds from the sample. Simple absorbance spectra, usually obtained using a post-column photodiode array detector, do not provide sufficient evidence to confirm structure. There are numerous literature examples where absorbance spectral characterization alone has been used which has produced spurious results (Day and Williamson 2001). Although careful use of electrochemical detectors may provide better selectivity than absorbance detectors, they are not sufficient on their own to confirm structures.

There are a number of deconjugating enzymes (β-glucuronidase, sulphatase) available for purchase, and these vary in source and purity. One of the problems encountered by researchers using these enzymes has been their lack of purity and specifically the presence of contaminating activities that can make their use as identification tools problematic. In particular, many of the commercially available sulphatases and β-glucuronidases are derived from molluscs (e.g. *Helix pomatia, Helix aspera, Patella vulgata*) and are crude or partially purified preparations that usually contain a mixture of both these activities. They can also contain other glycosyl hydrolase activities (e.g. β-glucosidase) that may be of relevance. A further issue that should be considered is the specificity of the deconjugating enzymes and the incubation conditions. Some flavonoid conjugates may not be substrates for particular sulphatases or β-glucuronidases. For example, using the *Helix pomatia* H-5 sulphatase preparation (Sigma S 3009; partially purified), an authentic sample of quercetin-3'-O-sulphate was not hydrolysed even though other sulphate conjugates were cleaved (PA Kroon, SM Dupont and RN Bennett, unpublished data). Clearly, it is imperative that due care is taken when using enzymes to aid in identification of flavonoid conjugates. Nevertheless, when used properly, deconjugating enzymes can provide useful information. As an example, treatment of plasma samples obtained from volunteers following ingestion of orange juice (containing naringenin and hesperetin glycosides) with either buffer, β-glucuronidase alone, sulphatase alone or a mixture of β-glucuronidase and sulphatase, showed that none of the hesperetin was present as aglycone, but that it was all glucuronidated and around 13% was also sulphated (i.e. present as mixed sulfo-glucurono-conjugates (Manach et al. 2003).

On-line MS provides the single most informative method for analysing flavonoid conjugates. A number of variations exist, with differences in the sample ionization method and the ionization mode is important. The choice depends on the chemical nature of the analytes of interest and is usually arrived at empirically. Data are collected in one or more modes–full scan, zoom scan, selected ion monitoring or in tandem (i.e. MS-MS). MS can provide the mass of molecular ions and of fragments (fragmentation patterns). For example, mass spectral analysis of a putative flavonoid glucuronide could provide the mass of the molecular ion as well as the ion masses for the flavonoid aglycone and the glucuronide moiety. Although this information is useful alone, mass spectrometry is most powerful when used in combination with other techniques including absorbance spectral analysis, coularray analysis and specific enzyme hydrolysis. Further, the data obtained from MS (and the other techniques) are most powerful when they are employed in studies where authentic standards are available.

Only a handful of studies reported to date have employed authentic flavonoid conjugates as standards during analysis. The reason for this is simple; flavonoid glucuronides and sulphates are generally not available commercially and their synthesis is not trivial. Nevertheless, flavonoid conjugates can be obtained by synthetic (Barron and Ibrahim 1987; Day et al. 2001; Needs and Williamson 2001; Bouktaib et al. 2002; Clarke et al. 2002; O'Leary et al. 2003) or biosynthetic (Wittig et al. 2001; Manach et al. 2003; O'Leary et al. 2003; Plumb et al. 2003) routes, and studies where they have been available have provided data with credence and depth. Appropriate flavonoid conjugates can be used to confirm retention times, absorbance spectra, shift reagent response and mass spectrum. In addition, access to these materials provides the means to monitor extraction/processing yields. As an example, Day and co-workers were interested in the structures of quercetin present in the plasma of volunteers who had been fed onions (Day et al. 2001). Some of the predicted conjugates were synthesized chemically and used to confirm the identity of some of the absorbance peaks with flavonol-like spectra. In this way, they were able to confirm that three of the four most abundant quercetin conjugates in plasma were quercetin-3'-O-sulphate, quercetin-3-O-glucuronide and 3'-O-methyl-quercetin-3-O-glucuronide (isorhamnetin-3-O-glucuronide). An additional nine quercetin conjugates were tentatively identified. This study is noteworthy in the number of complementary approaches used to establish the structure (or likely structure) of flavonoid conjugates in plasma. In addition to HPLC retention times, absorbance spectra and mass spectral data, the authors examined the sensitivity of peaks to β-glucuronidase and/or sulphatase treatment and used 'shift' reagents to provide information on the conjugation position. An extensive investigation identifying quercetin metabolites was recently conducted using six subjects fed red onions (Mullen et al. 2004). Several of the metabolites identified by Day et al. (2001) were available to use as reference compounds, and structural elucidation of newly identified metabolites was facilitated by using HPLC-PDA coupled to full scan MS-MS. The authors identified twenty-three distinct compounds in either plasma or urine including (methylated)quercetin mono- and di-glucuronides, quercetin sulphates as well as quercetin glucuronide sulphates. Interestingly, samples from one of the volunteers also contained trace amounts of quercetin aglycone, quercetin-3,4'-O-diglucoside, quercetin-3-O-glucoside and isorhamnetin-3-O-glucoside. Whether this individual has a polymorphism in one of the deconjugating enzymes such as LPH or CBG which could have accounted for the appearance of these compounds remains to be studied.

A couple of examples of studies that have used labelled materials to enhance investigations of physiological structures are worth mentioning. Feeding of radio-labelled flavonoids provides additional sensitivity in the analysis, and confirms that what is observed was derived only from the ingested material. Using [2-^{14}C]quercetin-4'-O-glucoside to feed rats, Mullen and co-workers were able detect a total of 18 radiolabelled compounds. They putatively identified 17 of these as glucuronidated and sulphated conjugates of quercetin or methylquercetin on the basis of mass spectral data; ten of these were present in plasma (Mullen et al. 2002). It is worth comparing these findings with those from one of the first reports concerned with the identification of quercetin conjugates in plasma that used only MS and (single wavelength) absorbance detection (Wittig *et al.* 2001). In this study, the authors identified five HPLC peaks that gave flavonol-like absorbance spectra and the quercetin glucuronide molecular ion (m/z 479). A mixture of quercetin glucuronides was obtained by biosynthetic means as standards for comparison (using rabbit liver as a source of UDP-glucuronosyl transferase activity). However, no sulphates or glucuronosulphates were reported, and it is likely this was due to the extraction conditions used (Day and Morgan 2003). Clearly, great care needs to be taken and appropriate controls used when attempting to identify and quantify flavonoid conjugates in plasma and urine samples.

8.2.3.2 Flavonoid conjugates identified in plasma and urine

A summary of the data in the various reports concerned with the structures of flavonoid conjugates in vivo has been published recently (Kroon *et al.* 2004); therefore, only a brief summary will be provided here, and the reader is advised to refer to this citation for a more detailed account. As mentioned, more detailed investigations are emerging as technology permits (Mullen *et al.* 2004, 2006; Jaganath *et al.* 2006).

The major dietary flavonols, quercetin and kaempferol, show some interesting differences with regard to the forms present in plasma following a flavonol-rich meal. As has been detailed above, quercetin is present in plasma only in conjugated forms in most individuals, comprising at least 12 glucuronide and sulphate conjugates of quercetin or methylquercetin (Day *et al.* 2001; Mullen *et al.* 2004). The major forms in plasma are quercetin-3'-O-sulphate (comprising around 50% of total quercetin; PA Kroon, SM Dupont and RN Bennett, unpublished data), quercetin-3-O-glucuronide, isorhamnetin-3-O-glucuronide and quercetin-3'-O-glucuronide (Mullen *et al.* 2006) (Figure 8.4). In contrast, a reasonable portion (~20%) of absorbed kaempferol is present in plasma as the aglycone, with the majority of the remainder accounted for by kaempferol-3-glucuronide with possibly some kaempferol monosulphate also present (Dupont *et al.* 2004).

For the flavones, there is some evidence to indicate a luteolin monoglucuronide is obtained (Shimoi *et al.* 1998, 2000). Chrysin-7-O-sulphate and chrysin-7-O-glucuronide

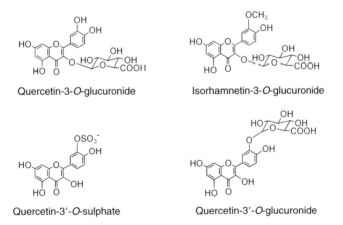

Figure 8.4 The structures of the four major quercetin conjugates present in plasma following ingestion of a quercetin glucoside-rich meal.

have been positively identified (Walle et al. 2001a). For the flavanones hesperetin and naringenin, it has been established that all the flavonoid is present as conjugates in plasma, and that the majority of the conjugates are monoglucuronides with a reasonable amount of mixed sulphoglucuronide(s) also present. However, the exact structures of flavanone conjugates have not been reported (Manach et al. 2003).

Of all the polyphenols, the bioavailability of the isoflavones is best understood, and arguably the most complete data available concerning the kinetics of absorption and the identity of the conjugates in vivo. The most likely reason for this was the fact that isoflavones were found to interact with known cellular receptors (i.e. estrogen receptors) and cause estrogenic effects in model systems and in humans, which provided the impetus for substantial research on these compounds. The interest in isoflavones as phytoestrogens with likely effects (positive or detrimental) on human health resulted in the funding of large research programmes in various countries around the world and, as a consequence a large body of literature has been produced. In terms of studies concerned with the identification of structures in vivo, and their quantification, the work described by Clarke et al. (2002) was extremely useful. Using [3-^{13}C]isoflavone internal standards and isotope dilution liquid chromatography and tandem mass spectrometry, it was shown that following ingestion of daidzein, 54% of daidzien conjugates were present in urine as 7-O-glucuronide, 25% as 4'-O-glucuronide, 13% as 7- and 4'-O-sulphates, 0.4% as 4',7-O-diglucuronide, 0.9% as sulphoglucuronides and 7% as unconjugated daidzein. Similar profiles were obtained for genistein. One study used isotope dilution-GC in combination with MS to identify both intact isoflavones and their microbial metabolites (e.g. equol, O-desmethyl-angolensin) in human feces (Adlercreutz et al. 1995). There is considerable interest in the microbial metabolites formed from isoflavones, particularly equol, which appears to be even more potent than the soya isoflavones in modulating estrogenic function in some models and in certain studies has shown a stronger association with reductions in disease risk.

The flavan-3-ols and their gallate esters have also received considerable attention regarding their bioavailability. A useful study was published by the group of Junji Terao who were able to identify epicatechin-3'-O-glucuronide, 4'-O-methyl-epicatechin-3'-O-glucuronide and 4'-O-methyl-epicatechin-5 or 7-O-glucuronide in human urine using HPLC-MS and NMR. Further, they were able to detect absorbance peaks of similar retention time and with the same molecular mass (using HPLC-MS) in human plasma (Natsume et al. 2003). Information on the conjugation position of (+)-catechin conjugates in plasma is currently not available. However, only traces (<2 nM) of the aglycone were detected in human plasma after red wine consumption (Donovan et al. 1999), and the vast majority of (+)-catechin in plasma is present as a mixture of catechin sulphates, catechin sulphoglucuronide and 3'-O-methyl-catechin glucuronide for which the stereochemistry has not been determined (Donovan et al. 1999, 2002a). Similarly, the conjugated structures of (−)-epicatechin gallate, (−)-epigallocatechin and (−)-epigallocatechin gallate in humans are not known. (−)-Epicatechin gallate from black tea has been found in human plasma exclusively as conjugates (Warden et al. 2001), but the exact positions of the conjugates are not known. Meng et al. (2001) reported that 4'-O-methyl-epigallocatechin was present five times higher in plasma than non methylated epigallocatechin. More recently the same group of researchers demonstrated the presence of a dimethylated metabolite of (−)-epigallocatechin gallate, 4',4''-di-O-methyl-epigallocatechin gallate (Meng et al. 2002). This metabolite was methylated both on the flavonoid ring and the gallate moiety.

The concentration of 4′,4″-di-O-methyl-epigallocatechin gallate was about 15% that of unmethylated epigallocatechin gallate in human plasma. It is likely that the gallated flavan-3-ols produce numerous combinations of conjugated metabolites. The structures of the predominant forms of these metabolites in plasma after green tea consumption need to be determined.

The anthocyanins are probably the least well understood in terms of the nature of the structures in vivo. This is not due to a lack of interest in anthocyanin bioavailability but due to technical difficulties in anthocyanin analysis. Before 2002, data arising from most studies concerned with anthocyanin bioavailability, and using mass spectrometry, indicated that anthocyanins were very poorly bioavailable, with <0.1% of oral dose reaching the urine, and that the anthocyanins were present in plasma and urine as the glycoside forms typically present in anthocyanin-containing foods and beverages such as berries, currants and red wine (Miyazawa *et al.* 1999; Cao *et al.* 2001; Matsumoto *et al.* 2001; Mazza *et al.* 2002; Nielsen *et al.* 2003). These reports indicated that if anthocyanins were going to affect cells/tissues beyond the gastrointestinal tract, they would need to be active at extremely low concentrations. Although an earlier study had reported much higher urinary yields (1.5–5.1% of ingested dose) and indicated the presence of unknown metabolized forms of red wine anthocyanins (Lapidot *et al.* 1998), the most revealing study concerned with bioavailability of strawberry anthocyanins reported a total urinary yield of 1.8% of the oral dose of pelargonidin-3-O-glucoside. This study provided some evidence pertaining to the structures in urine (the anthocyanidin itself, three distinct monoglucuronides, one sulpho conjugate and free pelargonidinin) and indicated that special procedures were required to prevent substantial degradation of pelargonidin and its conjugates from urine samples (Felgines *et al.* 2003). The major metabolite present in urine was a pelargonidin monoglucuronide. Another study has shown that cyanidin can be methylated to form a peonidin glucuronide conjugate (Wu *et al.* 2002).

There is little indication that the polymeric flavan-3-ols, such as proanthocyanidins and tannins, are absorbed intact to any great extent by humans or other mammals. There are only two reports that provide evidence supporting the presence of proanthocyanidins in human plasma, one for procyanidin B_2 from chocolate (Holt *et al.* 2002) and one for procyanidin B_1 from a grape seed extract (Sano *et al.* 2003). In both studies, plasma concentrations were reported after the hydrolysis of glucuronide and sulphate conjugates. The proportion of these proanthocyanidins that exist in their native versus conjugated form is not known. The plasma concentrations reported in both these studies were in the low nM range. As discussed in the next section, the plasma concentrations should be an important factor in designing studies to investigate possible biological activity.

8.2.4 *Pharmacokinetics of flavonoids in humans*

Approximately one hundred studies have been published to date on the bioavailability and pharmacokinetics of individual polyphenols following a single dose of pure compound, plant extract or whole food/beverage to healthy volunteers. We recently reviewed the pharmacokinetic data available for each class to estimate average pharmacokinetic parameters including the maximum concentration in plasma (C_{max}), T_{max}, the area under the plasma concentration versus time curve (AUC), elimination half-life ($T_{1/2}$) and percent of dose excreted in urine (Manach and Donovan 2004). Here, we present a summary of that data along with the pharmacokinetic curves that have been complied from various studies. Because flavonoids are present largely, if not exclusively, as conjugated metabolites, pharmacokinetic values generally represent the total amount of flavonoid including all known conjugated forms present. This is in contrast to other disciplines where only the native or 'parent' compound is of interest.

It is a challenge to make generalizations and to compare data obtained from different studies, because most studies have not used the same amounts of polyphenols, and have had different populations with different background diets. Furthermore, most studies have administered different amounts of polyphenols in different food sources. To facilitate the comparison between polyphenols, data have been converted to correspond to the same supply of polyphenols, a single 50 mg dose of aglycone equivalent. For this analysis, we assumed that the plasma concentrations increase linearly with doses that could be present in foods; however, this relationship has only been demonstrated for (−)-epigallocatechin gallate in humans (Ullmann et al. 2003). We have only included studies that used well-characterized sources and doses of polyphenols and appropriate analytical methods. The number of studies available for each selected polyphenol ranged from 4 to 12. It should be noted that a high variability has been observed between individuals and between the findings of several studies. Clifford and Brown (2006) have drawn attention to some specific examples of inter-individual variation.

The data presented in Figure 8.5 show that isoflavones are the best absorbed flavonoids. The maximum plasma concentrations reach about 2 μM after an intake of 50 mg dose. For the flavonoids, in general, quercetin glucosides, flavan-3-ols and anthocyanins show maximum plasma concentrations at about 1.5 h, reflecting an absorption in the small intestine

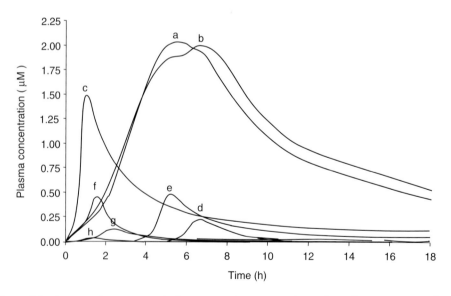

Figure 8.5 Average plasma concentration versus time curves drawn from literature survey. Data were converted to correspond to a single dose of 50 mg aglycone equivalent. (a) isoflavone aglycones (b) isoflavone glucosides (c) quercetin glucosides (d) quercetin rutinoside (e) flavanone glycosides (f) (epi)catechins (g) (−)-epigallocatechin gallate (h) anthocyanins.

or in the stomach, whereas maximum concentrations are reached at around 6 h for rutin and the flavanone rhamnoglucosides. The later T_{max} is consistent with an absorption in a more distal part of the intestine after hydrolysis into aglycones by the microflora. As discussed above, glycosylation has a dramatic influence on the bioavailability of polyphenols. Quercetin glucosides are obviously far better and faster absorbed than rutin. The C_{max} differs markedly between (−)-epigallocatechin gallate and (−)-epigallocatechin. By giving pure flavan-3-ols individually, Van Amelsvoort et al. (2001) demonstrated that galloylation of flavan-3-ols appears to reduce the C_{max} concentrations.

Anthocyanins appear to be poorly absorbed compared with the other polyphenols. When single doses of 150 mg to 2 g total anthocyanins were given to the volunteers, generally in the form of berries, berry extracts or concentrates, concentrations of anthocyanins measured in plasma were in the order of few tens of nmoles/L. However, as mentioned in the previous section, the bioavailability of anthocyanins may have been underestimated, because a significant portion of the metabolites were not quantified by the available methods of analysis. Felgines and colleagues (2003) showed that the conjugated metabolites of the strawberry anthocyanins were unstable and were extensively degraded when acidified urine samples were frozen for storage. This explains why such metabolites have not been observed in previous studies. Future studies using new methods for preservation and analysis of all anthocyanin metabolites may reveal a better absorption of these polyphenols than currently stated.

Only two studies are available in humans on proanthocyanidins (Holt et al. 2002; Sano et al. 2003). They both showed that the C_{max} in plasma for dimers B_1 and B_2 was in the 10–20 nM range even after consumption of two of the richest food sources, chocolate and grape seed extract. Due to the low plasma concentrations it has not been possible to determine other pharmacokinetic parameters such as elimination half-life. As discussed above, absorption of trimers or proanthocyanidins with higher degree of polymerization has never been reported and is unlikely to occur according to the size and polarity of the compounds and their strong binding capacity to proteins.

Based upon the elimination half-lives, it appears that flavan-3-ols or anthocyanins are unlikely to accumulate in plasma even with repeated intakes. However, some of their metabolites may have longer elimination half-lives and slower systemic clearance resulting in appreciable accumulation. One characteristic feature of quercetin is that the elimination of its metabolites is quite slow, with reported half-lives ranging from 11 to 28 h. Quercetin was shown to accumulate in plasma after multiple doses of supplements or onions (Conquer et al. 1998; Boyle et al. 2000; Moon et al. 2000).

It should be noted that the representation we used here does not take into account the mean dietary intake of each polyphenol. For example, even if isoflavones are efficiently absorbed, they are not likely to be the major circulating polyphenols in western populations because for these populations isoflavone intake is far lower than 50 mg per day. In contrast, a single glass of orange juice easily provides 50 mg hesperidin, and hesperetin metabolites in plasma reached 1.3–2.2 μM following an intake of 130–220 mg given as orange juice (Erlund et al. 2001; Manach et al. 2003), and up to 6 μM naringenin metabolites following a 200 mg dosing with grapefruit juice (Erlund et al. 2001). The relative urinary excretion of flavonoid conjugates (expressed as a percentage of the flavonoid intake) is shown in Figure 8.6. These data cannot be considered an accurate estimation of the amount absorbed because most of the flavonoids studied to date have been shown to be extensively

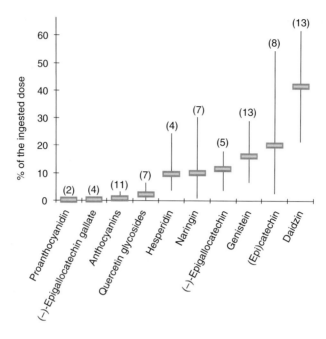

Figure 8.6 Mean urinary excretion of flavonoids calculated from literature survey. The horizontal bars represent the mean % of the ingested dose recovered in the urine for each polyphenol, the vertical bars represent the range between the lowest and the highest values of the literature, and the number above the green bars is the number of studies taken into account.

eliminated in bile (as discussed above) and metabolized further to phenolic acids in the colon (Sfakianos *et al.* 1997; Kohri *et al.* 2001). The percentage excreted in urine ranged from 0.3 to 43% of the dose for the flavonoids that have been studied. This clearly shows the high variability between flavonoids as well as different species in the subclasses. The relative urinary excretion of daidzein is markedly higher than that of the other polyphenols (43% of intake). The total urinary excretion of conjugated flavanones accounted for 8.6% of the intake for hesperidin and 8.8% for naringin. In the case of quercetin, mean urinary excretion was higher for glucosides (2.5%) than for rutin (0.7%). It is worth noting that concomitant consumption of foods may be important since urinary excretion reached higher values (3.6%) when purified glucosides were given in hydroalcoholic solution to fasting volunteers (Olthof *et al.* 2000). Although the significance of the differences between urinary excretion of (+)-catechin, (−)-epicatechin and (−)-epigallocatechin gallate need to be further investigated, it is clear that urinary excretion of (−)-epigallocatechin gallate is markedly lower (0.06%), because of preferential excretion in bile. Anthocyanins are rapidly excreted in urine and maximum concentrations were observed by 2.5 h in urine. Most studies reported low relative urinary excretions, ranging from 0.004 to 0.1% of the intake, although higher levels of anthocyanin excretion (up to 5%) have been reported after red wine (Lapidot *et al.* 1998) and strawberry consumption (Felgines *et al.* 2003).

It should be noted that the plasma concentration of flavonoids may be underestimated for some flavonoids because not all metabolites have been identified and quantified.

We have already discussed the difficulties in the analysis of the anthocyanins as described by Felgines *et al.* (2003) and the likely underestimation in the plasma levels of gallated flavan-3-ols due to extensive methylation (Meng *et al.* 2001, 2002). These challenges exist in addition to possible inefficiencies in enzymatic hydrolysis of the conjugate forms for all types of flavonoids. Thus, the levels that are thought to be present in plasma now represent minimum levels. There is also a large variability in the plasma concentrations obtained from different studies. One contributing factor may be the influence of food matrix. Some diet constituents may significantly affect polyphenol bioavailability. Inter-individual variation is also an important parameter. Most of the studies have used relatively small numbers of subjects and outlying values may have significantly altered the mean values. Some individuals could be better polyphenol absorbers than others. Analysis of the inter-individual variations may lead to identification of some polymorphisms or differential expression for some crucial enzymes or transporters involved in the absorption and metabolism of polyphenols. In addition, almost all data on flavonoid pharmacokinetics have been derived from young, healthy volunteers. Some changes in the physiology, induced by ageing or chronic diseases, may affect the metabolism and pharmacokinetics of polyphenols. For example, the mean serum concentration of isoflavones was shown to increase along with the severity of diabetic nephropathy (von Hertzen *et al.* 2004). The profile of specific conjugates may be affected by inflammation (Shimoi *et al.* 2001). Future studies should investigate the impact of ageing and major diseases on polyphenol bioavailability in humans. These aspects have also been discussed by Clifford and Brown (2006).

In summary, the pharmacokinetic characteristics vary widely between the classes of flavonoids as well as the specific compounds in some of the classes. The plasma concentration of total metabolites ranges from 0 to 4 µM for an intake of 50 mg aglycone equivalent. The polyphenols that are present at the highest postprandial concentrations are the isoflavones, monomeric flavan-3-ols, flavanones and flavonol glucosides. The flavonols, however, have the longest elimination half-lives and thus have more potential to accumulate with less frequent dosing. The anthocyanins and galloylated flavan-3-ols appear to be less bioavailable but may have been underestimated, because of metabolism into non detected compounds or poor stability in biological samples. The flavonoids that appear to be the least bioavailable, at least in the forms containing the flavonoid ring structure, are the proanthocyanidins, especially those with higher degrees of polymerization.

8.3 Hydroxycinnamic acids

Hydroxycinnamic acids such as caffeic, ferulic and coumaric acids occur in a large variety of fruits (see Chapter 7), in concentrations up to 2 g/kg fresh weight (Macheix *et al.* 1990). Caffeic acid, free or esterified, accounts for 75–100% of the total hydroxycinnamic acid content of most fruits. The most abundant hydroxycinnamic acid in food is 5-O-caffeoylquinic acid, the ester of caffeic acid with quinic acid, widely referred to as chlorogenic acid. Coffee is the major dietary source of chlorogenic acids and in the daily intake coffee drinkers may ingest up to 800 mg (Clifford 2000).

When ingested in the free form, hydroxycinnamic acids are rapidly absorbed from the stomach or the small intestine and are glucuronidated and sulphated in the same way as flavonoids (Clifford 2000; Cremin *et al.* 2001). Ferulic acid and caffeic acid were reported to be transported across human intestinal Caco-2 cells by the monocarboxylic acid transporter (Konishi et al. 2003; Konishi and Shimizu 2003).

Absorption is markedly reduced when caffeic acid is given in esterified rather than in free form (Azuma et al. 2000; Olthof et al. 2001; Gonthier et al. 2003b). In patients who have undergone colonic ablation, caffeic acid was much better absorbed than 5-O-caffeoylquinic acid, with 11% and 0.3% of the ingested dose excreted in urine respectively (Olthof et al. 2001). Similarly, when 5-O-caffeoylquinic acid was given by gavage to rats, no intact compound could be detected in plasma in the following 6 h, and the maximum concentrations of metabolites (various glucuronidated/sulphated derivatives of caffeic and ferulic acids) were 100-fold lower than those reached after administration of caffeic acid in the same conditions (Azuma et al. 2000).

The mechanism and site of absorption of 5-O-caffeoylquinic acid is still unclear. No esterase activity able to hydrolyse 5-O-caffeoylquinic acid into caffeic acid was detected in human tissues (intestinal mucosa, liver) or in biological fluids (plasma, gastric juice, duodenal fluid) in rats or humans (Plumb et al. 1999; Azuma et al. 2000; Andreasen et al. 2001; Olthof et al. 2001; Rechner et al. 2001). Several authors have detected 5-O-caffeoylquinic acid in urine after ingestion of coffee or pure 5-O-caffeoylquinic acid, with recoveries ranging from 0.3% to 2.3%, suggesting small ester absorption without hydrolysis (Cremin et al. 2001; Olthof et al. 2001; Gonthier et al. 2003a; Ito et al. 2004). On the other hand, rapid appearance of caffeic acid in plasma after 5-O-caffeoylquinic acid ingestion suggests that 5-O-caffeoylquinic acid may be hydrolysed in the upper part of the gastrointestinal tract before absorption (Azuma et al. 2000; Nardini et al. 2002). A low concentration of caffeic acid (1.2% of the perfused dose) was recently found in the effluent during in situ perfusion of 5-O-caffeoylquinic acid in a segment of the upper intestinal tract of rats (Lafay et al. 2005). This clearly indicates that hydrolysis of 5-O-caffeoylquinic acid may occur to a very low extent in the gut mucosa; however the main site of hydrolysis is the large intestine with microbial esterases. A number of colonic bacterial species capable of carrying out this hydrolysis have been identified (Couteau et al. 2001; Rechner et al. 2004). Finally, a minor portion of 5-O-caffeoylquinic acid could be absorbed in the proximal part of the gut, but the majority of 5-O-caffeoylquinic acid reaches the caecum or colon, where it is hydrolysed and metabolized before absorption.

The metabolites detected in most studies after 5-O-caffeoylquinic or caffeic acid ingestion were conjugated forms of caffeic, ferulic and isoferulic acids (Azuma et al. 2000; Rechner et al. 2001; Nardini et al. 2002; Gonthier et al. 2003a; Wittemer et al. 2005). Dihydroferulic has also been detected (Rechner et al. 2001; Wittemer et al. 2005). However, the metabolites of microbial origin, namely m-coumaric acid, derivatives of phenylpropionic (3,4-dihydroxyphenylpropionic, 3-hydroxyphenylpropionic acid), benzoic and hippuric acids (3-hydroxyhippuric and hippuric acids) may be of major importance (Gonthier et al. 2003a; Rechner et al. 2004). In rats, these microbial metabolites accounted for 57.4% (mol/mol) of the 5-O-caffeoylquinic acid intake (Gonthier et al. 2003a). Hippuric acid largely originated from the transformation of the quinic acid moiety and other microbial metabolites from the caffeic acid moiety. Proportions of the various metabolites may differ when caffeic or 5-O-caffeoylquinic acid are ingested. 5-O-caffeoylquinic acid which is poorly absorbed in the small intestine provided higher yields of microbial metabolites than caffeic acid but lower concentrations of caffeic and ferulic acid conjugates (Gonthier et al. 2003b).

In situ small intestine perfusion of hydroxycinnamic acids in rats showed that ferulic acid is better absorbed than caffeic acid (56.1% vs 19.5% of the perfused flux was absorbed, respectively), whereas caffeic acid is better absorbed than 5-O-caffeoylquinic acid (8%) (Adam *et al.* 2002; Lafay *et al.* 2005). This model also revealed that biliary excretion is low for hydroxycinnamic acids (Adam *et al.* 2002; Lafay *et al.* 2005).

Ferulic acid is not as abundant in common foods as caffeic acid. The main source of dietary ferulic acid is likely to be coffee which provides approximately 10 mg of its quinic acid ester per cup (Clifford 2000). When present in free form in tomato or in beer, ferulic acid is rapidly and efficiently absorbed in humans (up to 25% of the dose) (Bourne and Rice-Evans 1998; Bourne *et al.* 2000). Ferulic acid is also found in cereals, in which it is esterified to arabinoxylans and hemicelluloses in the aleurone layer and pericarp of the grains. This binding has been reported to hamper the absorption of ferulic acid in animals (Adam *et al.* 2002; Zhao *et al.* 2003). Ferulic acid metabolites excreted in urine represented only 3% of the ingested dose when ferulic acid was provided to rats as wheat bran, compared with 50% of the dose when provided as pure compound (Adam *et al.* 2002). Feruloyl esterases were shown to be present throughout the entire gastrointestinal tract, particularly in the intestinal mucosa. Ester bonds between ferulic acid and polysaccharides in cell walls may thus be theoretically hydrolysed in the small intestine (Andreasen *et al.* 2001). However, analysis of rat intestinal contents after ingestion of free or esterified ferulic acid suggests that their role may be very limited compared with that of the microbial xylanases and esterases (Zhao *et al.* 2003). In plasma, ferulic acid was mainly recovered as sulphoglucuronides or sulphates, with only 5–24% as free form (Rondini *et al.* 2002; Zhao *et al.* 2003). Diferulic acids from cereal brans were also shown to be absorbed in rats (Andreasen *et al.* 2001).

Kern *et al.* (2003) measured the urinary excretion and plasma concentration of ferulic acid metabolites after ingestion of breakfast cereals by humans. They deduced from the kinetic data that absorption of ferulic acid from cereals mainly took place in the small intestine, from the soluble fraction present in cereals. Only a minor absorption of ferulic acid linked to arabinoxylans was absorbed after hydrolysis in the large intestine.

In summary, hydroxycinnamic acids are well absorbed when they are present in free forms in food. Esterification markedly reduces their intestinal absorption and turns their metabolism towards microbial metabolism.

8.4 Gallic acid and ellagic acid

Red fruits such as strawberries, raspberries and blueberries, black tea, red wine and nuts are the main sources of gallic acid and ellagic acid (Tomas-Barberan and Clifford 2000). These phenolic acids exist in free form or as components of complex structures such as hydrolysable tannins (gallotannins in mangoes and ellagitannins in red fruits (Clifford 2000; also see Chapter 7).

Gallic acid appears to be very well absorbed in humans compared with other polyphenols (Shahrzad and Bitsch 1998; Shahrzad *et al.* 2001; Cartron *et al.* 2003). The main plasma metabolites are glucuronidated forms of gallic acid and 4-O-methylgallic acid. Plasma concentrations of total gallic acid metabolites rapidly reached $4\,\mu M$ after ingestion of

50 mg pure gallic acid. Total urinary excretion accounted for 37% of the dose (Shahrzad and Bitsch 1998; Shahrzad et al. 2001). On the basis of a large study carried out on 344 Australian volunteers, 4-O-methylgallic acid measured in 24 h urine was proposed as a reliable biomarker for black tea intake (Hodgson et al. 2004).

Several studies have recently investigated ellagic acid bioavailability. After consumption of pomegranate juice providing ellagic acid and ellagitannins, intact ellagic acid was detected in human plasma, with a maximum concentration 1 h after intake (Seeram et al. 2004). In contrast, ellagic acid was not recovered in urine from volunteers challenged with strawberries, raspberries, walnuts or oak-aged red wine (Cerda et al. 2005). A microbial metabolite urolithin B conjugated with glucuronide acid was found in the urine of all volunteers, which may become a biomarker of exposure to ellagitannins and ellagic acid.

8.5 Dihydrochalcones

Phloretin glycosides are the main dietary dihydrochalcones (Tomas-Barberan and Clifford 2000). They are characteristic of apples and are found in all apple derived products, such as juices, pomace and ciders. Phloridzin (phloretin-2'-O-glucoside) is a well-known inhibitor of the sodium-dependent glucose transporter SGLT1 and has been shown to be transported by SGLT1 (Walle and Walle 2003). Very little is known, however, about the possible effects of phloridzin on glucose absorption when consumed orally at concentrations found in the diet. Such an effect could be important in improving glucose tolerance in patients with non-insulin dependent diabetes mellitus. It was recently shown that glucose absorption was significantly delayed when consumed in apple juice compared with water (Johnston et al. 2002). While the apple juice contained other phenolics including flavonoids and hydroxycinnamates, phloridzin appears to be at least partially if not mostly responsible for this effect (Johnston et al. 2002, 2003).

The bioavailability of phloretin and its glucoside phloridzin was shown to be similar in rats, except that the kinetics of absorption was delayed for phloridzin (Crespy et al. 2001). About 10% of the ingested dose was excreted in the urine in 24 h, and plasma concentrations were returned to baseline at this timepoint. Plasma metabolites were glucuronidated and/or sulphated forms of phloretin. Intact phloridzin was not recovered. Phloretic acid was also detected in rat urine after phloretin gavage (Monge et al. 1984).

8.6 Betalains

Betalains are a class of phytochemicals contained in some families of the Caryophyllales order of plants, including the edible red beet (Tesoriere et al. 2005). In comparison with other types of plant secondary metabolites, little is known regarding the absorption and metabolism of betalains. These pigments were shown to incorporate into red blood cells and LDL of healthy volunteers after ingestion of cactus pear fruit pulp (Tesoriere et al. 2004, 2005). Maximum concentrations were reached by 3 h after intake and betalains disappeared from plasma by 12 h after intake. Indicaxanthin was excreted to a higher extent than betanin by urine.

8.7 Glucosinolates

Glucosinolates are sulphur-containing plant secondary metabolites that are present in cruciferous plants many of which are consumed as vegetables (e.g. broccoli, kale, Brussels sprouts, cabbage, watercress, salad rocket, turnip, mustard and radish) (see Chapter 2). Glucosinolates are β-thioglucoside N-hydroxysulphates that contain a variable side chain and a β-glucopyranose moiety linked through sulphur (Figure 8.7). These compounds are responsible for the pungent odour and bitter/biting taste of cruciferous vegetables. The biting taste/bitterness are actually due to the products of myrosinase-catalysed glucosinolate breakdown that are formed when the thioglucosidic bond is hydrolysed. These products include isothiocyanates (R–N=C=S) which contribute to the characteristic aroma/taste attributes and are often referred to as 'mustard oils'. More than 120 glucosinolates structures (i.e. with different 'R' groups) have been identified but only around 16 are common in plant foods consumed by humans (see Chapter 2 and Fahey *et al.* (2001)). Figure 8.7 shows the structures of glucosinolates and isothiocyanates from commonly consumed crop plants.

In many instances, plant breeding programmes have led to reductions in the levels of glucosinolates in cruciferous crop plants compared with their wild parents. The impetus for this has been the potential toxicity of glucosinolates in animals and humans, which has lead to the development of 'double-zero' rape seed to remove progoitrin. In addition there has also been a desire to reduce the pungency and/or bitterness of cruciferous plants grown commercially for human consumption. In contrast, the current interest in glucosinolates as health-promoting components of the diet has resulted in the development of cruciferous plants with increased levels of glucosinolates. For example, broccoli with enhanced levels of 4-(methylsulphinyl)butyl glucosinolate (glucoraphanin; the precursor of the isothiocyanate sulphoraphane) has been developed through a conventional breeding programme (Faulkner *et al.* 1998) and broccoli sprouts have been promoted as an alternative to mature broccoli florets on the basis of their very high glucoraphanin/sulphoraphane content (Fahey *et al.* 1997).

The current interest in glucosinolates is largely focused on their ability to protect against cancer. There is good evidence from epidemiological studies showing an inverse relationship between cruciferous vegetable consumption and cancer risk. In addition, some of the breakdown products arising following hydrolysis of the parent glucosinolate have been shown to have a number of activities in cell and animal models that would explain this anti-cancer action, including down-regulation of phase-I 'activation' enzymes, induction of phase-II 'detoxification' enzymes, induction of a cellular antioxidant response, inhibition of cellular proliferation, induction of apoptosis (programmed cell death) and cell cycle arrest. For example, sulphoraphane, the most studied isothiocyanate which was first isolated from broccoli (Zhang *et al.* 1992b), is a potent inducer of phase-II enzymes such as glutathione S-transferases (GSTs) and is able to block chemically-induced carcinogenesis in several animal models (Chung *et al.* 2000).

(a) *Basic glucosinolate (GLS) structure*

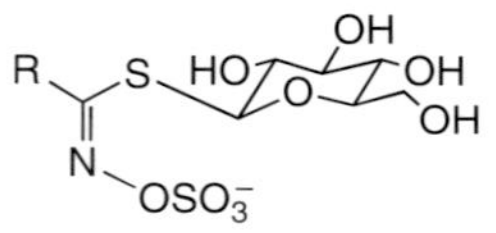

(b) *Glucosinolates and corresponding isothiocyanates in commonly consumed crop plants*

Glucosinolate structure	Trivial name	Corresponding isothiocyanate (R–N=C=S)	Example crop plants
3-(Methylsulphinyl)propyl-GLS	Glucoiberin	$(CH_2)_3S(=O)CH_3$-ITC (Iberin)	Brussels sprouts, Savoy cabbage, white cabbage, green cabbage
4-(Methylsulphinyl)butyl-GLS	Glucoraphinin	$(CH_2)_4S(=O)CH_3$-ITC (Sulphoraphane)	Broccoli, swede, Brussels sprouts, various cabbage varieties
3-Indolylmethyl-GLS	Glucobrassicin	3-Indolylmethyl-ITC	Brussels sprouts, broccoli, various cabbage varieties, swede, curly kale
Allyl (2-propenyl)-GLS	Sinigrin	$CH_2=CH-CH_2$-ITC (Allyl isothiocyanate)	Brussels sprouts, Savoy cabbage, red cabbage, green cabbage, white cabbage, mustard, cress
2-Phenethyl-GLS	Gluconasturtiin	$(C_6H_{12})-CH_2-CH_2$-ITC (Phenylethyl-ITC)	Watercress, swede, turnip, radish

Figure 8.7 Basic glucosinolate structure (a) and commonly consumed glucosinolates and their respective isothiocyanates (b). ITC, isothiocyanate.

The ability of glucosinolate to break down products from the diet to influence cancer risk will depend on how efficiently they are absorbed from the intestine, what metabolism they undergo, whether they accumulate in tissues/cells and how rapidly they are excreted. Readers are referred to a detailed review on the bioavailability of glucosinolates and related compounds by Holst and Williamson (2004) that provides more detail than can be covered here.

8.7.1 Hydrolysis of glucosinolates and product formation

Although intact glucosinolates have been detected in samples of plasma and urine of poultry fed rapeseed meal (Slominski *et al.* 1987), the amounts were small and the absorption of intact glucosinolates is likely to be a minor pathway. The general consensus is that intact glucosinolates are not efficiently absorbed across the intestinal barrier, probably due to their hydrophilic nature. Further, there is no evidence that the thioglucosidase activity required to hydrolyse glucosinolates and facilitate the formation of the various breakdown products exists in human cells including those lining the gut. Two mechanisms exist for the hydrolysis of glucosinolates. Glucosinolate-containing plants also contain significant quantities (often a large excess) of myrosinase, a thioglucosidase, which is maintained in a different cellular compartment to the glucosinolates. Disruption of the tissue, such as would occur with slicing, crushing and chewing, leads to contact between the enzyme and substrates and facilitates rapid deglycosylation. Myrosinase enzymes are covered further in Chapter 2 of this book. An alternative route for deglycosylation is facilitated by the colonic flora which has been shown to produce a thioglucosidase activity and is able to hydrolyse glucosinolates. Hydrolysis of the thioglucoside moiety of glucosinolates generates an unstable product (thiohydroxamate-*O*-sulphonate) which rapidly and spontaneously gives rise to several products (Figure 8.8; see also Chapter 2)

Commonly, isothiocyanates are formed from thiohydroxamate-*O*-sulphonate, and these have been most extensively studied in terms of bioavailability and anti-cancer activity. Once formed, most isothiocyanates are stable unless the parent glucosinolate contains a β-hydroxylated side chain or an indole moiety. However, other compounds are also formed; the number of products and the amount of each produced depends on various factors including the pH and whether active epithiospecifier protein (ESP) is present. ESP activity is also important since it favours the formation of (biologically inactive) nitrile products rather than isothiocyanates. Low pH also favours nitrile formation. In the case of raw vegetables, there may be considerable nitrile formation at the expense of isothiocyanate formation due to the presence of active ESP. In lightly cooked vegetables, the relatively heat-labile ESP can be denatured while the comparatively heat-stable myrosinase activity remains, facilitating isothiocyanate formation (Matusheski *et al.* 2004). In overcooked vegetables myrosinase is also inactivated and intact glucosinolates are ingested. Conaway *et al.* (2000) have shown that the urinary excretion of total isothiocyanates from overcooked broccoli (i.e. steamed for 15 min) was only 30% of that for fresh broccoli, which was related with the heat-induced loss of myrosinase activity from the cooked material. The total urinary excretion of isothiocyanates from fresh broccoli sprouts (isothiocyanate dose) was about six-fold higher than for an equivalent dose of cooked broccoli sprouts (glucosinolate dose) (Shapiro *et al.* 2001). Further, the peak plasma concentration following consumption

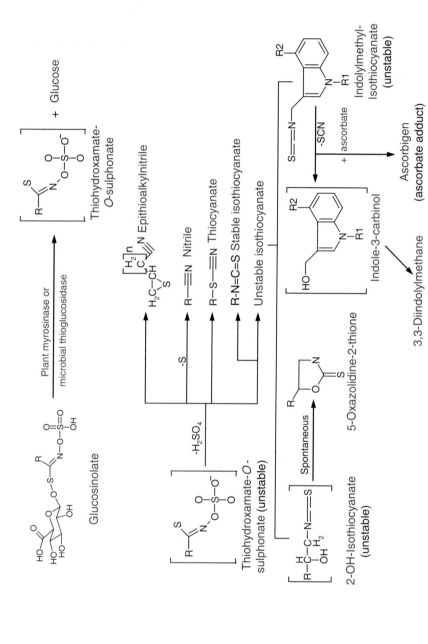

Figure 8.8 Scheme for glucosinolate breakdown and product formation.

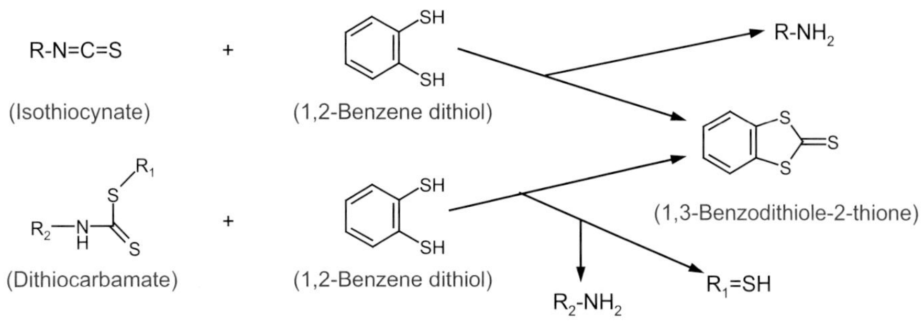

Figure 8.9 Reaction of an isothiocyanate with 1,2-benzenedithiol which gives rise to a cyclocondensation product (1,3-benzodithiole-2-thione) and an amine (R-NH$_2$).

of fresh broccoli occurred at about 2 h compared with 6 h for steamed broccoli because hydrolysis of glucosinolates in the cooked broccoli required the action of the gut microflora. Hence, domestic processing is a major factor determining the exposure to isothiocyanates from cruciferous vegetable consumption.

8.7.2 Analytical methods

Levels of total glucosinolates in foods can be quantified by methods that measure the amount of glucose released in the presence of myrosinase (Heaney *et al.* 1988). This approach requires care in accounting for the levels of endogenous glucose. Another approach has been to quantify the levels of bisulphate released following myrosinase hydrolysis, since this reaction is equimolar with respect to starting glucosinolate and product bisulphate, but these are rarely used today. Since quality controls need to include an assessment to ensure there is no remaining glucosinolates (e.g. following a chromatographic separation), and there is interest in data on glucosinolate profiles as well as quantity, the preference is for methods that include a chromatographic step such as HPLC or GC. HPLC is the current method of choice because it is reproducible and avoids the high temperatures associated with GC, which can lead to degradation of some glucosinolates products, such as indoles. This method of separation can be combined with MS, which is powerful in terms of sensitivity and selectivity, and modern methods facilitate the quantification of glucosinolates in crude plant extracts in a single chromatographic separation (Mellon *et al.* 2002).

The analysis of glucosinolates breakdown products is rather more complicated due to the volatility of many of the compounds and the lack of absorbance exhibited by some of the products, namely thiocyanates and nitriles. A particular analytical shortcoming is the lack of a suitable method for measuring the levels of a number of different glucosinolate hydrolysis products in plasma and urine samples, and this should be a priority for future research.

Many of the reports concerned with measuring organic isothiocyanates in plasma have employed a method described by Zhang *et al.* (1992a). This method makes use of the quantitative reaction that occurs between isothiocyanates (and their dithiocarbamate metabolites) and an excess of vicinal dithiols, resulting in the formation of condensation products with five-membered rings and release of the free amines (R-NH$_2$)

(Figure 8.9). Various studies have used this approach with plasma (and urine) samples to quantify total isothiocyanates (Getahun and Chung, 1999; Conaway et al. 2000; Shapiro et al. 2001; Ye et al. 2002). Samples of plasma/urine are first reacted with vicinal dithiols such as 1,2-benzenedithiol, the cyclocondensation reaction products (in this case, 1,3-benzenedithiole-2-thione) are extracted, and samples analysed by HPLC with absorbance detection. Although this method has been validated and has proved useful in the quantification of total isothiocyanates and their dithiocarbamate metabolites, it does have drawbacks. In particular, no information is provided as to the nature of the isothiocyanates in the samples, and other glucosinolates breakdown products (e.g. nitriles, indoles, thiocyanates) cannot be quantified using this approach. There is a need for a method or methods allowing quantification of a number of glucosinolates breakdown products in plasma, since there is a lack of information with regard to peripheral exposure to glucosinolates breakdown products following ingestion of glucosinolates.

The major forms of isothiocyanates present in urine are N-acetyl-cysteine conjugates (mercapturic acids) and these have been measured using HPLC with MS detection (Chung et al. 1992; Conaway et al. 2000; Fowke et al. 2003; Petri et al. 2003; Rouzaud et al. 2004). Appropriate standards can be readily synthesized. Figure 8.9 shows the reaction of an isothiocyanate with 1,2-benzenedithiol which gives rise to a cyclo-condensation product (1,3-benzodithiole-2-thione) and an amine (R-NH$_2$).

8.7.3 Absorption of isothiocyanates from the gastrointestinal tract

Whereas glucosinolates are effectively not bioavailable, the isothiocyanate breakdown products are rapidly and efficiently absorbed from the lumen of the upper gastrointestinal tract. Data obtained from jejunal in situ perfusion experiments with healthy volunteers showed that on average 74% of the sulphoraphane in the perfusate was absorbed from the lumen compared with less than 9% of a flavonoid glycoside (Petri et al. 2003). In the same experiments, the mean effective jejunal permeability was estimated at $18.7 \pm 12.6 \times 10^{-4}$ cm/s, indicating a high luminal disappearance rate. Pharmacokinetic studies in rodents and humans indicate that isothiocyanate appearance in peripheral blood occurs soon after ingestion of isothiocyanates and peaks within 1–3 h of oral dosing (Bollard et al. 1997; Conaway et al. 2000). In contrast, the appearance of isothiocyanates in blood following ingestion of intact glucosinolates (i.e. in the absence of myrosinase, as found in overcooked plant material) occurs significantly later and correlates with the time that glucosinolates would be expected to reach the microflora in the distal gut (Conaway et al. 2000).

8.7.4 Intestinal metabolism and efflux

The small intestine is an important site for xenobiotic metabolism. When the jejunum of healthy individuals was perfused with a solution of sulphoraphane, about 75% of the sulphoraphane was absorbed (i.e. taken up by the enterocytes) but a significant fraction (43%) of this absorbed material was excreted back into the gut lumen as a sulphoraphane-glutathione conjugate (Petri et al. 2003). Although isothiocyanate-glutathione conjugates are hydrophilic and therefore likely to be poorly absorbed by the gut wall, they would be delivered to the distal gut and the isothiocyanate may become bioavailable if released by microbial action.

Isothiocyanates accumulate in cells, and can reach quite remarkable concentrations. Studies utilizing cultured mammalian cells derived from various tissues demonstrated that isothiocyanates such as sulphoraphane are accumulated in cells at concentrations several hundred-fold above those outside the cells, and can reach millimolar levels (Zhang and Talalay 1998; Zhang 2000). Further, it was shown that the isothiocyanates are present in the cells predominantly as glutathione conjugates, indicating that very rapid conjugation occurred. Glutathione transferases are abundant in many mammalian tissues including the intestine, liver and kidney, and intracellular glutathione levels are considerable (~6 mM; plasma levels are ~0.004 mM). Similar studies have shown that a portion of the cellular isothiocyanates can also be present as cysteinylglycine conjugates (Callaway *et al.* 2004). Further, the isothiocyanate conjugates appear to be rapidly exported from cells in which they have accumulated, and there is evidence that this is an active process mediated by cellular export pumps including MRP-1 and P-glycoprotein-1 (Pgp-1) (Zhang and Callaway 2002; Callaway *et al.* 2004). The efflux of conjugated isothiocyanates from cells occurs rapidly. The maintenance of a high intra-cellular concentration of glutathione conjugates may be required for at least some of the anticarcinogenic actions of isothiocyanates (Zhang and Talalay 1998), hence extracellular isothiocyanate concentrations and cellular export of isothiocyanate conjugates are important factors determining the potency of isothiocyanates in cancer prevention. Figure 8.10 illustrates the cellular uptake, metabolism and export of isiothiocyanates.

8.7.5 Distribution and elimination

Whereas much is known of the absorption and metabolism of isothiocyanates, virtually nothing is known of the absorption and metabolism of the other products arising from the breakdown of glucosinolates (e.g. nitriles, thiocyanates). This is probably because of two main reasons: first, the majority of research has focused on isothiocyanates due to their potent anti-cancer activities, and second, there is a lack of methods for measuring other breakdown products. In particular, the dithiol condensation reaction is the most reported approach but is specific for isothiocyanates.

The major metabolic route for isothiocyanates in humans is via the mercapturic acid pathway which gives rise to N-acetyl-cysteine (NAC) derivatives (Figure 8.10). In the first step, isothiocyanates are conjugated with glutathione, catalysed by GSTs. The glutathione conjugates can undergo hydrolysis to yield the cysteine derivatives which are subsequently N-acetylated. Several studies have used HPLC-MS to analyse NAC-derivatives in urine collected after the ingestion of glucosinolates or isothiocyanates, and the amounts are substantial. For example, NAC-derivatives excreted in urine accounted for 30–67% of the oral dose for sulphoraphane, allyl isothiocyanate, phenethyl isothiocyanates and benzyl isothiocyanates (Mennicke *et al.* 1988; Chung *et al.* 1992; Jiao *et al.* 1994; Shapiro *et al.* 1998).

The urinary yield of isothiocyanates depends on the form in which the dose is given. The urinary yield of total isothiocyanates (determined using the dithiol cyclocondensation method) from an oral dose of sulphoraphane was 80% but for the parent glucosinolate

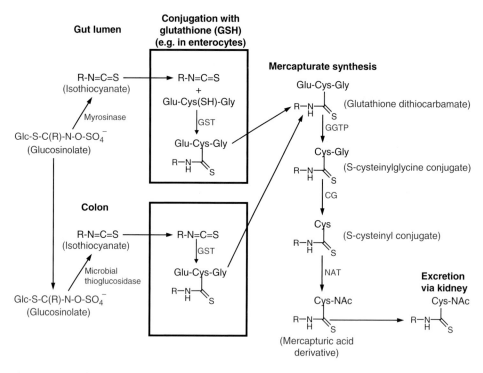

Figure 8.10 Cellular uptake, metabolism and export of isothiocyanates. Glu, glutamate; Cys, cysteine; Gly, glycine; GST, Glutathione-*S*-transferase; GGTP, γ-glutamyltransferase; CG, cysteinylglycinase; NAT, *N*-acetyltransferase.

and glucoraphanin, the total isothiocyanate yield was only 12% (Shapiro *et al.* 2001). When cruciferous vegetables are consumed, the urinary yield is strongly dependent on the degree of heating during cooking and on the amounts of crushing or chewing of the vegetable tissues. Thoroughly chewed broccoli sprouts generated a higher urinary yield of isothiocyanates compared with sprouts that were swallowed whole (Shapiro *et al.* 2001), while the bioavailability of total isothiocyanates from fresh broccoli (Conaway *et al.* 2000), broccoli sprouts (Shapiro *et al.* 2001) and watercress (Getahun and Chung 1999) were considerably lower if the vegetables were cooked prior to consumption compared with when consumed fresh.

8.8 Carotenoids

Humans do not have the ability to synthesize retinol (vitamin A) and have thus evolved to derive this compound from the diet, either directly as preformed retinol from foods of animal origin or from the metabolism of carotenoids primarily derived from plant tissues. Several hundred carotenoids have been isolated, named and their structures elucidated, but only a few have a known biochemical role in human metabolism or are present in the diet in sufficient quantities to be detected in plasma (Khachik *et al.* 1992). The chemical

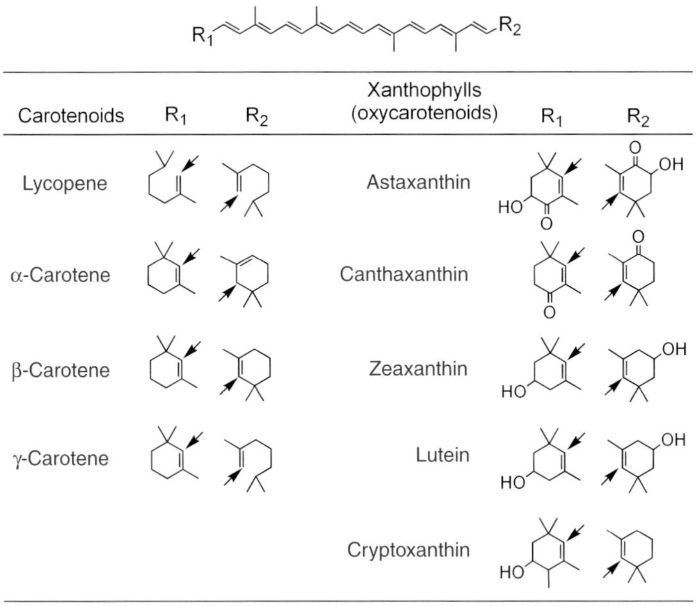

Figure 8.11 Chemical structures of dietary carotenoids.

structure of carotenoids means that they range in colour from pale yellow to intense red-orange and are most abundant in green photosynthetic plant tissues where their natural colour is masked by chlorophyll or in fruits, roots, flowers, seeds, leaves, etc where the colour provides visual cues as to ripeness, suitability as food, guidance for pollinating insects, photoprotectants and indicators of senescence. Because of the natural occurrence of carotenoids some are also permitted food colourants and may be added to foods either as the isolated compound, ester or as stabilized dispersible formulations. The amounts added are usually small because some of the carotenoids are very intensely coloured.

The carotenoids generally found in foods are linear all-*trans* (E form) C^{40} polyenes formed from eight isoprenoid units. The structures of common carotenoids are shown in Figure 8.11. The linear conjugated polyene structure has the ability to delocalize an unpaired electron and hence the capacity to act to terminate free radical reactions with the production of resonance stabilized free radical structures. Thus, carotenoids may potentially (a) provide retinol and (b) act as antioxidants.

The main dietary carotenoids are lycopene (linear, no substitutions), β-carotene and α-carotene (ring closure at both ends, no substitutions), β-cryptoxanthin (ring closure at both ends, substitution in the 3 position), lutein (ring closure at both ends, substitutions in the 3 and 3′ positions) and canthaxanthin (ring closure at both ends, [O] substitutions in the 4 and 4′ positions). In some tissues, particularly flower petals, the hydroxylated carotenoids may also be present as mono- or di-acyl esters, most commonly with C16 fatty acids. Further oxidation of the terminal ring may occur to produce the mono- and di-epoxides. For an exhaustive list of carotenoids, the 'Key to Carotenoids' (Straub 1987) is a recommended reference source.

The hydrocarbon carotenoids are apolar lipophylic molecules and show no solubility in water but are readily soluble in organic solvents and to some extent in fats and oils. The presence of hydroxy or keto groups gives the molecule some polarity but such compounds are still predominantly hydrophobic; nevertheless, these substitutions may significantly affect their distribution in biological systems. The all-*trans* structure is frequently subject to isomerization giving a *cis*-configuration (Z form) at various positions on the polyene backbone. Isomerization generally occurs as a result of high temperature processing but some β-carotene supplements derived from algae may naturally contain a high proportion (50%) of 11-*cis*-β-carotene (Mokady *et al.* 1990). Such isomeric change may have a significant effect on the physical and biochemical properties of the molecule.

8.8.1 Mechanisms regulating carotenoid absorption

About 90% of a large but physiological amount of β-carotene (10 mg), when dispersed in oil can be absorbed (Faulks *et al.* 1997), about 65% from cooked purée carrot but only about 40% from raw carrots (Liveny *et al.* 2003). About 25% of β-carotene was absorbed from cooked leaf spinach as measured by mass balance in ileostomy volunteers (Faulks *et al.* 2004). It would seem, therefore, that in normal healthy volunteers there is little inhibition to absorption of modest doses of free β-carotene and that the poor absorption from some foods is a result of the form of the carotenoid and the absence of an effective system of extraction from the food matrix during digestion.

It is also recognized that co-ingested dietary fat improves carotenoid bioavailability (Dimitrov *et al.* 1988; Prince and Frisoli, 1993; Reddy *et al.* 1995; Jayarajan *et al.* 1980). Because of their hydrophobic nature the carotenoids are constrained to the hydrophobic domains of plant tissues where they may be present as part of the photosynthetic mechanism in leaf chloroplasts (Cogdell 1988), as semicrystalline bodies in fruit and roots (Thelander *et al.* 1985) or dissolved in lipid droplets within plastids in ripe fruit (de Pee *et al.* 1998). Absorption studies would indicate that carotenoid present as semicrystalline bodies is most difficult to absorb, followed by carotenoids in leaves, with fruit carotenoids dispersed in oil droplets being most easily absorbed.

This hierarchy seems logical if we consider processing of food in the gastro-intestinal (GI) tract and how absorption can be manipulated by food processing. It is clear that carotenoids dissolved or dispersed in lipid are most efficiently absorbed and that the presence of lipid is essential for the absorption process. However, carotenoids in foods are generally embedded in a structured food matrix, the major component of which is water. With foods of plant origin, the structure of the matrix is made up of plant cell walls which are not digestible in the stomach or small intestine. The structure of much raw vegetable matter, although it may undergo some physical changes during passage through the upper GI tract, essentially leaves the terminal ileum intact. In order for the carotenoids to be absorbed they must be released from the plant cells through cell breakage or 'washed out' of this physically constrained aqueous environment by partitioning into lipid phases during the passage of the food through the upper GI tract, a period of around 6–8 h.

From such a model it will be appreciated that bulk lipid, even if it is emulsified in the gastric antrum or in the duodenum in the presence of bile salts cannot effectively penetrate pieces of vegetable tissue to wash out the carotenoid. However, once the emulsion droplets enter the duodenal environment the pH change alters the zeta-potential (oil emulsion droplet surface charge) and the addition of bile salts strips surface active proteins that may have contributed as a barrier to mass transfer of carotenoids. The 'clean' interface created by the bile salts permits rapid digestion of the lipid emulsion droplets by pancreatic lipase. The products of lipid digestion, mainly fatty acids, mono- and di-acylglycerols together with native triacylglycerols and bile salts change the surface characteristics of the emulsion droplet and give rise to mixed micelles. The mixed micelles also have the potential to dissolve the carotenoids. Because this is an interface process and the carotenoids are partitioned between the bulk oil and the interface, depending on their polarity, it may give rise to selective absorption or a temporal shift in absorption between the xanthophylls and the hydrocarbon carotenoids or between different isomers of the same carotenoid (Gaziano *et al.* 1995; Gartner *et al.* 1997).

8.8.2 Effects of processing

Most foods that we consume have undergone at least one of two main processes in either order: (1) particle size reduction (milling, grinding and shredding) and (2) thermal treatment, commonly with other food components. These processes generally make the food microbiologically safe, soften and disperse tissue architecture and improve palatability and digestibility. The disintegration of carotenoid containing foods helps to reduce the 'encapsulating' effect of the food structure and, if processed in the presence of lipid, helps to transfer some of the carotenoid to the lipid phase. Hence, it is common to observe yellow-orange fat globules when, for example, meat is cooked with carrot or tomato. Improved absorption of lycopene has been observed if tomato is puréed or heat treated in the presence of lipid (Stahl and Sies 1992; Gartner *et al.* 1997). Medium chain triglycerides are not re-synthesized into triacylglycerols but are absorbed directly into the portal vein if not metabolized in the enterocyte. In volunteers fed medium chain triglycerides and carotenoids a much reduced plasma excursion in carotenoid concentration is observed (Borel *et al.* 1998) but it is not clear whether there is a reduced absorption of carotenoids or simply a reduced transport to the serosal side. The carotenoids may be absorbed and retained in the enterocytes and hence not detected in the plasma. Figure 8.12 illustrates the known essential features of carotenoid mass transfer to absorbable lipid species in the upper GI tract (Rich *et al.* 2003a,b).

8.8.3 Measuring absorption

In the past most carotenoid 'availability' studies have been based on oral–faecal mass balance but over the last 25 years other methods have been used to assess their absorption. Methods vary from the simple plasma response following an oral dose (Johnson and Russell 1992) to analyzing plasma fractions, most commonly, the Triglyceride Rich Lipoproteins (TRL, chylomicron fraction) (van Vliet *et al.* 1995; Borel *et al.* 1996; Faulks *et al.* 2004), the use of radio (Goodman *et al.* 1966; Blomstrand and Werner, 1996) and stable isotopes (Parker *et al.* 1993; Dueker *et al.* 1994) and mass balance techniques (Shiau *et al.*, 1994; Faulks *et al.* 1997). It was soon noted, on the basis of plasma response, that humans tended to fall into groups of 'responders' and 'non-responders' (Johnson and

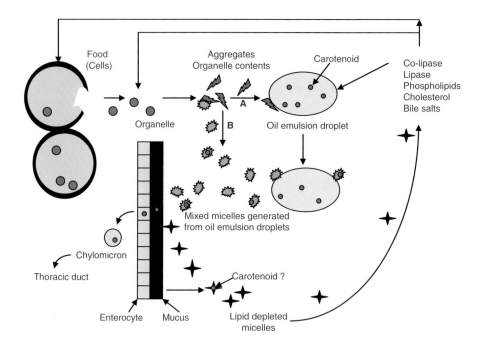

A - Carotenoid dissolution in bulk lipid
B - Carotenoid dissolution in mixed micelles

Figure 8.12 This figure illustrates the known essential features of carotenoid mass transfer to absorbable lipid species in the upper GI tract (Rich et al. 2003): (1) Carotenoids need to be released from the plant cells through cell breakage. (2) Carotenoid aggregates or carotenoid-containing plastids need to adsorb to emulsion droplet interface (Route A) or to mixed micelles (Route B) to provide a 'hydrophobic continuum'. This process is subject to the quality of the interface for example its composition and charge (zeta-potential). (3) The apolar (hydrocarbon) carotenoids are located in the core of the droplet while those with polar end groups (xanthophylls) are at the interface (Borel et al. 1996). (4) The carotenoid 'carrying capacity' of the structured mixed micelles will depend on their composition but it is generally thought to be much less than the carotenoid solubility in bulk lipid. (5) Mixed micelles are believed to dissociate at the enterocyte brush-border where the lipid component is absorbed. The carotenoids are also transferred to the enterocyte during this process but the mechanism is not well understood. Unabsorbed carotenoid may remain with the lipid-depleted bile salt micelles. (6) Lipid-depleted micelles (bile salt micelles) are recycled in the upper GI tract. This recycling is essential to provide sufficient bile salts for lipolysis and lipid transport. (7) Lipid absorbed by the enterocyte is re-synthesised into triglyceride. The triglyceride, lipid-soluble compounds, including carotenoids, are packaged into chylomicrons and released into the mesenteric lymph. (8) Carotenoids are rapidly dispersed to plasma lipoproteins and hence to tissues.

Russell 1992), and volunteers who habitually have a low plasma carotenoid concentration show a much smaller (but proportional) response than volunteers who have a high fasting plasma carotenoid concentration. However there is a high degree of intra-individual consistency (Borel et al. 1998). These results have been interpreted as reflecting differences in absorption.

However, a more detailed study (Borel *et al.* 1998) concluded that the plasma response also depends on the rate of clearance of the chylomicrons. Inter-individual differences in plasma response therefore cannot be interpreted as differences in the rate and extent of absorption unless chylomicron clearance rates are taken into account. It must also be emphasized that the absence of a detectable change in plasma concentration following a meal/dose cannot be interpreted as no, or very low, absorption, particularly bearing in mind that the carotenoids may be temporally stored in the enterocyte. In a mass balance study in ileostomy volunteers, coupled with plasma carotenoid excursions (Faulks *et al.* 1997) none of the volunteers showed a plasma response, yet they all apparently absorbed significant amounts of carotenoid, indicating that both the size of the dose and the clearance kinetics are crucial to observing a significant change in plasma concentration on top of the endogenous plasma carotenoids.

Newly absorbed carotenoid is most easily detected in the chylomicron or TRL plasma fraction, even in the absence of a whole-plasma response (Faulks *et al.* 1997) partly because this fraction is small and devoid of carotenoid in fasting volunteers, it is the most responsive pool and does not contain carotenoids sequestered and re-exported by the liver. The only disadvantage in this method is the necessity of obtaining fairly large plasma samples (5 ml) and the time taken to density adjust the plasma, ultracentrifugation and the quantitative recovery of the chylomicron fraction. A shorter and less demanding protocol for the isolation of chylomicrons has now been described (Borel *et al.* 1998).

8.8.4 Transport

The chylomicrons have a biologically controlled composition and consist of about 2% proteins made up of apoproteins A-1, A-2 and B-48 and 98% dietary lipid and carry the carotenoids and other hydrophobic compounds in the mesenteric lymph and hence into the thoracic duct. Unlike water soluble compounds that are transported via the portal vein to the liver, the thoracic duct empties into the inferior vena cava and hence into the main blood circulation. The first main metabolic site 'seen' by the chylomicrons is the extrahepatic capillary bed. The blood capillaries have an endothelial lipoprotein lipase which hydrolyses the chylomicron lipid, the products of which are absorbed by the endothelial cells, removing much of the lipid from circulation and leaving lipid depleted chylomicron remnants. The half-life of the chylomicrons is very brief in normal individuals, 11 min, but can be much longer in those with hyperlipidaemia, especially hyperchylomicronaemia, where it can be up to several hours (Grundy and Mok, 1976; Cortner *et al.* 1987). It is not known how much of the carotenoid is transferred from the plasma chylomicrons to the tissues in the extrahepatic capillary bed or if there is selectivity to absorb one or more of the carotenoids or carotenoid isomers (Gartner *et al.* 1996). The chylomicron remnants are avidly scavenged by the liver with any associated carotenoid sequestered and re-excreted in very low density lipoproteins (VLDL) and hence into the low density and high density lipoproteins (LDL, HDL).

There are very small numbers of chylomicrons present in plasma from fasting volunteers and they are generally carotenoid free, the bulk of carotenoids being carried by the LDL and HDL. The apolar and polar carotenoids show different distributions between these two lipoproteins. LDL carries around 80% of plasma β-carotene and lycopene but only about

50% of lutein and zeaxanthin, the remainder being carried by HDL (Krinsky et al. 1958; Lepage et al. 1994; Vogel et al. 1996). There is as yet no convincing explanation as to why the different classes of carotenoid partition in this way, and in the absence of demonstrable binding sites it is assumed that this is simply a reflection of a thermodynamically favourable distribution.

Human blood plasma contains mainly all-*trans* forms of the common dietary carotenoids but 5-cis-lycopene (up to 50% of the total plasma lycopene) and 9-cis- and 9'-cis-lutein and 9-cis-β-carotene (Khachik et al. 1992) are also commonly found. In some cases 5-cis-lycopene appears in plasma in a much greater proportion than in the food (Schierle et al. 1997). This could suggest that the 5-cis-lycopene is preferentially absorbed, or less rapidly cleared from the plasma, or that all-*trans*-lycopene undergoes isomerization as a result of some biochemical interaction. Simulated digestion *in vitro* does not cause significant acid-catalysed isomerization.

It has been reported that all *trans*-β-carotene preferentially accumulates in the lipoprotein carriers (Stahl et al. 1995) and (Gaziano et al. 1995) it was found that there was a marked preferential absorption of the all *trans*-β-carotene in males. Supplementation of female volunteers with 15 mg/day of palm oil carotenoids (a mixture of *trans* and *cis*-β-carotene) elicited a plasma response where the ratio of *cis:trans* forms was much lower than in the supplement but consistent across all volunteers, irrespective of the total plasma concentration of β-carotene (Faulks et al. 1998). This would indicate that the t*rans* form is better absorbed than the *cis* form, or that the *cis* form is cleared from the plasma more rapidly. However, in ileostomists given an acute oral dose of all-*trans*-β-carotene and 9-cis-β-carotene, both isomers appeared to be equally well absorbed from the gut and *cis-trans* isomerization did not occur during passage of the β-carotene through the GI tract (Faulks et al. 1997). Figure 8.13 illustrates how carotenoids are absorbed and distributed to the body.

8.8.5 Tissue distribution

There is not much information on the tissue concentration of carotenoids in relation to the long term plasma concentrations although there are a number of studies which clearly indicate that tissue concentrations increase during periods where the diet is enriched or supplemented with carotenoids for example breast milk (Lietz et al. 2001; Alien et al. 2002; Gossage et al. 2002), buccal mucosa (Paetau et al. 1999; Allen et al. 2003 ; Reifen et al. 2003), eye (Bone et al. 2003), adipose tissue (El-Sohemy et al. 2002; Gomez-Aracena et al. 2003; Walfisch et al. 2003) and skin (Walfisch et al. 2003). However, although the concentration of carotenoids may differ between tissues (Kaplan et al. 1990) there is no evidence of accumulation of any of the carotenoids in any tissue other than perhaps the eye (canthaxanthin retinopathy) which is known to contain membrane proteins with a high affinity for the xanthophyll carotenoids (Yemelyanov et al. 2001). In the absence of high affinity binding systems the carotenoids in tissues are in dynamic equilibrium with the transport system (plasma), and therefore tissue concentrations may be expected to fluctuate more or less with plasma concentration although some delays and attenuation can be expected. It is most likely that the carotenoids partition throughout the body 'pools' in a way that is thermodynamically favourable rather than being driven by energy dependent

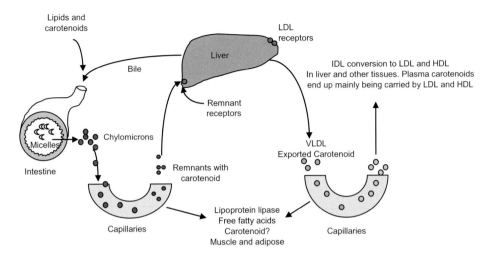

It is not clear how much of the carotenoid is dropped off each time the lipoprotein carrier is sequestered. Tissues with large numbers of LDL receptors tend to be the tissues with the highest carotenoid content, eg. liver.

Figure 8.13 Scheme depicting how carotenoids are absorbed and distributed to the body.

processes. The sudden (transient) appearance of a high concentration of a carotenoid in a tissue 'pool' during acute or chronic supplementation should not be interpreted as 'accumulation' unless there is clear evidence of a change in the partitioning ratio being dependent on the total concentration. Manipulation of tissue carotenoid concentration is therefore wholly dependent on plasma concentration which in turn is dependent on intake and absorption. It should be noted that any pharmacological action, for example mitigation of risk of cancer or cardiovascular disease and any hypothesized anti-oxidant effect or regulation of gene expression will be dependent on the effective carotenoid being delivered to the tissue, cell and sub-cellular target in sufficient quantity in a timely manner.

8.8.6 Metabolism

Once absorbed into the enterocyte, the carotenoids, which may act as retinol precursors, may be cleaved; β-carotene, the main precursor of vitamin A, theoretically producing two molecules of retinal (central cleavage), which is reduced to retinol or oxidized (irreversibly) to retinoic acid, while eccentric cleavage only gives rise to apocarotenals in which the conjugated polyene is successively shortened by β-oxidation to one molecule of retinoic acid. Such theoretical yields are not normally observed and the average retinol equivalent of β-carotene is taken as 6 : 1 (β-carotene : retinol, wt : wt) or 3 : 1 as a molar ratio (FAO/WHO, 1967). It is unclear what the relative contributions of absorption and conversion make to this figure, but recent studies with small doses of stable isotope (^{13}C) labelled β-carotene give a conversion ratio of 2.4 : 1 (wt:wt) (Van Lieshout *et al.* 2001). This would indicate an absorption efficiency of around 40%. Conversion to retinol is dependent on the presence of a β-ionone ring and all carotenoids possessing such a structure have potential vitamin A activity. Minor carotenoids with only one β-ionone ring (α-carotene) are assigned a retinol equivalent of 12:1 (wt:wt). Retinol produced in the enterocyte is esterified before being exported in the chylomicrons and then into the blood from which it is sequestered and stored by the liver. During this process there is generally no change in plasma retinol. It should, however, be noted that the plasma concentration of retinol is strictly controlled to the extent that carotenoderma (yellowing of the skin) may occur with chronic high β-carotene intake without significantly altering the plasma retinol concentration. Conversely, in retinol-deficient individuals, both the absorption and conversion of β-carotene to retinol may be much higher unless the deficiency is so severe as to affect absorption. The use of the 6:1 conversion factor for retinol equivalents of β-carotene should, therefore, be treated only as a very rough guide (Scott and Rodriquez-Amaya 1999; Mackerras 2004).

The 9-*cis*-β-carotene gives rise to 9-*cis*-retinoic acid which has retinoid activity and 11-*cis*-β-carotene yields 11-*cis*-retinol which is central to the regeneration of rhodopsin in the eye. It is not known if other β-carotene *cis*-isomers produce biologically active retinoids. There is currently some debate as to whether some of the other carotenoids can give rise to other biologically active retinoid analogues.

β-Carotene and other carotenoids with retinol potential that are not cleaved in the enterocyte may be cleaved in the liver, kidney and possibly other tissues that have the necessary enzyme, 15,15'-β-carotenoid dioxygenase (EC 1.13.11.21), and there is some debate as to the relative contribution of the different possible sites of cleavage.

8.8.7 Toxicity

As far as can be ascertained, the carotenoids are not cyto- or genotoxic in either large acute doses or in chronic supplementation studies lasting several years. However, chronic supplementation in smokers has been found to lead to a higher incidence of lung cancer (Goodman *et al.* 2004). Carotenoderma is commonly encountered in supplementation studies and in individuals who consume large quantities of carrot or carrot products, particularly if they have a low Body Mass Index. The condition is rapidly reversible after ceasing intake. Conversion to retinol is controlled so that vitamin A toxicity does not occur. However, consumption of high levels of lutein, zeaxanthin and canthaxanthin, which are accumulated in the eye, may lead to problems, for example canthaxanthin retinopathy.

8.8.8 Other metabolism

There is very little information on the metabolic fate of the non-pro-vitamin A carotenoids or the pro-vitamin A carotenoids that are not metabolized to retinol. It is assumed that they undergo oxidation (photo bleaching in the skin?), cleavage and polyene chain shortening by a process analogous to the β-oxidation of fats, and the unmetabolizable remnants are detoxified in the liver and are then eliminated in the faeces (enterohepatic circulation) and urine. Excretion of carotenoids via enterohepatic circulation has not been observed in ileostomy volunteers in that the ileal effluent is virtually free of carotenoids when volunteers are fed a carotenoid free diet. The observation that large intakes of carotenoids for example, β-carotene, canthaxanthin (oral bronzing agent) can result in yellowing of the skin suggests that the skin is a significant excretory mechanism.

8.9 Conclusions

The human diet consists of numerous classes of plant secondary metabolites that may potentially benefit human health. These compounds have complex underlying mechanisms controlling their uptake, distribution and metabolism. Flavonoids are partially bioavailable in humans when consumed in foods, with a proportion reaching the plasma and allowing exposure to tissues beyond the gut through peripheral circulation. The nature of the metabolites present in plasma depends to some extent on the structure of the flavonoid itself, but in most cases, flavonoids are present as combinations of sulphate, glucuronate and methyl conjugates. Hydroxycinnamic acids are rapidly absorbed from the stomach or the small intestine and are methylated, glucuronidated and sulphated in the same way as flavonoids. Gallic acid appears to be very well absorbed in humans compared with other polyphenols with conjugated forms reaching 4 μM after ingestion of 50 mg gallic acid. The main plasma metabolites are glucuronidated forms of gallic acid and 4-O-methylgallic acid.

Glucosinolates themselves are poorly absorbed, if at all. However, hydrolysis of the β-thioglucosidic bond releases an unstable intermediate which spontaneously rearranges to form a number of metabolites. The profile and quantities of breakdown products formed is highly dependent on whether the plant tissue is fresh or has been cooked, on the levels of certain proteins in the parent plants (e.g. ESP) and on a number of other factors, some of which may be modulated in order to control product formation. Isothiocyanates, which are potent anticancer compounds, are generally produced as major products. They are highly bioavailable and significant proportions (more than 50%) of an oral dose can be measured in urine (mainly as N-acetylcysteine conjugates) within 24 h of consuming cruciferous vegetables. There is very little information regarding the fate of other glucosinolate breakdown products such as nitriles, thiocyanates and epithioalkylnitriles, and this is an area that deserves attention in future studies.

Despite the obvious bioactivity of carotenoids and related compounds in model systems, human intervention studies have not given convincing evidence that the incidence of chronic disease is significantly affected by carotenoid intake *per se*. It may be that with adults the primary neoplastic and atherosclerotic changes have already occurred or that the course of the disease is not generally amenable to treatment by large dose and short-term supplementation. Focusing on children and young adults and the prevention of initiation of chronic disease would be a more fruitful exercise.

Studies evaluating the potential for plant secondary metabolites to alter disease risk through effects on relevant cells and tissues other than those associated with the oral cavity and the gastrointestinal tract should assess the impact of the known metabolites present in vivo rather than the chemical forms that exist only in foods. Furthermore, the pharmacokinetic characteristics of the metabolites should be taken into consideration and efforts made to establish their biological activities at relevant physiologic concentrations.

References

Abia, R. and Fry, S.C. (2001) Degradation and metabolism of ^{14}C-labelled proanthocyanidins from carob (*Ceratonia siliqua*) pods in the gastrointestinal tract of the rat. *J. Sci. Food Agric.*, **81**, 1156–1165.

Adam, A., Crespy, V., Levrat-Verny, M.A. *et al.* (2002) The bioavailability of ferulic acid is governed primarily by the food matrix rather than its metabolism in intestine and liver in rats. *J. Nutr.*, **132**, 1962–1968.

Adlercreutz, H., Fotsis, T., Kurzer, M.S. *et al.* (1995) Isotope dilution gas chromatographic-mass spectrometric method for the determination of unconjugated lignans and isoflavonoids in human feces, with preliminary results in omnivorous and vegetarian women. *Anal. Biochem.*, **225**, 101–108.

Alien, C.M., Smith, A.M., Clinton, S.K. *et al.* (2002) Tomato consumption increases lycopene isomer concentrations in breast milk and plasma of lactating women. *J. Am. Diet. Assoc.*, **102**, 1257–1262.

Allen, C.M., Schwartz, S.J., Craft, N.E. *et al.* (2003) Changes in plasma and oral mucosal lycopene isomer concentrations in healthy adults consuming standard servings of processed tomato products. *Nutr. Cancer*, **47**, 48–56.

Amelsvoort, J.M.M., van het Hof, K.H., Mathot, J.N.J.J. *et al.* (2001) Plasma concentrations of individual tea catechins after a single oral dose in humans. *Xenobiotica*, **31**, 891–901.

Andreasen, M.F., Kroon, P.A., Williamson, G. *et al.* (2001) Esterase activity able to hydrolyse dietary antioxidant hydroxycinnamates is distributed along the intestine of mammals. *J. Agric. Food Chem.*, **49**, 5679–5684.

Azuma, K., Ippoushi, K., Nakayama, M. *et al.* (2000) Absorption of chlorogenic acid and caffeic acid in rats after oral administration. *J. Agric. Food Chem.*, **48**, 5496–5500.

Barron, D. and Ibrahim, R.K. (1987) Synthesis of flavonoid sulphates. 1. Stepwise sulfation of position-3, position-7, and position-4' using n,n'-dicyclohexylcarbodiimide and tatrabutylammonium sulphate. *Tetrahedron*, **43**, 5197–5202.

Blomstrand, R. and Werner, B. (1996) Studies on the intestinal absorption of radioactive β-carotene and vitamin a in man. *Scand. J. Clin. Lab. Invest.*, **37**, 250–261.

Bollard, M., Stribbling, S., Mitchell, S. *et al.* (1997) The disposition of allyl isothiocyanate in the rat and mouse. *Food Chem. Toxicol.*, **35**, 933–943.

Bone, R.A., Landrum, J.T., Guerra, L.H. *et al.* (2003) Lutein and zeaxanthin dietary supplements raise macular pigment density and serum concentrations of these carotenoids in humans. *J. Nutr.*, **133**, 992–998.

Borel, P., Grolier, P., Armand, M. *et al.* (1996) Carotenoids in biological emulsions: solubility, surface-to-core distribution, and release from lipid droplets. *J. Lipid Res.*, **37**, 250–261.

Borel, P., Tyssandier, V., Mekki, N. *et al.* (1998) Chylomicron β-carotene and retinyl palmitate responses are dramatically diminished when men ingest beta-carotene with medium-chain rather than long-chain triglycerides. *J. Nutr.*, **128**, 1361–1367.

Bouktaib, M., Atmani, A. and Rolando, C. (2002) Regio- and stereoselective synthesis of the major metabolite of quercetin, quercetin-3-O − β-D-glucuronide. *Tetrahedron Lett.*, **43**, 6263–6266.

Bourne, L., Paganga, G., Baxter, D. *et al.* (2000) Absorption of ferulic acid from low-alcohol beer. *Free Radic. Res.*, **32**, 273–280.

Bourne, L.C. and Rice-Evans, C. (1998) Bioavailability of ferulic acid. *Biochem. Biophys. Res. Commun.*, **253**, 222–227.

Boyle, S.P., Dobson, V.L., Duthie, S.J. *et al.* (2000) Bioavailability and efficiency of rutin as an antioxidant: a human supplementation study. *Eur. J. Clin. Nutr.*, **54**, 774–782.

Callaway, E.C., Zhang, Y., Chew, W. *et al.* (2004) Cellular accumulation of dietary anticarcinogenic isothiocyanates is followed by transporter-mediated export as dithiocarbamates. *Cancer Lett.*, **204**, 23–31.

Cao, G., Muccitelli, H.U., Sanchez-Moreno, C. *et al.* (2001) Anthocyanins are absorbed in glycated forms in elderly women: a pharmacokinetic study. *Am. J. Clin. Nutr.*, **73**, 920–926.

Cartron, E., Fouret, G., Carbonneau, M.A. *et al.* (2003) Red-wine beneficial long-term effect on lipids but not on antioxidant characteristics in plasma in a study comparing three types of wine – description of two O-methylated derivatives of gallic acid in humans. *Free Radic. Res.*, 37, 1021–1035.

Cerda, B., Tomas-Barberan, F.A. and Espin, J.C. (2005) Metabolism of antioxidant and chemo-preventive ellagitannins from strawberries, raspberries, walnuts, and oak-aged wine in humans: identification of biomarkers and individual variability. *J. Agric. Food Chem.*, 53, 227–235.

Cheng, Z., Radominska-Pandya, A. and Tephly, T.R. (1999) Studies on the substrate specificity of human intestinal UDP-glucuronosyltransferases 1A8 and 1A10. *Drug Metab. Dispos.*, 27, 1165–1170.

Chung, F.L., Morse, M.A., Eklind, K.I. *et al.* (1992) Quantitation of human uptake of the anticarcinogen phenethyl isothiocyanate after a watercress meal. *Cancer Epidemiol. Biomarkers Prev.*, 1, 383–388.

Chung, F.L., Conaway, C.C., Rao, C.V. *et al.* (2000) Chemoprevention of colonic aberrant crypt foci in fischer rats by sulphoraphane and phenethyl isothiocyanate. *Carcinogenesis*, 21, 2287–2291.

Clarke, D.B., Lloyd, A.S., Botting, N.P. *et al.* (2002) Measurement of intact sulphate and glucuronide phytoestrogen conjugates in human urine using isotope dilution liquid chromatography-tandem mass spectrometry with [3-^{13}C]isoflavone internal standards. *Anal. Biochem.*, 309, 158–172.

Clifford, M.N. (2000) Chlorogenic acids and other cinnamates – nature, occurence, dietary burden, absorption and metabolism. *J. Sci. Food Agric.*, 80, 1033–1043.

Clifford, M.N. and Brown, J.E. (2006) Dietary flavonoids and health-broadening the perspective. In O. Andersen and K.R. Markham (eds) *Flavonoids: Chemistry, Biochemistry and Applications*. CRC Press, Beca Raton.

Cogdell, R. (1988) The function of pigments in chloroplasts. In T.W. Goodwin (ed.), *Plant Pigments*. Academic Press, London.

Conaway, C.C., Getahun, S.M., Liebes, L.L. *et al.* (2000) Disposition of glucosinolates and sulphoraphane in humans after ingestion of steamed and fresh broccoli. *Nutr. Cancer*, 38, 168–178.

Conquer, J.A., Maiani, G., Azzini, E. *et al.* (1998) Supplementation with quercetin markedly increases plasma quercetin concentration without effect on selected risk factors for heart disease in healthy subjects. *J. Nutr.*, 128, 593–597.

Cortner, J.A., Coates, P.M., Le, N.A. *et al.* (1987) Kinetics of chylomicron remnant clearance in normal and in hyperlipoproteinemic subjects. *J. Lipid Res.*, 28, 195–206.

Couteau, D., McCartney, A.L., Gibson, G.R. *et al.* (2001) Isolation and characterization of human colonic bacteria able to hydrolyse chlorogenic acid. *J. Appl. Microbiol.*, 90, 873–881.

Cremin, P., Kasim-Karakas, S. and Waterhouse, A.L. (2001) LC/ES/MS detection of hydroxycinnamates in human plasma and urine. *J. Agric. Food Chem.*, 49, 1747–1750.

Crespy, V., Morand, C., Manach, C. *et al.* (1999) Part of quercetin absorbed in the small intestine is conjugated and further secreted in the intestinal lumen. *Am. J. Physiol.*, 277, G120–G126.

Crespy, V., Aprikian, O., Morand, C. *et al.* (2001) Bioavailability of phloretin and phloridzin in rats. *J. Nutr.*, 131, 3227–3230.

Crespy, V., Morand, C., Besson, C. *et al.* (2003) The splanchnic metabolism of flavonoids highly differed according to the nature of the compound. *Am. J. Physiol.*, 284, G980–G988.

Day, A.J. and Morgan, M.R.A. (2003) Methods of polyphenol extraction from biological fluids and tissues. In C. Santos-Buelga and G. Williamson (eds), *Methods in Polyphenol Analysis*. Royal Society of Chemistry, Cambridge, pp. 17–47.

Day, A.J. and Williamson, G. (2001) Biomarkers for exposure to dietary flavonoids: a review of the current evidence for identification of quercetin glycosides in plasma. *Br. J. Nutr.*, 86, S105–S110.

Day, A.J., DuPont, M.S., Ridley, S. *et al.* (1998) Deglycosylation of flavonoid and isoflavonoid glycosides by human small intestine and liver β-glucosidase activity. *FEBS Lett.*, 436, 71–75.

Day, A.J., Canada, F.J., Diaz, J.C. et al. (2000) Dietary flavonoid and isoflavone glycosides are hydrolysed by the lactase site of lactase phlorizin hydrolase. *FEBS Lett.*, **468**, 166–170.

Day, A.J., Mellon, F., Barron, D. et al. (2001) Human metabolism of dietary flavonoids: identification of plasma metabolites of quercetin. *Free Radic. Res.*, **35**, 941–952.

Day, A.J., Gee, J.M., DuPont, M.S. et al. (2003) Absorption of quercetin-3-glucoside and quercetin-4′-glucoside in the rat small intestine: the role of lactase phlorizin hydrolase and the sodium-dependent glucose transporter. *Biochem. Pharmacol.*, **65**, 1199–1206.

de Pee, S., West, C.E., Permaesih, D. et al. (1998) Orange fruit is more effective than are dark-green, leafy vegetables in increasing serum concentrations of retinol and β-carotene in schoolchildren in indonesia. *Am. J. Clin. Nutr.*, **68**, 1058–1067.

Déprez, S., Mila, I., Scalbert, A. et al. (2001) Transport of proanthocyanidin dimer, trimer and polymer across monolayers of human intestinal epithelial caco-2 cells. *Antioxid. Redox Signal.*, **3**, 957–967.

Dimitrov, N.V., Meyer, C., Ullrey, D.E. et al. (1988) Bioavailability of β-carotene in humans. *Am. J. Clin. Nutr.*, **48**, 298–304.

Donovan, J.L., Bell, J.R., Kasim-Karakas, S. et al. (1999) Catechin is present as metabolites in human plasma after consumption of red wine. *J. Nutr.*, **129**, 1662–1668.

Donovan, J.L., Crespy, V., Manach, C. et al. (2001) Catechin is metabolized by both the small intestine and liver of rats. *J. Nutr.*, **131**, 1753–1757.

Donovan, J.L., Kasim-Karakas, S., German, J.B. et al. (2002a) Urinary excretion of catechin metabolites by human subjects after red wine consumption. *Br. J. Nutr.*, **87**, 31–37.

Donovan, J.L., Manach, C., Rios, L. et al. (2002b) Procyanidins are not bioavailable in rats fed a single meal containing a grape seed extract or the procyanidin dimer B_3. *Br. J. Nutr.*, **87**, 299–306.

Dueker, S.R., Jones, A.D., Smith, G.M. et al. (1994) Stable isotope methods for the study of β-carotene-d_8 metabolism in humans utilizing tandem mass spectrometry and high-performance liquid chromatography. *Anal. Chem.*, **66**, 4177–4185.

Dunn, R.T.I. and Klaassen, C.D. (1998) Tissue-specific expression of rat sulfotransferase messenger rnas. *Drug Metab. Dispos.*, **26**, 598–604.

Dupont, M.S., Day, A.J., Bennett, R.N. et al. (2004) Absorption of kaempferol from endive, a source of kaempferol-3-glucuronide, in humans. *Eur. J. Clin. Nutr.*, **58**, 947–954.

El-Sohemy, A., Baylin, A., Kabagambe, E. et al. (2002) Individual carotenoid concentrations in adipose tissue and plasma as biomarkers of dietary intake. *Am. J. Clin. Nutr.*, **76**, 172–1779.

Erlund, I., Meririnne, E., Alfthan, G. et al. (2001) Plasma kinetics and urinary excretion of the flavanones naringenin and hesperetin in humans after ingestion of orange juice and grapefruit juice. *J. Nutr.*, **131**, 235–241.

Fahey, J.W., Zhang, Y., Talalay, P. (1997) Broccoli sprouts: an exceptionally rich source of inducers of enzymes that protect against chemical carcinogens. *Proc. Natl. Acad. Sci.*, **94**, 10367–10372.

Fahey, J.W., Zalcmann, A.T., Talalay, P. (2001) The chemical diversity and distribution of glucosinolates and isothiocyanates among plants. *Phytochemistry*, **56**, 5–51.

FAO/WHO (1967) *Requirements of Vitamin A, Thiamine, Riboflavin and Niacin.* WHO technical report series No. 362. Geneva: World Health Organisation,

Faulkner, K., Mithen, R. and Williamson, G. (1998) Selective increase of the potential anticarcinogen 4-methylsulphinylbutyl glucosinolate in broccoli. *Carcinogenesis*, **19**, 605–609.

Faulks, R.M., Hart, D.J., Wilson, P.D. et al. (1997) Absorption of all-*trans*- and 9-*cis*-β-carotene in human ileostomy volunteers. *Clin. Sci.*, **93**, 585–591.

Faulks, R.M., Hart, D.J., Scott, K.J. et al. (1998) Changes in plasma carotenoid and vitamin E profile during supplementation with oil palm fruit carotenoids. *J. Lab. Clin. Med.*, **132**, 507–511.

Faulks, R.M., Hart, D.J., Brett, G.M. et al. (2004) Kinetics of gastro-intestinal transit and carotenoid absorption and disposal in ileostomy volunteers fed spinach meals. *Eur. J. Nutr.*, **43**, 15–22.

Felgines, C., Talavera, S., Gonthier, M.-P. et al. (2003) Strawberry anthocyanins are recovered in urine as glucuro- and sulfoconjugates in humans. *J. Nutr.*, **133**, 1296–1301.

Fowke, J.H., Chung, F.L., Jin, F. *et al.* (2003) Urinary isothiocyanate levels, brassica, and human breast cancer. *Cancer Res.*, **63**, 3980–3986.

Fuhr, U. and Kummert, A.L. (1995) The fate of naringin in humans: a key to grapefruit juice–drug interactions? *Clin. Pharmacol. Ther.*, **58**, 365–373.

Gartner, C., Stahl, W. and Sies, H. (1996) Preferential increase in chylomicron levels of the xanthophylls lutein and zeaxanthin compared to β-carotene in the human. *Int. J. Vitam. Nutr. Res.*, **66**, 119–125.

Gartner, C., Stahl, W. and Sies, H. (1997) Lycopene is more bioavailable from tomato paste than from fresh tomatoes. *Am. J. Clin. Nutr.*, **66**, 116–122.

Gaziano, J.M., Johnson, E.J., Russell, R.M. *et al.* (1995) Discrimination in absorption or transport of β-carotene isomers after oral supplementation with either all-*trans*- or 9-*cis*-β-carotene. *Am. J. Clin. Nutr.*, **61**, 1248–1252.

Gee, J.M., DuPont, S.M., Day, A.J. *et al.* (2000) Intestinal transport of quercetin glycosides in rats involves both deglycoslylation and interaction with the hexose transport pathway. *J. Nutr.*, **130**, 2765–2771.

Getahun, S.M. and Chung, F.L. (1999) Conversion of glucosinolates to isothiocyanates in humans after ingestion of cooked watercress. *Cancer Epidemiol. Biomarkers Prev.*, **8**, 447–451.

Ghazali, R.A. and Waring, R.H. (1999) The effects of flavonoids on human phenolsulphotransferases: potential in drug metabolism and chemoprevention. *Life Sci.*, **65**, 1625–1632.

Gomez-Aracena, J., Bogers, R., Van't Veer, P. *et al.* (2003) Vegetable consumption and carotenoids in plasma and adipose tissue in Malaga, Spain. *Int. J. Vitam. Nutr. Res.*, **73**, 24–31.

Gonthier, M.P., Cheynier, V., Donovan, J.L. *et al.* (2003a) Microbial aromatic acid metabolites formed in the gut account for a major fraction of the polyphenols excreted in urine of rats fed red wine polyphenols. *J. Nutr.*, **133**, 461–467.

Gonthier, M.P., Verny, M.A., Besson, C. *et al.* (2003b) Chlorogenic acid bioavailability largely depends on its metabolism by the gut microflora in rats. *J. Nutr.*, **133**, 1853–1859.

Goodman, D.S., Blomstrand, R., Werner, B. *et al.* (1966) The intestinal absorption and metabolism of vitamin A and β-carotene in man. *J. Clin. Invest.*, **45**, 1615–1623.

Goodman, G.E., Thornquist, M.D., Balmes, J. *et al.* (2004) The β-carotene and retinol efficacy trial: incidence of lung cancer and cardiovascular disease mortality during 6-year follow-up after stopping β-carotene and retinol supplements. *J. Natl. Cancer Inst.*, **96**, 1729–1731.

Gossage, C.P., Deyhim, M., Yamini, S. *et al.* (2002) Carotenoid composition of human milk during the first month postpartum and the response to β-carotene supplementation. *Am. J. Clin. Nutr.*, **76**, 193–197.

Grundy, S.M. and Mok, H.Y. (1976) Chylomicron clearance in normal and hyperlipidemic man. *Metabolism*, **25**, 1225–1239.

Heaney, R.K., Spinks, E.A. and Fenwick, G.R. (1988) Improved method for the determination of the total glucosinolate content of rape seed by determination of enzymatically released glucose. *Analyst*, **113**, 1515–1518.

Hodgson, J.M., Chan, S.Y., Puddey, I.B. *et al.* (2004) Phenolic acid metabolites as biomarkers for tea- and coffee-derived polyphenol exposure in human subjects. *Br. J. Nutr.*, **91**, 301–306.

Hollman, P.C., Bijsman, M.N., van Gameren, Y. *et al.* (1999) The sugar moiety is a major determinant of the absorption of dietary flavonoid glycosides in man. *Free Radic. Res.*, **31**, 569–573.

Holst, B. and Williamson, G. (2004) A critical review of the bioavailability of glucosinolates and related compounds. *Nat. Prod. Rep.*, **21**, 425–447.

Holt, R.R., Lazarus, S.A., Sullards, M.C. *et al.* (2002) Procyanidin dimer B2 [epicatechin-(4β-8)-epicatechin] in human plasma after the consumption of a flavanol-rich cocoa. *Am. J. Clin. Nutr.*, **76**, 798–804.

Ito, H., Gonthiera, M.P., Manach, C. *et al.* (2004) High-throughput profiling of dietary polyphenols and their metabolites by HPLC-ESI-MS-MS in human urine. *Biofactors*, **22**, 241–243.

Jaganath, I.B., Mullen, C. and Crozier, A. (2006) The relative contribution of the small and large intestine to the absorption and metabolism of rutin in man. Free Radical Research, in press.

Jayarajan, P., Reddy, V. and Mohanram, M. (1980) Effect of dietary fat on absorption of b-carotene from green leafy vegetables. *Indian J. Med. Res.*, **71**, 53–56.

Jiao, D., Ho, C.T., Foiles, P. *et al.* (1994) Identification and quantification of the n-acetylcysteine conjugate of allyl isothiocyanate in human urine after ingestion of mustard. *Cancer Epidemiol. Biomarkers Prev.*, **3**, 487–492.

Johnson, E.J. and Russell, R.M. (1992) Distribution of orally administered β-carotene among lipoproteins in healthy men. *Am. J. Clin. Nutr.*, **56**, 128–135.

Johnston, K.L., Clifford, M.N. and Morgan, L.M. (2002) Possible role for apple juice phenolic compounds in the acute modification of glucose tolerance and gastrointestinal hormone secretion in humans. *J. Sci. Food Agric.*, **82**, 1800–1805.

Johnston, K.L., Clifford, M.N. and Morgan, L.M. (2003) Coffee acutely modifies gastrointestinal hormone secretion and glucose tolerance in humans: glycemic effects of chlorogenic acid and caffeine. *Am. J. Clin. Nutr.*, **78**, 728–733.

Kaplan, L.A., Lau, J.M. and Stein, E.A. (1990) Carotenoid composition, concentrations, and relationships in various human organs. *Clin. Physiol. Biochem.*, **8**, 1–10.

Kern, S.M., Bennett, R.N., Mellon, F.A. *et al.* (2003) Absorption of hydroxycinnamates in humans after high-bran cereal consumption. *J. Agric. Food. Chem.*, **51**, 6050–6055.

Khachik, F., Beecher, G.R., Goli, M.B. *et al.* (1992) Separation and identification of carotenoids and their oxidation products in extracts of human plasma. *Anal. Chem.*, **64**, 2111–2112.

Kohri, T., Matsumoto, N., Yamakama, M. *et al.* (2001) Metabolic fate of $(-)$-[4-^{3}H]epigallocatechin gallate in rats after oral administration. *J. Agric. Food Chem.*, **49**, 4102–4112.

Konishi, Y. and Shimizu, M. (2003) Transepithelial transport of ferulic acid by monocarboxylic acid transporter in caco-2 cell monolayers. *Biosci. Biotechnol. Biochem.*, **67**, 856–862.

Konishi, Y., Kobayashi, S. and Shimizu, M. (2003) Transepithelial transport of *p*-coumaric acid and gallic acid in caco-2 cell monolayers. *Biosci. Biotechnol. Biochem.*, **67**, 2317–2324.

Krinsky, N.I., Cornwell, D.G. and Oncley, J.L. (1958) The transport of vitamin A and carotenoids in human plasma. *Arch. Biochem. Biophys.*, **73**, 233–246.

Kroon, P.A., Clifford, M.N., Crozier, A. *et al.* (2004) How should we assess the effects of exposure to dietary polyphenols in vitro? *Am. J. Clin. Nutr.*, **79**, 1145–1151.

Kuhnle, G., Spencer, J.P.E., Schroeter, H. *et al.* (2000) Epicatechin and catechin are *O*-methylated and glucuronidated in the small intestine. *Biochem. Biophys. Res. Commun.*, **277**, 507–512.

Lafay, S., Manach, C., *et al.* (2006) Absorption and metabolism of caffeic acid and chlorogenic acid in the small intestine of rats. *Br. J. Nutr.*, **136**, 1192–1197.

Lapidot, T., Harel, S., Granit, R. *et al.* (1998) Bioavailability of red wine anthocyanins as detected in human urine. *J. Agric. Food Chem.*, **46**, 4297–4302.

Lee, M.J., Wang, Z.-Y., Li, H. *et al.* (1995) Analysis of plasma and urinary tea polyphenols in human subjects. *Cancer Epidemiol. Biomarkers Prev.*, **4**, 393–399.

Lee, M.J., Maliakal, P., Chen, L. *et al.* (2002) Pharmacokinetics of tea catechins after ingestion of green tea and $(-)$-epigallocatechin-3-gallate by humans: formation of different metabolites and individual variability. *Cancer Epidemiol. Biomarkers Prev.*, **11**, 1025–1032.

Lepage, S., Bonnefont-Rousselot, D., Assogba, U. *et al.* (1994) Distribution of β-carotene in serum lipoprotein fractions separated by selective precipitation. *Ann. Biol. Clin.*, **52**, 139.

Lietz, G., Henry, C.J., Mulokozi, G. *et al.* (2001) Comparison of the effects of supplemental red palm oil and sunflower oil on maternal vitamin a status. *Am. J. Clin. Nutr.*, **74**, 501–509.

Liveny, O., Reifen, R., Levey, I. *et al.* (2003) β-Carotene bioavailability from differently processed carrot meals in human ileostomy volunteers. *Eur. J. Nutr.*, **42**, 338–345.

Macheix, J.-J., Fleuriet, A. and Billot, J. (1990) *Fruit Phenolics*. Boca Raton, FL, CRC Press.

Mackerras, D. (2004) Calculation of vitamin A activity from provitamin A carotenoids: What factor should we use? *Asia Pac. J. Clin. Nutr.*, **13**, S133.

Manach, C. and Donovan, J.L. (2004) Pharmacokinetics and metabolism of dietary flavonoids in humans. *Free Radic. Res.*, **38**, 771–785.

Manach, C., Texier, O., Morand, C. *et al.* (1999) Comparison of the bioavailability of quercetin and catechin in rats. *Free Radic. Biol. Med.*, **27**, 1259–1266.

Manach, C., Morand, C., Gil-Izquierdo, A. *et al.* (2003) Bioavailability in humans of the flavanones hesperidin and narirutin after the ingestion of two doses of orange juice. *Eur. J. Clin. Nutr.*, **57**, 235–242.

Manach, C., Scalbert, A., Morand, C. *et al.* (2004) Polyphenols: food sources and bioavailability. *Am. J. Clin. Nutr.*, **79**, 727–747.

Matsumoto, H., Inaba, H., Kishi, M. *et al.* (2001) Orally administered delphinidin 3-rutinoside and cyanidin 3-rutinoside are directly absorbed in rats and humans and appear in the blood as the intact forms. *J. Agric. Food Chem.*, **49**, 1546–1551.

Matusheski, N.V., Juvik, J.A. and Jeffery, E.H. (2004) Heating decreases epithiospecifier protein activity and increases sulphoraphane formation in broccoli. *Phytochemistry*, **65**, 1273–1281.

Mazza, G., Kay, C.D., Cottrell, T. *et al.* (2002) Absorption of anthocyanins from blueberries and serum antioxidant status in human subjects. *J. Agric. Food Chem.*, **50**, 7731–7737.

Mellon, F.A., Bennett, R.N., Holst, B. *et al.* (2002) Intact glucosinolate analysis in plant extracts by programmed cone voltage electrospray LC/MS: performance and comparison with LC/MS/MS methods. *Anal. Biochem.*, **306**, 83–91.

Meng, X., Lee, M.J., Li, C. *et al.* (2001) Formation and identification of $4'$-O-methyl-$(-)$-epigallocatechin in humans. *Drug Metab. Dispos.*, **29**, 789–793.

Meng, X., Sang, S., Zhu, N. *et al.* (2002) Identification and characterization of methylated and ring-fission metabolites of tea catechins formed in humans, mice, and rats. *Chem. Res. Toxicol.*, **15**, 1042–1050.

Mennicke, W.H., Gorler, K., Krumbiegel, G. *et al.* (1988) Studies on the metabolism and excretion of benzyl isothiocyanate in man. *Xenobiotica*, **18**, 441–447.

Miyazawa, T., Nakagawa, K., Kudo, M. *et al.* (1999) Direct intestinal absorption of red fruit anthocyanins, cyanidin-3-glucoside and cyanidin-3,5-diglucoside, into rats and humans. *J. Agric. Food Chem.*, **47**, 1083–1091.

Mokady, S., Avron, M. and Ben-Amotz, A. (1990) Accumulation in chick livers of 9-*cis* versus all-*trans* β-carotene. *J. Nutr.*, **120**, 889–892.

Monge, P., Solheim, E. and Scheline, R.R. (1984) Dihydrochalcone metabolism in the rat: phloretin. *Xenobiotica*, **14**, 917–924.

Moon, J.H., Nakata, R., Oshima, S. *et al.* (2000) Accumulation of quercetin conjugates in blood plasma after short-term ingestion of onion by women. *Am. J. Physiol.*, **279**, R461–R467.

Mullen, W., Graf, B.A., Caldwell, S.T. *et al.* (2002) Determination of flavonol metabolites in plasma and tissues of rats by HPLC-radiocounting and tandem mass spectrometry following oral ingestion of [2-^{14}C]quercetin-$4'$-glucoside. *J. Agric. Food Chem.*, **50**, 6902–6909.

Mullen, W., Boitier, A., Stewart, A.J. *et al.* (2004) Flavonoid metabolites in human plasma and urine after the consumption of red onions: analysis by liquid chromatography with photodiode array and full scan tandem mass spectrometric detection. *J. Chromatogr. A*, **1058**, 163–168.

Mullen, W., Edwards, C.A. and Crozier, A. (2006) Absorption, excretion and metabolite profiling of methyl-, glucuronyl-, glucosyl- and sulpho-conjugates of quercetin in human plasma and urine after ingestion of onions. *Br. J. Nutr.*, in press.

Nakagawa, K., Okuda, S. and Miyazawa, T. (1997) Dose-dependent incorporation of tea catechins, $(-)$-epigallocatechin-3-gallate and $(-)$-epigallocatechin, into human plasma. *Biosci. Biotechnol. Biochem.*, **61**, 1981–1985.

Nakamura, Y. and Tonogai, Y. (2003) Metabolism of grape seed polyphenol in the rat. *J. Agric. Food Chem.*, **51**, 7215–7225.

Nardini, M., Cirillo, E., Natella, F. *et al.* (2002) Absorption of phenolic acids in humans after coffee consumption. *J. Agric. Food Chem.*, **50**, 5735–5741.

Natsume, M., Osakabe, N., Oyama, M. et al. (2003) Structures of (−)-epicatechin glucuronide identified from plasma and urine after oral ingestion of (−)-epicatechin: Differences between human and rat. *Free Radic. Biol. Med.*, **34**, 840–849.

Needs, P.W. and Williamson, G. (2001) Syntheses of daidzein-7-yl-β-D-glucopyranosiduronic acid and daidzein-4′,7-yl-di-β-D-glucopyranosiduronic acid. *Carbohydr. Res.*, **330**, 511–515.

Nielsen, I.L Dragsted, L.O., Ravn-Haren, G. et al. (2003) Absorption and excretion of black currant anthocyanins in humans and watanabe heritable hyperlipidemic rabbits. *J. Agric. Food Chem.*, **51**, 2813–2820.

O'Leary, K.A., Day, A.J., Needs, P.W. et al. (2003) Metabolism of quercetin-7-and quercetin-3-glucuronides by an in vitro hepatic model: the role of human beta-glucuronidase, sulfotransferase, catechol-O-methyltransferase and multi-resistant protein 2 (mrp2) in flavonoid metabolism. *Biochem. Pharmacol.*, **65**, 479–491.

Olthof, M.R., Hollman, P.C., Vree, T.B. et al. (2000) Bioavailabilities of quercetin-3-glucoside and quercetin-4′-glucoside do not differ in humans. *J. Nutr.*, **130**, 1200–1203.

Olthof, M.R., Hollman, P.C.H. and Katan, M.B. (2001) Chlorogenic acid and caffeic acid are absorbed in humans. *J. Nutr.*, **131**, 66–71.

Pacifici, G.M., Franchi, M., Bencini, C. et al. (1988) Tissue distribution of drug-metabolizing enzymes in humans. *Xenobiotica*, **18**, 849–856.

Paetau, I., Rao, D., Wiley, E.R. et al. (1999) Carotenoids in human buccal mucosa cells after 4 weeks of supplementation with tomato juice or lycopene supplements. *Am. J. Clin. Nutr.*, **70**, 490–494.

Parker, R.S., Swanson, J.E., Marmor, B. et al. (1993) Study of β-carotene metabolism in humans using ^{13}C-β-carotene and high precision isotope ratio mass spectrometry. *Ann. N. Y. Acad. Sci.*, **691**, 86–95.

Petri, N., Tannergren, C., Holst, B. et al. (2003) Absorption/metabolism of sulphoraphane and quercetin, and regulation of phase II enzymes, in human jejunum in vivo. *Drug. Metab. Dispos.*, **31**, 805–813.

Piskula, M.K. and Terao, J. (1998) Accumulation of (−)-epicatechin metabolites in rat plasma after oral administration and distribution of conjugation enzymes in rat tissues. *J. Nutr.*, **128**, 1172–1178.

Plumb, G.P., O'Leary, K., Day, A.J. et al (2003) Enzymatic synthesis of quercetin glucosides and glucuronides. In C. Santos-Buelga and G. Williamson (eds), *Methods in Polyphenol Analysis*. Royal Society of Chemistry, Cambridge, pp. 177–186.

Plumb, G.W., Garcia-Conesa, M.T., Kroon, P.A. et al. (1999) Metabolism of chlorogenic acid by human plasma, liver, intestine and gut microflora. *J. Sci. Food Agric.*, **79**, 390–392.

Prince, M.R. and Frisoli, J.K. (1993) β-Carotene accumulation in serum and skin. *Am. J. Clin. Nutr.*, **57**, 175–181.

Rechner, A.R., Spencer, J.P., Kuhnle, G. et al. (2001) Novel biomarkers of the metabolism of caffeic acid derivatives in vivo. *Free Radic. Biol. Med.*, **30**, 1213–1222.

Rechner, A.R., Smith, M.A., Kuhnle, G. et al. (2004) Colonic metabolism of dietary polyphenols: influence of structure on microbial fermentation products. *Free Radic. Biol. Med.*, **36**, 212–225.

Reddy, V., Underwood, B.A. and Pee, S.D. (1995) Vitamin a status and dark green leafy vegetables. *Lancet*, **346**, 1634–1635.

Reifen, R., Haftel, L., Faulks, R. et al. (2003) Plasma and buccal mucosal cell response to short-term supplementation with all-*trans*-β-carotene and lycopene in human volunteers. *Int. J. Mol. Med.*, **12**, 989–993.

Rich, G.T., Bailey, A.L., Faulks, R.M. et al. (2003a) Solubilization of carotenoids from carrot juice and spinach in lipid phases: I. Modeling the gastric lumen. *Lipids*, **38**, 933–945.

Rich, G.T., Faulks, R.M., Wickham, M.S. et al. (2003b) Solubilization of carotenoids from carrot juice and spinach in lipid phases: II. Modeling the duodenal environment. *Lipids*, **38**, 947–956.

Richelle, M., Tavazzi, I., Enslen, M. et al. (1999) Plasma kinetics in man of epicatechin from black chocolate. *Eur. J. Clin. Nutr.*, **53**, 22–26.

Rios, L., Bennet, R.N., Lazarus, S.A. *et al.* (2002) Cocoa procyanidins are stable during gastric transit in humans. *Am. J. Clin. Nutr.*, **76**, 1106–1110.

Rondini, L., Peyrat-Maillard, M.N., Marsset-Baglieri, A. *et al.* (2002) Sulfated ferulic acid is the main in vivo metabolite found after short-term ingestion of free ferulic acid in rats. *J. Agric. Food Chem.*, **50**, 3037–3041.

Rouzaud, G., Young, S.A. and Duncan, A.J. (2004) Hydrolysis of glucosinolates to isothiocyanates after ingestion of raw or microwaved cabbage by human volunteers. *Cancer Epidemiol. Biomarkers Prev.*, **13**, 125–131.

Sano, A., Yamakoshi, J., Tokutake, S. *et al.* (2003) Procyanidin B1 is detected in human serum after intake of proanthocyanidin-rich grape seed extract. *Biosci. Biotechnol. Biochem.*, **67**, 1140–1143.

Santos-Buelga, C. and Williamson, G. (2003). *Methods in Polyphenol Analysis*. Cambridge, Royal Society of Chemistry.

Schierle, J., Bretzel, W., Buhler, I. *et al.* (1997) Content and isometric ratio of lycopene in food and human plasma. *Food Chem.*, **59**, 459–465.

Scott, J.K. and Rodriquez-Amaya, D. (2000) Pro-vitamin A carotenoid conversion factors: retinol equivalents – fact or fiction. *Food Chem.*, **69**, 125–127.

Seeram, N.P., Lee, R. and Heber, D. (2004) Bioavailability of ellagic acid in human plasma after consumption of ellagitannins from pomegranate (*Punica granatum l.*) juice. *Clin. Chim. Acta*, **348**, 63–68.

Setchell, K.D., Brown, N.M., Zimmer-Nechemias, L. *et al.* (2002) Evidence for lack of absorption of soy isoflavone glycosides in humans, supporting the crucial role of intestinal metabolism for bioavailability. *Am. J. Clin. Nutr.*, **76**, 447–453.

Sfakianos, J., Coward, L., Kirk, M. *et al.* (1997) Intestinal uptake and biliary excretion of the isoflavone genistein in rats. *J. Nutr.*, **127**, 1260–1268.

Shahrzad, S. and Bitsch, I. (1998) Determination of gallic acid and its metabolites in human plasma and urine by high-performance liquid chromatography. *J. Chromatogr. B*, **705**, 87–95.

Shahrzad, S., Aoyagi, K., Winter, A. *et al.* (2001) Pharmacokinetics of gallic acid and its relative bioavailability from tea in healthy humans. *J. Nutr.*, **131**, 1207–1210.

Shapiro, T.A., Fahey, J.W., Wade, K.L. *et al.* (1998) Human metabolism and excretion of cancer chemoprotective glucosinolates and isothiocyanates of cruciferous vegetables. *Cancer Epidemiol. Biomarkers Prev.*, **7**, 1091–1100.

Shapiro, T.A., Fahey, J.W., Wade, K.L. *et al.* (2001) Chemoprotective glucosinolates and isothiocyanates of broccoli sprouts: metabolism and excretion in humans. *Cancer Epidemiol. Biomarkers Prev.*, **10**, 501–508.

Shaw, I.C. and Griffiths, L.A. (1980) Identification of the major biliary metabolite of (+)-catechin in the rat. *Xenobiotica*, **10**, 905–911.

Shiau, A., Mobarhan, S., Stacewicz-Sapuntzakis, M. *et al.* (1994) Assessment of the intestinal retention of β-carotene in humans. *J. Am. Coll. Nutr.*, **13**, 369–375.

Shimoi, K., Okada, H., Furugori, M. *et al.* (1998) Intestinal absorption of luteolin and luteolin-7-O-β-glucoside in rats and humans. *FEBS Lett.*, **438**, 220–224.

Shimoi, K., Saka, N., Kaji, K. *et al.* (2000) Metabolic fate of luteolin and its functional activity at focal site. *Biofactors*, **12**, 181–186.

Shimoi, K., Saka, N., Nozawa, R. *et al.* (2001) Deglucuronidation of a flavonoid, luteolin monoglucuronide, during inflamation. *Drug Metab. Dispos.*, **29**, 1521–1524.

Slominski, B.A., Cambell, L.D. and Stanger, N.E. (1987) Influence of cecectomy and dietary antibiotics on the fate of ingested intact glucosinolates in poultry. *J. Sci. Food Agric.*, **80**, 967–984.

Spencer, J.P., Chaudry, F., Pannala, A.S. *et al.* (2000) Decomposition of cocoa procyanidins in the gastric milieu. *Biochem. Biophys. Res. Commun.*, **272**, 236–241.

Spencer, J.P.E., Schroeter, H., Shenoy, B. *et al.* (2001) Epicatechin is the primary bioavailable form of the procyanidin dimers B2 and B5 after transfer across the small intestine. *Biochem. Biophys. Res. Commun.*, **285**, 558–593.

Stahl, W. and Sies, H. (1992) Uptake of lycopene and its geometrical isomers is greater from heat-processed than from unprocessed tomato juice in humans. *J. Nutr.*, **122**, 2161–2166.

Stahl, W., Schwarz, W., von Laar, J. *et al.* (1995) All-*trans*-β-carotene preferentially accumulates in human chylomicrons and very low density lipoproteins compared with the 9-*cis* geometrical isomer. *J. Nutr.*, **125**, 2128–2133.

Straub, O. (1987) Key to carotenoids. In H. Pfander, M. Gerspacher, M. Rychener and R. Schwabe (eds). Verlag, Birkhauser.

Tesoriere, L., Allegra, M., Butera, D. *et al.* (2004) Absorption, excretion, and distribution of dietary antioxidant betalains in LDLs: potential health effects of betalains in humans. *Am. J. Clin. Nutr.*, **80**, 941–945.

Tesoriere, L., Butera, D., Allegra, M. *et al.* (2005) Distribution of betalain pigments in red blood cells after consumption of cactus pear fruits and increased resistance of the cells to ex vivo induced oxidative hemolysis in humans. *J. Agric. Food Chem.*, **53**, 1266–1270.

Thelander, M., Narito, O.J., and Gruissen, W. (1985) Plastid differentiation and pigment synthesis during tomato fruit ripening. *Curr. Topics Plant Biochem. Physiol.*, **5**, 137–147.

Tomas-Barberan, F.A. and Clifford, M.N. (2000) Dietary hydroxybenzoic acid derivatives – nature, occurrence and dietary burden. *J. Sci. Food Agric.*, **80**, 1024–1032.

Tsang, C., Auger, C., Mullen, W. *et al.* (2005) The absorption, metabolism and excretion of flavan-3-ols and procyanidins following the ingestion of a grape seed extract by rats. *Br. J. Nutr.* **94**, 170–181.

Ullmann, U., Haller, J., Decourt, J.P. *et al.* (2003) A single ascending dose study of epigallocatechin gallate in healthy volunteers. *J. Int. Med. Res.*, **31**, 88–101.

Vaidyanathan, J.B. and Walle, T. (2001) Transport and metabolism of the tea flavonoid (−)-epicatechin by the human intestinal cell line caco-2. *Pharm Res.*, **18**, 1420–1425.

Vaidyanathan, J.B. and Walle, T. (2002) Glucuronidation and sulfation of the tea flavonoid (−)-epicatechin by the human and rat enzymes. *Drug Metab. Dispos.*, **30**, 897–903.

Vaidyanathan, J.B. and Walle, T. (2003) Cellular uptake and efflux of the tea flavonoid (−)-epicatechin-3-gallate in the human intestinal cell line caco-2. *J. Pharmacol.Exp. Ther.*, **307**, 745–752.

Van Amelsvoort, J.M., Van Hof, K.H., Mathot, J.N. *et al.* (2001) Plasma concentrations of individual tea catechins after a single oral dose in humans. *Xenobiotica*, **31**, 891–901.

Van Lieshout, M., West, C.E., Mulhial *et al.* (2001) Bioefficacy of β-carotene dissolved in oil studied in children in Indonesia. *Am. J. Clin. Nutr.*, **73**, 949–958.

van Vliet, T., Schreurs, W.H. and van den Berg, H. (1995) Intestinal β-carotene absorption and cleavage in men: response of β-carotene and retinyl esters in the triglyceride-rich lipoprotein fraction after a single oral dose of β-carotene. *Am. J. Clin. Nutr.*, **62**, 110–116.

Vogel, S., Contois, J.H., Couch, S.C. *et al.* (1996) A rapid method for separation of plasma low and high density lipoproteins for tocopherol and carotenoid analyses. *Lipids*, **31**, 421–426.

von Hertzen, L., Forsblom, C., Stumpf, K. *et al.* (2004) Highly elevated serum phyto-oestrogen concentrations in patients with diabetic nephropathy. *J. Intern. Med.*, **255**, 602–609.

Walfisch, Y., Walfisch, S., Agbaria, R. *et al.* (2003) Lycopene in serum, skin and adipose tissues after tomato-oleoresin supplementation in patients undergoing haemorrhoidectomy or peri-anal fistulotomy. *Br. J. Nutr.*, **90**, 759–766.

Walgren, R.A., Karnaky, K.J., Jr., Lindenmayer, G.E. *et al.* (2000) Efflux of dietary flavonoid quercetin-4′-β-glucoside across human intestinal caco-2 cell monolayers by apical multidrug resistance-associated protein-2. *J. Pharmacol. Exp. Ther.*, **294**, 830–836.

Walle, T. and Walle, U.K. (2003) The β-D-glucoside and sodium-dependent glucose transporter 1 (sglt1)-inhibitor phloridzin is transported by both sglt1 and multidrug resistance-associated proteins 1/2. *Drug Metab. Dispos.*, **31**, 1288–1291.

Walle, T., Otake, Y., Brubaker, J.A. et al. (2001a) Disposition and metabolism of the flavonoid chrysin in normal volunteers. *Br. J. Clin. Pharmacol.*, **51**, 143–146.

Walle, T., Walle, U.K. and Halushka, P.V. (2001b) Carbon dioxide is the major metabolite of quercetin in humans. *J. Nutr.*, **131**, 2648–2652.

Walle, U.K., Galijatovic, A. and Walle, T. (1999) Transport of the flavonoid chrysin and its conjugated metabolites by the human intestinal cell line caco-2. *Biochem. Pharmacol.*, **58**, 431–438.

Warden, B., Smith, L.S., Beecher, G.R. et al. (2001) Catechins are bioavailable in man and women drinking black tea throughout the day. *J. Nutr.*, **131**, 1731–1737.

Wittemer, S.M., Ploch, M., Windeck, T. et al. (2005) Bioavailability and pharmacokinetics of caffeoylquinic acids and flavonoids after oral administration of artichoke leaf extracts in humans. *Phytomedicine*, **12**, 28–38.

Wittig, J., Herderich, M., Graefe, E.U. et al. (2001) Identification of quercetin glucuronides in human plasma by high-performance liquid chromatography-tandem mass spectrometry. *J Chromatogr. B.*, **753**, 237–243.

Wu, X., Cao, G. and Prior, R.L. (2002) Absorption and metabolism of anthocyanins in elderly women after consumption of elderberry or blueberry. *J. Nutr.*, **132**, 1865–1871.

Yang, B., Arai, K. and Kusu, F. (2000) Determination of catechins in human urine subsequent to tea ingestion by high-performance liquid chromatography with electrochemical detection. *Anal. Biochem.*, **283**, 77–82.

Yang, C.S., Lee, M.J. and Chen, L. (1999) Human salivary tea catechin levels and catechin esterase activities: implication in human cancer prevention studies. *Cancer Epidemiol. Biomarkers Prev.*, **8**, 83–89.

Ye, L., Dinkova-Kostova, A.T., Wade, K.L. et al. (2002) Quantitative determination of dithiocarbamates in human plasma, serum, erythrocytes and urine: pharmacokinetics of broccoli sprout isothiocyanates in humans. *Clin. Chim. Acta*, **316**, 43–53.

Yemelyanov, A.Y., Katz, N.B. and Bernstein, P.S. (2001) Ligand-binding characterization of xanthophyll carotenoids to solubilized membrane proteins derived from human retina. *Exp. Eye Res.*, **72**, 381–392.

Zhang, Y. (2000) Role of glutathione in the accumulation of anticarcinogenic isothiocyanates and their glutathione conjugates by murine hepatoma cells. *Carcinogenesis*, **21**, 1175–1182.

Zhang, Y. and Callaway, E.C. (2002) High cellular accumulation of sulphoraphane, a dietary anticarcinogen, is followed by rapid transporter-mediated export as a glutathione conjugate. *Biochem. J.*, **364**, 301–307.

Zhang, Y. and Talalay, P. (1998) Mechanism of differential potencies of isothiocyanates as inducers of anticarcinogenic phase II enzymes. *Cancer Res.*, **58**, 4632–4639.

Zhang, Y., Cho, C.G., Posner, G.H. et al. (1992a) Spectroscopic quantitation of organic isothiocyanates by cyclocondensation with vicinal dithiols. *Anal. Biochem.*, **205**, 100–107.

Zhang, Y., Talalay, P., Cho, C.G. et al. (1992b) A major inducer of anticarcinogenic protective enzymes from broccoli: isolation and elucidation of structure. *Proc. Natl. Acad. Sci.*, **89**, 2399–2403.

Zhao, Z., Egashira, Y. and Sanada, H. (2003) Ferulic acid sugar esters are recovered in rat plasma and urine mainly as the sulfoglucuronide of ferulic acid. *J. Nutr.*, **133**, 1355–1361.

Index

Aberrant crypt foci, 183, 191, 192
(−)-Abietadiene, 68–70
Abietadiene synthase, 68
(−)-Abietic acid, 69, 70
Abscisic acid, 75–7
Abscisic acid aldehyde, 77
Acetaminophen, 106
Acetophenones, 2
Acetyl coenzyme A: salutaridinol-7-O-acetyltransferase, 104, 105
Acetyl-CoA, 14
AcetylCoA carboxylase, 14
Acetyl-CoA thiolase, 51
Acetylenase, 138
Acetylenes, 138–55
   aliphatic, 138–40, 144–7, 150, 152
   allergenicity, 150–151
   antibacterial activity, 151–2
   anticancer, 153
   antifungal activity, 147–9
   antiinflammatory activity, 151–2
   antiplatelet aggregation, 151–2
   aromatic, 138–40, 145
   carrots, 153–5
   cytotoxicity, 152–3
   distribution and biosynthesis, 138–47
   dithiophene, 140, 144, 146
   falcarinol, 153–5
   furano, 147, 148
   isocoumarin, 138–40, 145
   neurotoxicity, 149–50
   thiophen, 138, 139, 145
Acetylsalicylic acid (aspirin), 210
N-Acetyl-S-allyl-L-cysteine, 36
7-O-Acetylsalutaridinol, 104
Acholeplasma-anaeroplasma, 178
α-Acids, 283, 284
β-Acids, 283, 284
Acrinylisothiocyanate, 217, 218
Adenosine triphosphate-binding cassette of transporters, 304, 308
S-Adenosylmethionine decarboxylase, 122
S-Adenosylmethionine, 119, 120, 122
S-Adenosylmethionine:7-methylxanthosine synthase, 119, 120

Adhupulone, 283, 284
Adlupulone, 283, 284
Aegopodium podagraria L., *see* Ground elder
(+)-Afzelechin, 7
Ajawain, *see* Ajowan
Ajmaline, 102, 117
(E)-Ajoene, 215
(Z)-Ajoene, 215
Ajowan (*Trachyspermum ammi*), 142, 150
Alanine aminotransferase, 72
Alfalfa (*Medicago sativa*), 21
Algae
   brown algae
      Ascophyllum spp., 262
      Laminaria spp., 262
      Macrocystis pyrifera, 262
      Sargassum spp., 262
   red algae
      Chondus crispus, 262
      Eucheuma spp., 262
      Gelidium spp., 262
      Gigartina spp., 262
      Gloiopeltis spp., 262
      Gracilaria spp., 262
      Iridaea spp., 262
      Porphyra tenera, 262
      Porphyra umbilicalis, 262
      Porphyra yessoensis, 262
      Pterocladia spp., 262
Alginates, 262
Alk(en)yl L-cysteine sulphoxides, 25
5,5′-(Alkadiyl) diresorcinols, 259
Alkaloids, 102–36
5-Alkenyl resorcinols, 259
Alkenyl-substituted L-cysteine sulphoxides, 35
S-Alkyl cysteine sulphoxides, 214
Alkyl resorcinols, 259
5-Alkylpyrazines, 276
Allergies, 150–151, 180, 184
Allicin, *see* Diallyl thiosulphinate
Alliin, *see* S-Allyl cysteine sulphoxide
Alliin-allinase system, 25, 34–7
Alliinase, 25, 34–6, 214, 215
Allium cepa, *see* Onions
Allium porrum, *see* Leeks

Allium sativum, see Garlic
S-Allyl cysteine sulphoxide (*aka* alliin and 2-propenyl-L-cysteine sulphoxide), 25, 35, 36, 214
Allyl sulphenic acid, 215
Allylisothiocyanate, 218, 326
Almonds (*Prunus dulcis*), 260, 261
Aloe-emodin, 249, 250
Amanita muscaria, see Red fly agaric mushroom
American cranberry (*Vaccinium macrocarpon*), 240
γ-Aminobutyric acid receptor, 64
  porphyria inducer, 64
Ammonia, 175, 183
Amorpha-4,11-diene, 81, 82
α-Amylase synthesis, 287
β- Amyrin, 86, 89
Anacardic acids, 261
Anethole (1-propenyl-4-methoxybenzene), 254, 255
Anethum graveolens, see Dill
Angel's trumpet (*Datura suaveolens*), 111
Angelica archangelica L. (*aka A. officinalis* Hoffm.), *see* Garden angelica
Angelicins, *see* Furanocoumarins, angular
Angular Furanocoumarins, *see* Furanocoumarins, angular
Anise (*Pimpinella anisum*), 130, 158, 253, 254
Anthemideae, 139, 144, 145, 149
Anthocyanidin reductase, 15, 19
Anthocyanidins, 3, 5, 8, 19
Anthocyanins, 8, 17, 19–21, *see also individual entries*
Anthracnose, 149
Anthriscus cerefolium (L.) Hoffm., *see* Chervil
Anthriscus sylvestris Hoffm., *see* Cow parsley
Antibacterial, 151–2
Anti-cancer, 152, 153
Antifungal, 147–9, 162, 163
Anti-inflammatory, 150–152
Anti-platelet aggregation, 151, 152
Anti-tubercular, 152
Apiaceae, 138, 141, 142, 149–51, 153, 156, 157
Apigenin, 4, 5, 15, 238
Apigenin-7-*O*-(2″-*O*-apiosyl)glucoside, 223
Apigenin-7-*O*-rutinoside, 225, 226
Apigenin-*C*-glycosides, 266
Apium graveolens L. var. *dulce, see* Celery
Apium graveolens L. var. *rapaceum, see* Celeriac
Apocarotenals, 339
Apoproteins, 337
Apples (*Malus* × *domestica*), 229–31
  cider apples, 229
Apricot (*Prunus armeniaca*), 231
Arabica coffee (*Coffea arabica*), *see* Coffee
Arabidiopsis (*Arabidopsis thaliana*), 16, 20, 21, 29–31, 55, 56, 58, 60, 70
Arabidopsis thaliana, see Arabidopsis
Araliaceae, 138, 150, 152, 153
Arbutin, *see* Hydroquinone glucoside
Arctic acid, 142, 146
Arctic acid-b, 142, 146
Arctic acid-c, 142, 146
Arctinal, 142, 144, 146
Arctinol-a, 142, 146

Arctinol-b, 142, 146
Arctinone-a, 142, 146
Arctinone-b, 142, 146
Arctium lappa, see Burdock
Area under the plasma concentration versus time curve, 317
L-Arginine, 110
epi-Aristolochene, 79, 82
Armoracia rustica, see Horseradish
Arrow poisons, 89, 90, 106, 114
Artemesia dracunculus, see Tarragon
Artemisin (Qinghaosu), 81, 82
  anti-malarial activity, 81
Artemisinic acid, 82
Artemisinin, 48, 81
Artichoke (*Cynara scolymus*), 224–5
Artichoke, *see* Globe Artichoke *and* Jerusalem Artichoke
Asafoetida (*Ferula assa-foetida* L.), 142
Ascorbic acid, *see* Vitamin C
Ascorbigen, 28, 33, 38, 39, 328
8*C*-Ascorbyl-epigallocatechin gallate, 266
Asian pumpkin (*Cucurbita moschata*), 229
Asiatic Pennywort (*Centella asiatica* L.), 141
Asparagus (*Asparagus officinalis*), 223, 224
Assamaicin A, 264
Assamaicins, 266, *see also individual entries*
Asteraceae, 138–40, 142–5, 149, 153
trans-Astringin, 246, 248
Atopobium, 178
Atractyligenin, 275, 276
Atropa belladonna, see Deadly nightshade
Atropine, 107, 108, 111
Aubergine (*Solanum melongena*), 127, 138, 143, 149, 227–8
Aurones, 3
Australian cashew (*Semecarpus australiensis*), 261
Autism, 184, 185
Avenanthramides, 258, 259
Avenulamides, 259
Avocado (*Persea americana*), 143, 149, 224, 225
Axillarin-4′-*O*-glucoside, 219, 220
Azadirachtin, 48, 87–9, 210

Bacteriocins, 182
Bacteroides spp., 175–8, 187, 188, 190, *see also individual entries*
Balloon flower (*Platycodon grandiflorum* (Jacq.) A. DC), 143, 145
Bananas (*Musa cavendishii*), 250, 251, 229
Barberry (*Berberis stolonifera*), 103
Barley (*Hordeum vulgare*), 258, 259, 281, 284, 287
Basil (*Ocimum basilicum*), 253, 254, 256, 257
Bay leaf (*Laurus nobilis*), 257
Beer, 281–5
  production, 281
Beet, 324
Beetroot (*Beta vulgaris*), 212, 213
Bellis perennis L., *see* Common daisy
Benzoic acid, 14, 17, 322
Benzoic acid 2-hydroxylase, 14, 17

Benzophenanthridine alkaloids, 103–5
Benzylisoquinoline alkaloids, 102–7
  biosynthesis, 103–5
  cellular location, 105
Berbamunine synthase, 104
Berberine, 102–7, 131
Berberine, bridge enzyme, 104, 105
Berberis stolonifera, see Barberry
Bergamot (*Mentha citrata*), 55
Bergamottin, 156, 158, 160
Bergapten, 156–2, 223
Bergaptol, 156, 158–60
Berries
  anthocyanin, ellagitannin and anthocyanin content, 242–3
  flavan-3-ols and proanthocyanidin content, 246
Betalains, 130, 213, 324–5
  as colouring agents, 130
Betalamic acid, 130
Betanin, 213, 304, 325
Betaxanthins, 130
Beverages, 263–88, see also individual entries
Bifidobacterium bifidum, 180, 181
Bifidobacterium animalis, 190
Bifidobacterium infantis, 182
Bifidobacterium lactis, 193
Bifidobacterium longum, 181, 182, 193
Bifidobacterium spp., 175–9, 182, 183, 186–8, 189–93, 197, 198, see also individual entries
Bifidobactrium breve, 182
Bilberry (*Vaccinium myrtillus*), 244
Bile acids, 174, 185
Bilobalide, 260, 262
Bioavailability of betalains, 324–5
Bioavailability of dihydrochalcones, 324
Bioavailability of flavonoids, 304–11
  absorption, 304–8
    glycosylated flavonoids, 305–6
    non-glycosylated flavonoids (flavan-3-ols), 306–8
  biliary excretion, 310
  elimination, 310–311
  hepatic metabolism, 310
  intestinal efflux of absorbed flavonoids, 308–9
  metabolism, 309–10
  small intestinal metabolism, 308
Bioavailability of carotenoids, 332–40
  absorption, 335–7
  effects of processing, 335
  mechanisms regulating absorption, 334–5
  metabolism, 339–40
  tissue distribution, 338–9
  toxicity, 340
  transport, 337–8
Bioavailability of gallic acid and ellagic acid, 323–4
Bioavailability of glucosinolates, 325–32
  absorption, 330
  distribution and elimination, 331–2
  hydrolysis and product formation, 327–9
  intestinal metabolism and efflux, 330–331
Bioavailability of hydroxycinnamic acids, 321–3
  urinary excretion, 322–3

Biological activity of in vivo flavonoid metabolites, 311–15
Biosynthesis of
  flavonoids, 15, 17–19
    anthocyanins, 15, 19
    flavan-3-ols, 15, 19
    flavones, 15, 18
    flavonols, 15, 19
    isoflavonoids, 15, 18
    proanthocyanins, 15, 19
  phenolics and hydroxycinnmates, 14, 16, 17
  stilbenes, 5, 19
γ-Bisabolene, 256, 257
$(-)$-α-Bisabolol, 81, 82, 256, 257
  anti-inflammatory properties, 81
Bisabolol oxides, 81–4
Bisabolyl cation, 79, 81, 82
Bisdemethoxycurcumin, 255
Bishop's weed, see Ground Elder
Bitter orange (*Citrus aurantium*), 9, 158, 232, 233
Black grapes (*Vitis vinifera*), 7
Black pepper (*Piper nigrum*), 254
Black walnut (*Juglans nigra*), 262
Blackberry (*Rubus* spp.), 240–242, 244, 246
Blackcurrant (*Ribes nigrum*), 240–242, 244, 246
Blood-brain barrier, 308
Bloodroot (*Sanguinaria canadensis*), 106
Blueberry aka blaeberry (*Vaccinium corymbosum*), 240, 241, 243, 244, 246, 323
Blumenol, 77, 78
Bog myrtle (*Myrica gale*), 281
Bog whortleberry (*Vaccinium uliginosum*), 244
Bok choi (*Brassica rapa*), 27, 29
Borage (*Borago officinalis*), 111, 254
Borago officinalis, see Borage
Bornanes, 61, 63
Borneol, 62, 63, 256
Borneol acetate, 256
Bornyl cation, 63
(+)-Bornyl pyrophosphate, 62, 63
Botanical supplements, see Dietary supplements
Botrytis cinerea, 147
Botrytis fabae, 147
Botrytis spp. 148, see also individual entries
Brassica juncea, see Mustard, brown
Brassinosteroids, 86
Broad bean (*Vicia faba*), 143, 147–9, 221
Broccoli (*Brassica oleracea*), 25, 27, 29, 31, 40, 217, 218, 220, 325–7, 329, 332
Bromelain, 235
Broomrape (*Orobanche* spp.), 94
Brugmansia sanguinea, see Red Angel's trumpet
Bruicine, 114
Brussels sprout (*Brassica oleracea*), 27, 29, 33, 38, 217, 325, 326
Bunium bulbocastanum L., see Great Earthnut
Burdock (*Arctium lappa*), 140, 142, 144–6
Burnet Saxifrage (*Petropselinum saxifraga* L.), 142
Buttermilk walnut (*Juglans cinerea*), 262
Butyriovibrio spp., 176
Byakangelicin, 159, 160
Byakangelicol, 159, 160

C1 transcriptional factor, 20
Cabbage (*Brassica oleracea*), 25, 27, 29, 211, 217, 325, 326
  goitrogenic effect, 218
  progoitrin, 218
Cabbage family and greens, 217–19
Cacao (*Theobroma cacao*), 118, 119
  chocolate, 119
Caco-2 cells, 307–9, 322
Cactus pear, 324
Cadaverine, 126
Cadinene, 82, 83, 256, 257
Cadinyl cation, 83
Cafestol, 72, 74, 275, 276
Caffeic acid, 12, 14, 17, 229, 230, 257, 258, 275, 276, 283, 284, 304, 321–3
Caffeic acid glycerol ester, 235, 236
Caffeic/5-hydroxyferulic acid *O*-methyltransferase, 14
Caffeine (1,3,7-trimethylxanthine), 102, 109, 118–21, 131, 264, 265, 269, 271, 272, 274–8
  physiological effects, 121
Caffeine synthase, 120
3-*O*-Caffeoyl-1,5-quinide, 275
4-*O*-Caffeoyl-1,5-quinide, 275
N-Caffeoyl-3-*O*-hydroxytyrosine, 278, 279
Caffeoylferuloylquinic acid, 275
Caffeoyl-L-glutamic acid, 278, 279
Caffeoyl-L-malic acid, 222
1-*O*-Caffeoylquinic acid, 225, 226
3-*O*-Caffeoylquinic acid (neochlorogenic acid), 12, 13, 212, 213, 218, 231–3, 270, 272, 274
4-*O*-Caffeoylquinic acid (crytochlorogenic acid), 12, 13, 272, 274
5-*O*-Caffeoylquinic acid (chlorogenic acid), 12–14, 17, 213, 222, 225–32, 244, 264, 270, 272, 274, 286, 321–3
Caffeoylshikimic acid glucosides, 236
Caffeoyltartaric acid, 222
Caftaric acid, 12, 13, 246–8, 280, 283
Calabar bean (*Physostigma venenosum*), 113
Calabrese (*Brassica oleracea*), 217
Calystegia sepium, see Hedge bindweed, 109
Calystegines, 109–11
  toxic effects, 109
Camellia irrawadiensis, 119
Camellia japonica, 266
Camellia oleifera, 266
Camellia sasanqua, 266
Camellia sinensis, see Tea
Camellia sinensis var. *assamica*, 263, 265
Camellia sinensis var. *sinensis*, 263, 265
Campanulaceae, 138, 143
Campestanol, 92
Campesterol, 86, 88, 224, 225
(+)-Camphene, 63
Camphor, 62, 63, 256
Camptothecin, as an anticancer drug, 117
Campylobacter spp., 198
Canadian wild lupins (*Lupinus reflexus*), 127
(*S*)-Canadine, 104, 105
(*S*)-Canadine synthase, 104, 105

Cancer
  gastric, 152
  genital, 161
  lung, 153
  skin, 137
Candida albicans, 174, 184
Canthaxanthin, 333, 338, 340
  canthaxanthin retinopathy, 338, 340
Capsaicins and piperines, 255
  role in taste, pain and analgesia, 255
Capsanthin, 228
Capsicum annum, see Peppers
Capsidiol, 79, 80, 82
Caraway (*Carun carvi*), 138, 141, 157, 254
Cardamon (*Elettara cardamonum*), 255
Cardols, 261
Carnosic acid, 256
Carnosol, 256
ζ-Carotene desaturase, 75
α-Carotene, 74–6, 153, 213, 224, 225, 251, 333, 339
β-Carotene, 48, 52, 74–6, 93, 153, 211–13, 219, 220, 222–7, 229, 231, 233, 234, 237, 251, 304, 333, 334, 337–40
γ-Carotene, 333
Carotenoderma, 340
Carotenoids, 332–40, see also individual entries
  food colourants, 333
  protection against heart disease and cancer, 74
  structures, 333–4
15,15'-β-Carotenoid dioxygenase, 340
Carotyl cation, 79
Carrageenans, 262
Carrot (*Daucus carota*), 138, 141, 147, 149, 151, 153–8, 164, 212–13, 334, 335, 340
Carum carvi L., see Caraway
Carvacrol, 61, 62, 256
Carvone, 60, 62, 256, 257
β-Caryophyllene, 81, 83
Casbene, 67, 68
Casbene synthase, 67
Cashew (Anacardium occidentale), 260, 261
Cassia (*Cinnamonum cassia*), 254
Castor bean (*Ricinus communis*), 67
(−)-Catechin, 5, 271
(+)-Catechin, 5, 6, 15, 19, 231, 232, 244, 246, 248, 264, 265, 270, 271, 278–84, 286, 305–11, 316, 320
Catechin sulphates, 316
Catechin sulphoglucuronide, 316
Catechol-*O*-methyltransferases, 309
Catenabacterium catenaeforme, see *Lactobacillus ruitinis*
Catharanthine, 114–16
Catharanthus roseus, see Madagascar periwinkle
Catmint (*Nepeta racemosa*), 65, 94
Cauliflower (*Brassica oleracea*), 217–19
Celandine (*Chelidonium majus*), 106
Celeriac (*Apium graveolens* var. *rapaceum*), 138, 141, 156, 157, 223
Celery (*Apium graveolens* var. *dulce*), 4, 138, 141, 151, 156, 157, 161–3, 213, 223
  accumulation of furocoumarins, 223
  diseased, 161, 162

fungal infection, 223
Celery root, *see* Celeriac
Centaur X$_3$, 139, 143
Centella asiatica, see Asiatic Pennywort
Cephaelis ipecacuanha, see Ipecac
Cephaline, 106
Cereals, 258–60
Chaerophyllum bulbosum, see Turnip-rooted Chervil
Chalcone isomerase, 15, 17
    type I, 17
    type II, 17
Chalcone reductase, 15
Chalcone synthase, 15, 17
Chalcones, 3
Chamazulene, 80, 83, 84
Chamomile
    sesquiterpene biosynthesis, 84
Chamomile
    Anthemis nobilities aka *Chamaemelum nobile*, 55, 83
    Matricaria chamomilla also, *see* German Chamomile, 139, 143, 257
Chelerythrine, 106
Chelidonine, 106
Chelidonium majus, see Celandine
Cherries
    sour (Prunus cerasus), 231
    sweet (*Prunus avium*), 231
Chervil (*Anthriscus cerefolium* Hoffm.), 141, 157
Chervin, *see* Skirret
Chicory (*Cichorium intybus*), 142, 145, 192
Chinese "happy" tree (*Camptotheca accuminata*), 113
Chinese Bellflower, *see* Balloon flower
Chinese cabbage (*Brassica pekinensis* aka *Brassica rapa*), 29, 33, 219
Chinese hamster ovary cells, 308
Chinese prickly ash (*Zanthoxylum bungeanum* Maxim.), 254
Chives (*Allium schoenoprasum*), 214
Chlorogenic acid, *see* 5-*O*-Caffeoylquinic acid
Chlorogenic acids, 12, 212, 213, 222, 226, 232, 254, 257, 266, 269, 272, 274, 275, *see also individual entries*
    carrot root fly attractant, 213
    mono- and diacyl derivatives incorporating 3,4-dimethyoxycinnamic acid, 275
Chocolate, 277–8, 307, 317, 319
Cholesterol, 48, 50–52, 72, 86, 87, 90–92, 121, 127, 128
Chondrodonendron tomentosum, see Tubo curare
Chromoplasts, 74
Chrysanthemic acid, 65, 67, 74
Chrysanthemum coronarium, see Garland chrysanthemum
Chrysanthemyl diphosphate synthase, 67
Chrysanthemyl pyrophosphate, 67, 74
Chrysin-7-*O*-glucuronide, 315
Chrysin-7-*O*-sulphate, 315
Chrysoeriol-6-*C*-glucoside, 257
Chrysoeriol-7-*O*-(2″-*O*-apiosyl)glucoside, 223
Chrysophanol, 249, 250
Chylomicrons, 335–7, 339, 340

Cichorium endivia, see Endive
Cichorium intybus, see Chicory
Cicuta maculata, see Spotted water hemlock
Cicuta virosa, see Water hemlock
Cicutoxin, 149, 150
Cider, 285–7
    bitterness and astringency, 287
    production, 285
Cilantro, *see* Caraway
Cinchona alkaloids, 65, 66
Cinchona calisaya, 114
    quinine content of bark, 114
    anti-malarial treatment, 114
Cinchonain Ib, 266, 267
1,8-Cineole, 62, 63
Cinnamaldehyde, 254
Cinnamate 4-hydroxylase, 14, 16
Cinnamic acid, 12, 14, 16
Cinnamon (*Cinnamomom zeylandicum*), 81, 210, 253, 254, 257
Cinnamoyl-L-aspartic acid, 278, 279
Citronellal, 61
Citronellol, 61
Citrus aurantiifolia, see Lime
Citrus aurantium, see Bitter orange
Citrus fruits, 156, 162, 232–5, *see also individual entries*
Citrus grandis (L.) Osbeck, *see* Pummelo
Citrus limon, see Lemon
Citrus paradisi, see Grapefruit
Citrus sinensis Osbeck, *see* Sweet Orange
Cladosporium fulvum, 149
Clastogenic changes induced by
    2-propenyl isothiocyanate, 38
    phenylethyl isothiocyanate, 38
Cleopatra and tropane alkaloids ,107
Clostridium butyricum, 179
Clostridium coccoides, 176, 178
Clostridium difficile, 176, 180, 181
Clostridium histolyticum, 185
Clostridium leptum, 176, 178
Clostridium perfringens, 176, 185
Clostridium spp., 175–8, 184–8, *see also individual entries*
Cloudberries (*Rubus chamaemorus*), 240
Clovamide, *see N*-Caffeoyl-3-*O*-hydroxytyrosine
Clover (*Trifolium* spp.), 9
Clove(s) (*Syzygium aromaticum*), 81, 254, 257, 258
Cnidilin, 158, 160
Cocaine, 102, 108, 109
    Amazonian coca (*Erythroxylum coca* var. *ipadu*), 109
    coca cola, 109
    Peruvian coca (*Erythroxylum coca*), 108
    physiological effects, 109
    Trujillo coca (*Erythroxylum novogranatense* var. *truxillense*), 108, 109
(*S*)-Coclaurine, 104
Coclaurine-*N*-methyltransferase, 103
Cocoa (*Theobroma cocoa*), 7, 277–8, 308
    criollo, 277
    trinitario, 277

Cocoa (*Theobroma cocoa*) (*Contd.*)
　var. *Forastero*, 277
Cocoa beans and the manufacture of cocoa and
　　chocolate, 277–8
Cocoa tea (*Camellia ptilophylla*), 119
Codeine, 102–6
　pharmacological effects, 106
Codeinone, 104, 105
Codeinone reductase, 104, 105
Coffea arabica aka Arabica coffee, *see* Coffee
Coffea canephora, *see* Robusta coffee
Coffea dewevrei, 119, 273
Coffea excelsa, 119
Coffea liberica, 119, 273
Coffea pseudozanguebariae, 273
Coffea racemosa, 273
Coffee, 12, 118–22, 131, 273–7, 321–3
　aroma, 276
　consumption, 119
　daily intake, 321
　diterpenes, 72, *see also individual entries*
　　increased cholesterol, 72, 276
　production, 273–4, 275
　　roasting beans, 275
Cohumulone, 283, 284
Colchicine, 102, 116
Coliforms, 188, *see also individual entries*
Colletotrichum gloeosporides, 149
Colon cancer, 183, 189, 191–3, 198
Colupuline, 283, 284
Comfrey (*Symphytum officinale*), 124
Common cow parsnip (*Heracleum sphondylium*),
　　142, 157
Common daisy (*Bellis perennis*), 143
Common giant fennel (*Ferula communis*), 142
Common lupin (*Lupinus polyphyllus*), 126
Common ragwort (*Senecio jacobaea*), 123, 124
Common sage (*Salvia officinalis*), 60, 61, 63, 91
Composition of herbal extracts, 91
Concord grapes (*Vitis labrusca*), 245
Condensed tannins, 5, 11, 15, *see also individual
　　entries*
δ-Coniceine, 129
Coniferaldehyde, 288
Coniine, 129–30
Conium maculatum, *see* Hemlock
Contact dermatitis, *see* Dermatitis, contact
Convallatoxin, 90
Convolvulus arvensis, *see* Field bindweed
(−)-Copalyl pyrophosphate aka (*ent*-Copalyl
　　pyrophosphate), 68, 71
(+)-Copalyl pyrophosphate, 47, 68–70
9,10-*syn*-Copalyl pyrophosphate, 71
ent-Copalyl pyrophosphate, *see* (−)-Copalyl
　　pyrophosphate
syn-Copalyl pyrophosphate, 71
ent-Copalyl pyrophosphate synthase, 70, 72
Copapyl diphosphate-methylerythritol kinase, 51
Copapyl diphosphate-methylerythritol synthase, 51
Coptisine, 106
Coriander (*Coriandrum sativum*), 141,157, 167, 254,
　　257, 258, 281

Coriandrin, 257, 258
Coriandrum sativum L., *see* Caraway
Coronary heart disease, 72
Cotton (*Gossypium* spp.), 82
Coumarate:CoA ligase, 14, 16
p-Coumaric acid glycerol ester, 235, 236
p-Coumaric acid, 2, 12, 14, 16, 17, 156, 159, 321
Coumarins, 2, 3
p-Coumaroyl-CoA, 14–17, 19
p-Coumaroyl-CoA:(+)-epilupinine
　　O-*p*-coumaroyltransferase, 126
(+)-*p*-Coumaroylepilupinine, 126
N-*trans*-*p*-Coumaroyloctopamine, 216
3-*O*-*p*-Coumaroylquinic acid, 212, 213, 231, 233
4-*O*-*p*-Coumaroylquinic acid, 229, 230, 270
5-*O*-*p*-Coumaroylquinic acid, 14, 17, 264
Coumestan, 9,18
Coumestrol, 9, 10, 18
Coutaric acid, 12, 13, 247, 248
Cow parsley (*Anthriscus sylvestris* Hoffm.), 141
Cow parsnip (*Heracleum lanatum*), 157
Cranberry (*Vaccinium oxycoccus*), 240, 241, 243,
　　244, 246
Crepenynic acid, 138–40, 148
Cresols, 183
Crithmum maritimum L., *see* Samphire
Crohn's disease, 182
Cryptotaenia canadensis, *see* Hornwort
Cryptotaenia japonica Hassk., *See* Japanese hornwort
β-Cryptoxanthin, 233, 234, 237, 238, 333
Crytochlorogenic acid, *see* 4-*O*-Caffeoylquinic acid
Cumin (*Cuminum cymimum*), 257
Curcumin, 255
Curcuminoids
　pharmacological properties, 255
Currants, 247–8
Cyanidin, 8, 15, 19
Cyanidin-3,5-*O*-diglucoside, 252
Cyanidin-3-*O*-(2^G-*O*-xylosylrutinoside), 241
Cyanidin-3-*O*-(6″-malonyl)glucoside, 216, 217, 222
Cyanidin-3-*O*-(6″-malonyl)laminaribioside, 216, 217
Cyanidin-3-*O*-arabinoside, 241
Cyanidin-3-*O*-galactoside, 230, 237, 241
Cyanidin-3-*O*-glucoside, 213, 231–3, 241, 244, 252
Cyanidin-3-*O*-rutinoside, 231–3, 241
Cyanidin-3-*O*-sambubioside, 241, 244
Cyanidin-3-*O*-sophoroside, 241
1-Cyano-2,3-epithiopropane, 26
1-Cyano-2-hydroxy-3-butene, 38
Cyclase, 75
Cyclic diketopiperazines, 277
Cycloalliin, 35, 36
Cycloartenol, 86–8
Cycloeucalenol, 88
Cyclopropyl carbinyl cation, 85
Cymbopogon citrates, *see* Lemongrass
p-Cymene, 256
Cynara scolymus, *see* Globe artichoke
Cynarin, *see* 1,3-*O*-Dicaffeoylquinic acid
CYP P450, 162
CYP P450 3A, 162
CYP79 enzymes, 30, 31

CYP79 gene family, 30
Cytosolic β-glucosidase, 305, 306, 311, 314
Cytotoxicity, 152, 153

Daffodil (*Narcissus pseudonarcissuss*), 93
Daidzein, 9, 10, 15, 17, 18, 221, 316, 320
  effects on human cancers, 10
Daidzein-4′-O-glucuronide, 316
Daidzein-4′,7-O-diglucuronide, 316
Daidzein-4′-O-sulphate, 316
Daidzein-7-O-(6″-O-acetyl)glucoside, 221
Daidzein-7-O-glucoside, 221
Daidzein-7-O-glucuronide, 316
Daidzein-7-O-(6″-O-malonyl)glucoside, 221
Daidzein-7-O-sulphate, 316
Dammarenyl cation, 89
Danieleone, 237, 238
Databases, 21, 288
  flavonoids in fruits, vegetables, beverages and foods, 288
  flavonol and flavan-3-ol content of Dutch produce, 288
  flavan-3-ol content of Spanish foodstuffs and beverages, 288
  proanthocyanidin content of foods in the USA, 288
Dates (*Phoenix dactylifera*), 235–6
Datura metel, see Hindu datura
Datura stramonium, see Thorn apple
Datura suaveolens, see Angel's trumpet
Daucus carota, see Carrots
Deacetylvindoline 4-O-acetyltransferase, 115, 116
Deadly nightshade (*Atropa belladonna*), 107, 109
Dehydrocrepenynic acid, 138–40, 147, 148
Dehydrofalcarindiol, 152, 153
Dehydrofalcarinol, 139, 152, 153
Dehydrofalcarinone, 139, 143
4,21-Dehydrogeissoschizine, 114, 115
3-Dehydroshikimic acid, 14, 16
Dehydrotheasinensin, 269
Delphinidin, 8
Delphinidin-3,5-O-diglucoside, 252
Delphinidin-3-O-galactoside, 241, 244
Delphinidin-3-O-glucoside, 246, 248, 252
Delphinidin-3-O-rutinoside, 241
Demethoxycurcumin, 255
Demethylsuberosin, 156, 159
Dendropanax arboreus (L.) Derne. & Planchon., 152
11-Deoxojervine (cyclopamine), 128
Deoxyclovamide (*N-p*-coumaroyl-tyrosine), 278, 279
Deoxyhypusine synthase, 123
1-Deoxyxylulose 5-phosphate, 49, 51, 52, 84
1-Deoxyxylulose 5-phosphate reductoisomerase, 51, 52
1-Deoxyxylulose 5-phosphate synthase, 51
Derived polyphenols, 1, 2, 8, see also individual entries
Dermatitis, contact, 150, 151
Desacetoxyvindoline, 115, 116
Desacetoxyvindoline 4-hydroxylase, 115, 116, 131
Diallyl disulphide, 36, 37, 215
Diallyl thiosulphinate (allicin), 36, 40, 215
Diallyl trisulphide, 215

Diamine oxidase, 109
Diarrhoea, 176, 180, 181
Dibenztropolones, 267
1,3-O-Dicaffeoylquinic acid, 225, 226
  formed from 1,5-O-dicaffeoylquinic acid, 225
1,4-O-Dicaffeoylquinic acid, 274
3,4-O-Dicaffeoylquinic acid, 274
3,5-O-Dicaffeoylquinic acid, 13, 213, 222, 274
4,5-O-Dicaffeoylquinic acid, 274
Dicaffeoylquinic acids, 272, see also individual entries
Dicaffeoyltartaric acid, 222
Dietary supplements, 91, 213, 233, 257, 308, see also individual entries
1,2-O-Diferuloylgentiobiose, 218, 220
2-O-Digalloyl-tetra-O-galloylglucose, 11
Digitoxin, 90
  cardiac stimulatory activity and high toxicity, 90
Dihydro-4′-methoxy-isoflavonol dehydratase, 18
Dihydrochalcones, 3, see also individual entries
Dihydroflavonol reductase, 15, 19
Dihydroflavonols, 3, 19, see also individual entries
Dihydrohydroxywyerol epoxide, 148
Dihydrokaempferol, 15, 19
3,6-Dihydronicotine, 110, 112
1,2-Dihydronicotine dehydrogenase, 112
2,3-Dihydrooenanthetol acetate, 139, 142
2,3-Dihydrooenanthetol, 139, 142
1,2-Dihydropyridine, 112
Dihydroquercetin, 15, 19
Dihydrosanguinarine, 105
Dihydrowyerol, 143, 148
Dihydrowyerone, 143, 148
Dihydrowyerone acid, 143, 148
3,4-Dihydroxy-6-(*N*-ethylamino)benzamide, 256
3,4-Dihydroxyphenylpropionic acid, 322
Diindolylmethane, 28, 33
3,3-Diindolylmethane, 39, 328
Dill (*Anethum graveolens*), 141, 150, 157, 253, 257
5,7-Dimethoxyisoflavone, 221, 260
Dimethylallyl pyrophosphate, 47, 49–52, 54–7, 65, 82, 84
1,2-Dimethylpyridinium, 275, 276
3,7-Dimethylxanthine, see Theobromine
Diplotaxis spp., see Rockets
Dithiophenes, acetylenic, 140, 144, 146
DNA adducts, 159, 161
Dopa, 103, 104, 130
Dopa 4,5-dioxygenase, 130
Dopamine, 103, 104, 251
Duboisia hopwoodii, see Pituri bush

Eczema, 159
Edible Burdock, see Burdock
Eggplant, see Aubergine
Elderberry (*Sambucus nigra*), 240, 241, 243, 244
Elenolic acid, 239
Elimination half life, 317–19, 321
  quercetin metabolites, 319
Ellagic acid, 11, 14, 16, 244, 247, 252, 288, 323, 324
Ellagitannins, 11, 14, 16, 241, 242, 244, 247, 323, 324, see also individual entries
  hexahydroxydiphenoyl moieties, 11

Emetine, 106
Emodin, 249, 250
Endive (*Cichorium endivia*), 143, 145
Enterobacteriaceae, 175, 177, 178, 187, see also individual entries
Enterococcus spp., 174–6, 179, 187
(−)-Epiafzelchin gallate-4-β-6-epigallocatechin gallate, 264
(−)-Epiafzelechin, 7, 264, 266
(−)-Epicatechin, 5, 6, 15, 19, 229, 230–232, 244, 246, 248, 250, 264, 265, 270, 271, 278–81, 283, 284, 286, 306–310, 320
(+)-Epicatechin, 5
(−)-Epicatechin gallate, 6, 7, 263–5, 306–8, 316
Epicatechin-[7,8-bc]-4-(4-hydroxyphenyl)-dihydro-2(3H)-pyranone, 266, 267
(−)-Epicatechin-3′-O-glucuronide, 316
(−)-Epigallocatechin, 6, 7, 264, 265, 267, 270
(−)-Epigallocatechin gallate, 6, 7, 263–5, 267, 270, 306, 307, 316, 318–20
   3-D structure, 7
(+)-Epirosmanol, 256
Epitheaflavic acid, 267, 268
Epithiospecifier protein, 26, 33, 38, 327
Epoxybergamottin, 158, 160
Epoxybergamottin hydrate, 158, 160, 162
9-*cis*-Epoxycarotenoid dioxygenase, 78
Equol (*O*-desmethyl-angolensin), 316
Eriodicytol, 308, 310
Eruca sativa, see Rockets
Erucin, 219
Escarole, *see* Endive
Escherichia coli, 176, 182, 183, 198
Eschscholzia californica, see Golden poppy
Esculetin, 2
Esdragon, *see* Tarragon
Estragole (1-allyl-4-methoxybenzene), 254
2-Ethyl-3,5-dimethylpyrazine, 276
N-Ethylglutamine, *see* Theanine
8′-Ethylpyrrolidinonyltheasinensin A, 269
Eucalyptol, 256
Eudesmyl cation, 80
Eugenol (1-allyl-3-methoxy-4-hydroxybenzene), 254, 255
Euphol, 86, 89
European Angelica, *see* Garden angelica
European grapes (*Vitis vinifera*), 245

Faecalibacterium prausnitzii, 176, 177
Falcarindiol, 139, 141–3, 147, 149, 150, 153, 164
Falcarindiol-3-acetate, 139, 141
Falcarindione, 139, 141, 142
Falcarinol (*aka* Panaxynol), 138, 139, 141–3, 147, 149–55, 163, 164, 213
  anticancer properties, 213
  pharmacokinetics, 155
Falcarinolone, 139, 141, 142
Falcarinone, 139, 141, 142, 150
Farnesyl cation, 78, 79, 81, 83

Farnesyl pyrophosphate, 49, 50, 57, 78, 79, 81, 82, 84–6
Farnesyl pyrophosphate synthase, 56–8, 84
Fatty acid acetylenase, 138
Fatty acyl serotonin derivatives, 276
Fennel (*Foeniculum vulgare*), 138, 142, 157, 254, 257
Fenugreek (*Trigonella foenum-graecum*), 121
Fermented soya products, 221
Fertaric acid, 12, 13, 248, 280
Ferula assa-foetida, *see* Asafoetida
Ferula communis, *see* Common giant fennel
Ferulate 5-hydroxylase, 14
Ferulate dimers 258
Ferulic acid, 12, 17, 235, 236, 238, 258, 284, 321–3
Ferulic acid esterified to arabinoxylans and hemicelluloses, 323
Feruloyl esterases, 323
N-trans-Feruloyloctopamine, 216
3-*O*-Feruloylquinic acid, 274
4-*O*-Feruloylquinic acid, 274
5-*O*-Feruloylquinic acid, 212, 213, 274
Feverfew (*Tanacetum parthenium*), 79
  health effects, 79
Field bindweed (*Convolvulus arvensis*), 109
Fig (*Ficus caria*), 238
  anti-tumour activity, 238
Fish poisons, 149
Flavan-3,4-diols, 3
Flavan-3-ol gallates, 268, *see also individual entries*
Flavan-3-ols, 221, 244, 246, 257, 260, 261, 263–7, 269–71, 27–81, 283, 284, 286, 288, *see also individual entries*
  diastereoisomers, 7
  monomers, 3, 5–7
  proanthocyanidins, 5–7
    procyanidins, 7
    prodelphinidins, 7
    propelargonidins, 7
    type A, 5, 6
    type B, 5, 6
Flavanone-3-hydroxlase, 15, 19
Flavanones, 5, 8–9, 10, 13, 17, 18
Flavone synthase, 15, 18
Flavones, 3, 4–5
Flavonoids
  basic structure, 3, 16
  data bases, 21
Flavonoid biosynthesis pathway, 14
Flavonoid conjugates in plasma and urine, 315–21
  pharmacokinetics, 317–21
Flavonol 3′-hydroxylase, 15, 19
Flavonol synthase, 15, 19
Flavonols, 3, 4, 5, 17, 21, *see also individual entries*
Flax (*Linum usitatissimum*), 259
Fluoxetine, 106
Foeniculum vulgare, *see* Fennel
Food safety, 163, 164
  acetylenes
    allergenicity, 150, 151
    neurotoxicity, 149, 150
    psorslens, 163
      DNA adducts, 159

P450 inhibition, 162
phototoxicity, 159–62
reproductive toxicity, 162
Formononetin, 18
Foxglove (*Digitalis purpurea*), 90
French beans (*Phaseolus vulgaris*), 221
French salad, *see* Chervil
Fructans, 186, 189, 192, *see also individual entries*
Fructo-oligosaccharides, 186–92
Fruit and vegetable consumption, dietary guidelines, 208
Fruits, 229–52, *see also individual entries*
Furanocoumarins
   angular, 155, 161
   linear, 137, 155–63
      geranylated, 156
      prenylated, 156
Furanones, 276
Furfuryl mercaptan, 276
2-Furfuryl-thiol, 276
Fusobacterium spp., 175, 187

Galacto-oligosaccharides, 186–8
Galileo, 278
Gallacetophenone, 2
Gallagic acid, 252
   toxicity, 252
Gallic acid, 2, 7, 11, 14, 16, 244, 246–9, 252, 253, 267, 270, 280, 283, 288, 304, 323, 324, 341
(+)-Gallocatechin, 6, 251, 264, 270, 271
(+)-Gallocatechin gallate, 6, 7
   3-D structure, 7
Gallotannins, 11, 14, 16, 237, 249, 257, 323, *see also individual entries*
5-O-Galloylquinic acid, 264, 267, 270
Galloylquinic acids (theogallins), 266, *see also individual entries*
Galloyltransferase, 14
Garden angelica (*Angelica archangelica* aka *A. officinalis*), 157
Garden lovage, *see* Lovage
Garland chrysanthemum (*Chrysanthemum coronarium*), 139, 143
Garlic (*Allium sativum*), 25, 34–6, 39, 40, 214–16
   bad breath and diallyl disulphide and diallyl trisulphide, 215
   cooking and formation of diallyl sulphides and ajoenes, 215
   protective effects, 215–16
Gastric cancer, 152
Gastric pH value, 174, 178, 186
Gastric ulcer, 175
Genetic engineering
   for defense against insects, 93
   of flavonoid biosynthetic pathway
      constraints, 21
      manipulation, 20–21
   to attract insects, 94
   to produce
      berberine, 131
      caffeine deficiency, 131

carotenoids, 93
columbamine, 131
enhanced oil quality in mints, 60–61
sanguinarine, 131
scopolamine, 131
terpenoid indole alkaloids, 131
valuable terpenes, 94
to suppress germination of parasitic weeds, 94
Genistein, 9, 10, 221
   effects on human cancers, 10
Genistein glycosides, 257, *see also individual entries*
Genistein-7-O-(6″-O-acetylyl)glucoside, 221
Genistein-7-O-glucoside, 221
Genital cancer, 161
Gentio-oligosaccharides, 186
Geranial, 61
Geraniol 10-hydroxylase, 114, 116, 131
Geraniol, 48, 60, 61, 65, 66
Geraniol/linalool family, 60, 61
Geranyl pyrophosphate, 49, 50, 56, 57, 60, 61, 63, 65, 68, 83, 84
Geranyl pyrophosphate synthase, 60, 56, 93
Geranylgeranyl pyrophosphate, 49, 50, 57, 65, 67–75, 93
   cyclization modes, 68
Geranylgeranyl pyrophosphate synthases, 56, 57
6-Geranylnaringenin, 284, 286
Germacrene, 78–80, 82
Germacrene synthases, 78
Germacryl cation, 78–81, 83
German chamomile (*Matricaria chamomilla*), 80, 81
   anti-inflammatory properties, 80
Germinated barley foodstuff, 191
Gibberellin A_1, 48, 72, 73
Gibberellin A_{12}, 73
Gibberellin A_{12} aldehyde, 73
Gibberellin A_3 (*aka* Gibberellic acid), 72, 73, 287
Gibberellin A_7, 72, 73
Gibberellin biosynthesis, 72, 73
Gibberellins, 56, 65, 68, 72, 76, *see also individual entries*
Gin, 257
Ginger (*Zingiber officinale*), 255, 257
Gingkgo (*Gingko biloba*), 69, 261–2
   medicinal properties, 69
   pharmacological effects, 262
Gingkolides, 69, 70, 260, 262
Ginseng (*Panax ginseng*), 152
Ginsenosides, 86, 88, 89, 152
Glandular trichomes, 59, 60
Globe Artichoke (*Cynara scolymus*), 143
Glucobrassicin, 209, 218, 219, 258, 326
Glucoerucin, 219
β-Glucogallin, 14, 16
Glucoiberin, 326
Gluconasturtiin, 326
Gluco-oligosaccharides, 186
Glucoraphanin, 218, 219
Glucoraphanin, *see* 4-(Methylsulphinyl)butyl glucosinolate
$\bar{\beta}$-Glucosidase, 305, 306, 311, 313

Glucosinolates, 163
  affects of cooking, 33, 327, 329, 330
  anti-nutritional effects, 38
  beneficial effects, 38–41
    anti-inflammatory activity, 40
    anti-proliferative activity, 40
    epidemiological evidence, 38–9
    induction of phase II enzymes, 39–49
    inhibition of phase I CYP450, 39
    mechanisms of action, 39–41
    reduction in *Helicobacter pylori*, 40–41
  biological activity, 37–8
  biosynthesis, 29–31
  degradation, 26–8
  down-regulation of phase-I 'activation' enzymes, 325
  factors affecting content, 31–2
  hydrolysis products, 33–4
  human metabolism, 34
    N-acetylcysteine-isothiocyanate conjugates, 34
  induction of phase-II 'detoxification' enzymes, 325
  structures in crucifers, 27–9
    side chains, 27–9
Glucosinolates-myrosinase system, 26–7, 32
β-Glucuronidase, 310, 313, 314
N-L-γ-Glutamyl-S-sinapyl-L-cysteine, 235, 236
Glutathione S-transferase, 325, 331, 332
Glyceraldehyde 3-phosphate, 49, 51, 52, 84
Glycine max, see Soya bean
Gobo, see Burdock
Goitre, 4, 218, 259
Goitrin, 209, 218
Golden poppy (*Eschscholzia californica*), 105, 131
Goldenrod (*Solidago canadensis*), 78
Goldenseal (*Hydrastis canadensis*), 106, 107
Gooseberry (*Ribes grossularia*), 240
Gossypol, 82, 83
  activity as a male contraceptive, 82
Grape seed extract, 307, 308, 317, 319
Grapefruit (*Citrus paradisi*), 9, 158, 162, 232–4
Grapes, 245–8,
Great earthnut (*Bunium bulbocastanum*), 141
Greater burnet saxifrage (*Pimpinella major*), 142, 158
Ground elder (*Aegopodium podagraria*), 141, 157
Guaiacylglycerol-β-caffeic acid ether, 216
Guaiacylglycerol-β-ferulic acid ether, 216
Guaianolides, 144, 146
Guaiyl cation, 80
Gut microflora, 174–207
  activity, 193, 194
  composition, 174–7, 193, 194
  DNA microarrays, 195
  diet, 178, 197, 198
  fluorescent in situ hybridization, 195
  host age, 177, 178
  in vitro models, 193, 194
  molecular methods, 194–7
  proteomics, 196, 197

Hamburg parsley (*Petoselinum crispum* var. *tuberosum*), 142
Hazelnut (*Corylus avellana*), 260, 261
Heart disease, 177, 193
Hedge bindweed (*Calystegia sepium*), 109
Helianthae, 149
Helianthus tuberosus, see Jerusalem artichoke
Helicobacter pylori, 175, 176
Heliotridine, 122
Hellebore (*Helleborus niger*), 90
Hemigossypol, 83
Hemlock (*Conium maculatum*), 103, 129, 130
Hemlock water dropwort (*Oenanthe crocata*) 150
Henbane (*Hyoscyamus niger*), 107, 109, 131
HepG2 cells, 310
5-(12-Heptadecenyl)-resorcinol, 237, 261
Heracleum lanatum, see Cow parsnip
Heracleum sphondylium, see Common cow parsnip
Herbal remedies, risk assessment, 91
Herbal tea, 12, 123, 257, 271, 272, see also individual entries
Herbs and spices, 252–8, see also individual entries
Heroin, 102, 106
Hesperetin-7-O-neohesperidoside (neohesperidin), 9, 10, 233, 234
Hesperetin-7-O-rutinoside (hesperidin), 9, 10, 233, 234, 250, 305, 306, 319, 320
Hesperidin, see Hesperetin-7-O-rutinoside
Hexahydroxydiphenic acid, 11, 16
High performance liquid chromatography-tandem mass spectrometry, 306, 313–6, 329, 330, 331
Hindu Datura (*Datura metel*), 107
Hippuric acid, 322
Hogweed, see Common cow parsley
Homospermidine, 122, 123
Homospermidine synthase, 122, 123
  localisation, 122
Hops (*Humulus lupulus*), 81, 281, 283–4
Hormesis, 153, 154, 162–4
Hornwort (*Cryptotaenia canadensis*), 141
Horseradish (*Armoracia rusticana*), 29, 217, 258
Hululone
  inhibition of angiogenesis, 283–4
α-Humulene, 81, 83
Humulinic acids, 283
Humulone, 283–5
Humulyl cation, 79–81, 83
Hydrastis canadensis, see Goldenseal
Hydrogen sulphide, 183
Hydrolysable tannins, 11, 14, 16, 252, 264, 266, 272, see also individual entries
Hydroquinone glucoside (arbutin), 230
2-Hydroxy-3-butenyl glucosinolate, see Progoitrin
(E)-4-Hydroxy-3-methylbut-3-enyl pyrophosphate, 51, 54, 55
Hydroxy-3-methyl-coenzyme A, 51
3-Hydroxy-3-methylglutaryl CoA synthase, 51
3-Hydroxy-4,5-dimethylfuran-2-one, 276
($2'$,5-Hydroxyalkenyl) resorcinol, 259
($4'$,5-Hydroxyalkyl) resorcinol, 259
6-Hydroxyapigenin, 257
Hydroxybenzoates, see Phenolic acids
Hydroxybenzoic acid glycosides, 253
$2'$-(4-Hydroxybenzyl)-rosmarinic acid, 253, 254
2-Hydroxycinnamaldehyde, 254

Hydroxycinnamates, 2, 8, 12, 14, 16, 20, see also individual entries
5-Hydroxyferulic acid, 14, 17
3-Hydroxyhippuric acid, 322
13α-Hydroxylupanine, 127
5-(Hydroxymethyl)-2-furaldehyde, 232, 233, 288
13α-Hydroxymultiflorine, 126, 127
(S)-3'-Hydroxy-N-methylcoclaurine, 104, 105
3'-Hydroxy-N-methylcoclaurine 4'-O-methyltransferase, 103, 104
4-Hydroxyphenylacetaldehyde, 103, 104
p-Hydroxyphenylacetic acid, 2
3-Hydroxyphenylpropionic acid, 322
15-Hydroxyprostaglandin dehydrogenase, 151
8-Hydroxy-psoralen, see Xanthotoxol
8-Hydroxy-5-methoxy-psoralen, 156, 158, 160
16-Hydroxytabersonine, 114, 115
5-Hydroxytryptamine (serotonin), 251
Hydroxytyrosol, 239
Hygrine, 109, 110
Hyoscine, see Scopolamine
Hyoscyamine 6β-hydroxylase, 110, 111, 131
Hyoscyamine, 107, 109–11, 131
Hyoscyamus niger, see Henbane
Hyperchylomicronaemia, 337

Iberin, 326
Ichthyothereol, 149, 150
Ichthyothereol acetate, 149, 150
Ilex paraguariensis, see Maté
Imperatorin, 156, 158, 160
Increasing flavonoid levels in food crops, 20–21
Indian Pennywort, see Asian pennywort
Indian snakeroot (*Rauwolfia serpentaria*), 113
Indicaxanthin, 325
Indole-3-carbinol, see Indolyl-3-carbinol
Indole-3-methanol, see Indolyl-3-carbinol
Indolyl glucosinolate(s), 28, 29, 32, 33, 38, 39
Indolyl-3-acetonitrile, 28, 33, 38
Indolyl-3-carbinol, 28, 38, 218, 219, 328
3-Indolylmethyl isothiocyanates, 326
Inflammatory bowel disease, 182–3, 190, 198
Interleucin IL8, 183
Inulin, 186, 187, 189–93, see also fructans
Ipecac (*Cephaelis ipecacuanha*), 106
  medicinal properties, 106
Ipomoea batatas, see Sweet potato
IQ, 183
Iridodial, 66
Iridoids, 64–6, see also individual entries
Iriflopheone, 237
Irritable bowel syndrome, 184, 198
Isoalliin, see S-Allyl cysteine sulphoxide
Isobetanin, 213
Isochlorogenic acid, see 3,5-O-Dicaffeoylquinic acid
Isocoumarins, 138–40, 145
Isoflavone reductase, 18
Isoflavone synthase, 15, 18
Isoflavone-O-methyltransferase, 18
Isoflavones, 9–10, 303–6, 316, 318, 319, 321, see also individual entries
reduced prostate and breast cancer, 10
Isoformononetin, 18
Isohumulones, 283
Isoimperatorin, 156–8, 160
Isoliquiritigenin, 15, 17, 18
Isomalto-oligosaccharides, 18–28
Isomangiferin, 237
Isomyricetin-3-glucoside, 267
Isopentenyl pyrophosphate, 47–59, 82–4
Isopentenyl pyrophosphate and dimethylallylpyrophosphate biosynthesis, 49–55
  biosynthesis of cholesterol, 51–2
  1-deoxyxylulose 5-phosphate (*aka* methylerythritol 4-phosphate) pathway, 52–5
    transit peptides, 53
  interconversion of isopentenyl pyrophosphate and dimethylallylpyrophosphate, 54–5
  isopentenyl pyrophosphate and dimethylallylpyrophosphate biosynthesis, 55
  mevalonic acid pathway, 49–52
Isopentenyl pyrophosphate isomerase, 51
Isopentenyl pyrophosphate/dimethylallyl pyrophosphate synthase, 51, 56–9
Isopimpinellin, 156–8, 160–162
(−)-Isopiperitenol, 61, 62
(−)-Isopiperitenone, 62
Isoprene, 47, 49, 50, 83, 84
Isoprenoid biosynthesis in the cytosol, 78–90
  sesquiterpenes, 78–85
  triterpenes, 85–90
Isoprenoid biosynthesis in the plastids, 59–78
  (+)-abscisic acid, 75–8
  carotenoids, 74–8
  diterpenes, 65–74
    gibberellins, 72, 73
    ginkolides, 69, 70
    rice phytoalexins, 70–71
    taxol, 67, 69
  monoterpenes, 59–65
    iridoids, 64–6
    "irregular" monoterpenes, 65,
    menthanes, 60–62
Isoprenyl diphosphate isomerase, 54, 55
Isoprenyl pyrophosphate synthases, 56–8
  active sites, 57–8
  cis/trans enzymes, 57
4-Isopropylphenol, see Thymol
(+)-*cis*-Isopulegone, 62
Isorhamnetin, 4
Isorhamnetin-3-O-glucoside, 314
Isorhamnetin-3-O-glucuronide (*aka* 3'-methylquercetin-3-O-glucuronide), 314, 315
Isorhamnetin-4'-O-glucoside, 216
Isothiocyanates, 325–7, 329, 330–332, 341
  absorption from the gastrointestinal tract, 330–332
  N-acetyl-cysteine conjugates, 330, 331, 341
Isothiocyanate sulphoraphane, 325
Isoveraldehyde, 276
Isoxanthohumol, 284, 286
Itadori plant (*Polygonum cuspidatum*), 12–13

Jacobine, 123, 124
Japanese golden seal (*Coptis japonica*), 131
Japanese hornwort (*Cryptotaenia japonica*), 141
Japanese knotweed, *see* Itadori plant
Jerusalem artichoke (*Helianthus tuberosus*), 138, 143, 212
Juglone (5-hydroxy-1,4-naphthoquinone), 2, 260, 262
Juniper (*Juniperus communis*), 82, 256, 257

Kaempferol, 4, 15, 19, 21, 308, 310, 315
Kaempferol glycosides, 218, 237, 266, *see also* individual entries
Kaempferol monosulphate, 315
Kaempferol-3,7-*O*-diglucoside, 219, 220
Kaempferol-3-*O*-glucuronide, 315
Kaempferol-3-*O*-rutinoside, 250
Kaempferol-3-*O*-sophoroside, 218, 220
Kaempferol-3-*O*-sophorotrioside-7-*O*-sophoroside, 218, 220
Kahweofuran, 276
Kahweol, 72, 74, 275, 276
Kale (*Brassica oleracea*), 29, 217, 325, 326
ent-Kaurene *aka* (−)-Kaurene, 54, 68, 72, 73
ent-Kaurene synthase, 72
Kiwi fruit (*Actinidia deliciosa*), 249–50
  potentially protective effects, 250
Knob celery, *see* Celeriac
Kohlrabi (*Brassica oleracea*), 217
Kor tongho, *see* Garland chrysanthemum
Kucha (*Camellia assamica* var. kucha), 119

Lachrymatory factor in onions, 214
Lachrymatory factor synthase, 35, 36
Lactase phloridzin hydrolase, 305, 306, 311, 314
Lactobacillus acidophilus, 180–182, 190, 193
Lactobacillus bulgaricus, 180, 182
Lactobacillus casei, 181, 182, 190
Lactobacillus gasseri, 177
Lactobacillus plantarum, 182, 184
Lactobacillus reuteri, 177, 181, 184
Lactobacillus rhamnosus, 180, 184, 190, 193
Lactobacillus ruminis, 177
Lactobacillus spp., 174–83, 186–8, 190–192, 197, 198, *see also individual entries*
Lactose maldigestion, 180
Lactosucrose, 186
Lactuca sativa, *see* Lettuce
Lactulose, 186, 188, 189, 193
Lambertianin C, 242–4, 247
Lampranthin I and II, 130
Lappaphen-a, 142, 146
Lappaphen-b, 142, 146
Lauraceae, 143, 149
LC transcriptional factor, 20, 21
Leaf mould, 149
Leeks (*Allium porrum*), 25, 34, 35, 39, 214, 217
Legumes, 219–22
Leguminosae, 143, 149, 156
Lemongrass (*Cymbopogon citratus*), 59, 257
Lemons (*Citrus limon*), 156, 158, 232, 233

Lens culinaris, *see* Lentil
Lentil (*Lens culinaris*), 143, 147–9
Lettuce (*Lactuca sativa*), 138, 143, 222–3
  Lollo roso, 222,
  Round, 222
  Iceberg, 222
Leucocyanidin 4-reductase, 15, 19
Leucocyanidin deoxygenase, 15, 19
Leucocyanidins, 15, 19
Leucoplasts, 60
Levisticum officinale, *see* Lovage
Levopimaradiene, 69, 70
Lily of the valley (*Convallaria majalis*), 90
Lime (*Citrus aurantifolia*), 156, 158, 232
Limonene, 60–62, 233, 235, 256
Limonene 3-hydroxylase, 60, 61
Limonene 6-hydroxylase, 60, 61
Limonene synthase, 60
Limonin, 86, 89, 234, 235
  anticancer properties, 86
Limonin-17-*O*-glucoside, 234, 235
Limonoids as
  insect anti-feedants, 234
  pharmacological properties, 234
Linalool, 60, 61, 93, 210
  as a signal to attract preditors of herbivores, 93
Linalyl acetate, 55, 60, 61
Linalyl cation, 60, 61
Linalyl pyrophosphate, 60, 61, 63, 78
Linear furanocoumarins, *see* Furanocoumarins, linear
Linoleic acid, 138, 140
(6′-*O*-Linoleoylglucosyl)sitosterol, 238
Lipase, 335–7, 339
Lipid metabolism, 189, 190, 193
Liquiritigenin, 15, 17
Littorine, 110
Lobetyol, 143, 144
Lobetyolin, 143, 144
Lobetyolinin, 143, 144
Lochnericine, 116
Loc-I-Gut® model, 308, 309
Loganberry (*Rubus loganbaccus*), 240
Loganin, 65, 66
L-Ornithine, 110, 111, 123
Lovage (*Levisticum officinale*), 142, 157
Lucerne (*Trifolium* spp.), 9
Lung cancer, 153
Lupanine, 125–7
Lupinine, 126
Lupinus angustifolius, *see* Narrow leaved lupin
Lupulone, 283–5
Luputriones, 283
Lutein, 74–6, 93, 211, 212, 219, 220, 222, 224–9, 251, 333, 338, 340
  in the light-harvesting complex of the chloroplasts, 74
Luteolin, 4, 5, 238, 306, 308–10
Luteolin-6-*C*-glucoside, 257
Luteolin-7-*O*-(2″-*O*-apiosyl)glucoside, 223
luteolin-7-*O*-(2″-*O*-apiosyl-6″-*O*-malonyl)glucoside, 228
Luteolin-7-*O*-glucoside, 225, 226, 305, 306

Luteolin-7-O-rutinoside, 225, 226
Luteolin-8-C-glucoside, 257–9, 264
Luteolin-8-C-rutinoside, 257
Lycopene, 74–6, 93, 226, 227, 229, 245, 333, 335, 337, 338
Lycopersicon esculentum, see Tomato
Lysine, 126
Lysine decarboxylase, 126

Mace, *see* Nutmeg
Maclurin, 237
Madagascar periwinkle (*Catharanthus roseus*), 113, 114, 116–17
  anti-cancer drugs, 117
Madin-Darby canine kidney cells, 308
Maillard products, 277
Maize (*Zea mays*), 20, 21, 78, 94, 258, 259, 287
Malarial parasites
  Plasmodium falciparum, P. ovale, P. vivax, and *P. malariae*, 117
Malonic acid pathway, 17
Malonyl CoA, 14, 15, 17, 19
Malvidin, 8
Malvidin-3,5-O-diglucoside, 9, 246, 248
Malvidin-3-O-(6″-O-acetyl)glucoside, 9, 246, 248
Malvidin-3-O-(6″-O-p-coumaroyl)glucoside, 9, 246, 248, 280
Malvidin-3-O-(6″-O-p-coumaroyl)glucoside, linked to
  (+)-catechin, 281, 282
  (−)-epicatechin, 81, 282
  procyanidin dimer B$_3$, 281, 282
Malvidin-3-O-glucoside, 9, 246, 248, 280, 281, 283
Malvidin-3-O-glucoside, linked to
  (+)-catechin, 281, 282
  (−)-epicatechin, 81, 282
  procyanidin dimer B$_3$, 281, 282
Mandragora autumnalis (Autumn Mandrake), 111
Mandragora officinarum, see Mandrake
Mandrake (*Mandragora officinarum*), 107, 281
Mangiferin, 2, 237
  inhibition of carcinogenesis, 237
Mango (*Mangifera indica*), 236–7, 261, 323
  pharmacological properties, 237
Marine Fennel, *see* Samphire
Marjoram (*Origanum marjoram*), 253, 254, 257
Marmesin, 156, 159
Masterwort, *see* Cow Parsnip
Maté (*Ilex paraguariensis*), 12, 119, 271–3
  cancer of the oesophagus, 119, 273–3
Matesaponins, 272
Matricaria chamomilla, see Chamomile and German Chamomile
Matricin, 80, 83, 84
Maximum concentration in plasma, 317, 319, 320, 322, 324
Meadow rue (*Thalictrum flavum*), 105
Medicago sativa, see Alfalfa
Medicarpin, 18
Melissa (*Melissa officinalis*), 253
Melons (*Cucumis melo*), 228, 245

Mentha citrata, see Bergamot
Mentha spicata, see Spearmint
Mentha x *piperita, see* Peppermint
(+)-Menthofuran, 61, 62
(−)-Menthol, 48, 60, 62, 256
(−)-Menthone, 62
Metabolic channeling, 19, 21
Metabolic engineering, *see* Genetic engineering
7-Methoxy-luteolin-6-C-glucoside, 257
16-O-Methyl cafestol, 275, 276
S-Methyl cysteine sulphoxide, 25, 214
Methyl jasmonate, 33
N-Methyl nucleosidase, 120
N-Methyl-Δ^1-pyrrolium cation, 109, 110, 112
4-Methylaminobutanal, 109, 110
3′-O-Methylcatechin, 310
3′-Methyl-catechin glucuronide, 316
Methylcoclaurine, 103, 104
N-Methylcoclaurine 3′-hydroxylase, 104
24-Methylene-cycloartanol, 88
4′-O-Methyl-epicatechin-3′-O-glucuronide, 316
4′-O-Methyl-epicatechin-5-O-glucuronide, 316
4′-O-Methyl-epicatechin-7-O-glucuronide, 316
4′-O-Methyl-epigallocatechin, 316
4′,4″-di-O-Methyl-epigallocatechin gallate, 315, 316
2-C-Methylerythriol 4-phosphate, 51–3
Methylerythritol cyclodiphosphate synthase, 51
4-O-Methylgallic acid, 323, 324, 341
  biomarker for tea intake, 324
N-Methylputrescine, 109, 110
1-Methylpyridinium, 275, 276
(S)-cis-N-Methylstylopine, 105
4-(Methylsulphinyl)butyl glucosinolate (glucoraphanin), 304, 325, 332
6-O-Methyltransferase, 103, 104
7-Methylxanthine, 120
7- Methylxanthosine, 119, 120
7-Methylxanthosine nucleosidase, 120
7- Methylxanthosine synthase, 119, 120
N-Methyl-Δ^1-pyrrolinium cation, 109, 112
Mevalonate 5-phosphate decarboxylase, 51
Mevalonate kinase, 51
Mevinolin, 52
Mignonette (*Reseda odorata*), 29
Millet (*Pennisetum americanum*), 258, 259
Mineral absorption, 189, 190
Mint (*Mentha rotundifolia*), 256, 257
Mint family (*Mentha* spp.) 60, *see also individual entries*
Mitsuba, *see* Japanese hornwort
Momilactone A, 71
Monocarboxylate transporter, 308
Moraceae, 156
Morpheus, the creator of sleep and dreams in *Ovid*, 103
Morphinan alkaloids, 103, 105
Morphine, 102–6, 108
  as an analgesic, 103, 105
  pharmacological effects, 105–6
Multidrug resistance protein, 304, 308, 311, 331
(−)-Multiflorine, 126

Mustard, 325, 326
  black (*Brassica nigra*), 217, 218, 258
  brown (*Brassica juncea*), 29
  english, *see* Mustard, white
  white (*Sinapis alba*), 27, 29, 217, 218, 258
Mycocentrospora acerina, 149
Mycosis fungoides, 159
β-Myrcene, 61, 283, 284
Myricetin, 4
Myricetin-3-*O*-galactoside, 244, 245
Myricetin-3-*O*-glucoside, 244, 245
Myricetin-3-*O*-rutinoside, 244, 245
Myristicin (1-allyl-3,4-methylenedioxy-5-methoxybenzene), 254
Myrosin cells, 32
Myrosinase, 25, 26, 32, 33, 217–19, 327–30, 332
Myrosinase-associated proteins, 32–3
Myrosinase-binding proteins, 32, 33
Myrrhis odorata, *see* Sweet Cicely

Naphthoquinones, 2, *see also individual entries*
Naringenin, 2, 15, 17–21, 227
Naringenin-7-*O*-neohesperidoside (naringin), 9, 10, 233, 234, 251
Naringenin-7-*O*-rutinoside (narirutin), 225, 226, 233, 234, 306
Naringenin-chalcone, 15, 17, 19
Naringin, *see* Naringenin-7-*O*-neohesperidoside
Narirutin, *see* Naringenin-7-*O*-rutinoside
Narrow leaved lupin (*Lupinus angustifolius*), 125, 127
Nasturtium officinale, *see* Watercress
Necine base, 12–14
Nectarines (*Prunus persica* var. *nectarina*), 231
Neem (*Azadirachta indica*), 87, 88, 210
  as a folk medicine and insect repellent, 87, 88
Neochlorogenic acid, *see* 3-*O*-Caffeoylquinic acid
Neohesperidin, *see* Hesperetin-7-*O*-neohesperidoside
9-*cis*-Neoxanthin, 76–8
all-*tran*-Neoxanthin, 77
(4a*S*,7*S*,7a*R*)-Nepetalactone, 65, 66, 94
  as a pheromone, 65
  as an insect attractant, 94
Neral, 61
Nero, Roman Emperor, 103
Nerolidyl pyrophosphate, 59, 78, 79
Neurotoxicity, 149, 150
Niacin, *see* Nicotinic acid
Nicotiana benthamiana, 55
Nicotiana tabacum, *see* Tobacco
Nicotine, 102, 109–13
Nicotine biosynthetic pathway, 110, 112
  regulation and localization, 112, 113
Nicotinic acid, 112, 122, 275, 276
N-nitroso compounds, 183
Noah, 278
Nobel Prize, 50
Nobiletin, 4, 5, 233, 234
Nomilin, 235
Nomilin limonoate A-ring lactone, 234, 235
Norcoclaurine 6-*O*-methyltransferase, 103, 104
(*S*)-Norcoclaurine, 103, 104

central precursor of benzylisoquinoline alkaloids, 103
Norcoclaurine synthase, 103, 104
Norepinephrine (noradrenaline), 117, 251
Novel flower colours, 20
Nutmeg *aka* mace (*Myristica fragrans*), 254
Nuts, 260–262, 323
Nuts, concentration of flavan-3-ols and proanthocyanidins, 261
Nux-vomica (*Strychnos nux vomica*), 114

Oats (*Avena sativa*), 259, 287
Obtusifoliol, 88
Oenanthe crocata, *see* Hemlock water dropwort
Oenanthe javanica, *see* Water dropwort
Oenanthetol, 139, 142
Oenanthetol acetate, 139, 141, 142
Oenanthotoxin, 149, 150
Oestradiol, 9, 10
Olacaceae, 138
Olea europa, *see* Olive
Oleic acid, 138–40, 145–8, 238, 239
(−)-Oleocanthal, 239
  anti-inflammatory properties, 239–40
Oleuropein, 238, 239
Oleuropein aglycone, 239
Olive (*Olea europa*), 64, 238–40
  protective effects, 239–40
Onions (*Allium cepa*), 25, 34–6, 38–40, 208, 214, 216, 217, 225
Oolong homo-bis-flavans, 266
Oolongtheanin, 266
Ophthalmic alkaloids, 108, 109
Opium, 102, 103
Opium poppy (*Papaver somniferum*), 102, 105, 106
Opium wars, 103
Orange (*Citrus sinensis*), 232
Oregano (*Origanum vulgare*), 210, 253, 256
Organic anionic transport polypedtide-2, 310
Ornithine decarboxylase, 110, 123
Oryzalexins, 70, 71
Osteoporosis, 189–90
Otonecine, 122
Oxalic acid, 249
5-Oxazolidene-2-thione, 328
5-(2′-Oxoalkyl) resorcinol, 259
Oxypeucedanin, 158, 160
Oxypeucedanin hydrate, 158, 160

Pacific yew (*Taxus brevifolia*), 67
Pallidol, 246
Palmitic acid, 238
(6′-*O*-Palmitoylglucosyl)sitosterol, 238
Panax ginseng, *see* Ginseng
Panaxydol, 152
Panaxynol, *see* Falcarinol
Panaxytriol, 152
Papaver somniferum, *see* Opium poppy
Papaverine, 102
Papaya (*Carica papaya*), 237–8
Paradisin, 158–60

Parsley (*Petoselinum crispum*), 4, 120, 130, 138, 141, 142, 151, 156–8, 162
Parsnip (*Pastiuaca sativa*), 130, 138, 142, 156, 158, 162, 212, 213
Parthenolide, 79, 80
Passiflora incarnata, see Passion flower
Passion flower (*Passiflora incarnata*), 113
Pastinaca sativa, see Parsnip
Patchouli (*Pogestomon cablin*), 81
Patchoulol, 59, 80, 81
Pea (*Pisum sativum*), 72
Peaches (*Prunus persica*), 231, 232
Peanut (*Arachis hypogaea*), 221, 260, 261
Pears (*Pyrus communis*) 230–231
Pecans (*Carya illinoensis*), 260, 261
Pelargonidin, 8
Pelargonidin conjugates
    degradation in urine, 317
Pelargonidin-3-*O*-glucoside, 241, 317
5-($\Delta^{8,11,14}$-Pentadecatrienyl)-resorcinol, 261
Penta-*O*-galloyl-glucose, 14, 16
Peonidin, 8
Peonidin glucuronide, 317
Peonidin-3-rutinoside, 231, 233
Pepper, green, black and white, 255
Peppermint (*Mentha* x *piperita*), 52, 55, 60, 253
Peppers (*Capsicum annuum*), 74, 227–8
    coloration, 227–8
Peptostreptococcus spp., 175
Persea americana, see Avocado
Persenone A and B, 224, 225
    anti-tumour properties, 224
    inhibition of superoxide and nitric oxide generation, 224
Petoselinum. crispum var. *tuberosum*, see Hamburg parsley
Petroselinum crispum, see Parsley
Petunidin, 8
Petunidin-3-*O*-glucoside, 241, 244, 248
P-glycoprotein, 304, 308, 331
Phaseolic acid, see caffeoyl-L-malic acid
Phellopterin, 160
Phenolic acids, 2, 11–12, 14
Phenolics and antioxidant capacity of red table grapes and red wine grapes, 249
Phenolic compounds, classification of, 2–14
    flavonoids, 2–11
        anthocyanidins, 8
        flavan-3-ols, 5–8
        flavanones, 8–9
        flavones, 4–5
        flavonols, 4
        isoflavones, 9–10
    non-flavonoids, 11–14
        hydroxycinnamates, 2, 12, 13
        phenolic acids, 2, 11–12
        stilbenes, 2, 12–14
Phenolics in red and white wines, 283
Phenols, 183
Phenylacetic acid, 2
Phenylalanine ammonia-lyase, 14
L-Phenylalanine, 14, 16

Phenylethyl isothiocyanates, 326
Phenylpropanoid pathway, 12, 14
Phloretic acid, 324
Phloretin, 304, 324
Phloretin-2′-*O*-(2″-*O*-xylosyl)glucoside, 229, 230
Phloretin-2′-*O*-glucoside (phloridzin), 229, 230, 286, 324
    inhibitor of the sodium dependent glucose transporter, 324
Phloridzin, see phloretin-2′-*O*-glucoside
3′-Phosphoadenosine 5′-phosphosulphate:desulphoglucosinolate sulphotransferase, 31
Phosphomevalonate kinase, 51
Photoactivation, 137
Photoactive furoisocoumarins, 257
Photodermatitis, 137, 159–62
Phototoxicity, 161
Physostigmine, 113
Phytoalexins, 137, 138, 147, 149, 163
Phytocassane D, 71
Phytochemicals, function in planta, 209–11
15-*cis*-Phytoene, 50, 75
Phytoene, 47, 49, 74, 85
Phytoene desaturase, 75, 93
Phytoene synthas,e, 74, 75, 93
Phytoestrogens, 9, 163
    effects on reproduction of cows and sheep, 9
Phytol, 47, 48, 52, 55
Phytophotodermatitis, 223
Phytosterols, 86, 88, 91–3
Piceatannol, 249, 250
Piceid, see *trans*-Resveratrol-3-*O*-glucoside
Picroliv, 65
    liver regeneration, 65
Picrorhiza kurroa, 65
Pimpinella anisum, see Anise
Pimpinella major, see Greater burnet saxifrage
Pimpinella saxifraga, see Burnet saxifrage
Pineapple (*Ananas comosus*), 235
(−)-β-Pinene, 62, 64, 256
(−)-α-Pinene, 62, 63,
(+)-α-Pinene, 48, 63, 256
Pinto beans *aka* kidney beans (*Phaseolus vulgaris*), 221
Piperanine, 255
Piperidine, 255
Piperine, 255
Pisatin, 18
Pistachio (*Pistachio vera*), 260, 261
Pittosporaceae, 138
Pituri bush (*Duboisia hopwoodii*), 109
Plantains (*Musa cavendishii*), 250, 251
Plasmodium falciparum, malarial parasite, 52, 54
Platycodon grandiflorum, see Balloon flower
Plums (*Prunus domestica*), 231–2
Poison hemlock (*Conium maculatum*), 129, 130
Poison ivy (*Rhus toxicodendron*), 261
Polydatin, see *trans*-Resveratrol-3-*O*-glucoside
Polygonum cuspidatum, see Japanese knotweed
Polyphenol oxidase, 248, 263, 267, 277, 287
Pomegranate (*Punica granatum*), 251–2, 324

Pontica epoxide, 143, 144
Potato (*Solanum tuberosum*), 20, 21, 109, 127, 128
Pouchitis, 182, 183
Prebiotics, 177, 178, 186–92
　colon cancer, 189, 191, 192
　gastrointestinal disease, 190, 191
　health effects, 189, 190
　irritable bowel disease, 190, 191
　lipid metabolism, 189, 190
　mineral absorption, 189
　modulation of gut microflora, 186–9
　osteoporosis, 189, 190
　ulcerative colitis, 190, 191
6-Prenylnaringenin, 284, 286
8-Prenylnaringenin, 284, 286
Pre-phytoene pyrophosphate, 74, 75
Pre-squalene pyrophosphate, 85
Prickly pear (*Opuntia* spp.), 130
Proanthocyanidin A_2 dimer, 6
Proanthocyanidin B_1 dimer, *see* Procyanidin B_1 dimer
Proanthocyanidin B_2 dimer, *see* Procyanidin B_2 dimer
Proanthocyanidin B_2 dimer, *see* Procyanidin B_2 dimer
Proanthocyanidin B_5 dimer, *see* Procyanidin B_5 dimer
Proanthocyanidin C_1 trimer, *see* Procyanidin C_1 dimer
Proanthocyanidin C_2 trimer, *see* Procyanidin C_2 dimer
Proanthocyanidin gallates, 268, *see also individual entries*
Proanthocyanidins, (*see* also procyanidins), 5–7, 244, 257, 259, 260, 266, 278, 303, 307, 308, 317, 319–21, *see also individual entries*
　adtringency, 278
　depolymerisation, 307–8
Probiotics, 178–86
　allergies, 184
　autism, 184
　colon cancer, 183
　diarrhoea, 180–181
irritable bowel disease, 184
Procyanidin B_1 dimer, 229–32, 281, 307, 317, 319
Procyanidin B_2 dimer, 6, 229, 230, 250, 278, 279, 281, 307, 317, 319
Procyanidin B_3 dimer, 284, 286, 307
Procyanidin B_5 dimer, 6, 278, 279, 307
Procyanidin C_1 trimer, 278, 279, 281
Procyanidin C_2 trimer, 17, 281
Procyanidin oligomers, 229, 236, 246, 247, 278, *see also individual entries*
Procyanidins, 7, 231, 236, 246, 261, *see also individual entries*
Prodelphinidin B_3 dimer, 284, 286
Prodelphinidins, 7, 221, 261, *see also individual entries*
Progoitrin (2-hydroxy-3-butenyl glucosinolate), 26, 33, 38, 209, 218, 325
S-Propanethial oxide, 35, 36
　lachrymatory factor, 214
　onion lachrymatory factor, 35
Propelargonidins, 7, 221, 261, *see also individual entries*
S-Propenyl cysteine sulphoxide (*aka* isoalliin and 1-propenyl-L-cysteine sulphoxide), 35, 36, 214

2-Propenyl glucosinolate, *see* Sinigrin
1-Propenyl-L-cysteine sulphoxide (isoalliin), *see* S-Propenyl cysteine sulphoxide
2-Propenyl-L-cysteine sulphoxide (alliin), *see* S-Allyl cysteine sulphoxide
1-Propenylsulphenic acid, 35, 36
2-Propenylsulphenic acid, 36
Prostaglandins, 151
Protein prenyltransferases, 56
Protoberberine alkaloids, 103–6
Protocatechuic acid, 235, 236, 253, 284
Protodioscin, 223, 224
Protosteryl cation, 87
Pseudotropine (Ψ-tropine), 109–11
Psoralen, 156–62, 213, 223
　blistering, 213
　treatment of skin disorders, 223
Psoralenes, *see* Furanpocoumarins, angular, *also individual entries*
Psoralens
　bioactivity, 159–63
　　antibacterial, 162, 163
　　antifungal, 162, 163
　　P450 inhibition, 162
　　phototoxicity, 159–62
　　reproductive toxicity, 162
　distribution and biosynthesis, 155–9
Psoriasis, 159
5,8-Dihydroxy-psoralen, 156
Psychotrine, 106
Pterocarpins, 18, *see also individual entries*
Puerin A and B, 266, 267
(+)-Pulegone, 62
Pummelo (*Citrus grandis*), 158
Pumpkin (*Cucurbita maxima*), 72
Punicalagin, 252
Punicalin, 252
Purine alkaloids, 118–22
Purine alkaloid biosynthesis, 119, 120
　localization, 120
Putrescine, 109–12, 122, 123, 251
Putrescine N-methyltransferase, 109–12, 133
PUVA therapy, 160–162
Pyrethric acid, 65, 67
Pyrrolizidine alkaloids, 122–5
　biosynthesis, 122–3
　localization, 123
　toxicity, 123–4
　　metabolic activation, 124

Qinghaosu, *see* Aertemisinin
Quercetin, 4, 305, 306, 308–11, 314, 315, 319, 320
Quercetin glycosides, 230, 237, 246, 266, 267, *see also individual entries*
Quercetin-3,4'-O-diglucoside, 216, 306, 314
Quercetin-3-glucuronide, 221
Quercetin-3-O-(6''-malonyl)glucoside, 222
Quercetin-3-O-arabinoside, 278, 279
Quercetin-3-O-galactoside, 229, 230, 237, 244, 245, 264, 270

Quercetin-3-O-glucoside, 229–32, 237, 244, 245, 264, 270, 278, 279, 304–6, 314
Quercetin-3′-O-glucuronide, 308, 315
Quercetin-3-O-glucuronide, 314, 315, 308
Quercetin-3-O-rhamnoside, 228–30
Quercetin-3-O-rutinoside (rutin), 223, 224, 226, 227, 231, 232, 244, 250, 251, 264, 270, 272, 306, 318–20
   colonic deglycosylation, 306
Quercetin-3-O-sophoroside, 218, 220
Quercetin-3′-O-sulphate, 313–15
Quercetin-3-O-xylosylglucuronide, 244, 245
Quercetin-4′-O-glucoside, 216, 306, 314
Quercetin-7-O-glucuronide, 308
Quinidine, 106
Quinine, 65, 66, 114, 117
Quinolinate phosphoribosyltransferase, 112
Quinolinic acid, 112
Quinolizidine alkaloids, 126–8
   biosynthesis, 126–8
   toxicity, 128

Radish (*Raphanus sativus*), 27, 29, 33, 217, 325, 326
Raisins, 247–8
Raspberry (*Rubus idaeus*), 240, 243, 244, 246, 323, 324
Red angel's trumpet (*Brugmansia sanguinea*), 109
Red beet (*Beta vulgaris*), *see also* beet, 130
Red fly agaric mushroom (*Amanita muscaria*), 108
Red wine, 316, 317, 320, 323, 324
Red wine production, 278, 280
Redcurrant (*Ribes rubrum*), 240, 242, 244
Redox potential, 174, 177
Regulatory genes, 20
Reserpine, 113, 117
   pharmacological effects, 117
Resin ducts, 59
Resistant starches, 186, 189
Resorcinols and appetite suppression, 260
cis-Resveratrol, 12, 13
trans-Resveratrol, 12–15, 19, 249, 250
   inhibition of LDL oxidation, 14
cis-Resveratrol-3-O-glucoside, 13
trans-Resveratrol-3-O-glucoside (*aka* piceid, polydatin), 12, 13
trans-Resveratrol-3-O-glucoside, 246, 248
Reticuline, 103–5
Retinoic acid, 339, 340
Retinoids, 340
Retinol (vitamin A), 74, 76, 332, 333, 339, 340
   deficiency, 74
Retinol precursors, 339, 340
Retronecine, 122, 124
Rhamnosidase, 306
Rhapontigenin, 249, 250
Rhein, 249, 250
Rheum rhizoma, 249
Rhodopsin, 74
Rhodopsin regeneration in the eye, 340
Rhubarb (*Rheum rhaponticum*), 248–9
Rice (*Oryza sativa*), 258, 259
   "Golden" rice, 93

   production of labdane diterpene phytoalexins, 70, 71
Risk assessment of herbal remedies, 91
16s rRNA profiles, 176, 178, 195–7
Robusta coffee (*Coffea canephora*), 12, 273
Rocket (*Eruca sativa*), 217, 219
Rockets (*Eruca sativa* and *Diplotaxis* spp.), 27, 29
Root crops, 212–13
Rosemary (*Rosmarinus officinalis*), 210, 253, 256, 257, 281
Rosmanol, 256
Rosmaridiphenol, 256
Rosmarinic acid, 253, 254
Rosmariquinone, 256
Rotenoids, 18
Rotenone, 18
Ruminococus spp., 176, 177
Rutaceae, 156, 158
Rutin, *see* Quercetin-3-O-rutinoside
Rye (*Secale cereale*), 259, 287

(+)-Sabinene, 62, 63, 65
Saccharomyces boulardii, 180–182
Safrole (1-allyl-3,4-methylenedioxybenzene), 254
Sage (*Salvia officinalis*), 61, 91, 253, 254, 256, 257
Sage family (*Salvia* spp.,), 60
Salad chervil, *see* Chervil
Salad rocket, 325
Salicylic acid, 14, 17, 210, 211, 253, 261
   blood clotting, 211
   disease resistance, 17
Salmonella spp., 198
(7S)-Salutaridinol, 104, 105
Salutaridinol 7-O-acetyltransferase, 104, 105
Samphire (*Crithmum maritimum* L.), 141
Sandaracopimaradiene, 70
Sandaracopimarenyl cation, 68, 70
Sanguiin H-6, 242–4, 247
Sanguiin H-10, 11
Sanguinaria canadensis, *see* Bloodroot
Sanguinarine, 102–6, 131
   in oral hygiene products, 106
   medicinal properties, 106
Santalaceae, 138
Saponins, 152, 89–90
   cholesterol-lowering and anticancer agents, 90
   toxic properties, 89
Sassafras (*Sassafras albidum*), 254
   carcinogenic effects, 254
Savory (*Satureja* spp.), 61
Schefflera arboricola, 150
Scopolamine, 102, 107–11, 131
   pharmacological effects, 111
Scotch whisky, 287–8
(S)-Scoulerine, 104
Scoulerine-9-O-methyltransferase, 104, 105
Scutellarein, 233, 234
Secologanin, 65, 66, 113–15
Secologanin synthase, 114
Secondary metabolites, function
   allelopathic agents, 1

Secondary metabolites, function (*Contd.*)
  attractants, 1
  disease resistance, 3
  signal molecules, 1
  UV protectants, 1, 3, 19
Senecionine, 122, 123
Sesamolins, 253
Sesquiterpene lactones, 144, 146
Shikimate pathway, 14, 16
Short chain fatty acids, 174, 176, 182, 190–192
Shungiku, *see* Garland chrysanthemum
Sinalbin, 217, 218, 258
Sinapaldehyde, 288
Sinapic acid, 14, 17
1-*O*-Sinapoyl-2-*O*-feruloylgentiobiose, 218, 220
S-Sinapylgluthathione, 235, 236
S-Sinapyl-L-cysteine, 235, 236
Sinensetin, 233, 234
Sinigrin (2-propenyl glucosinolate), 26, 217, 218, 258, 326
Sitostanol, 92
Sitostanol esters, 92
  in Benecol™ margarine, 92
    cholesterol lowering properties, 92
Sitosterol, 86, 88, 212, 213, 224, 225
Sium sisarum, *see* Skirret
Skin cancer, 137
Skirret (*Sium sisarum*), 142
Snake wood (*Strychnos coulbrina*), 114
Sodium-dependent glucose transporter, 306, 311, 324
Soft fruits 240–247, *see also individual entries*
Solanaceae, 143, 149
Solanidine-uridine diphosphate-glucose glucosyltransferase, 127
Solanum melongena, *see* Aubergine
Solanum tuberosum, *see* Potato
Sorbic acid, 232, 233
Sour cherries (*Prunus cerasus*), 231
Sour Orange, *see* Bitter Orange
Soya *aka* soyabean (*Glycine max*), 9, 10, 12, 125, 221, 260, 303, 316
Soya oligosaccharides, 186–8
Soyabean, *see* Soya
Sparteine, 127
Spearmint (*Mentha spicata*), 60, 61, 253
Spermidine, 122, 123, 251
Spermidine synthase, 122, 123
Spinacetin-3-*O*-gentobioside, 219, 220
Spinach (*Spinaceae oleraceae*), 211, 219, 334
Spinach-beet (*Beta vulgaris*), 219
Spiroacetal enol ethers, 138–40, 145
Sporomusa spp., 178
Spotted water hemlock (*Cicuta maculata*), 149
Spring groundsal (*Senecio vernalis*), 123, 125
Squalene, 47, 49, 50, 74, 85, 87
Squalene biosynthesis, 85
Squalene oxide, 85–9
Squashes (*Cucurbita* spp.), 228, 229
St. Paul, 278
Staphylococci, 176, 177
Star anise (*Illicium verum*), 253, 255
Statin drugs, 52, 92

Steroid biosynthesis, 86
Steroidal glycoalkaloids, 127, 128
  pharmacological effects, 128
Steroidal saponins, 90
Stigmasterol, 86, 88
Stilbene synthase, 15, 19
Stilbenes, 12–14, 246, 249, 250, 280, *see also individual entries*
  as phytoalexins, 12
Stomach cancer, 175
Strawberry (*Fragaria* × *ananassa*), 240, 242, 244, 246, 317, 319, 320, 323, 324
Streptococcus faecium, 181
Streptococcus spp., 174–7, *see also individual entries*
Streptococcus thermophilus, 170, 180–183
Strictin, 264
Strictinin, 264, 266
Strictosidine, 114,115
Strictosidine aglycoside, 114, 115
Strictosidine synthase, 11–16, 131
Strictosidine β-D-glucosidase, 114–16
Strigol, 77, 78, 94
Structural genes, 20
Strychnine, 107, 113, 114, 117
(*S*)-Stylopine, 104
Sugar alcohols, 186
Sulphatase, 313, 314
Sulphoraphane, 218, 219, 325, 326, 330, 331
  accumulation in cells, 331
Sulphotransferases, 309, 310
Swede (*Brassica napus*), 27, 29, 212, 213, 217, 219
Sweet Cicely (*Myrrhis odorata*), 142
Sweet Orange (*Citrus sinensis*), 158
Sweet potato (*Ipomoea batatas*), 109
Sweet wormwood (*Artemisia annua*), 81
  anti-malarial activity, 81
Swiss chard (*Beta vulgaris*), 219
Synbiotics, 178, 192, 193
  lipid metabolism, 193
Syringaldehyde, 288
Syringic acid, 235, 236, 253, 288

Tabersonine, 114–16
Tabersonine 16-hydroxylase, 114, 115
Tangelo (*Citrus reticulata*), 233
Tangeretin, 4, 5, 233, 234
Tangerine (*Citrus reticulata*), 232
Tannins, and
  astringency, 11, 12
  demise of the red squirrel, 12
  herbivore nutrition, 12
  manufacture of leather, 8, 11
Tannin-protein interactions, 11, 12
Tarragon (*Artemesia dracunculus*), 140, 142, 153
Taxa-4(5),11(12)-diene, 67–9
Taxadiene synthase, 67, 69
Taxol, 47, 67, 69, 93
  as a potent anticancer drug, 67
  biosynthesis, 67
Tea (*Camellia sinensis*), 7, 10, 11, 13, 118–21, 263–71, 264, 265, 270, 272, 323

black tea, 7, 266–9
  tannins, 8
  thearubigins, 7–8
  theaflavins, 6, 7
  consumption, 119
  green tea, 7, 263–6
    flavan-3-ols and gallated derivatives, 7
    phenolics, effect of fermentation, 270
  manufacture, 263, 266–7
  Oolong, 266
  Pu'er, 267
Tea cream, 271
Tea scum, 271
Tea tree (*Melaleuca alternifolia*) oil
  acne, 90
Temperature gradient gel electrophoresis, 176, 177
Terpene biosynthesis enzymes, 55–9
  prenyltransferases, 55–6
    isoprenyl pyrophosphate synthases, 56
    mechanism of chain elongation, 56–8
    terpene synthases (*aka* terpene cyclases), 56, 58–9
Terpene cyclases, *see* Terpene synthases
Terpene indole alkaloids, 66
Terpenes
  in the environment and human health, 90–94
Terpene synthases (*aka* terpene cyclases), 56, 58–9, 74
Terpenoid indole alkaloids, 113–18
  biosynthesis, 114–16
    localization of enzymes, 116
    regulation, 116
γ-Terpinene, 256
α-Terpineol, 62, 63
Terpinyl cation, 62–4
α-Testosterone, 10
(6Z)-Tetradeca-6-ene-1,3-diyne-5,8-diol, 149
Tetrahydrocannabinol, 47
(S)-Tetrahydrocolumbamine, 104, 105
2,2',6,6'-Tetrahydroxydiphenyl, 266, 267
1,3-*cis*-Tetrahydroxyphenylindan, 275
Thalictrum flavum, see Meadow rue, 105
Theacitrin, 268
Theacitrin-3,3'-digallate, 268
Theacitrin-3'-gallate, 268
Theacitrin-3-gallate, 268
Theacitrins, 267, *see also* individual entries
Theadibenztropolone A, 268
Theadibenztropolones, 267, *see also* individual entries
Theaflagallin, 267, 268
Theaflavate A, 268
Theaflavin, 267–9, 271
Theaflavin gallates, 267, 268, 270, *see also* individual entries
Theaflavin-3,3'-digallate, 6, 268, 270
Theaflavin-3'-gallate, 6, 268, 270
Theaflavin-3-gallate, 6, 268, 270
Theaflavins, 7, *see also* individual entries
Theaflavonin, 268
Theaflavonins, 267, *see also* individual entries
Theanaphthoquinone, 267, 269
Theanine, 263–5, 269
Thearubigins, 7, 8, 267–9, 271
  astringency, 269

Theasinensin gallates, 268
Theasinensins, 266, 267
Thebaine, 104, 105
Theobroma cacao, *see* Cocoa
Theobromine (3,7-dimethylxanthine), 118–21, 131, 264, 269, 271, 278
Theobromine synthase (7-methylxanthine N-methyl transferase), 120, 131
Theogallin, 266, 269
Theogallinin, 267, 268
Theogallins, *see* Galloylquinic acid
Thiarubrines, 138
Thioglucosidase, 327, 328, 332
Thiohydroximate-O-sulphonate, 33, 327, 328
Thiophenes, acetylenic, 138–40, 145
Thiosulphanate degradation products, 37
Thiosulphinates, degradation, 37
Thorn-apple (*Datura stramonium*), 107, 108
Thujanes and pinanes, 61–5
(−)-α-Thujone, 64, 65, 256
  in absinthe, 64
    Vincent van Gogh, 64
Thyme (*Thymus* spp.), 61, 253, 256, 257
Thymol, 61, 62, 91, 256
  in toothpaste, 91
Tigloyl-CoA:(−)-13α-hydroxymultiflorine/ (+)-13α-hydroxylupanine-O-tigloyltransferase, 126, 127
(−)-Tigloyloxymultiflorine, 13α-, 126
Time to reach maximum plasma concentration, 306, 307, 317, 319
Tirucallol, 86, 89
Tobacco (*Nicotiana tabacum*), 20, 111–13
  smoking and health problems, 113
Tomatine, 226, 227
Tomato (*Lycopersicon esculentum*), 20, 21, 74, 78, 127, 138, 143, 149, 225–7, 323, 335
  types, 225–6
Trachyspermum ammi, *see* Ajowan
Tribenztropolones, 268
Tri-bromophenol 2, 4, 6-, 262
Tricetanidin, 267, 268
Trifolium spp., *see* Lucerne and Clover
Trigonelline (N-methylnicotinic acid), 121, 122, 275, 276
1,4,5-Trihydroxynaphthalene-4-β-D-glucoside, 262
1,3,7-Trimethylxanthine, *see* Caffeine
1,2,2'-O-Trisinapoylgentiobiose, 218, 220
Tropane alkaloids, 107–11
  biosynthesis, 109–11
  cellular localization, 111
  toxicity, 107
Tropic acid, 111
Tropine, 109, 110
Tropinone, 109, 110
Tropinone reductase, 109, 110
Tryptophan decarboxylase, 114–16, 131
Tubo curare (*Chondrodonendron tomentosum*), 106
  arrowhead poison, 106, 107
(+)-Tubocurarine, 102, 103, 106, 107
Turmeric (*Curcuma longa*), 255

Turnip (*Brassica campestris*), 27, 29, 212, 213, 217, 219, 325, 326
Turnip-rooted chervil (*Chaerophyllum bulbosum*), 141
Turnip-rooted parsley, *see* Hamburg parsley
Turpentine, 59
Tyrosinase, 118, 130
Tyrosine, 103, 104, 130
Tyrosine/dopa decarboxylase, 103, 104

Ulcerative colitis, 182, 183, 190, 191
Umbelliferone, 156, 159
Uridine diphosphate glucose:thiohydroximate glucosyltransferase, 30
Uridine-5'-diphosphate glucuronosyltransferases, 309, 315
Urolithin B glucuronide
   biomarker of exposure to ellagitannins and ellagic acid, 324
Urushiols, 261

Valerian (*Valeriana officinales*), 65
   tranquilliser, 65
Valeriana officinales, *see* Valerian
Valtrate, 65, 66
Vanillic acid, 235, 238, 288
Vanillin, 288
Vegetables, 211–29, *see also individual entries*
Verbascoside, 239
Verticillene, 67, 69
Verticillium alboatrum, 149
Vestotone reductase, 18
Vicia faba, *see* Broad bean
Vinblastine, 102, 113–17
Vincristine, 113, 116, 117
Vindoline, 114–16
ε-Viniferin, 246, 248
Viniferins, 12
4- Vinylguaiacol, 276
5-Vinyloxazolidine-2-thione, 26, 38
Violaxanthin, 75–7
Vitamin A, *see* Retinol
Vitamin B, 224
Vitamin C (ascorbic acid), 33, 130, 224, 237
Vitamin E, 224
Vitexin, *see* Luteolin-8-*C*-glucoside
Vitiligo, 159
Vitis vinifera, *see* Black grapes and White grapes
Vitisin A (malvidin-3-*O*-glucoside-pyruvic acid), 281, 282
Vitisin B (malvidin-3-*O*-glucoside-4-vinyl), 281, 282
Vitisins, *see also individual entries*

Vomilenine, 116
VSL#3, 182–4

Walnuts (*Juglans* spp.), 261, 262, 324
Wasabi (*Wasabia japonica*), 29
Wasabia japonica, *see* Wasabi
Water celery, *see* Water dropwort
Water chestnuts (*Eleocharis dulcis*), 259
Water dropwort (*Oenanthe javanica*), 142
Water hemlock (*Cicuta virosa*), 149
Watercress (*Nasturtium officinale*), 25, 27, 29, 33, 217, 219, 325, 326, 332
Watermelons (*Citrullus lanatus*), 245
Western false hellbore (*Veratrum californicum*), 128
Wheat (*Triticum vulgare*), 258, 259, 281, 287
Whisky production, 287–8
White Chervil, *see* Hornwort
White wine production, 280
Wild Chervil, *see* Hornwort
Wines, 278–83
Witchweed (*Striga* spp.), 94
Witloof Chicory, *see* Chicory
Wormwoodd (*Artemisia absinthum*), 64
   wormwood oil as herbal medicine, 64
Wort, 282–3, 285, 287
Wyerol, 143, 148
Wyerol epoxide, 143, 148
Wyerone, 143, 148
Wyerone acid, 143, 148
Wyerone epoxide, 143, 148

Xanthohumol, 284, 286
Xanthones, 2
Xanthophylls, 75, 76
Xanthosine, 119, 120
Xanthosine monophosphate, 120
Xanthotoxin, 156–8, 160–162, 223
Xanthotoxol, 156
Xanthoxin, 76, 77
Xylo-oligosaccharides, 186

Yams (*Dioscorea* spp.), 90
Yarrow (*Achillea millefolium*), 281
Yeast and beer
   bottom-feeding (*Saccharomyces carlsbergensis*), 281
   top-fermenting (*Saccharomyces cerevisiae*), 281
Yohimbine, 113, 117, 118

Zea mays, *see* maize
Zeaxanthin, 75, 76, 93, 224, 225, 228, 333, 338, 340
Zingerone, 255
(−)-Zingiberene, 256, 257